Fundamentals of Pharmacology for Veterinary Technicians

Janet Amundson Romich, DVM, MS

THOMSON
DELMAR LEARNING™

Australia Canada Mexico Singapore Spain United Kingdom United States

Fundamentals of Pharmacology for Veterinary Technicians
Janet Amundson Romich, DVM, MS

Vice President, Career Education Strategic Business Unit:
Dawn Gerrain

Director of Editorial:
Sherry Gomoll

Acquisitions Editor:
David Rosenbaum

Developmental Editor:
Gerald O'Malley

Editorial Assistant:
Christina Gifford

Director of Production:
Wendy A. Troeger

Production Manager:
Carolyn Miller

Production Editor:
Kathryn B. Kucharek

Director of Marketing:
Wendy Mapstone

Channel Manager:
Donna Lewis

Cover Images:
Getty One

Cover Design:
Dutton & Sherman Design

Printed in Canada
4 5 XXX 08

For more information contact Thomson Delmar Learning, 5 Maxwell Drive, PO Box 8007, Clifton Park, NY 12065-2919.

Or you can visit our Internet site at http://www.delmarlearning.com.

Library of Congress Cataloging-in-Publication Data
Romich, Janet Amundson.
Fundamentals of pharmacology for veterinary technicians / Janet Amundson Romich.
p. cm.
ISBN- 13: 978-1-4018-4293-2
ISBN- 10: 1-4018-4293-3
1. Veterinary pharmacology. I. Title.

SF915.R62 2005
636.089'51—dc22 2004046067

NOTICE TO THE READER

Publisher does not warrant or guarantee any of the products described herein or perform any independent analysis in connection with any of the product information contained herein. Publisher does not assume, and expressly disclaims, any obligation to obtain and include information other than that provided to it by the manufacturer.

The reader is expressly warned to consider and adopt all safety precautions that might be indicated by the activities herein and to avoid all potential hazards. By following the instructions contained herein, the reader willingly assumes all risks in connection with such instructions.

The Publisher makes no representation or warranties of any kind, including but not limited to, the warranties of fitness for particular purpose or merchantability, nor are any such representations implied with respect to the material set forth herein, and the publisher takes no responsibility with respect to such material. The Publisher shall not be liable for any special, consequential, or exemplary damages resulting, in whole or part, from the readers' use of, or reliance upon, this material.

Fundamentals of Pharmacology for Veterinary Technicians

CONTENTS

CHAPTER 7

Drugs Affecting the Nervous System 97

CHAPTER 8

Cardiovascular Drugs 123

CHAPTER 9

Respiratory System Drugs 140

CHAPTER 10

Hormonal and Reproductive Drugs 149

CHAPTER 11

Gastrointestinal Drugs 174

CHAPTER 12

Urinary System Drugs 191

CHAPTER 13

Drugs Affecting Muscle Function 204

CHAPTER 14

Antimicrobials 211

CHAPTER 18

Ophthalmic and Otic Medications 281

CHAPTER 19

Fluid Therapy and Emergency Drugs 298

CHAPTER 20

Antineoplastic and Immunosuppressive Drugs 318

CHAPTER 21

Vaccines 327

CHAPTER 22

Behavior-Modifying Drugs 340

CHAPTER 23

Herbal Therapeutics 350

PREFACE

INTRODUCTION

It is an exciting time to be in the veterinary medical profession! There are limitless opportunities to be part of a profession that continues to expand its field of knowledge. One of the fields of knowledge that is ever-expanding is the discipline of pharmacology. One reason there is increased interest in veterinary pharmacology is the expansion of drug choices available for animal use. New categories of drugs and new uses for existing drugs continue to be presented in the literature on a regular basis. Clients' willingness to treat their animals and manage clinical disease has also increased the use of pharmacological agents in the veterinary field.

Another cause of interest in veterinary pharmacology is that we have entered the age of consumer advertising in the veterinary profession. Ready or not, we are being dragged into the brave new world of pharmacology-savvy clients armed with their Internet research on the latest medications available to treat their pets. Newsstand magazines contain full-color ads for Frontline®, Cosequin®, and Rimadyl®. Television ads featuring friendly veterinarians discussing flea and tick control with not-so-compliant companion animals are now frequently seen by the viewing public. Veterinary pharmaceuticals are now being marketed and sold directly to the consumer. How are we, as veterinary professionals, preparing ourselves to respond to this change in the marketplace—a change that will require us to stay current in our knowledge of drugs and their applications?

The key ingredient in professional preparedness is a good understanding of the fundamentals of pharmacology. This text is designed to give veterinary professionals the solid foundation in pharmacology upon which is built the professional habit of *staying current* with emerging trends in pharmacology. Building a solid foundation of understanding in pharmacology requires a textbook in an easy-to-read format that provides practical applications of new information and reviews concepts, calculations, and critical thinking skills. These applications, during a course of study using this text, will give student technicians ample opportunity to develop their confidence and proficiency in the area of pharmacology. Developing the confidence in applying pharmacological agents to specific medical uses, and acquiring the language needed to explain these uses in both professional and lay terms, is the benchmark for veterinary professionals.

This textbook and the accompanying learning materials make the process of understanding, applying, and staying current with the changes in pharmacology as straightforward as possible. This textbook presents brief but thorough explanations and reviews of basic terminology, anatomy and physiology and disease processes—all necessary to understand how drug therapy is utilized. Case studies, practice calculations throughout the text and on CD-ROM, and interesting facts presented as "Tips" provide the learner with many ways to examine the material presented in pharmacology. Summary sections and chapter reviews guide the learner toward understanding the key points of each pharmacological topic. An introduction to how drugs are developed, made, and delivered to the veterinary community provides the learner with key insights into the world of pharmacology.

ORGANIZATION OF THE TEXTBOOK

This textbook begins with an introduction to the history of veterinary pharmacology, including a review of terminology and veterinary drug development and control. A discussion follows of the principles of the therapeutic range, routes of drug administration, and pharmacokinetic factors in drug therapy. Practical information regarding drug prescribing, prescription writing, and sources of drug information follow the pharmacokinetic section. Mathematical reviews and presentation of mathematical manipulations performed in a veterinary setting are presented in both text and CD-ROM formats to ensure that the learner has an adequate understanding of the math and skills that will be required in the workplace. The remainder of the text is organized according to the major drug classifications, identified either by their clinical use or by the body system they affect. For each classification of drug discussed in the text, the underlying anatomy and physiology are reviewed and the basic pharmacological principles of drug action are discussed. Material is presented in text, chart, and table formats throughout the text to reinforce the material. Practical applications, with case studies and additional dose calculations, as well as content review, are provided at the end of these chapters. Condensed tables at the end of chapters and at the end of the text allow the user to review material quickly and to utilize the text as a workplace reference. Review questions and summaries keep users focused on key pharmacological concepts, in addition to testing their knowledge. Images and line art help clarify more difficult concepts by providing a visual and varied representation of the information in text. The variety of presentation formats enhances content comprehension and retention.

FEATURES AND BENEFITS

- Two interactive CD-ROMs to aid in mathematical concepts are included with the textbook. The first CD-ROM is the *Delmar's Veterinary Dosage Accu-Calc® CD-ROM* that helps the student with a review of basic math skills, metric conversions, dose calculations, and dilution problems. The second CD-ROM is *Delmar's Fluids: Electrolytes for Veterinary Technicians CD-ROM* that helps the student to understand fluid calculations utilized in a clinical setting.
- Chapter objectives and key terms are listed at the beginning of each chapter to guide students in understanding the organization of that chapter.
- A case scenario in each chapter introduction helps students understand the importance of the chapter topic and lets them apply this information to a clinical setting. Answers to these introductory case scenarios are found on the Delmar Online Companion.
- A summary and chapter review are provided for every chapter, to allow students to review and track their understanding of a topic.
- Case studies are provided at the end of many chapters to give students a chance to apply the information they just learned.
- Easy-to-use tables and charts help students organize drugs into the various drug classes.
- Straightforward illustrations and photos help students comprehend difficult concepts.
- Appendices provide the student with quick references throughout the course and in a clinical setting.
- A comprehensive list of drugs, arranged alphabetically is provided in Appendix L. The action of each drug is provided in this chart.

- A glossary at the end of the text gives students a resource in which to quickly look up key terms in pharmacology.
- An online resource that provides discussion of case studies, Web addresses for researching relevant topics and resources, and critical thinking questions supplies students with additional resources and independent study options for learning pharmacology.
- An instructor's manual is available with answers to the end-of-the-chapter questions and case studies, as well as additional testing material.
- To view Online Companion instructor resources, go to <http://www.agriculture.delmar.com> and click on Instructor Center on the sidebar. Follow the steps displayed.

- To view student resources, go to <http://www.agriculture.delmar.com> and click on Resources. Click on Online Resources (or Online Companion) and select from the titles listed.

ABOUT THE AUTHOR

Dr. Romich received her Bachelor of Science degree in Animal Science from the University of Wisconsin–River Falls, and her Doctor of Veterinary Medicine and Master of Science degree from the University of Wisconsin–Madison. She worked as a pharmacy technician in both human and veterinary settings while attending veterinary school. Her master's thesis was based on FDA research for a veterinary pharmaceutical company. Currently Dr. Romich teaches at Madison Area Technical College in Madison, Wisconsin where she has taught and continues to teach a variety of science-based courses. Dr. Romich was honored with the Distinguished Teacher Award in 2004 for her use of technology in the classroom, advisory and professional activities, publication list, and fund raising efforts. She is a member of the Biosafety Committee for a biopharmaceutical company and an IACUC member for a hospital research facility. Dr. Romich authored the textbook *An Illustrated Guide to Veterinary Medical Terminology with FLASH! CD-ROM* and co-authored *Delmar's Veterinary Technician Dictionary*. Dr. Romich remains active in veterinary practice through her relief practice, where she works in both small animal and mixed animal practices.

ACKNOWLEDGMENTS

The author wishes to express her appreciation to all who contributed to and supported this project:

- the staff at Thomson Delmar Learning, for the opportunity to work on this project and their encouragement throughout the development of this project.
- fellow veterinarians who helped in reviewing content and content relevance, including Dr. Bob Garrison, Dr. Linda Kratochwill, Dr. Kenneth Brooks, Dr. Ron Fabrizius, Dr. Kelly Gilligan, and Dr. Jim Meronek.
- Grace A. Cooper, CVT, who read every page of the manuscript for clarity, relevance, and content. Without her help, this textbook would not be as clear, concise, and organized as it is. She also kept me focused during the editing process to ensure that this product would be a valuable resource for veterinary technician students.
- former and current students who continue to question this material and examine the content through a variety of viewpoints.
- my family, who supported me throughout the writing process and allowed me to complete this project in a timely fashion.

REVIEWERS

Textbook

Kenneth Brooks, DVM, Diplomate ABVP
Associate Veterinarian
Lodi Veterinary Hospital
Lodi, WI

Grace A. Cooper, CVT
Supplemental Educational Services
Madison Area Technical College
Madison, WI
Breeder/Trainer/Handler of traditional New England-type Labrador retrievers
Wanderer Kennels
Madison, WI

Robert Garrison, DVM, MS, Diplomate ABMM
Assistant Scientist, Communicable Disease Division
Wisconsin State Laboratory of Hygiene
Madison, WI

Kelly Gilligan, DVM
Associate Veterinarian
Four Paws Veterinary Hospital
Prairie du Sac, WI

Linda Kratochwill, DVM
Associate Veterinarian
Crow-Goebel Veterinary Clinic
Scanlon, MN

James Meronek, DVM
Field Veterinarian
American Breeder's Service (ABS) Global
DeForest, WI

CD-ROM Products

Thomas G. Curro, DVM, MS
Staff Veterinarian
Henry Doorly Zoo
Omaha, NE

Robert O. Erdman, DVM
Associate Veterinarian
Fairwood Animal Hospital
Spokane, WA

A. Bruce King, DVM
Associate Veterinarian
Kootenai Animal Hospital
Post Falls, ID

Cathy King, DVM, MS, PhD
Associate Veterinarian
Deer Park Veterinary Clinic
Deer Park, WA

Robert Taylor, DVM, MS
Instructor
Madison Area Technical College
Madison, WI

Terry Teeple, DVM
Instructor
Pierce College at Fort Steilacoom
Lakewood, WA

Delmar Reviewers

Thomas Colville, DVM
North Dakota State University
Fargo, ND

Richard Flora, DVM
Newberry College
Newberry, SC

James Kennedy, DVM
Colorado State University
Rocky Ford, CO

Paul Porterfield, DVM
Central Carolina Comunity College
Sanford, NC

NOTICE TO THE READER

Every attempt has been made to assure that the information included in this book is correct; however, errors may occur and it is suggested that the reader refer to original references or the approved labeling information of the product for additional information.

COMMENTS

Comments regarding this textbook and the accompanying material can be sent to Thomson Delmar Learning, Executive Woods, 5 Maxwell Drive, Clifton Park, NY 12065-2919 or relayed via the Delmar Web site at <http://www.delmarlearning.com>.

A Brief History of Veterinary Pharmacology

OBJECTIVES

Upon completion of this chapter, the learner should be able to

- describe the history of pharmaceuticals and their development.
- define key terms used in pharmacology.
- define the FDA's role in drug approval and drug monitoring.
- compare and contrast the use of prescription drugs, over-the-counter drugs, extra-label drugs and controlled substances.
- describe a veterinarian-client-patient relationship.
- explain the differences between C-I, C-II, C-III, C-IV, and C-V drugs.

KEY TERMS

pharmacology
therapy
pharmacotherapy
pharmacotherapeutics
kinetics
pharmacokinetics
pharmacodynamics
Food and Drug Administration (FDA)
biologic
prescription drugs
veterinarian/client/patient relationship
extra-label drug
over-the-counter drug
controlled substance
Drug Enforcement Administration (DEA)

INTRODUCTION

The owner of a sick dog comes into a veterinary clinic seeking treatment for the animal. The owner is depending on the veterinarian to prescribe drugs that will effectively treat the dog's illness, yet be safe for the dog and manufactured properly. How can this owner be sure that this is true? How can you convince this owner that this is true? Can you explain to the owner of that sick dog how drugs are manufactured and their safe use in animals is monitored? What agencies regulate drugs? Are all drugs regulated in the same way?

A HISTORY OF VETERINARY PHARMACOLOGY

Veterinary medicine has existed since ancient times; archeologists have discovered ruins in India of a hospital for horses and elephants that operated in 5000 B.C. However, *veterinary pharmacology*—the study and use of drugs in animal health care—is a much younger specialty. Its origins are traced to the early 1700s, when an epidemic in western Europe wiped out most of the cattle population there. The Europeans realized, in the wake of this catastrophe, how little anyone knew about animal diseases. In France, this concern stimulated the establishment of five veterinary colleges in the 1760s. Austria,

TABLE 1-1 Drug Sources

Drug Source	Example
Minerals	sulfur, iron, electrolytes
Botanical (from plants, molds, bacteria)	digitalis, antibiotics
Animal	insulin, thyroid hormone, lanolin
Synthetic (manmade or engineered)	aspirin, steroids, procaine

Germany, the Netherlands, England, and Scotland followed France's lead in the late 1790s. By the 1850s, veterinary colleges were being organized in America, with a veterinary college opening in Philadelphia in 1852 and another one opening in Boston in 1854. Since then, the number of American veterinary colleges has significantly increased.

What did all these colleges have to do with the birth of veterinary pharmacology? These veterinary colleges were founded as adjuncts to schools of medicine, the curricula of which included *materia medica*, the study of the physical and chemical characteristics of materials used as medicines. These materials were natural plant components. As scientists extracted and synthesized more sophisticated drugs from these plant components, *materia medica* gave way to a new field: pharmacology, and veterinary *materia medica* became veterinary pharmacology. Even today, many drugs come from plants, bacteria, and animal sources. Many anticancer drugs have been discovered in plants, most antibiotics have been discovered in soil bacteria, and many hormonal drugs are processed from animal tissue. These natural substances may now be made semisynthetically, with substances chemically modified from a natural source. The vast majority of drugs currently in use are made by synthetic means, through chemical reactions in a laboratory. These agents are synthesized after determination of how the chemical structure of a compound relates to its pharmacological properties. Because synthetic drugs are made in the laboratory, they tend to have greater purity than those that are naturally derived. Table 1-1 summarizes sources of drugs.

WHAT DID YOU SAY? UNDERSTANDING PHARMACOLOGICAL TERMS

Understanding pharmacological terms is fundamental to understanding pharmacology. The first part of the word **pharmacology** has the root *pharmaco-* which is Greek for "drug" or "medicine"; therefore any compound word with *pharmaco-* in it has something to do with drugs. Three terms that explain the workings of veterinary pharmacology: pharmacotherapeutics, pharmacokinetics, and pharmacodynamics all contain the root *pharmaco-*. Examining the other word components in these terms will provide an understanding of their meaning.

Therapy is the treatment of disease; therefore, **pharmacotherapy** is the treatment of disease with medicines and **pharmacotherapeutics** is the field of science that examines the treatment of disease with medicines. To study veterinary pharmacotherapeutics, then, is to study how a sick animal responds to drugs.

Kinetics is the scientific study of motion. In describing the motion of drugs, the term **pharmacokinetics** is used to indicate the study of the absorption, blood levels, distribution, biotransformation (or metabolism), and excretion of drugs. Knowing these factors, and how much of a drug is in the body, it can be determined how much of a drug gets to its desired site of action. To study veterinary pharmacokinetics, then, is to study how the motion of drugs in the body affects or brings about an animal's response to them.

Response to drugs must be studied in both healthy and diseased animals to determine if the drug used to treat one illness may cause another problem. **Pharmacodynamics** is the study of the mechanisms of action of a drug and its biological and physiological effects. To study veterinary pharmacodynamics, then, is to study a healthy animal's response to drugs to determine their effects on the physiological and biochemical systems of the body.

Once scientists know the pharmacotherapeutics, pharmacokinetics, and pharmacodynamics of a particular drug, they know how the drug might be modified to produce a greater effect. Pharmacology emerged from *materia medica* when scientists applied their knowledge to modify existing drugs which in turn resulted in the alteration of their molecular action. Through sustained trial and error, these scientists created more effective, more stable, and more predictable drugs. Something else occurred in the field of pharmacology at the turn of the last century that helped create more effective, stable, and predictable drugs.

WHO'S IN CHARGE? THE FOOD AND DRUG ADMINISTRATION

The **Food and Drug Administration**, or **FDA**, came into being as a government agency to enforce the federal Pure Food and Drug Act of 1906. Before 1906, drug manufacturers had no obligation to establish the safety, purity, or effectiveness of their drugs, so many questionable or even harmful products were legally sold. The Pure Food and Drug Act established standards for drug strength and purity, and guidelines for drug labeling.

For three decades, the FDA was a small agency with limited influence and authority. The drugs available in that era were primarily from botanical (plant) sources. Public concern focused on three botanical drugs: ergot, a derivative of rye fungus used to induce labor and treat migraines; quinine, a derivative of tree bark used to treat malaria; and digitalis, a derivative of the foxglove plant used to treat cardiac failure. All three drugs perform well if correctly dosed, but all are quite toxic if overdosed. At that time, dosing and overdosing were frequently a matter of trial and error. Until the late 1930s, the FDA had little power to determine and enforce correct dosage information.

Then Congress passed the federal Food, Drug, and Cosmetic Act of 1938, which required that a drug be adequately tested to demonstrate its safety when used as its label directs. The 1938 law greatly expanded the power and responsibilities of the FDA, and this expansion continued in the postwar years, as chemicals became widely used for drugs, cosmetics, food additives, and pesticides. In 1972, the Act was amended to include many more protections.

The FDA is headed by a commissioner and organized into a number of different centers, each performing a specific function. The FDA's Center for Veterinary Medicine (CVM) ensures that approved veterinary medicines will not harm animals, or at least that the harm a drug produces will be outweighed by its benefit. Take, for example, a dog with tachycardia (rapid heartbeat). The veterinarian gives the dog a pill that returns the heartbeat to normal within two hours. The dog then eats a bowl of food, but the medication suddenly causes him to vomit. The dog feels poorly again, but for a different reason. The drug that slowed the dog's heartbeat made him throw up and feel very uncomfortable.

The FDA would prohibit sale and use of a drug that would cause animals to suffer from serious health problems. In this case, though, the vomiting subsides; the drug's ability to slow the heartbeat far outweighs the unpleasant but tolerable side effect of vomiting. The FDA thus strives to protect consumers, health professionals, and animals by maximizing the benefit of drugs while minimizing their dangers.

TIP

It is important to note that FDA regulations do not cover certain medically significant compounds known as **biologics** (therapeutic agents derived from living organisms, such as vaccines, antibodies, and toxoids). Biologics are governed by regulations of the United States Department of Agriculture (USDA) and are brought to market under an entirely different system from that by which drugs are brought to market.

TIP

The U.S. system for new drug approvals is perhaps the most rigorous in the world.

TIP

An idiosyncratic reaction (an abnormal response to a drug that is peculiar to an individual animal) does not cause the FDA to disapprove a drug. FDA approval of a particular drug is based on the drug's effects and side effects in many test animals.

The FDA's power to protect veterinarians and animals is a result of the 1972 amendments to the Act concerning "New Animal Drugs." These amendments require a drug manufacturer to demonstrate that its drug is safe for animals and does what the label claims. However, the 1972 amendments also affect veterinarians in a completely different way: drug manufacturers must now provide both a reliable analytical method to detect drug residues in animal foodstuffs, and an acceptable drug withdrawal period following a food-producing animal's last dose. These provisions seek to assure consumers that the dairy, poultry, and meat products are drug-free. Since the 1972 amendments, veterinarians must take care to instruct livestock owners about the laws governing any drugs they feed their animals.

HOW DO YOU GET DRUGS?

Prescription drugs, which the FDA regulates, are limited to use under the supervision of a veterinarian or physician because of their potential danger, toxicity concerns, administration difficulty, or other considerations. Because every drug has the potential to cause harm, if given for the wrong reason, to the wrong animal or in the wrong amount, they must be regulated to ensure safe use. Prescription drugs for animals can be obtained only through a veterinarian or via a prescription. Before a prescription drug can be prescribed for an animal, a **veterinarian/client/patient relationship** must exist. Animals need to be seen and examined by a veterinarian, who assumes responsibility for making clinical assessments based on sufficient knowledge about the health of the animals, their need for treatment, and the need for follow-up. Prescription drugs must be labeled with the statement or legend: "Caution: Federal law restricts the use of this drug to use by or on the order of a licensed veterinarian."

Sometimes veterinarians utilize drugs in the treatment of animals that are not FDA approved for that particular disease or condition in that particular species. Many times this lack of FDA approval is due to the facts that the drug was tested for human use and that the cost to conduct experiments in animals for animal-use approval would not be financially rewarding for the drug manufacturer. Veterinarian discretion allows veterinarians to use drugs in a manner not indicated by the labeling. This is termed **extra-label drug** use and is defined as the use of a drug in a manner not specifically described on the FDA-approved label. Extra-label use of certain approved animal drugs, and approved human drugs for animals under certain conditions, is allowed under the Animal Medicinal Drug Use Clarification Act of 1994 (AMDUCA). The key constraints of AMDUCA are that any extra-label use must be by or on the order of a veterinarian within the context of a veterinarian/client/patient relationship. This extra-label use also must not result in drug residues in food-producing animals. The Code of Federal Regulations lists drugs that are subject to prohibitions on extra-label use. The drugs on this list cannot be used in animals.

TIP

Some controlled substances may be found at multiple schedule levels. For example, codeine is found under C-II, C-III, and C-V. This multiple schedule labeling is due to the level of controlled substance found in each product. Combination products are classified as lower-level controlled substances.

Some drugs do not have a significant potential for toxicity or do not require special administration. These drugs are referred to as **over-the-counter drugs** (OTC drugs) and may be purchased by the client without a prescription. Aspirin is a common over-the-counter drug used in veterinary medicine.

Some drugs are classified as **controlled substances**, or drugs considered dangerous because of their potential for human abuse or misuse. Controlled substances are regulated by the **Drug Enforcement Administration (DEA)** through the Comprehensive Drug Abuse Prevention and Control Act of 1970 (commonly known as the Controlled Substances Act).

The original law regulating drugs defined *drug abuse* as the illicit use of an illegal drug or the improper use of a legal prescription drug. However, in 1970, Congress decided that the law should reflect a much more detailed understanding of psychoactive (affecting the mind or behavior) controlled substances. The Controlled Substances Act classifies a drug into one of five schedules based upon the drug's potential for harm relative to its medical benefit. The higher the schedule, the lower the risk of abuse potential. For example, schedule V drugs (C-V) have less abuse potential than schedule II (C-II) drugs. The Act applies to veterinarians largely because they have access to controlled substances. Table 1-2 describes the schedules, lists their definitions, and gives examples of drugs in each category.

Veterinarians who wish to use or prescribe controlled substances must register annually with the DEA, keep it informed of all address changes, and receive a registration number to be used on all prescriptions and supply order forms. Controlled substances must be stored in a locked cabinet or, preferably, in a safe attached to a concrete floor. The registered veterinarian and other authorized handlers must keep records of orders, receipts, uses, discards, and thefts of controlled substances for two years following each transaction. This record is often in the form of a controlled substance log (Figure 1-1). The veterinarian must keep an inventory of controlled substances on hand and file it with the DEA every two years. Inventory of controlled substances on the premises is also needed for unannounced inspections by DEA agents.

Both the veterinarian and the pharmacist are responsible for the proper prescribing and dispensing of controlled substances. The manufacturer and distributor must identify a controlled substance on the label of the original container by a symbol corresponding to the above schedules specified by the DEA (Figure 1-2). Though federal, state, and municipal regulations governing purity, manufacture, sale, and dispensing of drugs may differ, the most stringent regulation takes precedence. The federal regulations of the Controlled Substances Act for recording and storage of controlled substances are often minimum requirements.

TIP

The FDA, a branch of the Department of Health and Human Services, regulates the development and approval of drugs. The laws and rules regarding the purchase, storage, and use of controlled substances are regulated by the DEA, which is a branch of the Justice Department.

TABLE 1-2 Controlled Substances Categories and Examples

Drug Schedule	Definition of Schedule	Examples
Schedule I (C-I)	Substance has high potential for abuse and has no currently accepted medical use; there is a lack of accepted safety for use (considered most dangerous, with virtually no medical benefit)	heroin, LSD, marijuana
Schedule II (C-II)	Substance has high potential for abuse but has currently accepted medical use, with severe restrictions	cocaine, morphine, amphetamines, pentobarbital, etorphine, fentanyl, codeine
Schedule III (C-III)	Substance has potential for abuse less than schedule I and II drugs and has accepted medical uses	acetaminophen/codeine combinations, ketamine, thiamylal, thiopental, hydrocodone
Schedule IV (C-IV)	Substance has low potential for abuse relative to drugs in schedule III and has accepted medical uses	diazepam (Valium®), phenobarbital, butorphanol
Schedule V (C-V)	Substance has low potential for abuse relative to drugs in schedule IV and has accepted medical uses	buprenophrine, diphenoxylate (Lomotil®), codeine cough syrups

Drug Name: Ketamine Strength: 100 mg/ml Bottle Number: 34

Date	Owner's Name	Pet's Name	Amount Drawn	Amount Used	Amount Discarded	Amount Remaining	Initials
01/10/03	Smith	Gilbert	1.0cc	1.0cc	0	9.0cc	JR

FIGURE 1-1 Controlled drug log

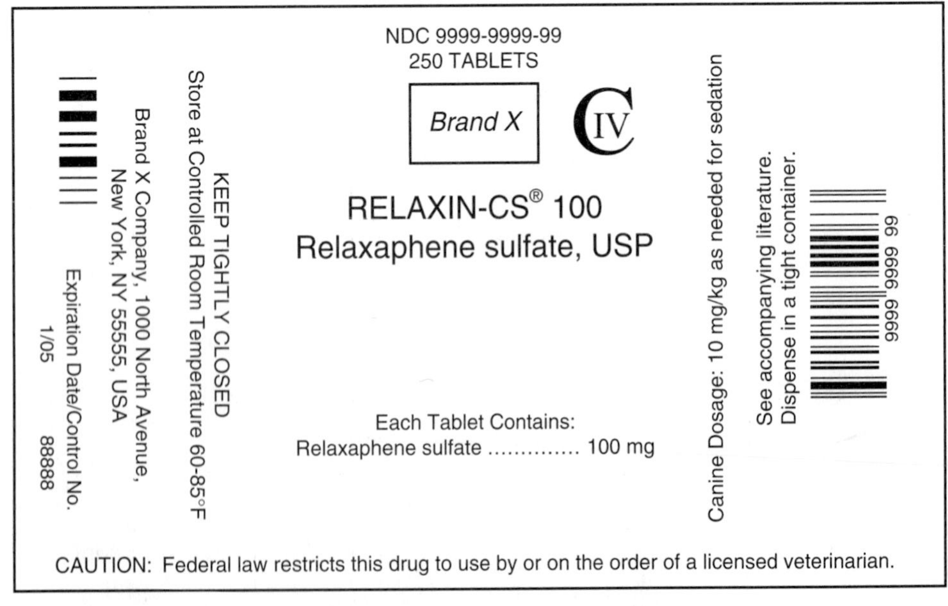

NDC 9999-9999-99
250 TABLETS
Brand X
CIV
RELAXIN-CS® 100
Relaxaphene sulfate, USP
Each Tablet Contains:
Relaxaphene sulfate .... 100 mg
KEEP TIGHTLY CLOSED
Store at Controlled Room Temperature 60-85°F
Brand X Company, 1000 North Avenue,
New York, NY 55555, USA
Expiration Date/Control No.
1/05 88888
Canine Dosage: 10 mg/kg as needed for sedation
See accompanying literature.
Dispense in a tight container.
9999 9999 99
CAUTION: Federal law restricts this drug to use by or on the order of a licensed veterinarian.

FIGURE 1-2 Sample controlled-substance medication label

SUMMARY

The history of treating sick animals can be traced to ancient times; however, the use of drug treatment in animals is more recent. Studying drug therapies and drug movement in animal bodies allows veterinarians to use drugs effectively. Drug regulation, for both people and animals, is accomplished through the FDA.

The FDA carries out varying levels of drug monitoring. Its oversight responsibilities include prescription drugs, extra-label drugs, over-the-counter drugs, and controlled substances. Controlled substances are the most heavily regulated substances, because of their abuse potential by humans, and are regulated by the DEA.

CHAPTER REFERENCES

Drug Enforcement Administration. Web site at <http://www.dea.org>

Food and Drug Administration. Web site at <http://www.fda.org>

Wynn, R. (1999). *Veterinary pharmacology*. Cambridge, MA: Harcourt Learning Direct.

CHAPTER REVIEW

Matching

Match the term or phrase with its proper definition

1. ______ drugs that can be purchased without a prescription
2. ______ drugs considered dangerous because of their potential for human abuse or misuse
3. ______ drugs that can be obtained only through a veterinarian or via a prescription
4. ______ drugs used in a manner not specifically described on the FDA-approved label
5. ______ study of a drug's mechanism of action and its biological and physiological effects
6. ______ study of absorption, blood levels, distribution, metabolism, and excretion of drugs
7. ______ the treatment of disease with medicines
8. ______ the study and use of drugs in animal health care
9. ______ the law that allows extra-label use of a drug under certain conditions
10. ______ agency that ensures that approved veterinary medicines are relatively safe for animals

a. pharmacodynamics
b. controlled substances
c. pharmacokinetics
d. over-the-counter drugs
e. pharmacotherapy
f. prescription drugs
g. extra-label drugs
h. veterinary pharmacology
i. FDA-CVM
j. Animal Medicinal Drug Use Clarification Act of 1994

Multiple Choice

Choose the one best correct answer.

11. The FDA became a government agency after the passage of the
   a. federal Food and Drug Act of 1906.
   b. Controlled Substances Act of 1970.
   c. Food, Drug, and Cosmetic Act of 1938.
   d. 1972 New Animal Drugs amendment to the Food and Drug Act.

12. A person studying how the body absorbs, uses, and gets rid of codeine is engaged in the pharmacological specialty called
   a. pharmacotherapeutics.
   b. pharmacodynamics.
   c. pharmacokinetics.
   d. pharmaconeurology.

13. Controlled substances must
   a. be kept in a locked cabinet or safe.
   b. have orders, receipts, uses, and thefts recorded.
   c. be ordered by veterinarians who register annually with the DEA.
   d. All of the above are true.

14. The higher the schedule number (for example, V versus I) of a controlled-substance drug,
   a. the higher the risk for human abuse potential.
   b. the more questionable its manufacture is.
   c. the lower the risk for human abuse potential.
   d. the less medical value it has.

True/False

Circle a. for true or b. for false.

15. Prescription drugs are limited to use under the supervision of a veterinarian or physician.
   a. true
   b. false

16. The majority of veterinary drugs in use during the early 1900s were found naturally in plants.
   a. true
   b. false

17. The major requirement of the Food, Drug, and Cosmetic Act of 1938 is the requirement of drug safety.
   a. true
   b. false

18. Valium® is an example of a schedule I drug.
   a. true
   b. false

19. Over-the-counter drugs are approved for human use only by the FDA.
   a. true
   b. false

20. All drugs are utilized on sick animals.
   a. true
   b. false

Case Study

21. An owner of a 12-year-old male/neutered (M/N) German Shepherd calls the clinic because her dog has been vomiting blood. She says the dog was fine yesterday and has been more active since the owner starting giving aspirin to relieve the pain associated with the dog's arthritis. You explain to the owner that aspirin can cause gastrointestinal upset, and that some signs the animal may show are vomiting and diarrhea. The owner says that it is impossible for the aspirin to be causing the dog to vomit blood, because aspirin can be purchased without a prescription.
   a. What do you tell this owner?
   b. What advice can you give this owner?

CHAPTER 2

Veterinary Drug Development and Control

OBJECTIVES

Upon completion of this chapter, the learner should be able to

- outline the stages of drug development in the United States.
- differentiate between effective dose and lethal dose.
- define margin of safety, calculate it, and understand how it relates to the appearance of toxicity signs.
- describe how side effects related to reproduction, carcinogenicity, and teratogenicity are monitored.
- describe the methods of drug marketing in the United States.

KEY TERMS

preliminary studies
preclinical studies
carcinogenicity
teratogenicity
clinical trial
shelf life
short-term tests
long-term tests
special tests
toxicity evaluation
parameter
effective dose-50 (ED_{50})
lethal dose-50 (LD_{50})
margin of safety
therapeutic index
systems-oriented screen
long-term toxicity test
chronic study
short-term toxicity test
direct marketing
distributor
wholesaler
generic company
bioequivalency

INTRODUCTION

A horse is diagnosed with a type of cancer that requires long-term treatment. The owner of this horse knows that drugs used to treat cancer have many side effects and may even be lethal. The owner wants to know how drugs like this can be approved and how the drug's side effects are monitored. The owner is also concerned about the cost of this treatment. Can you explain why some drugs are expensive and what sources may be available to lower the costs for this owner? Can you explain the drug approval process to the client in clear and concise terms that he can understand? Why are some drugs more expensive than other drugs that are very similar in composition? How are drug side effects monitored?

HOW DO VETERINARY DRUGS GET HERE? THE STAGES OF VETERINARY DRUG DEVELOPMENT

The development of a new veterinary drug for use in the United States requires evaluation of its effects on animals through a series of tests mandated by the FDA. These mandatory evaluations require the drug company to spend a great deal of resources, both time and money, to prove that the new drug is safe and effective. It can typically take an average of seven years of

TIP

Discovering and developing safe and effective new medicines is a long, difficult, and expensive process.

TIP

The FDA Center for Veterinary Medicine (CVM) is responsible for ensuring the safety and efficacy of animal drugs and medicated feeds.

testing and millions of dollars to bring a new veterinary drug to market. Table 2-1 identifies the regulatory agencies involved in animal health products. Figure 2-1 presents an overview of the animal test phases of drug development.

The four major steps in drug development, as illustrated in the figure, are

1. Synthesis/discovery of a new drug compound
   - Preliminary studies
2. Safety/effectiveness evaluation
   - Preclinical studies
   - Clinical trials
3. Submission and review of the new animal drug application (NADA)
   - Review by FDA (EPA or USDA)
   - Approved or rejected based on clinical trial data
4. A postmarketing surveillance stage
   - Product monitoring for safety and effectiveness
   - Listing in *Green Book*, a list of animal drug products published and updated monthly by the Drug Information Laboratory. Information on the Green Book can be found at <http://www.fda.gov>; go to the CVM section and use key search term: green book.

Overview of Drug Development

When scientists synthesize or discover a substance with potential therapeutic value (Step 1), it must be tested in a series of **preliminary studies**. Preliminary studies are performed to determine if the drug produces the intended effect(s) and whether it produces toxic side effects. These tests may include simulated testing via computer models, testing in laboratory media, or testing on simple organisms such as bacteria or fungi.

If the preliminary studies produce satisfactory results, **preclinical studies** begin. Preclinical studies are a series of tests performed on laboratory animals to determine safety and effectiveness (Step 2) of the drug (*target species*, those species in which the drug is intended for use, may also be used in the preclinical studies). Safety and effectiveness tests include short-term and long-term toxicity studies and special tests of immediate drug reactions, organ system damage, reproductive effects, **carcinogenicity** (the ability or tendency to produce cancer), and **teratogenicity** (the capacity to cause birth defects).

When sufficient animal data demonstrate the new drug's relative safety and effectiveness, researchers submit an Investigational New Animal Drug (INAD) application to the FDA. If the product is a pesticide, the scientists must file for an Experimental Use Permit (EUP) with the Environmental Protection Agency (EPA); if the product is a biologic, the scientists must file with the Animal and Plant Health Inspection

TABLE 2-1 Regulatory Agencies Involved in Animal Health Product Approval

Government Agency	Area of Regulation
FDA (Food and Drug Administration)	Development and approval of drugs
EPA (Environmental Protection Agency)	Development and approval of topical pesticides
USDA (United States Department of Agriculture)	Development and approval of biologics such as vaccines and antitoxins

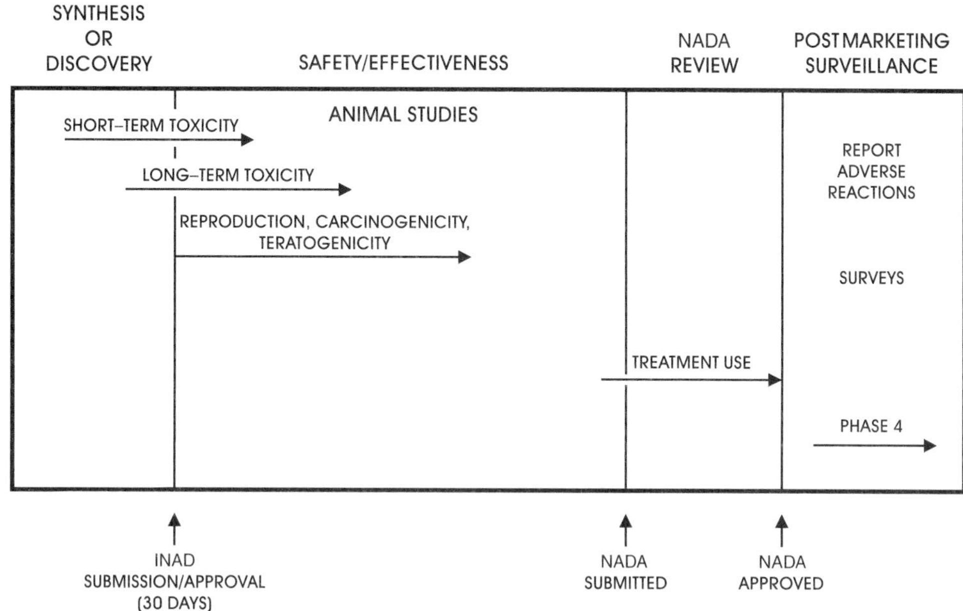

FIGURE 2-1 Animal test phases of drug development

Services (APHIS) of the USDA. The FDA reviews the INAD application and responds within 30 days.

If the FDA approves the application (the desired outcome of Step 3), researchers may proceed with **clinical trials** on the target species. Clinical trials are conducted in the target species and are done to evaluate the drug's safety and effectiveness in that species. Any toxic or adverse side effects in the target species are determined in the clinical trials as are tissue residue and withdrawal time information. Studies on **shelf life** (i.e., stability studies), to show how long a drug remains stable and effective for use, are also conducted.

Satisfactory clinical trial results allow the scientists to file a New Animal Drug Application (NADA) with the FDA (or other agencies for pesticides and biologics). All of the research studies are then submitted to the FDA, USDA, or EPA for review. If the results of testing are favorable, an approval and license for manufacture are granted. A new animal drug is deemed unsafe unless there is an approved NADA for that particular drug. Following approval of the drug, the company and the government monitor the product as long as the drug is manufactured. This monitoring ensures product safety and efficacy.

> **TIP**
>
> Preclinical studies involve laboratory testing to show the biological activity of the compound against the targeted disease and evaluation of the compound for safety.
>
> Clinical trials are conducted in different phases, each assessing a different aspect of the drug. Phase I looks at toxicity effects and may include absorption, distribution, metabolism, and excretion of the drug and its duration of action. Phase II monitors the drug's effectiveness. Phase III monitors efficacy and adverse reactions.

Safety and Effectiveness Evaluation

All drugs that are approved by the FDA must be tested to determine if they are effective and whether they cause side effects. If the drug is intended for use in food-producing animals, it must also be tested for safety to human consumers, and the edible animal products must be free of drug residues. The company developing the drug must also devise analytical methods to detect and measure drug residues in edible animal products. **Short-term tests** occur in the hours following a test dose, to check the animal for such obvious adverse reactions as convulsions, paralysis, depressed breathing, depressed heart rate, and death. **Long-term tests** typically run for 3 to 24 months of repeated dosing, to check the animal's various organ systems for toxicity damage. **Special tests**, such as tests done to determine reproductive effects, carcinogenicity, and teratogenicity, are both short- and long-term—hence the "special test" designation. Reproductive tests evaluate the drug's effect on conception, fertilization, and pregnancy. Carcinogenicity tests check to see if the drug causes cancerous tumors in such soft tissues as the urinary bladder or brain. Teratogenic tests check for development of fetal defects in pregnant test animals.

Toxicity Evaluation

Drugs must also be tested to determine if they cause toxic side effects. Typically conducted on mice, the **toxicity evaluation** is done to determine the dose at which a drug induces organ or tissue damage that may, at a high enough dose, result in permanent injury or death. Figure 2-2 shows a typical data sheet researchers would use to collect this information.

DATA SHEET
SHORT–TERM TOXICITY EVALUATION
GENERAL PROFILE SCREEN IN ANIMAL

COMPOUND # ________ DOSE ________ DATE ________
ANIMAL # ________
ANIMAL WT. ________
NORMAL HR ________
NORMAL BP ________
NORMAL RR ________

PARAMETER — TIME IN MINUTES AFTER DRUG

	1	5	10	15	30	45	60	120	180
RESPIRATION RATE									
HEART RATE									
BLOOD PRESSURE									
RELAXATION									
SEDATION/SLEEP									
SEIZURES									
CONVULSIONS									
OTHER (LIST)									

NOTES/COMMENTS

____
INVESTIGATOR

____
WITNESS

____
DATE

FIGURE 2-2 Data sheet used to collect toxicity evaluation information

The top areas of the sheet are for the compound (drug) number, the animal's number, the animal's weight, the dose of the drug given to the animal, the date of the report, and the animal's blood pressure, and its heart and respiration rates (control rates) before drug dosing. The left-hand column lists the **parameters** (the intensity of effect measured on a subjective scale of either 1 through 3 or 1 through 10) observed and the times of observation. The bottom is reserved for investigator notes and the signatures of the investigator and a witness. The data sheet records information vital to determining everything researchers want to know about a drug, including the key questions: "Does it do anything at all?" and "How much does it take to do it?"

Effective and Lethal Dose Evaluation

Researchers must also determine the amount of a drug, or *dose*, that produces a desired effect. This is called the *effective dose*. A dose can be called effective only if the amount of the test drug causes a defined effect in 50 percent of the animals that receive it. For example, if 50 out of 100 mice display a slower heart rate at a certain dose of the drug, that dose is the amount of drug that causes the desired effect. This term is abbreviated **ED_{50}** for **effective dose-50**.

In addition to finding a drug's effective dose, researchers must also determine the lethal dose. The *lethal dose* is the dose of a test drug that kills 50 percent of the animals that receive it. This figure is important for two reasons: (1) drugs that kill with a small dose are rarely evaluated further; and (2) every drug marketed must include lethal dose information in the accompanying official drug monograph (a written account of a single drug). The 50 percent figure means, for example, that if 26 mice receive the same dose of a test drug and 13 of them die, then the dose that all 26 received is the test drug's lethal dose. The lethal dose is abbreviated **LD_{50}** for **lethal dose-50**. Table 2-2 lists some typical LD_{50} values.

The Therapeutic Index

Another important part of the short-term toxicity phase is determining the test drug's **margin of safety**, also referred to as the **therapeutic index**. The therapeutic index is the drug dosage or dose that produces the desired effect with minimal or no signs of toxicity. This value is determined by comparing the drug's lethal dose (LD_{50}) and its effective dose (ED_{50}). A wide therapeutic index or margin of safety means that the drug can produce its desired effect without approaching toxicity. The wider this therapeutic index or margin of safety, the better. The lethal dose is used as the measure of toxicity because death is easily observed, whereas other toxic effects can be ambiguous.

If the LD_{50} and ED_{50} values are known for a given drug, its therapeutic index is determined by dividing the LD_{50} value by the ED_{50} value. For example, if Drug A has an LD_{50} of 100 milligrams per kilogram and an ED_{50} of 2 milligrams per kilogram, the therapeutic index is

$$\text{Drug A } LD_{50}/ED_{50} = 100 \text{ mg per kg}/2 \text{ mg per kg} = 50$$

TABLE 2-2 LD_{50} Values of Some Well-Known Drugs

Drug	LD_{50} in Mice
Valium® (tranquilizer)	720 mg/kg given orally
Cardizem® (blood pressure)	500 mg/kg given orally
Vasotec® (blood pressure)	2000 mg/kg given orally

If Drug B has an LD_{50} value of 10 milligrams per kilogram and an ED_{50} value of 5 milligrams per kilogram, the therapeutic index is

$$\text{Drug B } LD_{50}/ED_{50} = 10 \text{ mg per kg}/5 \text{ mg per kg} = 2$$

TIP

The larger the therapeutic index, the safer the drug.

Drug A's therapeutic index, 50, means that one would have to take 50 times the effective dose to ingest the lethal dose. Drug B's therapeutic index, 2, means that one would have to take only twice the effective dose to ingest the lethal dose. Therefore, Drug A has a substantially larger therapeutic index or margin of safety.

Systems-Oriented Screen

After the toxicity evaluation, researchers conduct a **systems-oriented screen**, a test of the drug's effect on a particular physiological system. This test evaluates effects of interest from the toxicity evaluation as they affect a specific body system. For example, a systems-oriented screen of the cardiovascular system might ask: How low did blood pressure go, and with what dose? How much was the heart rate reduced? Did any dose have no effect on blood pressure or heart rate? Body systems typically tested include the cardiovascular, respiratory, muscular, nervous, and endocrine systems. Around this time, researchers also start tests that will go on for many months. These tests are done to find out if the drug has any long-term side effects.

Evaluation of Long-Term (Chronic) Effects

Over the years, drugs have found their way to the marketplace before anyone knew of their dangerous long-term effects. Researchers conduct **long-term toxicity tests**, also called **chronic studies**, that run anywhere from three months to two years, to prevent such disastrous consequences of a drug therapy. Throughout long-term toxicity studies, test animals are given regular doses of the test drug. At the end of the study, researchers observe the animals for toxic effects that did not appear in the short-term studies. Researchers check each animal's behavior, as well as its blood, urine, and eyes. They then euthanize the animal and examine its tissues. Typical toxic effects monitored include liver, heart, and nervous system disorders. Many of these toxic effects appear only after repeated dosing over a longer period of time.

A **short-term toxicity test** that produces adverse reactions may prompt the manufacturer to terminate the drug testing. Further tests sometimes exonerate the drug as the cause of short-term adverse reactions, but usually the study of the drug is terminated because so many other drug candidates await testing. In any case, the next tests do not happen at all unless the bulk of the toxicity tests confirm a degree of safety and effectiveness in the test drug. These tests are too expensive and time-consuming to perform on any drug that does not show significant promise.

Evaluating Reproductive Effects, Carcinogenicity, and Teratogenicity

Just as a drug that lowers the heart rate of a mouse may lower a dog's heart rate, a drug that impairs the fertility of a laboratory rat may do the same to other animal species. Drugs linked to cancer and birth defects have found their way into human and veterinary medicine in the past. Researchers and the FDA want to do everything possible to make sure that drugs with the potential to cause serious disease and side effects are not marketed and used on animals in a clinical setting.

Reproductive tests are carried out in test animals to answer the following questions:

1. Does the test drug prevent egg cell production?
2. Does the test drug prevent fertilization of the egg cell by the sperm?

3. Does the test drug prevent uterine growth of the fertilized egg cell?
4. Does the drug cause early expulsion of the embryo from the uterus?

If the answer to any of these questions is yes, testing on the drug usually ends.

FDA-approved drugs must be free of any cancer-causing potential. Researchers give large daily doses of a test drug to thousands of test animals for six months to see if cancerous tumors appear. After six months they inspect each animal, particularly its urinary bladder and related organs, for tumors. If they find any, testing on the drug usually ends.

The FDA requires that all drugs be tested to see if they cause fetal defects in animals. Researchers give the test drug to hundreds of pregnant laboratory animals. Researchers then inspect the embryos for such defects as abnormal bone formation, shortened limbs, cleft palate, open skulls, and open spinal columns. Because these defects increase the risk to all pregnant animals taking the drug, the FDA will typically not approve it for general use.

TIP

Once the FDA approves a drug, the company must continue to submit periodic reports assessing adverse reactions, as well as the appropriate quality-control records.

HOW DO DRUGS GET TO YOU? DRUG MARKETING SYSTEMS

Drugs purchased by veterinarians can be marketed in a variety of ways. **Direct marketing** is when a drug is purchased directly from the company that manufactures it. Another way to purchase pharmaceuticals is through distributors. **Distributors** or **wholesalers** are agencies that purchase the drug from the manufacturing company and resell it to veterinarians. Distributors can purchase products from many different companies and make these products available to veterinarians via sales representatives or telephone order systems. **Generic companies** sell drugs that are no longer under patent protection. Once a drug patent expires, other drug companies can apply to the FDA to sell a generic equivalent drug, a drug determined to be the therapeutic equal of a drug on which the patent has expired. The application for generic equivalent approval is called an Abbreviated New Drug Application (ANDA). Although no preclinical or clinical data need be included on this application if the generic drug is identical to the original in potency, dosage form, and product labeling, the applicant must demonstrate the generic drug's **bioequivalency**, or ability to produce similar blood levels after administration as the original patented drug.

TIP

Drugs intended for administration to animals fall under the same current good manufacturing practices (cGMPs) requirements as do drugs for humans.

Some companies advertise prescription products for sale to non-veterinarians. These companies do require a prescription from a veterinarian before they will allow clients to purchase drugs for their animals.

SUMMARY

The development of new veterinary drugs in the United States takes time and extensive testing to ensure drug effectiveness and low-level or zero toxicity. Agencies involved in drug testing include the FDA (for drug approval), the EPA (for pesticide approval), and the USDA (for biologics approval). The stages of drug approval include preliminary and preclinical studies, clinical trials, FDA (or other agency) review, product approval or rejection, and product monitoring after the drug is marketed.

An effective dose and lethal dose are determined in the process of developing drugs. The effective dose (ED_{50}) is the amount of a drug that produces a desired effect in 50 percent of the population. The lethal dose (LD_{50}) is the dose of drug that kills 50 percent of the animals that take it. Both the effective dose and the lethal dose are used to calculate the drug's therapeutic index or margin of safety, which is simply a comparison between the lethal and effective doses.

A wide therapeutic index or margin of safety means that the drug can produce the desired effect without approaching toxicity. Systems-oriented screens are used to test a drug's effect on a particular physiological system. Long-term studies, including examination of reproductive effects, carcinogenicity, and teratogenicity, are also conducted and are important in monitoring the safety of individual drugs.

Drugs are marketed by direct marketing, distributors or wholesalers, and generic companies. All these marketing techniques have advantages and disadvantages for the veterinarian and staff.

CHAPTER REFERENCES

Beary, J. (Fall 1997). The drug development and approval process. Newsletter of Aplastic Anemia Foundation of America, Inc.

Wynn, R. (1999). *Veterinary pharmacology.* Cambridge, MA: Harcourt Learning Direct.

Zaret, E. (August/September 2001). Equating Fido's drugs with human quality standards. *Pharmaceutical Formulation & Quality.*

CHAPTER REVIEW

Matching

Match the term or abbreviation with its proper definition.

1. ______ NADA
2. ______ FDA
3. ______ EPA
4. ______ USDA
5. ______ INAD
6. ______ clinical trials
7. ______ preclinical studies
8. ______ therapeutic index
9. ______ systems-oriented screen
10. ______ company that sells drugs that are no longer under patent protection

a. studies conducted in the target species that are done to prove that the drug is safe and effective in that species
b. new animal drug application
c. series of tests performed on laboratory animals to determine safety and effectiveness of the drug
d. government agency that develops and approves drugs
e. government agency that develops and approves biologics such as vaccines and antitoxins
f. government agency that develops and approves topical pesticides
g. test of a drug's effect on a particular physiological system
h. generic company
i. drug dosage or dose that produces the desired effect with minimal or no signs of toxicity
j. investigational new animal drug

Multiple Choice

Choose the one best correct answer.

11. Direct marketing of veterinary drugs
    a. is done by agencies that purchase the drugs from manufacturers and resell them to veterinarians.
    b. is done by generic companies that sell generic drugs under their own companies's names.
    c. occurs when a drug is purchased by the veterinarian directly from the company that manufactures it.
    d. is not utilized in the veterinary community.
12. The margin of safety is often referred to as
    a. the effective dose.
    b. the lethal dose.
    c. the safety parameter.
    d. the therapeutic index.
13. The LD_{50} / ED_{50} is the mathematical expression of what value?
    a. the lethal dose
    b. the effective dose
    c. the margin of safety
    d. the mortality dose
14. A drug that has a margin of safety of 75 is
    a. safer than a drug whose margin of safety is 5.
    b. less safe than a drug whose margin of safety is 5.
    c. more likely to cause toxic side effects.
    d. not marketable in the United States.
15. How long a drug remains stable and effective for use is known as its
    a. half-life.
    b. shelf life.
    c. effective life.
    d. special test life.
16. The term used to describe the capacity to cause birth defects is
    a. reproductivity.
    b. carcinogenicity.
    c. teratogenicity.
    d. theriogenicity.

True/False

Circle a. for true or b. for false.

17. The FDA is responsible for approval of all chemicals dispensed by veterinarians.
    a. true
    b. false
18. Once the FDA approves a drug, it is no longer monitored for safety and effectiveness because it has already undergone extensive testing prior to approval.
    a. true
    b. false
19. Satisfactory clinical trial results allow scientists to file a NADA with the FDA.
    a. true
    b. false
20. As the margin of safety decreases, the lower the level of drug needed to produce the lethal dose.
    a. true
    b. false

Case Study

21. A client calls your office to ask a question regarding his animal's medication. He is currently giving his dog one antibiotic tablet twice a day to treat a skin infection. The client is planning to go on vacation and is wondering if his dog sitter could give four antibiotic tablets once every two days instead of one antibiotic tablet four times over two days. Because the total dose over the two days would be the same, the client thinks that this would save the dog sitter some time and trouble.

    Is it recommended for this dog to receive its entire two-day dose at one time? Why or why not?

CHAPTER 3

Therapeutic Range and Routes of Administration

OBJECTIVES

Upon completion of this chapter, the learner should be able to

- describe the therapeutic range and its role in drug efficacy.
- describe graphically the components of therapeutic range, subtherapeutic range, and toxicity levels.
- describe the three components of the therapeutic range.
- compare and contrast the different routes of administration with regard to rate of onset, formulation, and use.
- differentiate between bolus administration, intermittent therapy, and infusion of fluid IV techniques.
- describe concepts of drug dosing.
- describe dosage and its use in drug dosing.

KEY TERMS

therapeutic range
parenteral
nonparenteral
intravenous (IV)
intramuscular (IM)
subcutaneous (SC or SQ)
intraperitoneal (IP)
intradermal (ID)
intra-arterial (IA)
epidural
subdural
intrathecal
intracardiac (IC)
intra-articular
intramammary
intramedullary
intraosseous
emulsion
immiscible
repository (depot) preparation
inhalation
volatilize
nebulize
topical
oral
solution
suspension
dose
loading dose
maintenance dose
total daily dose
dosage
dosage interval
dosage regimen

INTRODUCTION

A cat presented to your clinic is having extreme difficulty breathing. The cat is breathing with its mouth open and you notice that its gums and mucous membranes are not as pink as they should be. The veterinarian quickly begins a physical exam to assess the situation. The respiratory sounds are difficult to hear because of fluid in the cat's chest. The veterinarian asks you to get a drug that will remove some fluid from the cat's chest to help it breathe easier. You rush over to the drug cabinet and notice that there are different forms of the drug the veterinarian asked you to get. Which one do you reach for, and why? Are you be able to determine which route of drug administration the veterinarian would choose, and are you prepared to assist in the drug administration? In emergency situations, what are the most rapid routes of drug administration? What drug forms are available for these routes of administration?

KEEPING SAFE: THE THERAPEUTIC RANGE

Drugs are capable of producing a wide variety of effects in animals. All drugs should be considered potential poisons and should be dispensed and given with great care. Only appropriate administration of the appropriate drug at the appropriate dosage determines whether a compound benefits an animal or causes toxicity. Appropriate administration of a drug includes administration of the appropriate amount of drug, based on dosage and introduction of the appropriate amount of drug into the animal's body by the appropriate route of administration. The goal of drug therapy is to deliver the desired concentration of a drug within the target area of the body to ultimately achieve the desired effect. The **therapeutic range** of a drug is the drug concentration in the body that produces the desired effect in the animal with minimal or no signs of toxicity (Figure 3-1A and B). Laboratory and clinical evaluation determine the drug dosage used to get and keep the drug level in the therapeutic range. The concept that "more is better" does not hold true: more drug—either an increased amount or increased frequency of dosing—can produce damage to body organs. Similarly, the concept that "less is better" does not hold true: less drug will not produce toxic side effects, but it will not achieve proper levels in the body to enable the beneficial effects of the drug.

Many factors play a role in getting and keeping drugs in the therapeutic range. Some factors involve properties of the drugs; other factors involve the health and physiology of the animal. Maintaining drugs in the therapeutic range involves maintaining a balance among the rate of drug entry into the body, absorption of the drug, distribution of the drug, metabolism of the drug, and excretion of the drug.

TIP

The rights of proper drug administration serve as a checklist of activities to be followed by those giving medication. These rights include:

- right drug
- right dose
- right route
- right time
- right patient
- right documentation

STAYING IN THE SAFE ZONE

Three major drug factors involved in staying within a drug's therapeutic range include route of administration, drug dose, and dosage interval.

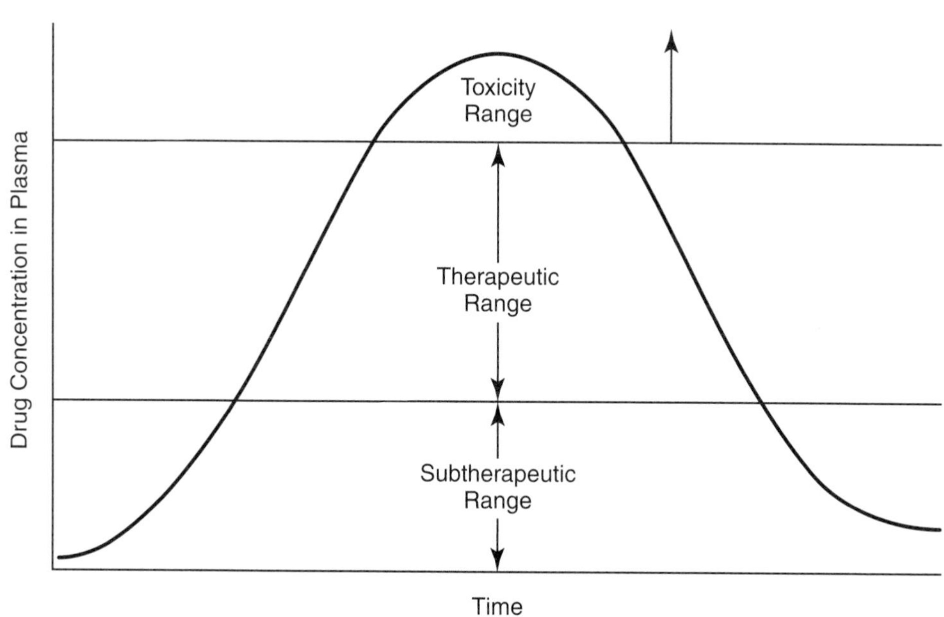

FIGURE 3-1A Therapeutic range

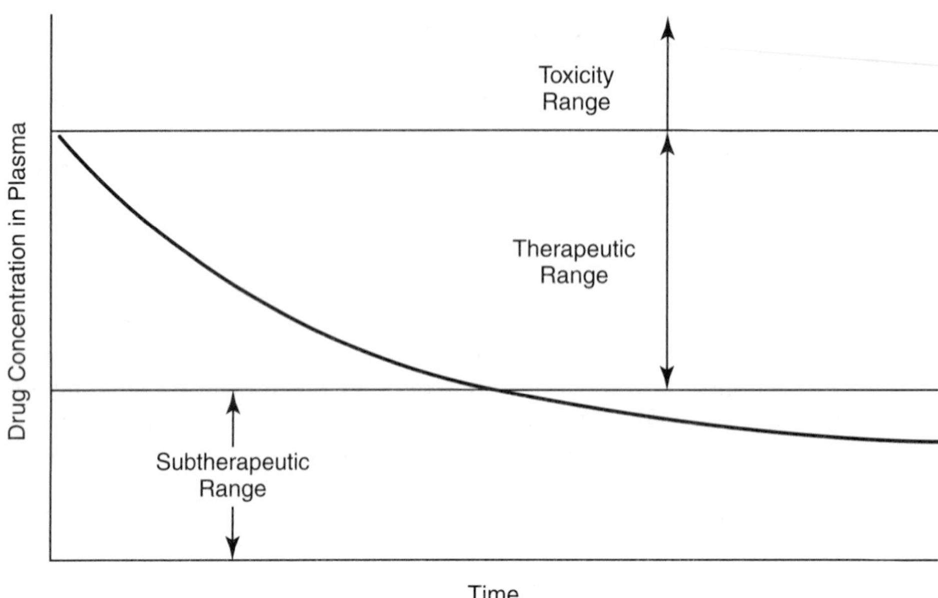

FIGURE 3-1B Therapeutic range for a drug given IV

Route of Administration

The route of administration is how you administer a drug or the manner in which a drug enters the body. Drugs can enter the body in two ways: parenterally and nonparenterally. **Parenteral** drugs are given by a route other than the gastrointestinal tract and **nonparenteral** drugs are given through the gastrointestinal tract (orally) (Figure 3-2).

Many factors are considered before the route of administration for a drug is selected. Some factors are based on the drug itself; other factors are based on the animal being treated. Drug factors that influence the route of administration include the following:

- Some drugs cause one effect when given parenterally and another effect when given nonparenterally. For example, magnesium sulfate causes muscle relaxation when given intravenously (IV) and diarrhea when given orally.
- Some drugs are insoluble in water and can be given intramuscularly (IM) but cannot be injected IV.
- Some drugs are destroyed by stomach acid and cannot be given orally.

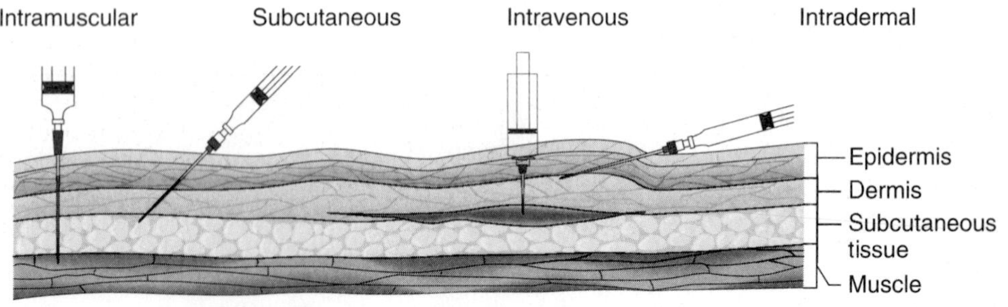

FIGURE 3-2 Some examples of parenteral routes of drug administration

Animal factors also influence the choice of route of drug administration. Animals that are actively vomiting cannot absorb drugs given orally and must receive drugs through another route of administration. Critically ill patients need to get therapeutic levels of drug in their bodies more rapidly than moderately ill patients do; therefore, critical patients tend to receive IV medication while moderately ill patients may be treated with oral medications.

Injectable Administration and Dose Forms Drugs administered by injectable routes are types of parenteral drugs. *Parenteral* literally means "excluding the intestines" (*para-* = apart from, *entero* = intestine). Injectable drugs are administered by needle and syringe (Figures 3-3, 3-4, and 3-5). A syringe consists of a barrel, plunger, flange, and tip. The barrel holds the medication and has graduated markings (calibrations) on its surface for measuring medication. The plunger is a movable cylinder that is inserted within the barrel and forms a tight-fitting seal. The plunger is the mechanism by which a medication is drawn into and pushed out of the barrel. The flange is at the end of the barrel where the plunger is inserted. It forms a rim around the end of the barrel against which one places the index and middle fingers when drawing up solution for injection. The tip is the end of the barrel where the needle is attached. Syringes come in many sizes and are chosen based on the animal, route of administration, and amount of drug to be given. For information on measuring doses in syringes, consult the Accu-Calc CD provided with this textbook.

Injectable drugs are usually supplied as sterilized solutions, prepackaged syringes with needles for injection, powders that must be reconstituted with sterile solution, or in vials to be drawn up into syringes for injection. Vials that store injectable

TIP

To help remember routes of administration, remember key word parts. *Intra-* means within, *inter-* means between, *sub-* means under, *epi-* means above, *trans-* means across.

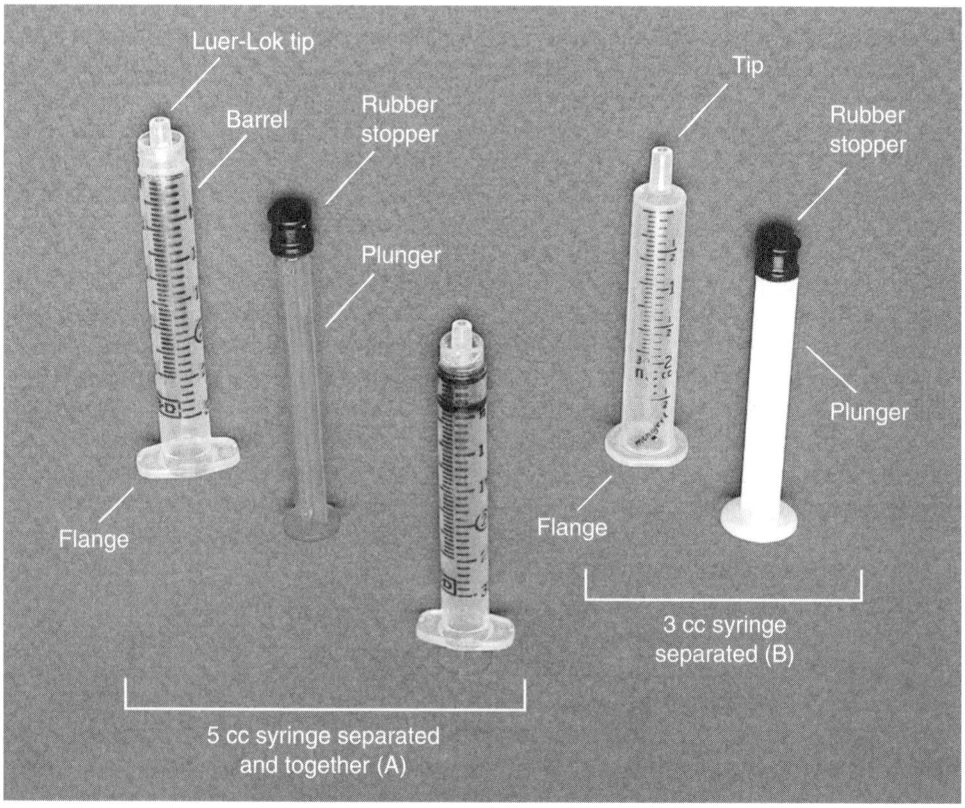

FIGURE 3-3 Parts of a syringe: (A) A 5 cc syringe separated and unseparated, with LUER-LOK® tip. (B) A 3 cc syringe separated with plain tip (Courtesy of Becton, Dickinson and Company)

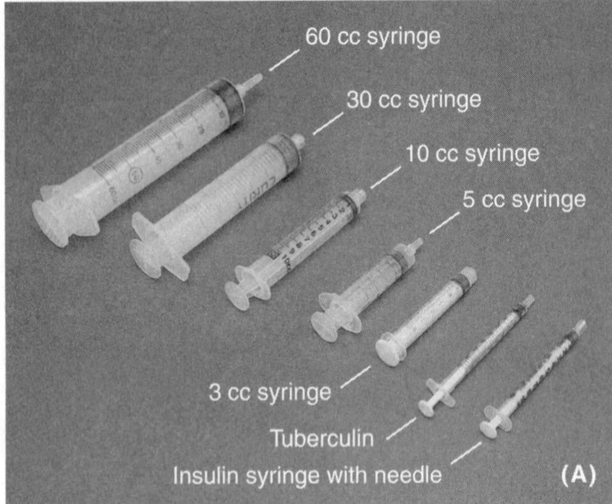

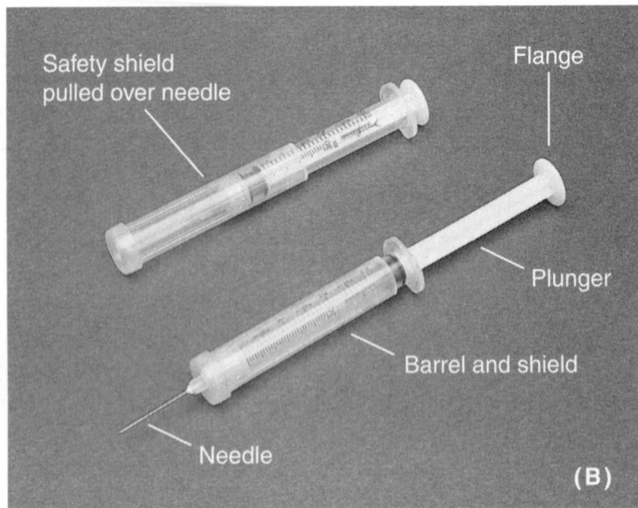

FIGURE 3-4 (A) Various sizes of disposable syringes. (B) A type of safety syringe (Courtesy of Becton, Dickinson and Company)

drugs can be multidose, meaning that the vials contain more than one dose, or single dose, meaning that the vials contain only one drug dose. Injectable drugs can also be stored in ampules, which are small, sterile, prefilled glass containers containing medication for injection (Figure 3-6).

The three most common injectable administration routes in animals are **intravenous**, or within a vein; **intramuscular**, or within the substance of a muscle; and **subcutaneous**, or under the skin. Other less common routes of injectable drug administration are **intraperitoneal (IP)**, **intradermal (ID)**, **intra-arterial (IA)**, **epidural/subdural/intrathecal**, **intracardiac (IC)**, **intra-articular**, **intramammary**, and **intramedullary/intraosseous**.

One of the fastest means of getting drugs into the bloodstream is intravenous injection. Figure 3-7 shows a dog receiving an IV injection. The IV route of administration gives a predictable concentration of drug and usually produces an immediate response. Most IV injections are aqueous solutions (solutions of drugs dissolved in liquid, usually water), but a few are emulsions. An **emulsion** is a mixture of two **immiscible** (incapable of mixing) liquids, one being dispersed throughout the other in small droplets. It's important that no foreign matter or particles be injected along with the medication, because particles can collide with the blood cells and cause blood clots.

TIP

When giving medication IV, remember that it should be given slowly in most cases. Also, make sure to remove all air bubbles in the substance to be injected before giving an IV injection. Injecting air bubbles can cause them to lodge in areas of the body, forming air emboli and causing tissue damage (and potentially death in small animal species).

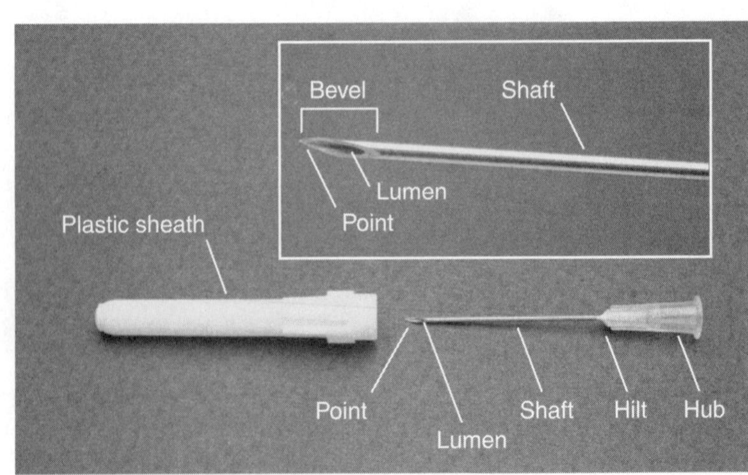

FIGURE 3-5 Parts of a needle and needle sheath. The insert shows point, lumen, bevel, and shaft.

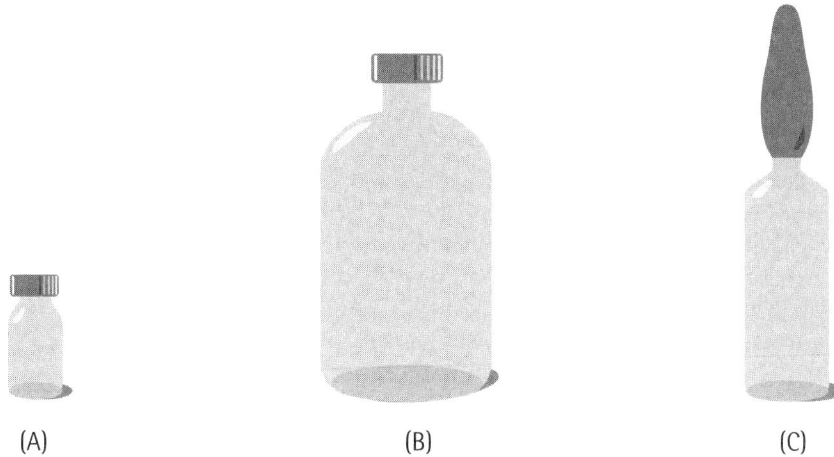

FIGURE 3-6 Vials and ampules store injectable drugs: (A) A single-dose vial. (B) A multidose vial. (C) An ampule

Filter needles, with filters in the needle hub, may be recommended when emulsions are given.

Veterinarians use three intravenous injection techniques: bolus administration, intermittent therapy, and infusion of fluid. Bolus intravenous administration involves injecting a drug in a minute amount of fluid, with a syringe and needle only. A *bolus* is a concentrated mass of pharmaceutical preparation. Veterinarians use this technique to achieve immediate high concentrations of drugs. Intermittent intravenous therapy involves diluting a drug dose in a small volume of fluid and administering it during a period of 30 to 60 minutes by means of an indwelling catheter (a catheter designed to remain in place). Intermittent intravenous therapy is the best way to maintain blood levels of antibiotics. Infusion of fluid involves the administration of large volumes of fluid continuously over extended periods of time. Veterinarians use this method to administer electrolytes and nutritional agents like amino acids.

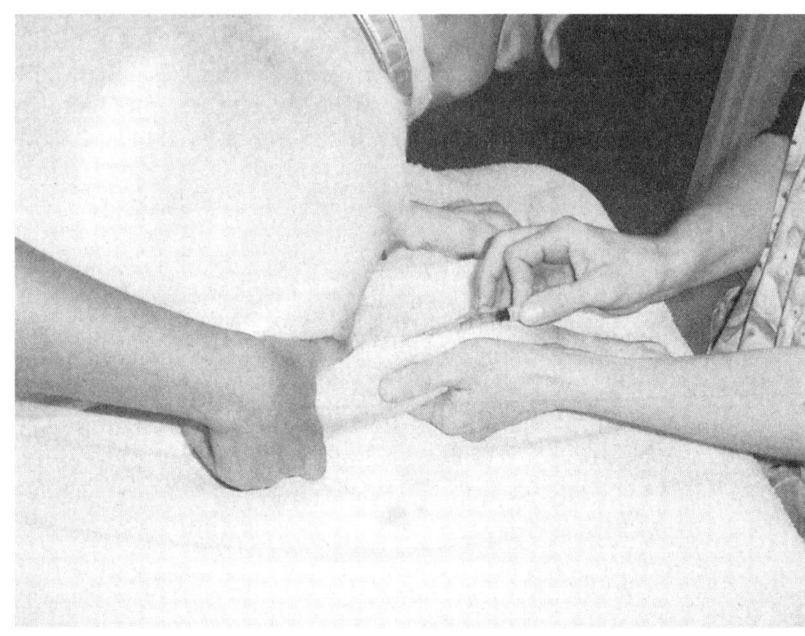

FIGURE 3-7 Dog receiving an IV injection

TIP

A *solution* is a clear liquid preparation that contains one or more solvents and one or more solutes. A *suspension* is a liquid preparation that contains solid drug particles suspended in a suitable medium.

TIP

Repository forms of drugs include sustained-release products (SR), implants, and spansules.

TIP

If you see the term *depo* associated with an injectable drug, think long-acting.

TIP

The rate of absorption of SQ medications can be manipulated to some degree by temperature. Warmer medications or heat applied to the area can increase the rate of absorption; conversely, cooler medications or cold applied to the area can decrease the rate of absorption.

Drugs for intramuscular injection come in aqueous (prepared in water) solutions, aqueous suspensions, oily suspensions, and injectable pellets. The body absorbs intramuscular aqueous solution injections rapidly; significant blood levels of the drug can appear within 5 minutes, but typically take about 30 minutes. The body absorbs aqueous suspensions more gradually, so introduction of the drug into the bloodstream is prolonged. Placing an injectable drug in a substance that delays absorption is called a **repository** or **depot preparation**. One means of prolonging drug action and further slowing the body's absorption of an intramuscular injection is to mix the drug with oil, creating an oily suspension. Examples of this are procaine penicillin G and methylprednisolone (Depo-Medrol®). Another way to prolong drug action is to add an ingredient that has limited absorption, which slows absorption of both drugs. An example of this is protamine zinc insulin injection.

When giving medication IM, remember:

1. Withdraw the plunger of the syringe to make sure you are not injecting directly into a blood vessel.
2. Give the injection within the muscle to make sure the drug will be absorbed properly. If given too shallowly, the drug may not actually have been given IM.
3. Keep in mind that some drugs cause pain when given IM.

A third injectable route of administration is subcutaneous (SC, SQ, or subQ). Subcutaneous injections place the drug into the connective tissue underneath the dermis of the skin. When the onset of drug effect desired is faster than oral administration and slower than intramuscular injection, subcutaneous injection is used. The vascularity of the subcutaneous space is less than that of skeletal muscle; hence, drugs given subcutaneously tend to be absorbed more slowly than IM administrations. Irritating or hyperosmotic solutions should not be given SQ. Larger amounts of solutions can be given SQ; however, the amount given should be based on the animal species involved and should not be so much as to cause tissue death or sloughing of skin. One way to further extend the duration of action of a particular drug given subcutaneously is to implant a pellet into the subcutaneous space. Here the drug is absorbed slowly and gradually. Steroids are examples of drugs that are administered with injectable pellets.

All injectable routes of administration are summarized in Table 3-1.

Inhalation Administration Drugs inhaled are rapidly absorbed into the bloodstream. **Inhalation** administration introduces the drug to the animal by having it breathe the drug into its lungs. After the animal breathes in the gas, the gas particles enter the alveoli of the lung. Here the particles diffuse (move from an area of high concentration to an area of low concentration) across the alveolar membrane. From the alveolar membrane the drug molecules enter the blood via capillaries surrounding the alveoli. Once in the capillary, the drug travels through the bloodstream to cause its effect. Gas anesthesia is one of the most common medications delivered by inhalation (Figure 3-8).

The inhalation route is also used to treat local respiratory conditions. Inhalation drugs include bronchodilators (drugs that open the airway for better breathing), mucolytic enzymes (drugs designed to liquefy thick mucous secretions for better removal from the lungs), antibiotics (for lung infections), and steroids (for inflammatory lung conditions).

Drugs that are administered by inhalation must be **volatilized** (turned into gases) in order to be inhaled into the lungs. Anesthetic gases are volatilized from liquids using a gas anesthetic vaporizer. Other drugs such as bronchodilators, mucolytics, antibiotics, and steroids, are **nebulized** (turned into a fine spray) and then inhaled into the lungs.

The inhalation route of administration is also summarized in Table 3-1.

TABLE 3-1 Parenteral Routes of Drug Administration

Route	General Rules
Intravenous (IV): within the vein	• rapid onset of action • higher initial body levels of drug • shorter duration of activity (need to be redosed more frequently) • smaller doses can be given • irritating drugs can be given (e.g., oxytetracycline) • increased risk of adverse effects (if drug given too rapidly, not sterile, or not properly mixed) • drug must be pure, sterile, and free of particles • drug must be water soluble
Intramuscular (IM): within the muscle	• relatively rapid onset of action (generally about 30 minutes) • rate of absorption depends on formulation (oil-based is slow, water-based is fast) • provides reliable blood levels • longer duration of action than IV (can dose less frequently) • shorter duration of action versus oral (generally) • absorption may be altered by vehicle present in preparation • cannot use irritating solutions • convenient route in fractious animals
Subcutaneous (SQ, subQ, or SC): beneath the skin into the subdermis	• slower onset of action than IM • less reliable blood levels (similar to oral) • longer duration of action than IM (can be given less frequently) • absorption may be altered by vehicle in preparation • cannot use irritating solutions • generally used for giving large volumes of solution
Intraperitoneal (IP): within the abdominal body cavity	• variable onset of action • variable blood levels • provides large surface area for drug absorption • irritating solutions may cause peritonitis • care must be taken so that the needle does not penetrate any organs; this could lead to peritonitis • first passes via portal system through liver, which could inactivate (or enhance) the drug's action

(continues)

TIP

Drugs that constrict blood vessels could delay drug absorption because they affect perfusion to the administration site. Blood vessels that are constricted or narrowed cannot absorb as much drug as blood vessels that are dilated or widened. This delay in absorption can prolong drug action. An example of this altered perfusion occurs when injecting lidocaine with epinephrine for local anesthesia. The lidocaine is absorbed more slowly because of the presence of epinephrine, thus lengthening its duration of action.

TABLE 3-1 *Continued*

Route	General Rules
Epidural/Subdural/Intrathecal: above the dura mater of the meninges, under the dura mater of the meninges, or into the subarachoid space of the meninges	• rapid onset of action localized to the central nervous system (CNS) • used for diagnostic procedures and administering some types of anesthetic agents • disadvantages include potential for misperformance resulting in spinal injection or drug moving cranially up the CNS
Intra-arterial (IA): within the artery	• used for treating a specific organ only, because very high drug levels are delivered to a specific site • may be accidental (e.g., given in carotid artery instead of jugular vein)
Intradermal (ID): within the skin	• injection given between dermis and epidermis • very slow absorption • low blood levels obtained • used for local treatments or allergy testing
Intracardiac (IC): within the heart	• rapid drug levels attained because drug passes from heart to systemic circulation • may be used in emergency situations or for euthanasia
Intra-articular: within the joint	• injection given in the synovial space of joints • aseptic technique is critical • drug can be absorbed systemically
Intramedullary or Intraosseous: within the medullary cavity of bone	• provides rapid blood levels • not commonly used and is painful • route of rapid fluid administration in smaller animals and birds • usually administered in the femur/ humerus
Inhalation: inhaled into the respiratory system	• examples include gas-masking of animals, endotracheal administration of gas anesthesia, and nebulization of drugs • establishes rapid blood levels because the lung provides a large surface area for absorption • may be used for anesthesia, emergency procedures, and treatment of respiratory disease
Topical: applied on top of a surface	• topical routes of administration include nasal, conjunctival, subconjunctival, intramammary, transdermal (by patch), rectal, and vaginal

TABLE 3-1 *Continued*

Route	General Rules
Topical (cont.)	• used mainly in dermatology and ophthalmology • may or may not be absorbed systemically • drug must first dissolve and then penetrate the skin by diffusion • good local effect • may be irritating • easy to administer • animal may chew/lick/rub off

Topical Applications and Dose Forms A **topical** medication goes on the surface of the skin or mucous membranes. Topical medications come in ointment, cream, gel, liniment, paste, lotion, powder, and aerosol dose forms (Table 3-2). Topical drugs must first dissolve and then penetrate the skin by diffusion. Topical applications are used for localized treatment of skin conditions such as localized skin infections (lesions), abrasions, and localized skin allergies. The body as a whole absorbs topical medication more slowly than any other application route; the localized site, however, benefits from a high concentration of the drug. Topical application is good for drugs needed externally that are toxic if injected. The disadvantage of veterinary topical application is that fur and feathers inhibit good skin contact (although clipping or plucking can eliminate the problem). Topical treatment also includes eye and ear medications. Eye preparations are available as liquid drops and ophthalmic ointments. Ear preparations are available as drops, ointments, and creams. Other routes of topical drug adminstration include nasal, rectal, vaginal, intramammary, and transdermal patches. All forms of topical routes of administration are also summarized in Table 3-2.

Oral Administration and Dose Forms **Oral** administration delivers the medicine directly to the animal's gastrointestinal (GI) tract (Table 3-3). Oral medications are the

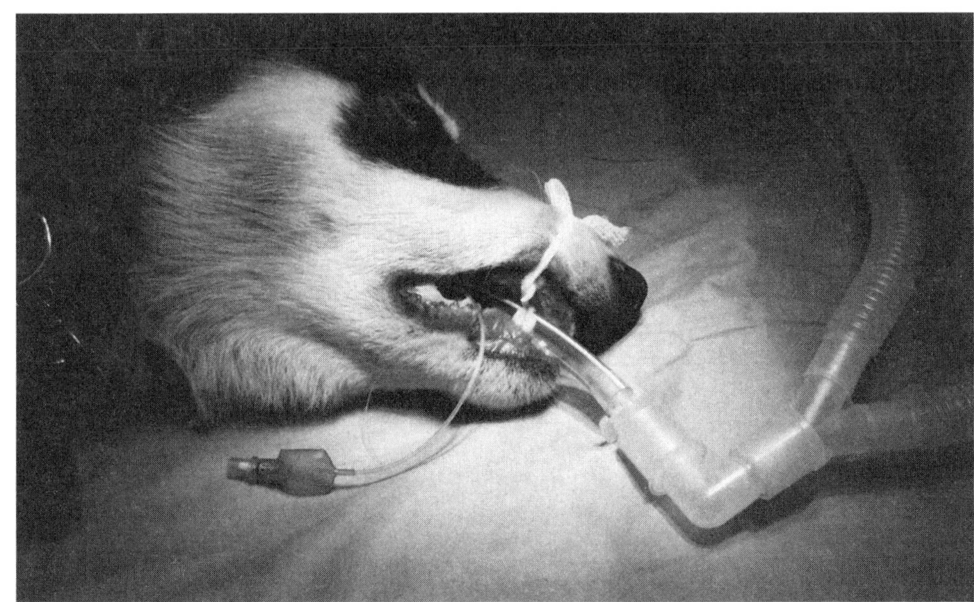

FIG 3-8 Dog receiving inhalant medication in the form of inhalant (gas) anesthesia. (*Source:* Linda Kratochwill, DVM)

TABLE 3-2 Topical Drug Forms

Topical Drug Form	Description
Aerosols	Drug suspended in solvent and packaged under pressure
Cream	Drug suspended in water-oil emulsion
Gel	Drug suspended in semisolid or jelly-like substance
Liniments	Drug suspended in oily, soapy, or alcohol-based substance applied with friction
Lotions	Drug suspensed in liquid for dabbing, brushing, or dripping on skin without friction
Ointments	Drug suspended in semisolid, greasy preparation that melts at body temperature
Paste	Drug suspended in semisolid preparation that retains its state at body temperature
Powder	Drug suspended in powder for external lubrication or absorption

most convenient to give and are less likely to cause adverse reactions. They do not have to be sterile because they enter the nonsterile environment of the gastrointestinal tract.

Before entering the bloodstream, an oral drug must go through a series of events, including release from the dose form (tablet or liquid), transport across the gastrointestinal tract, and passage through the liver. Each of these events can decrease the amount of drug that reaches the bloodstream.

TIP

Only scored tablets should be divided, because these are the only ones guaranteed by the manufacturer to distribute equal amounts of drug throughout the tablet (Figure 3-10).

The first step, dissolving the dose form, varies with each type of dose form. Oral dose forms are either solid or liquid. The animal either swallows the intact drug (the most accurate dose method) or eats food laced with the dose (less accurate, as the animal may leave some food uneaten).

Solid oral drug forms include tablets, capsules, boluses, lozenges, and powders (Figure 3-9). Tablets, capsules, and boluses can be given to an animal by hand or pilling gun. Tablets are medicines mixed with an inert binder and molded or compressed into a hard mass. The tablet disintegrates in stomach liquid, releasing the drug for

TABLE 3-3 Nonparenteral Routes of Administration

Route	General Concepts
Oral (po)	• most convenient route of administration for owner • slower onset of action • longer duration of activity • sometimes erratic and incomplete absorption because drug is affected by gastric fluids (acid) • absorption may be affected by gastrointestinal disease • relatively safe • drug must be able to get through gastrointestinal mucosa • drug need not be sterile • generally causes fewer adverse drug reactions • absorption in ruminants may be questionable with some medications given orally

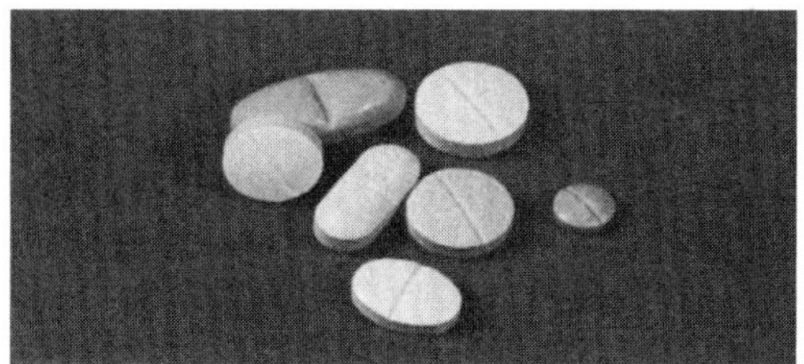

A. SCORED TABLETS

B. LAYERED TABLET

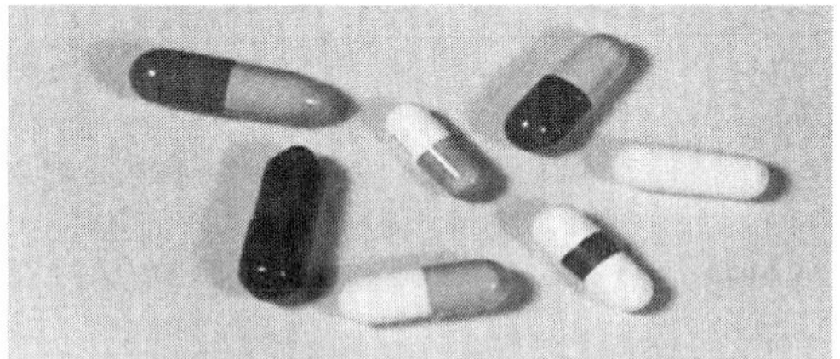

C. HARD GELATIN CAPSULES

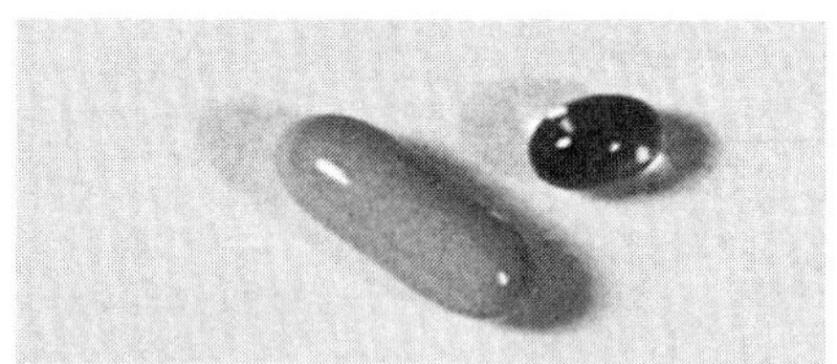

D. SOFT GELATIN CAPSULES

FIGURE 3-9 Examples of solid and semisolid preparations

absorption into the bloodstream. Coated or enteric-coated tablets are covered with a special coating that prevents the drug from dissolving in the stomach. Coated tablets are believed to cause less stomach irritation. Molded tablets are soft, chewable tablets that contain drug mixed with lactose, sucrose, or dextrose. Capsules are gelatin shells holding a powdered or liquid form of the drug. Most capsules are colored and may bear identifying product markings. The gelatin shell dissolves in the stomach liquids, releasing the drug for absorption into the bloodstream. Boluses are large, compressed, rectangular tablets that are typically used to dose large animals. Lozenges are drugs incorporated into a hard candy tablet that allows slow release of the drug. Lozenges are not utilized in veterinary medicine because of the obvious problems associated with getting animals to suck, rather than chew, the lozenges. Powders are dry, granulated versions of the drug mixed with inert bulking and flavoring agents (such as lactose) to enable dilution. Powders are easily mixed into the animal's food or drinking water.

Liquid oral drug forms include solutions, suspensions, and emulsions. **Solutions** are drug preparations dissolved in liquid; they do not settle out if left standing. Solutions often have flavoring added to mask the taste of the drug. Examples of solutions are syrups (drugs dissolved in 85 percent sucrose), elixirs (drugs dissolved in sweetened alcohol), and tinctures (alcohol solutions meant for topical application). An example of a solution is phenobarbital liquid (it is actually an elixir).

A finely divided, undissolved substance dispersed in water is called a **suspension**. Drugs that will not easily dissolve can be dispersed and suspended in liquid so that shaking the container distributes them uniformly. An example of a suspension is pyrantel pamoate oral suspension (a deworming medication).

An emulsion consists of fine droplets of oil in water or water in oil. They separate into layers after standing for long periods of time and must be shaken vigorously before they are ready for use. An example of an emulsion is castor oil.

FIGURE 3-10 Scored tablets are the only tablets that should be divided. Unless tablets are scored, the manufacturer cannot guarantee equal distribution of drug in each part.

Oral liquid drugs may be given by dropper, syringe, and drench (given by forcing the animal to drink) (Figure 3-11). Solutions, suspensions, and emulsions can be mixed with food. Although mixing drugs into food is not as accurate as a direct dose, it offers some advantages if the animal eats all the food.

Medications mixed with food are less likely to irritate the stomach and intestinal tract, as capsules and tablets can settle in a small area where any irritants in the drug then concentrate. Mixing drugs in food is also a safe way to medicate an unpredictable or aggressive animal. Getting the animal to eat all the food, however, can be a challenge, especially if the drug smells or tastes bad or the animal's appetite has decreased.

The second event that determines whether an orally administered drug gets into the bloodstream is transport across the gastrointestinal tract. Anatomical differences can affect drug absorption. A ruminant usually takes longer to respond to oral medication than does a monogastric animal. A drug mixed with food can remain in the rumen up to three days, where it is continually exposed to ruminal secretion. Consequently, some medications that are effective orally in monogastric animals biodegrade and become ineffective in ruminants. In general, the more complex the digestive tract, the longer it will take to attain therapeutic blood levels of an orally administered drug.

The third event, drug passage through the liver, affects drug concentration in the blood following oral administration, because the liver can alter the drug and render it less (or more) active. Animals with liver disease will need special consideration when receiving oral medications. General guidelines for drug administration are summarized in Table 3-4.

Drug Dose

A **dose** of a drug can also affect maintenance of a drug in the therapeutic range. A *dose* is the amount of a drug administered at one time to achieve the desired effect. Examples of doses include milliliters (ml or mL), cubic centimeters (cc), milligrams (mg),

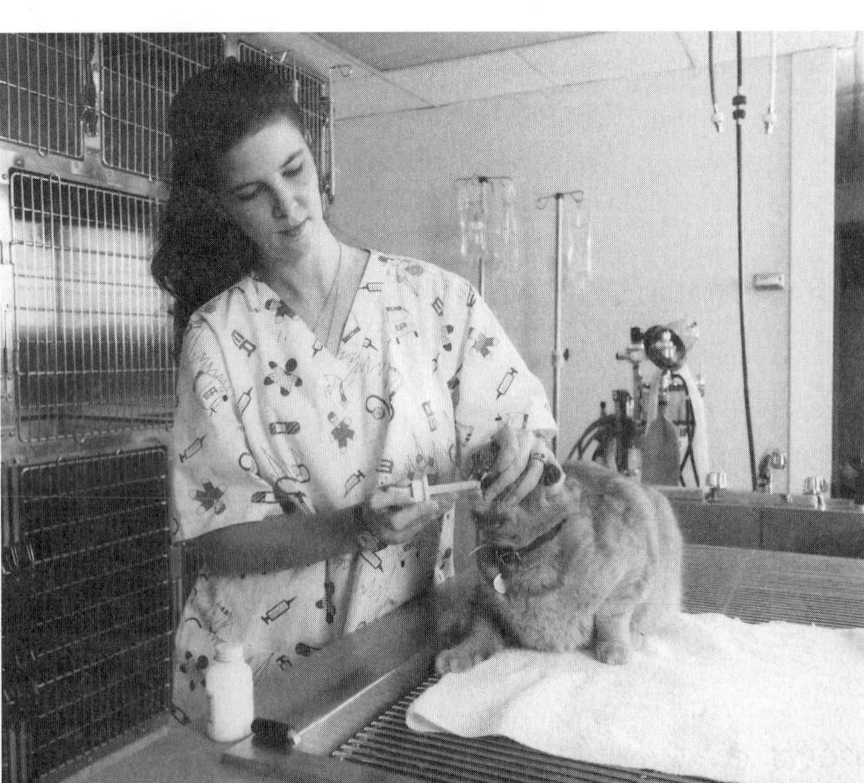

FIGURE 3-11 A liquid medication is administered orally to this cat.

TABLE 3-4 General Guidelines for Drug Administration (adapted from *Pharmacological Aspects of Nursing Care* 6th ed.)

1. Enteric-coated tablets should not be administered with antacids, milk, or other alkaline substances, because enteric-coated agents require the acid environment of the stomach to be effective.
2. Enteric-coated tablets should not be crushed before administration; crushing will alter absorption.
3. Suspensions and emulsions must be shaken thoroughly immediately before use, because the separation that occurs after standing for a short period will alter the dose if used in the separated form.
4. Suspensions should never be administered IV.
5. Solutions administered parenterally or in the eye must be sterile to prevent causing infection.
6. Solutions administered IV must be free of particulate matter that could serve as an embolus.
7. Proper storage of solutions is very important to prevent contamination and evaporation.
8. Skin integrity should be assessed for rashes or open areas before applying topical medications, as these condiditons will alter absorption time of the medication.
9. Transdermal therapeutic systems or patches allow drugs to pass through intact skin and care should be taken when applying these to animals to prevent self-medication.
10. Proper disposal of transdermal patches is important to prevent their ingestion by animals or improper exposure to people.

grams (g), or tablets (T). When recording a dose in an animal's record, it is more accurate to state the dose in mass units, such as grams or milligrams, because each manufacturer makes tablets in different sizes and liquids in different concentrations from other manufacturers.

Two terms used in reference to dose are **loading dose** and **maintenance dose**. The *loading dose* is the initial dose of drug given to get the drug concentration up to the therapeutic range in a very short period of time. A *maintenance dose* is the dose of drug that maintains or keeps the drug in the therapeutic range. The **total daily dose** is the total amount of drug delivered in 24 hours; for example, 30 mg given QID (four times daily) = 120 mg total.

Another term used in medicating animals is **dosage**, the amount of drug per animal species' body weight or measure. Examples of a drug dosage are 5 mg/kg or 25 g/lb.

Dosage Interval

Dosage is involved in the third factor in maintaining a drug in the therapeutic range, the **dosage interval**. The dosage interval is how frequently the dosage is given; for example, SID (*semel in die*) or once daily, BID (*bis in die*) or twice daily, TID (*ter in die*) or three times daily, or QID (*quarter in die*) or four times daily. The dosage interval and the dosage together represent the **dosage regimen**. An example of a dosage regimen is 30mg/kg TID.

OUT OF THE SAFE ZONE

Keeping the level of drug in the therapeutic range involves extensive studies of the drug's efficacy and toxicity signs. Drug toxicities may also occur due to human error and/or accident. Some causes of drug toxicity are summarized in Table 3-5.

Some ways to ensure that drugs stay within the therapeutic range and benefit the animal include the following:

- Use the proper dosage, frequency, and duration of treatment for each drug prescribed.
- Avoid combination drug treatment if possible, or allow a margin of error in dosing multiple drugs.
- Use less toxic drugs if available (for example, some antifungals are more toxic than others).
- Be aware of potential hazards and precautions. Sometimes fluid therapy may be given to an animal prior to and during treatment to enhance renal excretion of a drug.
- Use high-quality drugs, check expiration dates, check handling requirements, understand contamination possibilities, and make sure the drug is thoroughly mixed and not precipitating out of solution.
- Follow label directions carefully.
- Know the patient's history and keep in contact with the animal's owner.

If an animal develops drug toxicity, the veterinary staff needs to act quickly to counteract any problems caused by the drug treatment. Some ways to treat drug toxicities include:

- removal of the offending drug (for example, washing off flea products that are causing neurologic signs).

TABLE 3-5 Causes of Drug Toxicity

Cause of Drug Toxicity	Example
Outright Overdose	Dosing too frequently, dosing too high, administering too long
Relative Overdose	Recommended dose was too much for this animal, due to individual variation, impaired metabolism or excretion of drug by this individual animal, improper route of administration
Side Effects	Normal side effects associated with drug may occur at a higher level in this individual
Accidental Exposure	Exposure of animal to drug that is absorbed through skin or inhalation or accidental ingestion
Interaction with Other Drugs	If drug A is highly protein bound and drug B is protein bound as well, the competition for protein binding will affect the amount of free drug (active drug) available to the animal
Incorrect Treatment	Misdiagnosis: treatment causes toxicity levels of drug because the animal does not have the disease; for example, hormone replacement raises hormone levels above normal in healthy animal

- enhancing drug removal by the animal (for example, making the animal vomit or administering fluids to enhance drug excretion by the kidney).
- counteracting with an antidote (for example, giving naloxone for an overdose of morphine).
- providing symptomatic care or nursing care for the animal until the toxicity signs have diminished.

SUMMARY

Maintaining drugs within the therapeutic range is important in obtaining optimal drug efficacy. The therapeutic range of a drug is the drug concentration in the body that produces the desired effect in the animal with minimal or no signs of toxicity. Too little drug will keep the drug level in the subtherapeutic range; hence, there is little or no benefit to using the drug. Too much drug will push the drug level into the toxicity range, which will have adverse effects on the animal.

Three key aspects to keeping drug levels within the therapeutic range are route of administration, dose, and dosage interval. Route of drug administration plays a role in speed of drug absorption and utilization of the drug. Routes of drug administration include parenteral and nonparenteral. Various drug forms exist for both parenteral and nonparenteral drugs.

Dose is the amount of drug given to the animal and is determined based on the dosage of drug recommended by the manufacturer and the weight of the animal. Dosage is the amount of drug per animal species' body weight or measure. Dosage interval is how frequently we give the dosage.

Drug toxicities do develop in animals for a variety of reasons. If an animal experiences drug toxicities, the signs of these reactions must be identified quickly and treated with a variety of options, depending on the drug and what toxicities it has caused.

CHAPTER REFERENCES

Reiss, B., & Evans, M. (2002), *Pharamcological aspects of nursing care* (6th ed., revised by Bonita E. Broyles, pp. 2–77). Clifton Park, NY: Thomson Delmar Learning.

Rice, J. (1999). *Principles of pharmacology for medical assisting* (3d ed.). Clifton Park, NY: Thomson Delmar Learning.

Veterinary pharmaceuticals and biologicals (12th ed.). Lenexa, KS: Veterinary Healthcare Communications.

Wynn, R. (1999). *Veterinary pharmacology*. Cambridge, MA: Harcourt Learning Direct.

CHAPTER REVIEW

Matching

Match the definition with its proper term.

1. ______ injectable drug that is in a substance that delays its absorption
2. ______ small, sterile, prefilled glass containers containing medication for injection
3. ______ concentrated mass of drug
4. ______ administration of a large volume of fluid continuously over extended periods of time
5. ______ dilution of a drug dose in a small volume of fluid that is administered over a period of time (typically 30 to 60 minutes) via an indwelling catheter
6. ______ solution types that should not be given subcutaneously
7. ______ term for turned into a gas
8. ______ drug suspended in water-oil emulsion
9. ______ drug suspended in a semisolid preparation that retains its state at body temperature
10. ______ suspended in liquid for dabbing, brushing, or dripping on skin without friction

a. irritating or hyperosmotic solutions
b. ampule
c. infusion of fluids
d. cream
e. intermittent IV therapy
f. lotion
g. volatilized
h. repository or depot preparation
i. paste
j. bolus

Multiple Choice

Choose the one best answer.

11. The dose or dosage of a drug that produces the desired effect in the animal with minimal or no signs of toxicity is the
    a. toxicity subrange.
    b. therapeutic range.
    c. route of administration.
    d. subtherapeutic range.
12. An example of a parenteral drug route of administration is
    a. by mouth (po).
    b. rectally.
    c. IV.
    d. sublingual.
13. In general, which route of drug administration has a longer duration of action than IV, yet a shorter duration of action than oral?
    a. rectal
    b. transdermal
    c. inhalation
    d. IM
14. Which route of drug administration do gastric fluids affect?
    a. IM
    b. po
    c. IV
    d. SQ

15. An injectable drug placed into a substance that delays absorption is called a
    a. parenteral drug.
    b. repository preparation.
    c. suspension.
    d. solution.

16. Drugs administered via nebulization are
    a. volatilized for inhalation.
    b. mucolytic and administered orally.
    c. depot prepared and injected.
    d. turned into a fine mist for inhalation.

True/False

Circle a. for true or b. for false.

17. The loading dose of a drug is the initial dose given to get the drug concentration up to the therapeutic range in a very short period of time.
    a. true
    b. false

18. The dosage of a drug is the total amount of drug delivered in 24 hours and is based on the animal's weight.
    a. true
    b. false

19. 100 mg/kg BID is an example of a dosage regimen.
    a. true
    b. false

20. BID is the Latin abbreviation for twice daily.
    a. true
    b. false

Case Study

21. An elderly client on a limited budget brings her cat in for an examination. The owner tells you that the cat has been vomiting for two days and hiding under the bed. On physical examination, the cat appears lethargic, but its vital signs are normal. You collect blood from the cat for a chemistry panel and a complete blood count. Because the blood test results are not available immediately, the owner wants to give the cat something to make it stop vomiting. The owner states that cost is the only concern in medicating this cat.

    Cost may be an important factor determining treatment; however, it is not the only concern. What other concerns regarding treatment of this cat can you describe?

CHAPTER 4

Pharmacokinetics

OBJECTIVES

Upon completion of this chapter, the learner should be able to

- outline the four components of pharmacokinetics.
- describe the four mechanisms that allow drugs to move across cell membranes.
- describe factors that affect drug absorption.
- describe ion trapping.
- describe factors that affect drug distribution.
- describe the role protein binding plays in drug availability.
- describe factors that affect drug biotransformation.
- list sites of drug biotransformation and excretion.
- describe factors that affect drug excretion.
- describe the role of receptors in pharmacology.
- differentiate between agonist and antagonist.

KEY TERMS

pharmacokinetics
passive diffusion
lipophilic
hydrophilic
ionized
nonionized
facilitated diffusion
active transport
pinocytosis
phagocytosis
absorption
bioavailability
ion trapping
distribution
biotransformation
metabolism
induce
tolerance
elimination
excretion
withdrawal time
half-life
steady state
receptor
affinity
agonist
antagonist

INTRODUCTION

A horse is experiencing front-limb lameness from an injury that occurred during a race. The owner of this horse calls the veterinarian out to the ranch to have the horse examined. The vet determines that the horse needs an anti-inflammatory drug to decrease some of the front-limb joint swelling noted on the physical exam. The treatment includes an IM injection of an anti-inflammatory drug. The owner wonders why you are giving an injection in the horse's muscle when the horse has pain in its joint. How does the medication get from a muscle to another location in the body? How do medications get to where they are needed?

DRUG MOVEMENT

Pharmacokinetics refers to the physiological movement of drugs (*pharmaco* means drug and *kine* means motion). The term covers the way drugs move into, through, and out of the body. There are four steps in pharmacokinetics: absorption, distribution, metabolism or biotransformation, and excretion.

Pharmacokinetics also includes the movement of substances across cell membranes. Four basic mechanisms allow drugs to move across cell membranes: passive diffusion, facilitated diffusion, active transport, and pinocytosis/phagocytosis. These mechanisms are summarized in Table 4-1.

Passive Diffusion

Passive diffusion is the movement of drug molecules (particles) from an area of high concentration to an area of low concentration. Drug molecules move from an area where there are many such molecules (high concentration) to an area where there are

TABLE 4-1 Mechanisms of Drug Movement

Type of Movement	Energy	Carrier	Notes
Passive diffusion: random movement of molecules from an area of high concentration of molecules to an area of low concentration of molecules	--	--	• Rapid for nonionic, lipophilic, small molecules • Slow for large, ionic, hydrophilic molecules • Lower temperatures slow the rate of diffusion; higher temperatures speed the rate of diffusion • Thick cell membranes slow the rate of diffusion; thin cell membranes do not slow the rate of diffusion
Facilitated diffusion: passive movement with special molecules within the membrane that carry the molecules through the membrane	--	++	• Like passive diffusion except uses a carrier molecule • Cannot concentrate molecules on one side or the other • Entry of glucose into the cells occurs by facilitated diffusion with the help of insulin molecules

(continues)

TABLE 4-1 *Continued*

Type of Movement	Energy	Carrier	Notes
Active transport: movement of molecules across membranes involving a carrier molecule that pumps the molecule against a concentration gradient	++	++	• Like facilitated diffusion but needs energy (ATP) • Movement of strongly acidic or basic substances into urine usually occurs by active transport • pH gradient in body systems usually occurs by active transport
Pino/phagocytosis: cell drinking or cell eating in which cellular membrane surrounds the molecule and takes it within the cell	++	--	• Large drug molecules, such as proteins, are usually involved in these processes

few such molecules (low concentration) until they start to even out between the two sides (Figure 4-1A). When both sides are nearly equal in concentration, the molecules continue to exchange at an even rate to keep equal numbers of molecules both inside and outside the cell. Passive diffusion does not require energy, nor does it expend energy.

The drug must dissolve in the cell membrane (which is made primarily of phospholipids) and must pass readily through the membrane in order for passive diffusion to occur. Drugs that move through cell membranes by passive diffusion must be small in size, lipophilic, and nonionic. Keep in mind that pores in the cell membrane are small, therefore, drugs that pass through these pores must also be small. Large drugs cannot pass through small pores and so are left outside the cell.

Lipophilic means fat-loving (*lipo* means fat and *phil* means loving). Lipophilic drugs are chemicals that dissolve in fats or oils. Lipophilic drugs are able to dissolve in the phospholipid or fat-containing cellular membrane because like molecules dissolve in like molecules. If you remember that oil and water do not mix, the fact that lipophilic and hydrophilic substances do not mix makes sense. The intestinal mucosa has lipid-rich cell membranes, allowing lipophilic drugs to be well absorbed from the gut via passive diffusion. **Hydrophilic** means water-loving (*hydro* means water and *phil* means loving). Hydrophilic drugs are chemicals that dissolve in water. Hydrophilic drugs do not pass through lipid-rich cell membranes as easily as lipophilic drugs. Drugs that are already in fluid (body water) do not need to move through phospholipid membranes. For example, drugs injected intramuscularly are deposited in the fluid that surrounds the cells and must diffuse through fluid to reach the capillaries. Intramuscular drugs should ideally be in a hydrophilic form for rapid absorption. Intramuscular drugs may be in a lipophilic form to slow the rate of absorption.

TIP

Lipophilic drugs tend to pass through phospholipid cell membranes readily. Hydrophilic drugs have difficulty passing through phospholipid cell membranes.

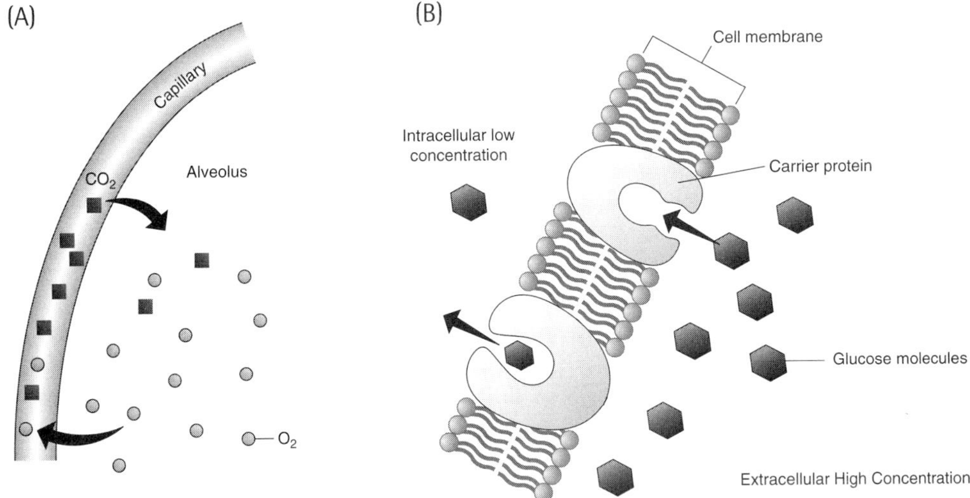

FIGURE 4-1A and B (A) *Diffusion* is the movement of atoms, ions, or molecules from an area of high concentration to an area of low concentration. In this example, O_2 and CO_2 diffuse in the lung. (B) Facilitated diffusion is a special kind of diffusion that utilizes a carrier molecule. Glucose moving from blood into cells is an example. Glucose molecules are too big and are not lipid-soluble enough to diffuse through the cell membrane; therefore, they combine with special carrier proteins so that they can pass through the cell membrane.

Drug ionization also affects movement of the drug. The ionization or charge of a drug depends on the pH of the liquid in which it is immersed and the pH of the drug. **Ionized** drugs have either a positive or a negative charge. **Nonionized** drugs have no charge or are considered neutral. Ionization or neutrality also affects the hydrophilic and lipophilic properties of a drug. Nonionized drugs tend to be in lipophilic form; ionized drugs tend to be in hydrophilic form. One goal of passive diffusion is to move molecules through a lipid-containing cell membrane. The molecules that move most effectively through the cell membrane are lipophilic; therefore, nonionized or neutrally charged particles would pass through the cell membrane more easily, as they tend to be lipophilic. When a drug is given, some of the drug molecules go into the hydrophilic form and other drug molecules go into the lipophilic form. The amount of drug that goes into a particular form depends on the chemistry of the drug and the environmental pH in which the drug is found (for example, oral drugs act in the acid environment of the stomach, whereas intravenous medications are placed in the more neutral environment within the blood vessels).

Facilitated Diffusion

The second type of cellular movement is **facilitated diffusion**, passive diffusion that utilizes a special carrier molecule (Figure 4-1B). This carrier molecule helps the drug across the cell membrane. When both sides are nearly equal in concentration, the molecules continue to exchange at an even rate to keep equal numbers of molecules both inside and outside the cell. No energy is needed for facilitated diffusion, molecules still move from an area of high concentration to an area of low concentration.

Active Transport

Active transport, the third type of cellular movement, involves both a carrier molecule and energy (Figure 4-1C). Energy is needed in active transport because drug molecules move against the concentration gradient, from areas of low concentration

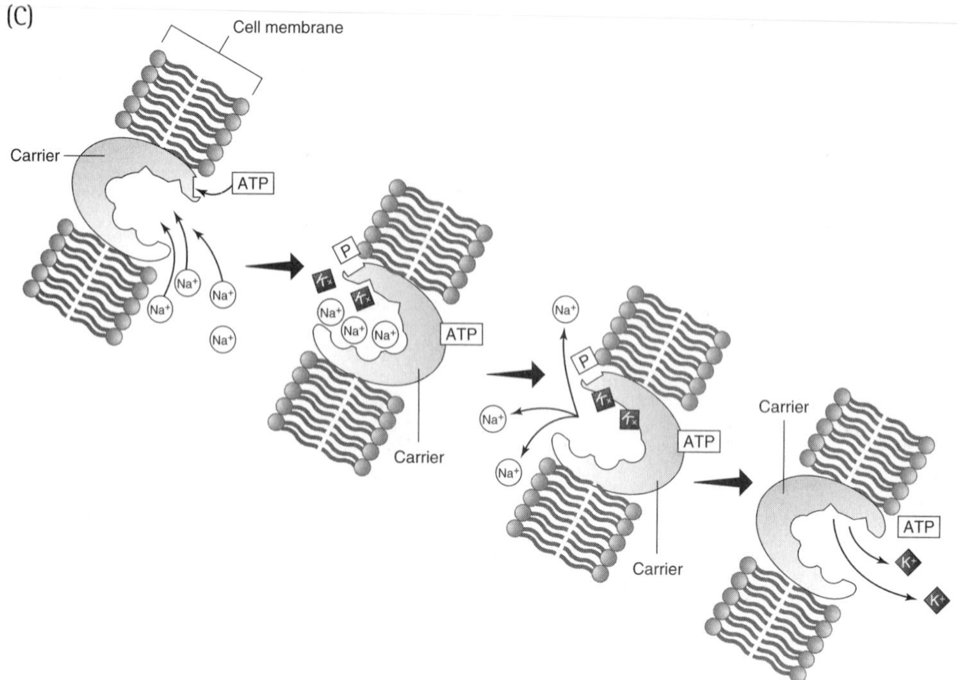

FIGURE 4-1C The active transport process moves particles against the concentration gradient from a region of low concentration to a region of high concentration. Active transport requires energy and a carrier molecule. An example is the Na^+-K^+ pump, which keeps sodium levels high outside the cell and potassium levels high inside the cell.

of molecules to areas of high concentration of molecules. Active transport allows drugs to accumulate in high concentration within a cell or body compartment. Sodium, potassium, and some other electrolytes move via active transport and accumulate in higher amounts on one side of a membrane than the other.

Pinocytosis and Phagocytosis

Pinocytosis and **phagocytosis** are mechanisms of molecule movement in which molecules are physically taken in or engulfed by a cell (Figure 4-1D). Pinocytosis (cell drinking) occurs when the cell membrane surrounds and engulfs liquid particles. Phagocytosis (cell eating) occurs when the cell membrane surrounds and engulfs solid particles. Engulfing liquid or solid particles into the cell requires energy. Pinocytosis and phagocytosis are important for the movement of large molecules such as proteins, that cannot enter a cell by passing intact through the cell membrane.

PHARMACOKINETICS PART I: GETTING IN

Drug **absorption** is the movement of drug from the site of administration into the fluids of the body that will carry it to its site(s) of action. Most drugs are carried in the systemic blood circulation. Drug factors and patient factors can affect the degree of drug absorption within an animal. Drug factors such as solubility, drug pH, and molecular size can affect drug absorption. Patient factors, such as the animal's age and health status, greatly affect how that individual animal can absorb a drug. Keep in mind that we do not want some drugs to be absorbed, such as topical anesthetics and activated charcoal.

(D)

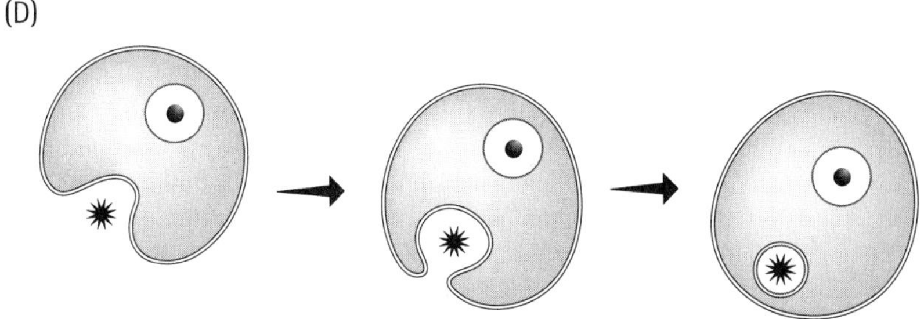

FIGURE 4-1D The processes of phagocytosis (engulfing solids) and pinocytosis (engulfing liquids) move very large particles into the cell by surrounding the particle and forming a vesicle.

Amount of Drug in the Body

The degree to which a drug is absorbed and reaches the circulation is an important component of drug absorption. This component is termed **bioavailability**. Bioavailability is the percent of drug administered that actually enters the systemic circulation. Intravenous or intra-arterial drugs immediately enter the systemic blood circulation; therefore, drugs given IV or IA are 100 percent bioavailable (they have a bioavailability number of 1). Drugs can be 100 percent bioavailable even if they do not enter the systemic blood circulation immediately. Drugs that are partially absorbed have a bioavailability of less than one. The lower the bioavailability of a drug, the less of it there is in the circulation and subsequently in tissue. Factors that affect bioavailability of a drug include the blood supply to the area (muscle has a greater blood supply than the subcutaneous space, so drugs given IM tend to have a higher bioavailability than those given SQ), surface area of absorption (increased surface area means more space for absorption to take place), mechanism of drug absorption, and dosage form of the drug (IV and IM routes of administration should yield higher bioavailabilities than oral routes of administration).

TIP

For a drug to be effective, it must get from where you administer it to the bloodstream. The percent of drug that does this determines the bioavailability of the drug.

pH and Ionization

The pH of the drug and the pH of the environment where the drug is administered play a role in drug absorption. *pH* is the measurement of the acidity or alkalinity of a substance. pH is based on a scale of 0–14, with the lower numbers indicating acid and the higher numbers indicating alkaline or base. Neutrality on the pH scale is 7.

Drugs exist in both ionized (charged) and nonionized (uncharged) forms. When weakly acidic drugs, such as aspirin, are swallowed, they enter the acidic environment of the stomach. Some of the drug molecules are ionized and some of the molecules are nonionized in the stomach. The drug particles then progress to the small intestine where the pH is higher. Most of the aspirin molecules are in ionized form in the small intestine. Recall that hydrophilic drugs are usually ionized (charged) and that lipophilic drugs are usually nonionized (uncharged). The hydrophilic form of aspirin is the charged portion in the small intestine and the lipophilic form of aspirin is the uncharged portion in the stomach. Because the lining of the gastrointestinal tract is composed mainly of phospholipid, the lipophilic or nonionized form is needed for the drug to be readily absorbed through the gastrointestinal mucosa. Weakly acidic drugs are more likely to be absorbed in the stomach, as there is a greater amount of nonionized drug in the stomach than in the small intestine. The pH environment of the small intestine favors the hydrophilic or ionized form of a weakly acidic drug and therefore limits its absorbance in the small intestine.

TIP

Ionized = charged; Nonionized = uncharged.

Hydrophilic drugs are usually ionized; lipophilic drugs are usually nonionized.

The acidic or alkaline nature of the drug itself also helps determine whether a drug is predominately in lipophilic or hydrophilic form. Weakly acidic drugs tend to be in hydrophilic form in an alkaline environment, whereas weakly alkaline drugs tend to be in hydrophilic form in an acid environment. Acidic drugs tend to take ionized form in an alkaline environment, whereas alkaline drugs tend to take ionized form in an acidic environment.

TIP

Acid drugs in an alkaline environment tend to be charged. Alkaline drugs in an acid environment tend to be charged.

Ion Trapping

Ion trapping occurs when a drug molecule changes from ionized to nonionized form as it moves from one body compartment to another. Drugs can pass from one body compartment to another body compartment with a different pH. When the drug changes compartments, it may change its ionization and become trapped in that new environment. For example, aspirin taken orally enters the acidic environment of the stomach. The majority of the acidic aspirin molecules are in nonionized form in the stomach (remember, some stay in ionized form). The phospholipid layer of the stomach wall readily absorbs the nonionized form. The aspirin molecules enter the stomach cells and are then in an environment with a pH near neutrality (7). The pH of the environment is becoming more alkaline, and thus the aspirin shifts to ionized form (however, some stay in nonionized form). When the aspirin shifts to ionized form, the molecules become trapped in the stomach cells. The equilibrium between ionized and nonionized molecules within the stomach cells continues to be maintained and some nonionized molecules pass out of the stomach cell and into the blood. Nonionized molecules are converted to their ionized forms in the blood. This process helps keep the absorbed aspirin molecules in the bloodstream where they can then be distributed to the body cells (Figure 4-2). Ion trapping is especially important in drug excretion, as alterations in the urine pH can allow drugs to be trapped in the urine and excreted.

Oral versus Parenteral Drug Forms

Absorption is further affected by the choice of oral or parenteral route of drug administration. Solid drugs administered orally must first dissolve in gastric or intestinal fluids before they can be absorbed. Administering fluids with solid drugs generally increases the rate at which a drug dissolves and may speed its rate of absorption. Drug size, gastric motility, and the lipophilic nature of the drug also affect orally administered drugs. Parenterally administered drugs are more affected by tissue blood flow and the hydrophilic nature of the drug than orally administered drugs. These effects are summarized in Table 4-2.

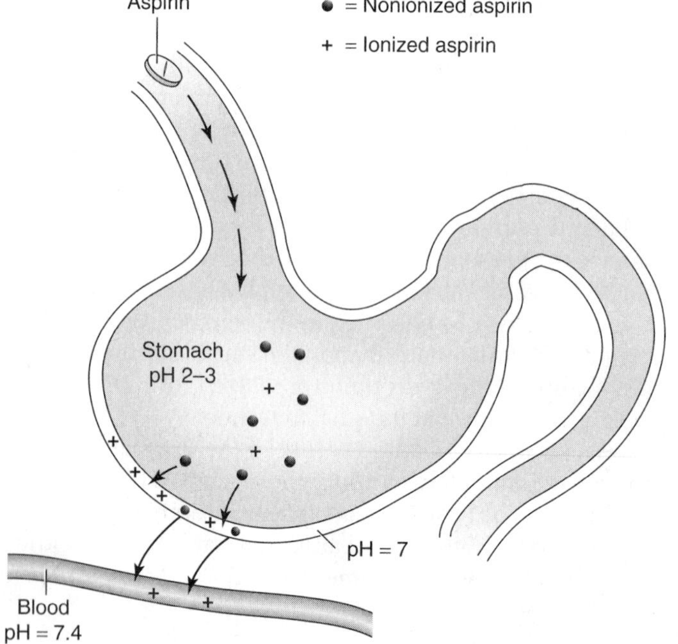

FIGURE 4-2 Ion trapping occurs when a drug molecule changes from ionized to nonionized form as it moves from one body compartment to another. Aspirin taken orally is nonionized in the stomach and is absorbed. When it enters the stomach cells, the aspirin shifts to its ionized form and cannot pass back into the lumen of the stomach. As a result of equilibrium, some nonionized aspirin molecules pass from the stomach cells into the blood, where they become ionized. These ionized aspirin molecules cannot pass effectively from the blood to the stomach cells.

TABLE 4-2 Drug Factors That Affect Drug Absorption

Drug Chemistry	• Lipophilic drugs dissolve in oil-based fluids. Lipophilic drugs are absorbed well across phospholipid-based cell membranes. • Hydrophilic drugs dissolve readily in water (tissue, fluid, and lymph). Tissue fluid is water soluble; therefore, drugs that are hydrophilic dissolve in and diffuse through tissue fluid quite well.
Drug Size	• Molecular size of the drug: Small molecules can pass more readily through cell membranes
Ionization of the Drug	• Nonionized or neutral drugs are lipophilic and can pass through phospholipid cell membranes. • Ionized or charged drugs are hydrophilic and dissolve in and diffuse through tissue fluid. • Ionization of the drug depends on the drug pH and the environmental pH
Acid-Base Characteristics	• pH of drug: Drugs may change their form when the pH of the environment changes. Weak acids become more ionized as the pH of the environment increases. Weak bases become more ionized as the pH of the environment decreases.
Ion Trapping	• Drugs can pass from one compartment to another compartment with a different pH. When the drug changes compartments, it may become ionized and become trapped in its new environment
Drug Form	*Oral* drugs must be in lipophilic form to penetrate the GI mucosa. They must be small to dissolve in the membrane. • Tablets must dissolve into smaller particles. Liquid drugs do not have a dissolution step; therefore, oral liquid drugs have a quicker onset of action than pills. • The drug (enteric coating) or special construction of the tablet (sustained-release) may alter dissolution and/or absorption. • Decreased gastric motility lengthens the time it takes for the drug to reach the absorption site. • Increased gastric motility shortens the time the drug remains in the GI tract. This time may not be long enough to allow drug dissolution, and the drug may pass unused in the feces. • The drug must be able to survive first-pass effect or detoxification by the liver. Remember that a drug is absorbed from the intestine and passes through the liver before it enters the systemic circulation. • The presence of food may interfere with the dissolution and absorption of certain drugs. *Parenteral* drugs must be in hydrophilic form. • Anything that interferes with diffusion of the drug from the administration site or alters blood flow to the injection site will delay absorption.

(continues)

TABLE 4-2 *Continued*

Drug Form (cont.)	• Some drugs are formulated to have a delayed absorption (repository or depot injections). • If there is limited blood flow in the injection site, absorption will be slowed (fat is poorly perfused, muscle is richly perfused). • Temperature may result in vasoconstriction or vasodilation, and affect blood flow to the administration site. • Other drugs may affect blood flow as well (e.g., lidocaine).

TABLE 4-3 Patient Factors That Affect Drug Absorption

Animal Factor	Example of effect
Age	• Young animals may not have well developed gastrointestinal tracts • Young animals may have less active enzyme systems
Health	• Fever may cause molecules to move faster and increase absorption • Disease signs like diarrhea may speed the drug through the gastrointestinal tract and not allow enough time for proper absorption • Disease signs like vomiting may affect the time the drug is in the stomach, hindering absorption
Metabolic Rate	• Animals with a high basal metabolic rate may metabolize and/or eliminate drugs more rapidly than those with a normal metabolic rate
Genetic factors	• Individual variation in response to drugs may occur because of genetic differences between animals. For example, one animal may metabolize a drug more slowly due to a genetically based enzyme deficiency.
Sex	• Male and female animals have different body compositions. The proportion of fat to lean body mass may influence the action and distribution of drugs throughout the animal's body.

Patient Factors

Patient factors affecting drug absorption are summarized in Table 4-3.

PHARMACOKINETICS PART II: MOVING AROUND

Drug **distribution** is the physiological movement of drugs from the systemic circulation to the tissues. The goal of drug distribution is for the drug to reach the target tissue or intended site of action. Factors that affect drug distribution include membrane permeability, tissue perfusion, protein binding, and volume of distribution.

Membrane Permeability

Membrane permeability has a great effect on drug distribution. Blood capillaries are only one cell layer thick and have *fenestrations* or small holes between cells to allow drug molecules to move in and out of the capillaries. Large molecules usually cannot pass through these fenestrations and are trapped in the bloodstream. One exception to this permeability is the blood-brain barrier. Capillaries in the central nervous system (CNS, brain and spinal cord) have no fenestrations. There are also supporting cells (glial cells) in the CNS that surround the capillaries, creating an extra barrier. The lack of fenestrations in the CNS capillaries and the glial cell barriers allow only the most lipophilic drugs to enter an undamaged CNS. Disease manifestations such as fever and inflammation may alter this blood-brain barrier, thus allowing drugs to penetrate the CNS.

Another potential barrier to membrane permeability is the placenta. However, the existence of the placental barrier does not prevent all drugs from passing from mother to fetus. Many drugs can pass through the permeable placental capillaries. Drugs that pass through the placental barrier will affect the fetus.

Tissue Perfusion

Tissue perfusion affects how rapidly drugs will be distributed. Tissue perfusion is the relative amount of blood supply to an area or body system. The level of tissue perfusion varies among body systems and animal species. Distribution will be rapid to well-perfused tissues like the brain and slow to poorly perfused tissues like fat. Some areas that are well perfused (such as the brain) may initially have high levels of drug, because of the rapid distribution of a particular drug from the bloodstream to the well-perfused area; other poorly perfused areas (such as fat) have lower levels of drug. An example of this occurs with the anesthetic agent thiopental. Thiopental is lipophilic and is stored in fat. It would seem reasonable that fat patients given thiopental should have high levels of the drug in the fat and less of the drug in the brain, resulting in delayed onset of anesthesia. Clinically, however, we see that patients given thiopental IV become anesthetized quickly. Why is this? Remember that tissue perfusion also plays a role in drug distribution. The well-perfused tissues, like the brain, rapidly receive high concentrations of thiopental when it is given IV. Thiopental in the circulation is simultaneously being slowly distributed to the poorly perfused fat tissue. The concentration of thiopental in the blood decreases as the drug moves from blood to fat. At some point, the concentration of thiopental becomes lower in the blood than its concentration in the brain. Lower drug levels in the bloodstream result in drug being moved to the bloodstream from the brain. The drug in the bloodstream then goes to the fat. As thiopental continues to be redistributed from the brain to the bloodstream to the fat, the animal regains consciousness. Typically, more thiopental is given to keep the animal anesthetized. This could eventually lead to high levels of drug in the fat, due to redistribution of thiopental between the brain, bloodstream, and fat. The animal will have too much anesthetic in fat storage, and over time the drug will continue to be redistributed to the brain as the brain levels of thiopental drop. This phenomenon may keep the animal anesthetized too long. This is especially true of fat patients given lipophilic drugs, because they have more fat available to store the lipophilic drugs.

Tissue perfusion can also be affected by alterations in blood flow rates, which may occur because of disease conditions or other drug treatments that cause blood vessel constriction or dilation. For instance, decreased blood flow to tissues, as a result of a patient being in heart failure, will decrease the rate and amount of drug delivered to various tissues.

Protein Binding

Protein binding also affects drug distribution. Some drugs bind to proteins in the blood. Proteins are large and cannot leave the capillaries, so the drug-protein complexes become trapped in the circulation. Free or unbound drugs are able to leave the capillaries.

TIP

Increased (↑) protein binding = less free drug available to tissue

Decreased (↓) protein binding = more free drug available to tissue

Albumin is the principal protein in systemic circulation and is produced in the liver. An animal that has liver disease or another protein-losing disease will have less protein available for protein binding. Less protein binding of drug will result in more free drug, which means more drug available to the target tissue. Liver disease and protein-losing diseases should alert the veterinary staff to closely monitor drug dosing, because of the decreased potential for drug protein binding. Less protein binding of drug results in more drug available to the target tissue(s), and potential toxic side effects from high tissue levels of the drug could occur.

Volume of Distribution

Volume of distribution is another factor in drug distribution. *Volume of distribution* is how well a drug is distributed throughout the body based on the concentration of drug in the blood. Volume of distribution assumes that the drug concentration in the blood is equal to the drug concentration dispersed throughout the rest of the body. Drug concentration in blood will be lower if the drug has a large volume to distribute itself through. The larger the volume of distribution, the lower the drug concentration in the blood after distribution.

An example of how volume of distribution affects drug concentration is in a dog with ascites (fluid in its abdomen). A dog with ascites has a greater volume to distribute the drug in (the abdominal fluid), and hence there is less drug concentration in the blood and other tissues. Distributing 10 milligrams of drug in a small volume (normal dog) will provide a certain concentration of drug in all body compartments. Distributing 10 milligrams of drug in a large volume (dog with ascites) will provide a decreased concentration of drug in all body compartments. Think of it as 10 milligrams of drug in a volume of 100 milliliters (normal dog) versus 10 milligrams of drug in a volume of 1000 milliliters (dog with ascites). The larger volume is less concentrated than the smaller volume. A less concentrated situation (dog with ascites) may keep a drug out of the therapeutic range, thus altering the effectiveness of the drug. Drug dosing may have to be altered to achieve drug levels in the therapeutic range in cases with increased volumes of distribution.

PHARMACOKINETICS PART III: CHANGING

Biotransformation is also called drug **metabolism**, *drug inactivation*, and/or *drug detoxification*. Biotransformation is the chemical alteration of drug molecules by the body cells of patients to a metabolite that is in an activated form, an inactivated form, and/or a toxic form. Usually the metabolite is more hydrophilic (therefore more ionized, less likely to be stored in fat, and less likely to pass through membranes) than the parent compound, and is more readily excreted in the urine or bile. The metabolite may also have less affinity for plasma proteins, resulting in wider distribution to tissues or excretion.

TIP

The goal of biotransformation is to make drugs more water-soluble so that they can be more easily excreted from the body.

There are four main pathways by which drugs undergo biotransformation: oxidation reactions (loss of electrons); reduction reactions (gaining of electrons); hydrolysis (adding of water molecules to a drug, causing it to split); and conjugation (addition of the glucuronic acid molecule, which makes the drug more water-soluble).

The primary site of biotransformation is the liver. Other sites of biotransformation include the kidneys, small intestine, brain and neurologic tissue, lungs, and skin. The liver is the primary organ where an enzyme called *cytochrome P450* is located. Cytochrome P450 is found in the hepatocytes (liver cells). Cytochrome P450 is actually a family of detoxifying enzymes that alter the structure of drug molecules. Many types of drug interactions are the result of either inhibition or induction of cytochrome P450.

Inhibition of cytochrome P450 generally involves competition with another drug for enzyme binding sites, which can prolong the activity of a particular drug. Cytochrome P450 induction occurs when one drug stimulates the production of more enzyme, thereby enhances its metabolizing capacity.

It is important to understand biotransformation because of drug interactions (how one drug's effect on the body affects other things). Some drug interactions are good and some are not so good. Ways drugs interact with each other include the following.

Altered absorption: One drug may alter the absorption of other drugs. For example, antacid medication alters the pH of the stomach and may affect how other drugs are absorbed through the gastrointestinal tract.

Competition for plasma proteins: Drug A and Drug B may both bind to plasma proteins. Drug A may have a higher affinity for binding to the plasma protein and so may displace drug B. Drug B may then reach toxic levels because it is free and able to cause its effect(s).

Altered excretion: Some drugs may act directly on the kidney and decrease the excretion of other drugs. Diuretics increase the production of urine and may affect drugs excreted via the kidneys.

Altered metabolism: The same enzymes may be needed for biotransformation of two drugs that are prescribed at the same time for an animal. This may cause the enzyme system to become saturated and decrease the rate of metabolism of both drugs. This altered metabolism may cause the parent drug to be broken down more slowly, resulting in lower levels of the active metabolite. The metabolite is produced upon metabolism of the parent drug and is then absorbed to cause the desired effect. If this process is slowed, the desired effect may not be seen. Veterinary staff may then inadvertently increase the animal's drug dose.

Another way some drugs may alter metabolism is by causing the liver enzymes to become more efficient. These drugs **induce** the enzyme system and hence are known as microsomal enzyme inducers (phenobarbital is one example). The enzyme system is referred to as induced if the rate of biotransformation by the enzyme system is increased. If other drugs that are metabolized by this same system are present, their biotransformation will also be increased. An increased rate of drug biotransformation by induction may require the dose to be increased to maintain adequate therapeutic levels.

Under some circumstances, the liver's ability to metabolize drugs may be impaired. Neonates with immature livers may not yet secrete adequate levels of microsomal enzymes. Old animals with liver damage may have diminished production of microsomal enzymes. Drug doses may have to be decreased in animals with altered liver function to avoid toxicity in these patients.

Another way metabolism may be altered is by animals developing tolerance. **Tolerance** is decreased response to a drug. There are two types of tolerance: metabolic (drug is metabolized more rapidly with chronic use) and cellular ("down regulation" or decreased cellular receptors with repeated use). Table 4-4 summarizes the primary factors affecting biotransformation.

PHARMACOKINETICS PART IV: GETTING OUT

Drug **elimination** is removal of a drug from the body. Drug elimination is also known as drug **excretion**. Routes of drug elimination include the kidneys, liver, intestine, lungs, milk, saliva, and sweat. The most important routes of drug elimination are the kidneys and liver.

Renal elimination of drugs involves a couple of mechanisms associated with urine production. Glomerular filtration occurs at the level of the nephron and works by pushing water and small molecules through a tuft of capillaries called the *glomerulus*,

TABLE 4-4 Factors Affecting Biotransformation

Factor	How Biotransformation Is Affected
Plasma protein binding	Less plasma protein binding allows excretion
Storage in tissue and fat depots	Fat and tissue storage decrease the rate of metabolism
Liver disease	Affects cytochrome P450 production
Age of patient	Young animals have decreased metabolic pathways (except horses), a blood-brain barrier that is not yet well established, and higher percent of body water that affects volume distribution
Nutritional status of patient	Poor nutrition yields inadequate plasma proteins
Species and individual variation	Cats have a reduced ability to biotransform aspirin
Body temperature	Increased body temperature increases rate of drug metabolism
Route of administration	Some drugs given parenterally have an effect, but when given orally have no effect (e.g., apomorphine)

similar to pushing things through a sieve (Figure 4-3). The rate of glomerular filtration depends on blood pressure. The higher the volume of blood going through the glomerulus, the higher the blood pressure of the capillaries of the glomerulus, and hence the greater the amount of particles that could potentially be filtered. Glomerular filtration is nonselective. If a particle is small, nonionic, and non-protein bound, it goes through.

The second mechanism by which drugs are eliminated by the kidney is tubular secretion. *Tubular secretion* is active transport across the convoluted tubule membrane that moves certain molecules from blood into the urine filtrate. Drugs such as penicillin are secreted into the tubule, allowing concentrations of these drugs to accumulate in the nephron and urine filtrate. Tubular secretion is generally a more rapid process of elimination than glomerular filtration and results in rapid elimination of drugs. Tubular secretion of drugs requires energy and any disease of the animal may hinder this function.

Another mechanism that influences how drugs are eliminated by the kidney is tubular reabsorption. Tubular reabsorption occurs in the loop of Henle of the nephron. Tubular reabsorption depends on lipid solubility and molecule size. A drug that is highly lipid soluble will have increased reabsorption (nonionized particles will have increased reabsorption). Ionization of molecules results in poor reabsorption. Figure 4-4 summarizes the renal elimination of drugs.

TIP

A disease condition of the liver and/or kidneys will affect elimination of drug from the body.

Urine pH can also affect the rate of drug excretion by changing the chemical form of a drug to one that is more readily excreted or to one that can be reabsorbed into the bloodstream. Drugs that are weak acids, such as barbiturates and other drugs available as sodium or potassium salts, tend to be better excreted in more basic urine, as this increases the proportion of drug in the ionized hydrophilic form. Weak bases, such as atropine and other drugs available as sulfate, hydrochloride, or nitrate salts, are better excreted in more acidic urine.

Hepatic elimination of drugs is also important. Drugs excreted by the liver usually move by passive diffusion from the blood into the liver cell. Once in the liver cell,

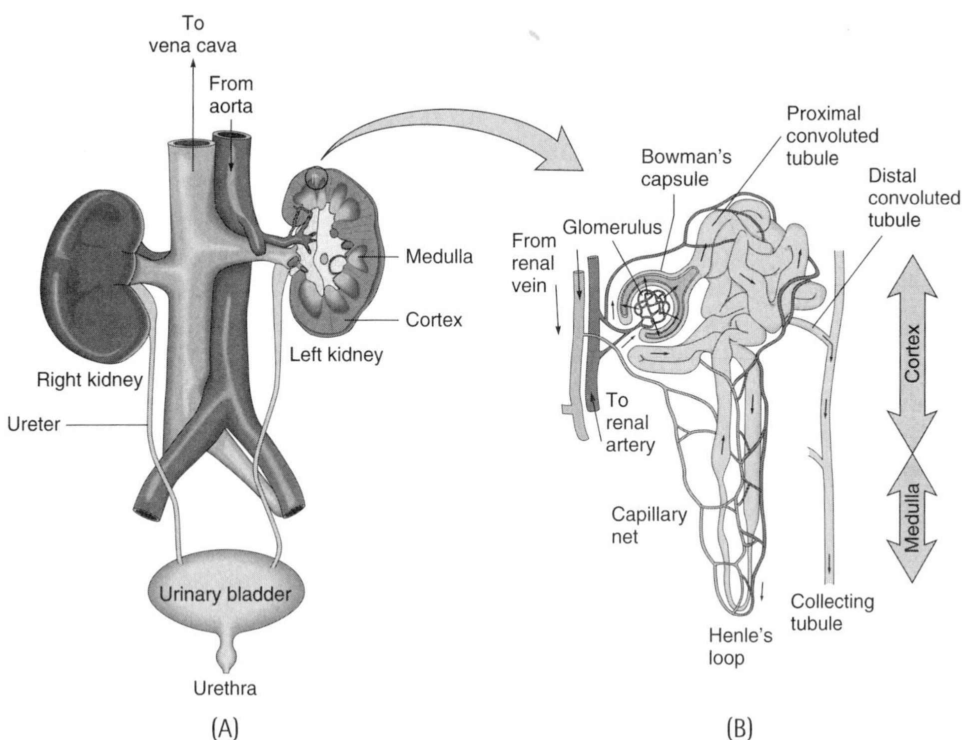

FIGURE 4-3 (A) Structures of the urinary tract. (B) Structures of a nephron

they are secreted into the bile. The bile is then secreted into the duodenum (either via the gall bladder or directly to the duodenum if the animal does not have a gall bladder). Lipophilic drugs entering the duodenum will reenter the bloodstream and go back to the liver. Hydrophilic drugs entering the duodenum will most likely become part of the feces and be eliminated by the body.

Intestinal elimination of drugs occurs when drugs are given orally and are not absorbed (allowing them to pass through the feces), when drugs are excreted in bile (allowing them to pass in the feces), or when drugs are actively secreted across mucous membranes into the gut. Pulmonary routes of drug excretion involve the movement of drug molecules out of blood and into the alveoli of the lungs to be eliminated in the expired air. Saliva and sweat also play a minor role in elimination of drugs from the body.

Another route of drug elimination is milk. Drugs or their metabolites, as well as nutrients, may pass directly from the blood to the milk via the mammary glands. Passage of nutrients into milk is important for the nourishment of offspring and human consumers of milk and milk products. Passage of drugs through the milk is an important consideration for veterinarians, as these drugs will affect the suckling offspring and also raise the issue of drug residues for consumers. Residues found in food products such as milk, meat, and eggs are important to people who have allergic reactions to the drug, for the development of antibiotic resistance, and for the development of disease secondary to drugs passing through these food sources. **Withdrawal time** is the period of time after drug administration during which the animal cannot be sent to market for slaughter and the eggs or milk must be discarded because of the potential for drug residues to persist in these products. Withdrawal times have been established for approved drugs that produce drug residues. The withdrawal time for drugs approved for use in food-producing animals is calculated using the drug's half-life. The **half-life** (abbreviated $T_{1/2}$) is the time required for the amount of drug in the body to be reduced by half of its original level. The half-life of a drug gives you an idea of how quickly a drug is eliminated by the body and the drug's steady state. The **steady state** of a drug is the point at which drug accumulation and elimination are balanced.

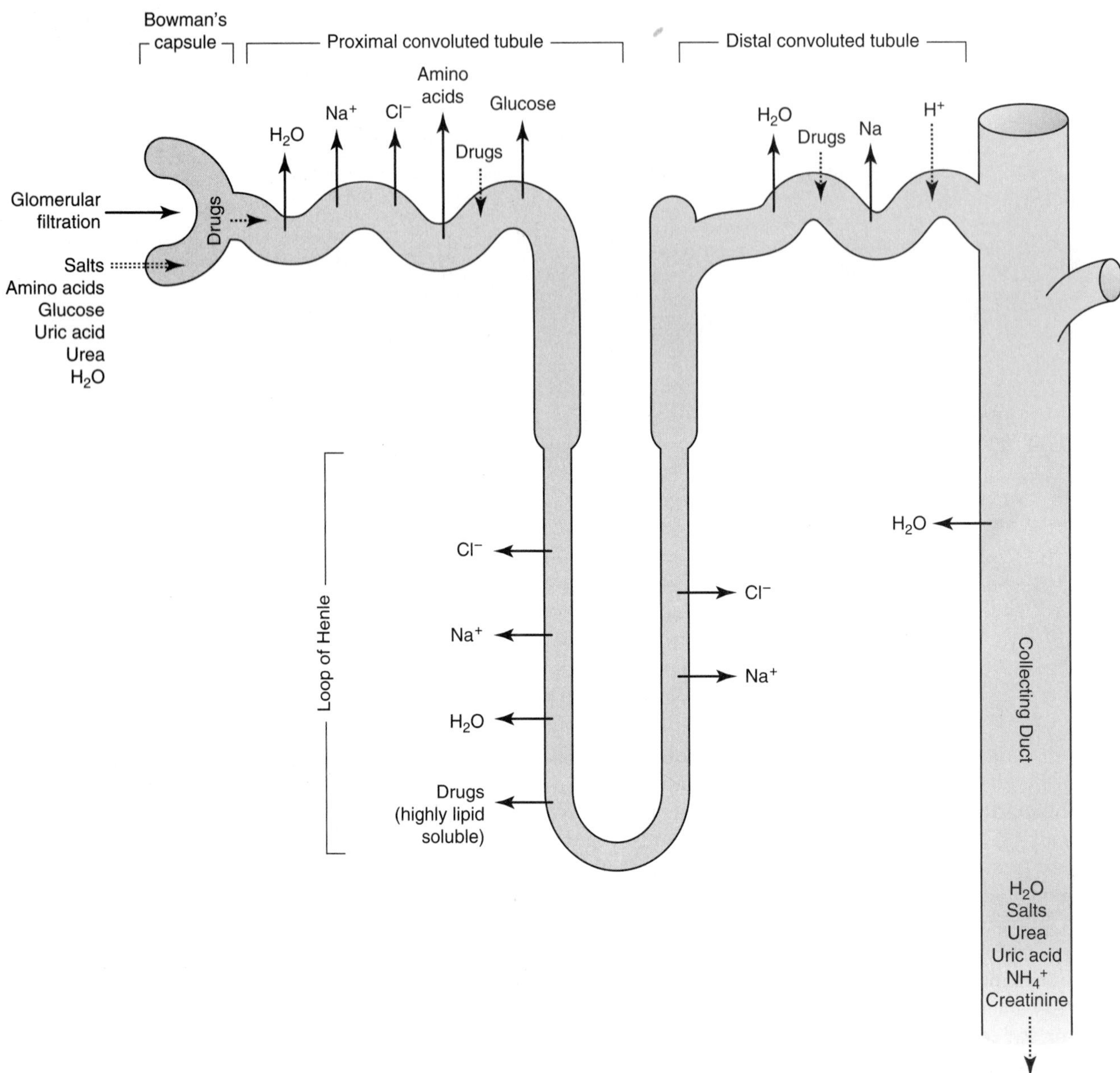

FIGURE 4-4 Renal elimination of drugs involves mechanisms associated with urine production. Glomerular filtration pushes small, nonionized, non-protein bound molecules through Bowman's capsule. Tubular reabsorption actively moves molecules from within the tubules into the capillaries surrounding the tubules. Tubular secretion actively moves molecules from capillaries into the tubules.

The patient's physiology also plays a role in the efficacy of drug excretion. Hydration status, age, and the disease or health status of the animal can all affect drug excretion. Hydration status affects the animal's blood volume and thus affects glomerular filtration of drug molecules excreted through the renal process. Dehydrated animals have less vascular fluid volume and so have decreased pressure in the vessels being filtered by the glomerulus. Age and disease may alter the levels of blood protein in an animal and affect the level of the drug's protein binding. Changes in protein binding can affect elimination rates. Animals with low blood protein levels will have less protein-bound drug (more free drug), which may be filtered through the glomerulus before the drug has had time to cause its effect.

HOW DO DRUGS WORK?

Once a drug is in the animal's body, it is absorbed and distributed by the bloodstream to the tissues. Drugs are capable of causing a variety of effects in animals.

Drugs alter existing cellular functions; they do not create new ones. For example, antibiotics slow the growth (reproduction) of bacteria. Laxatives increase the rate of peristaltic action of the gastrointestinal tract. Drug action is described relative to the physiological state that existed when the drug was administered.

Drugs may interact with the animal's body in several different ways. Some alter the chemical composition of body fluids (antacids alter the acidity of the stomach). Other drugs accumulate in tissues because of their affinity for a tissue component (gas anesthetics have an affinity for the lipid portion of the nerve cell membranes).

Drugs can form a chemical bond with specific cell components on target cells within the animal's body (this is the most common way in which drugs act). These specific cell components are called receptors. **Receptors** are three-dimensional proteins or glycoproteins that may be located on the surface, in the cytoplasm, or within the nucleus of cells (Figure 4-5). Animals have a finite or limited number of receptors. Not all cells have receptors for all substances, but some cells may multiple receptor sites.

This binding between drug and receptor will occur only if the drug and its receptor have a compatible chemical shape. This interaction between a drug and its receptor is often compared to the relationship between a lock and key. When the drug binds to the receptor, a series of reactions results in some cellular change, which may include the opening or closing of an ion channel; activation of secondary messengers such as cAMP, cGMP, or calcium; inhibition of normal cellular function; and/or activation of normal cellular function.

Affinity is the strength of binding between a drug and its receptor. Drugs whose molecules fit precisely into a given receptor can be expected to elicit a good response (for example, most penicillins); drugs that do not perfectly fit the receptor shape may produce only a weak response or no response at all. The better the fit of the drug with its receptor, the stronger the drug's affinity will be for the receptors and the lower the dose required to produce a pharmacological response. For example, hormone receptors are highly specific. Hormone response may often be elicited by the presence of only minute concentrations of an appropriate hormone, because it has a strong affinity for the receptor. The measure of a drug's affinity for its receptor is known as the dissociation constant K_D.

TIP

High-affinity drugs bind more tightly to a receptor than do low-affinity drugs.

An **agonist** is a drug that binds to a cell receptor and causes action (usually, more than one molecule of drug must bind to the receptor to cause action). A strong agonist is a drug that causes maximal cellular effect with only a few drug molecules occupying a few receptors. A weak agonist has to have many receptors bound with the drug before the effect occurs. A partial agonist is a substance that creates only a partial effect even if the drug occupies all of a cell's receptors.

An **antagonist** is a drug that inhibits or blocks the response of a cell when the drug is bound to the receptors. A competitive antagonist competes with the agonist for the same receptors. Competitive antagonism can be overcome by high doses of agonist (known as *surmountable* or *reversible antagonism*). A noncompetitive antagonist binds to a different site than the agonist's binding site and mechanically changes the agonist's receptor, resulting in a roadblock to the action of the agonist (known as *insurmountable* or *irreversible antagonism*). Figure 4-6 shows the different types of antagonism.

Some drugs work in ways other than binding to a receptor. One example is mannitol, an osmotic diuretic (a drug that causes increased urination). Mannitol prevents resorption of water back into the renal tubule by its presence in urine. Because mannitol is a large molecule, it sets up a concentration gradient (one side has more molecules than the other side) in the nephron and water moves into the nephron to try to equalize the concentration of fluid between the nephron and blood. Certain laxatives, such as milk of magnesia, raise the concentration of dissolved substances in the

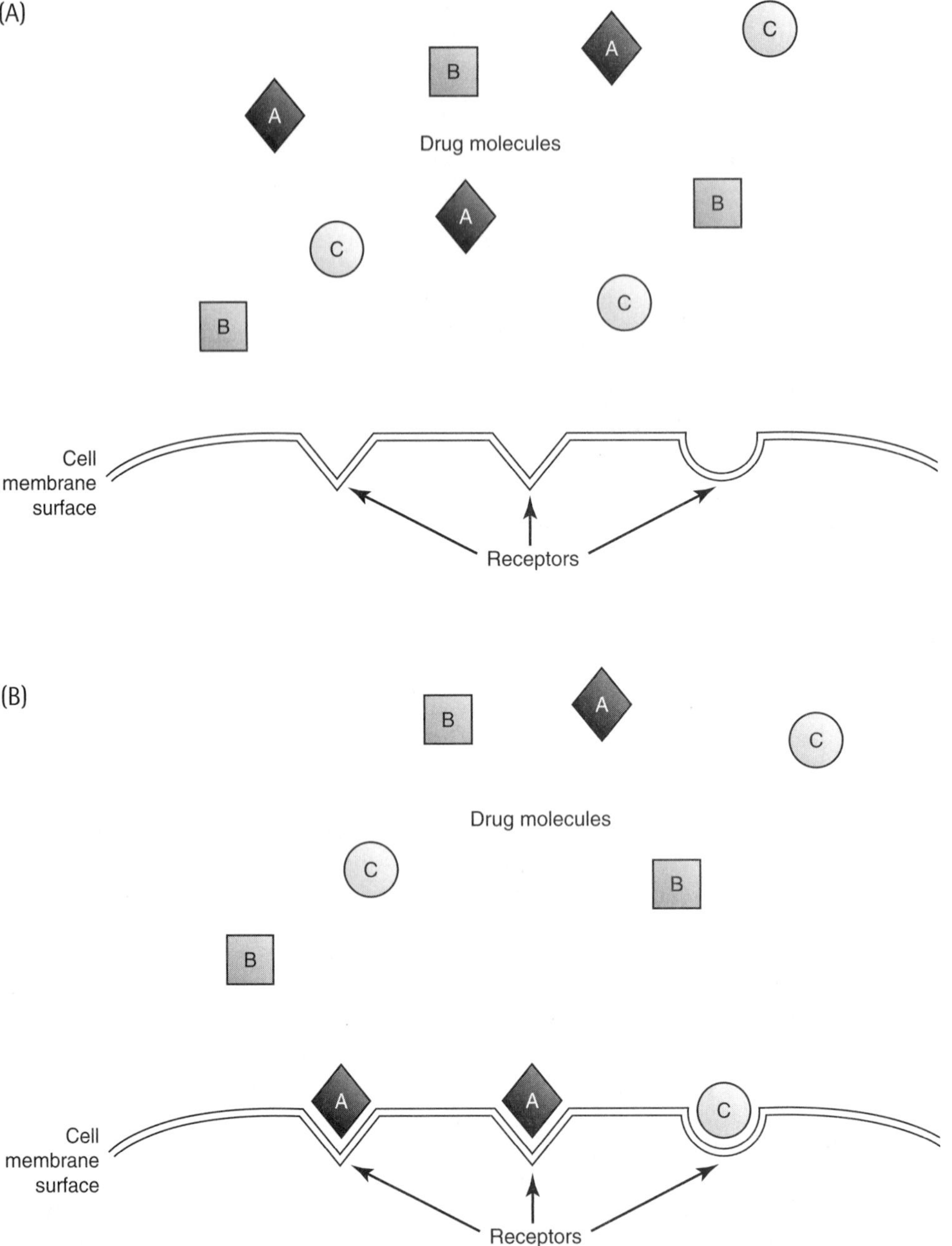

FIGURE 4-5 (A) Receptors are three-dimensional proteins located on the cell membrane surface. Receptors can only bind substances specific for that site. (B) This cell membrane can bind substances A or C.

gastrointestinal tract, thereby osmotically attracting fluid into the gut in an attempt to dilute the concentration of particles within the gastrointestinal tract.

SUMMARY

Pharmacokinetics describes the physiological movement of drugs. Passive diffusion, facilitated diffusion, active transport, or pino/phagocytosis are ways drugs move. Physiological drug movement involves four steps: absorption, distribution, biotransformation, and excretion. Absorption of drugs is affected by the

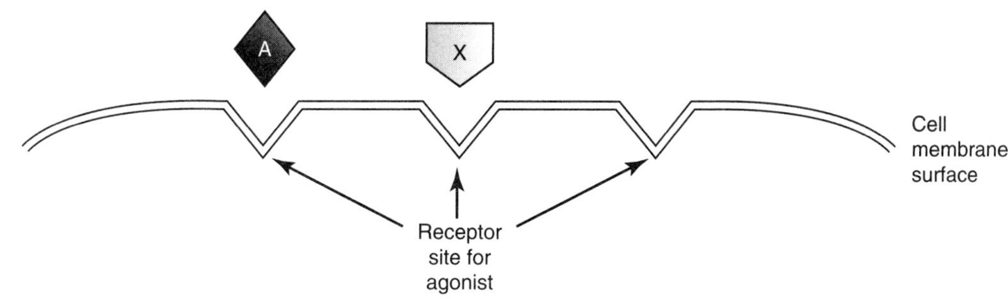

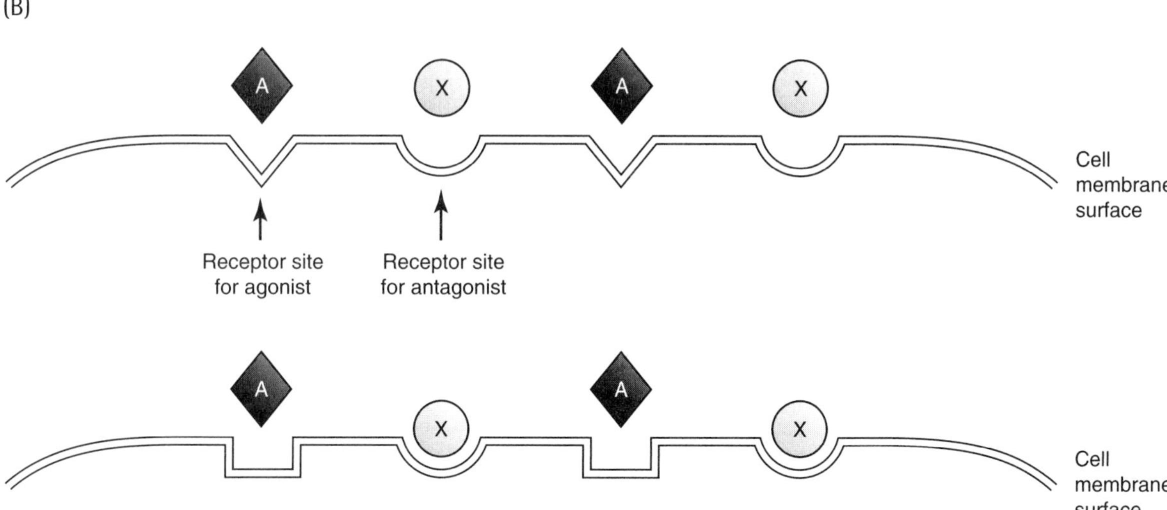

FIGURE 4-6 A = agonist, X = antagonist
(A) Competitive antagonism: The receptor is specific for A; however, X can compete for the same receptor because of its shape.
(B) Noncompetitive antagonism: When the antagonist binds to its receptor site, it changes the receptor so that the agonist can no longer bind to that site.

drug's bioavailability (the percentage of drug administered that actually enters the systemic circulation). Drugs with a bioavailability of 1 are 100 percent bioavailable. Other factors that affect drug absorption include the lipophilic or hydrophilic form of the drug, the degree of ionization, and size of the drug molecule. Acid-base status of the drug and the pH of the surrounding environment play a role in ionization of the drug. Ion trapping occurs when a drug molecule changes from ionized to nonionized form as it moves from one body compartment to another.

Drug distribution is the physiological movement of a drug to reach the target tissue or intended site of action. Membrane permeability, tissue perfusion, protein binding of the drug, and volume of distribution can affect drug distribution within an animal's body.

Biotransformation, or drug metabolism, occurs mainly in the liver and is affected by a variety of patient and drug parameters. Biotransformation occurs via oxidation reactions, reduction reactions, hydrolysis, and conjugation.

Drug elimination or excretion is the removal of a drug from the body. Routes of excretion include the kidney, liver, intestine, lungs, milk, saliva, and sweat. The two main routes of drug excretion are through the kidney and the liver.

The majority of drugs work by binding to receptors. Receptors are protein based, may be located in a variety of places on the cell, and are three-dimensional. Agonists bind to a cell receptor and cause action. Antagonists inhibit or block the response of a cell when bound to its receptors. Antagonists may be competitive or noncompetitive. Some drugs work without the use of receptors.

CHAPTER REFERENCES

Adams, H. R. (Ed.). *Veterinary pharmacology & therapeutics*, 8th ed. Iowa State University Press.

Johns Cupp, M., & Tracy, T. (January 1, 1998). *Cytochrome P450: New nomenclature and clinical implications*. Vol. 57: pp. 107–116. American Academy of Family Physicians.

Reiss, B., & Evans, M. (2002). *Pharmacological aspects of nursing care* (6th ed. revised by Bonita E. Broyles, pp. 2–31). Clifton Park, NY: Thomson Delmar Learning.

Veterinary pharmaceuticals and biologicals (12th ed.). (2001). Lenexa, KS: Veterinary Healthcare Communications.

Wynn, R. (1999). *Veterinary pharmacology*. Cambridge, MA: Harcourt Learning Direct.

CHAPTER REVIEW

Matching

Match the definition with its proper term.

1. ______ type of molecule or drug that easily passes through the cell membrane
2. ______ type of cellular movement in which molecules move against their concentration gradient
3. ______ type of cellular movement that allows intake of large molecules
4. ______ percent of drug administered that enters the systemic circulation
5. ______ term for charged
6. ______ term for water loving
7. ______ physiological movement of drugs from the systemic circulation to the tissues
8. ______ how well a drug is distributed throughout the body based on the concentration of drug in the blood
9. ______ primary site of biotransformation
10. ______ term used to describe an increase in the rate of biotransformation by an enzyme system

a. bioavailability
b. hydrophilic
c. pinocytosis/phagocytosis
d. active transport
e. volume of distribution
f. lipophilic, nonionized, small
g. liver
h. drug distribution
i. induce
j. ionized

Multiple Choice

Choose the one best answer.

11. Which drug has a greater percent that actually enters the systemic circulation?
    a. a drug with a bioavailability of 0.8
    b. a drug with a bioavailability of 0.2
12. The movement of molecules from an area of high concentration to an area of low concentration is known as
    a. passive diffusion.
    b. facilitated diffusion.
    c. active transport.
    d. pino/phagocytosis.
13. What type of drug is well absorbed from the gastrointestinal tract?
    a. hydrophilic
    b. lipophilic
    c. water-based
    d. charged drugs
14. Acidic drugs exist mostly in what form in an acidic environment?
    a. nonionized
    b. ionized
15. Based on blood perfusion, which body compartment will get adequate drug levels more quickly?
    a. fat
    b. subcutaneous tissue
    c. skeletal muscle
    d. smooth muscle
16. Which animal has a greater volume of distribution?
    a. a 7 percent dehydrated cat
    b. a normal cat
17. A dog has decreased renal perfusion. What will this do to the blood levels of a drug excreted through the kidneys?
    a. It will decrease.
    b. It will increase.
    c. It will remain normal.
    d. It will totally stop.
18. Drug affinity is the
    a. strength of binding between a drug and its receptor.
    b. the measure of the drug's action.
    c. number of receptors that must be occupied by the drug.
    d. binding of the drug to its receptor.
19. Which food-producing animal drug has the shorter withdrawal time?
    a. drug A with a half-life of 60 minutes
    b. drug B with a half-life of 120 minutes
20. Which of the following is true?
    a. Drugs with short half-lives need not be given more frequently.
    b. Noncompetitive antagonists are easily reversed in a clinical setting.
    c. Thin animals with low plasma protein levels require more drug than animals of normal weight.
    d. Giving fluids to an animal will increase the excretion of drug.

Case Study

21. A client brings a 13-year-old M/N Siamese cat into the clinic for an examination. The client states that the cat has been drinking more water and urinating more frequently in the litterbox. Because this is an older cat, the veterinarian collects blood to assess the cat's liver and kidney enzymes. A urine sample is also collected to assess whether the increased drinking and urinating are due to an infection or systemic disease. The blood results indicate that the cat has elevated liver and kidney enzymes. The urinalysis indicates that the cat also has a bacterial urinary tract infection. In addition to medication to treat liver and kidney disease, the client is given antibiotics to treat her cat's urinary tract infection.

    The client feels that because this cat has multiple health problems, it would be better to give a larger dose of antibiotics so that the infection can be cleared more quickly. Why is this not a good idea?

CHAPTER 5

Veterinary Drug Use and Prescribing

OBJECTIVES

Upon completion of this chapter, the learner should be able to

- compare and contrast generic and brand-name drugs.
- explain how veterinary drugs are used and prescribed.
- outline the parts of a drug label and explain the rationale for having this information on the drug label.
- describe the process used in prescribing drugs.
- describe proper drug dispensing techniques.
- list different drug resources and describe how they are used.
- explain drug package inserts and the information they provide.
- differentiate between prescribing and dispensing drugs.
- outline the essential components of a prescription label.
- define abbreviations used in prescription writing.

KEY TERMS

generic name
brand name
nonproprietary name
trade name
proprietary name
package insert
expiration date
prescription

INTRODUCTION

A woman calls your veterinary clinic and states that she would like some antibiotics to treat her cat's respiratory infection. She works at a discount store that has a pharmacy and she would like a prescription called in to that pharmacy for a human-labeled antibiotic that can be used in cats. Essentially, she is asking about the use of human-labeled products for veterinary purposes. Can drugs used for animals be dispensed at human pharmacies? What do you tell her?

BRAND NAME (®) OR NOT?

Drugs are named in a variety of ways (Table 5-1). The chemical name provides scientific and technical information because it is a precise description of the substance in accordance with chemical nomenclature rules established by the International Union of Pure and Applied Chemistry (IUPAC).

TABLE 5-1 Drug Identification

Term	Definition	Example
Chemical structure	Diagram of the chemical arrangement of the drug	H_3C O N N Cl
Chemical name	The name of the drug based on its chemical structure	7-chloro-1,3-dihydro-1-methyl-5-phenyl-2H-1, 4-benzodiazepin-2-one
Generic or nonproprietary name	The name picked by the company, but not exclusive to that company	diazepam
Trade or proprietary name	The name picked by the company that is registered	Valium^R or Valium®

The chemical name describes the chemical structure of the drug. Because they are long and complex, however, chemical names are rarely used in clinical medicine. An example of a chemical name is N-(2,6-dimethylphenyl)-5,6-dihydro-4H-1,3-thiazin-2-amine, which is more commonly known as xylazine.

Drugs are also named and marketed under two names, a **generic name** and a **brand name**. The generic name (sometimes referred to as the **nonproprietary name**) is written using lowercase letters. The generic name is the official identifying name of the drug, which is assigned by the U.S. Adopted Names (USAN) Council. The generic name is commonly used to describe the active drug(s) in the product and is easier to pronounce and remember than the chemical name. An example of a generic drug name is the animal sedative that goes by the generic name of xylazine. Another example of a generic drug name is the human-approved painkiller acetaminophen.

The brand name, **trade name**, or **proprietary name** of a drug establishes legal proprietary recognition for the corporation that developed the drug. The brand name of a drug is registered by the U.S. Patent Office, is approved by the U.S. Food and Drug Administration, and may be used only by the company that has registered the drug. For example, the Bayer Corporation calls its brand of xylazine by the brand name Rompun®. The acetaminophen produced by McNeil Pharmaceutical is commonly known by its brand name Tylenol®.

The brand name of a drug is usually written in capital letters or begins with a capital letter. It is considered a proper noun—the name of a drug. It may also have the superscript R or the symbol ®, both of which stand for "registered," next to the name. The brand name Rompun, for example, starts with a capital letter. With its superscript, the Rompun brand name looks like this: Rompun^R or Rompun®.

Once a manufacturer's patent for a drug has expired (usually 17 years from its registration date), other companies are free to market the drug under their own trademarked names or under the generic name of the drug. The therapeutic equivalence or bioequivalency of these products should be the same regardless of whether they are manufactured by the original patent-holding company or by another company. In most cases, no significant difference in response is noted when competing products are used. In some instances, however, different products containing identical drugs and drug doses have produced different pharmacological responses, even in the same patient. The vast price differences between some brand-name drugs and other brand-name drugs, and between brand-name drugs and generic drugs, have fueled the debate about whether "identical" drugs have the same therapeutic effectiveness as drugs in a formulation different from the original drug. No generalization can be made regarding the therapeutic effectiveness of competing drug products containing the same dose of a drug. The veterinary staff must carefully assess a patient's response to any drug product change, and the veterinarian should address any variation in therapeutic response.

GETTING INFORMATION

Drug Standards

In the United States, the Food and Drug Administration (FDA) of the Department of Health and Human Services is responsible for enforcement of the Food, Drug, and Cosmetic Act and its amendments, including the 1972 New Animal Drugs amendment. The FDA oversees adherence to drug standards and regulations established by physicians, pharmacists, dentists, and veterinarians. The standards for drugs are found in the *United States Pharmacopoeia* (USP). This publication is the legally recognized drug standard of the United States. The USP is revised and published periodically by a committee of the U.S. Pharmacopeial Convention, composed of delegates from all the major American health and medical science organizations. This official compendium describes the source, appearance, properties, standards of purity, and other requirements of the most important pure drugs. A drug may include the designation "USP" on its label if it meets the standards described in the compendium.

The FDA requires that all drugs meet USP standards of purity, quality, and uniformity. The FDA also requires that all drug containers be correctly and completely labeled; each container must identify the manufacturer or distributor and it must give directions for intended use. All accompanying advertisements and information must be true and correct statements of the indication, toxicity, and general usage of the drug.

Package Inserts

TIP

Drugs packaged or supplied in brown bottles are light-sensitive and should be stored in a dark area.

Drugs are packaged or supplied in bottles or vials that are labeled with specific information: the drug names (generic and trade), drug concentration and quantity, name and address of the manufacturer, the manufacturer's control or lot number, the expiration date of the drug, withdrawal times (if warranted), and the controlled substance status of the drug (if warranted). The label on the drug bottle or vial usually does not have enough space to list all the information required by FDA regulations. Manufacturers may reference **package inserts** provided with their drugs in order to meet regulatory requirements (Figures 5-1 and 5-2). Much of the information on package inserts is also found in drug references such as the *Physicians' Desk Reference (PDR), Compendium of Veterinary Products (CVP), Veterinary Pharmaceuticals and Biologicals (VPB),* and other drug reference books. Package inserts provide the following information (some package inserts may vary slightly from the terms listed here):

- Bold-faced letters with a circled R indicate a registered trade name of the product. The generic and/or chemical name of the drug follows the trade name. If there is no registered brand name (i.e., the product is a generic drug made by a manufacturer), the generic or chemical name will be given at the top of the insert. An Rx symbol may appear to the right of the brand name to signify that this drug is available through prescription only. The C symbol with an enclosed Roman numeral indicates that the drug is a controlled substance; the Roman numeral indicates one of several varying degrees of addiction potential. The manufacturer's name and trademark may also be on the insert.

- A *description or composition statement* usually follows the product name. This section describes the physical and chemical properties of the active drug. This section may include information about the drug's appearance, solubility, chemical formula and structure, and melting point. Other ingredients added to the product may also be included in this section. It is important to note any additional ingredients because these may make the product unsafe for some patients. This section may also be referred to as the *chemistry section.*
- *Clinical pharmacology* or *actions* or *mode of action statements* describe the activities and various properties of the main drug. The toxicology study data on the drug may be represented here. Metabolism and excretion of the drug may also be described in this section.
- The *indications and usage* section of the insert lists the specific uses for which the drug has been approved (indication) and describes how and for how long the drug is generally used (usage). Species-specific information will be listed in this section. Use in species not listed in the indications section and/or uses other than those described in this section are considered extra-label use (refer to Chapter 1 for a description of extra-label drug use).
- The *contraindications* section describes the situations in which the drug should not be used and the animals that should not receive the drug.
- The *precautions* section includes reasons to use the drug carefully. This does not imply that the drug use is contraindicated. Precautions usually describe conditions in which the drug is more likely to cause a problem (i.e., use of sulfa antibiotics in dehydrated animals) and suggests steps that should be taken in order for safe drug use.
- The *warnings* section relates situations in which potentially serious problems may occur if the drug is used (for example, during pregnancy or renal disease). The warnings section may lead into an animal toxicity section if additional studies were completed during development or if market use of the drug has indicated potential toxicity in certain cases.
- *Adverse reactions* or *side effects* are undesirable reactions to drugs that are significant enough to warrant extreme caution in use of the drug. A side effect is any normally occurring effect of the drug other than the intended therapeutic effect. Cautions are sometimes included in this section of the insert as well.
- *Overdosage* information lists the dangers of using excessive quantities of the drug. This section also describes the signs of overdose and how they should be handled.
- *Dosage and administration* information describes the parameters for use of the drug. Dosage refers to the amount of drug (usually per unit of body weight) that will produce the desired effect. Administration refers to the route that can be employed to deliver the drug into the animal's body.
- *Storage* information specifies the temperature and conditions under which the drug must be stored to maintain its integrity and viability.
- *How supplied* information states the dosage forms, strengths, and container size in which the drug is sold.

Additional information found in package inserts may include any or all of the following: how to prepare a suspension if mixing is needed, application methods for topical products, special disposal information, expiration dates on reconstituted powders, residue and withdrawal information, name and address of the manufacturer and distributor, and a *references* section for information on how the drug was tested and dosing information obtained.

NADA 141-108, Approved by FDA

EtoGesic®

(etodolac)

TABLETS FOR ORAL USE IN DOGS ONLY

Caution: Federal (U.S.A.) law restricts this drug to use by or on the order of a licensed veterinarian.

DESCRIPTION

Etodolac is a pyranocarboxylic acid, chemically designated as (±) 1,8-diethyl-1,3,4,9-tetrahydropyrano-[3,4-b] indole-1-acetic acid. The structural formula for etodolac is shown:

The empirical formula for etodolac is $C_{17}H_{21}NO_3$. The molecular weight of the base is 287.37. It has a pKa of 4.65 and an *n*-octanol:water partition coefficient of 11.4 at pH 7.4. Etodolac is a white crystalline compound, insoluble in water but soluble in alcohols, chloroform, dimethyl sulfoxide, and aqueous polyethylene glycol. Each tablet is biconvex and half-scored and contains either 150 or 300 mg of etodolac.

PHARMACOLOGY

Etodolac is a non-narcotic, nonsteroidal anti-inflammatory drug (NSAID) with anti-inflammatory, anti-pyretic, and analgesic activity. The mechanism of action of etodolac, like that of other NSAIDs, is believed to be associated with the inhibition of cyclooxygenase activity. Two unique cyclooxygenases have been described in mammals[1]. The constitutive cyclooxygenase, COX-1, synthesizes prostaglandins necessary for normal gastrointestinal and renal function. The inducible cyclooxygenase, COX-2, generates prostaglandins involved in inflammation. Inhibition of COX-1 is thought to be associated with gastrointestinal and renal toxicity, while inhibition of COX-2 provides anti-inflammatory activity. In *in vitro* experiments, etodolac demonstrated more selective inhibition of COX-2 than COX-1[2]. Etodolac also inhibits macrophage chemotaxis *in vivo* and *in vitro*[3]. Because of the importance of macrophages in the inflammatory response, the anti-inflammatory effect of etodolac could be partially mediated through inhibition of the chemotactic ability of macrophages.

Pharmacokinetics in healthy beagle dogs: Etodolac is rapidly and almost completely absorbed from the gastrointestinal tract following oral administration. The extent of etodolac absorption (AUC) is not affected by the prandial status of the animal. However, it appears that the peak concentration of the drug decreases in the presence of food. As compared to an oral solution, the relative bioavailability of the tablets when given with or without food is essentially 100%. Peak plasma concentrations are usually attained within 2 hours of administration. Though the terminal half-life increases in a nonfasted state, minimal drug accumulation (less than 30%) is expected after repeated dosing (i.e., steady-state). Pharmacokinetic parameters estimated in a crossover study (fed vs. fasted) in eighteen 5-month old beagle dogs are summarized in the following table:

Mean Pharmacokinetic Parameters Estimated in 18 Beagle Dogs After Oral Administration of 150 mg of Etodolac (approximately 12-17 mg/kg)

Pharmacokinetic Parameter	Tablet/ Fasted	Tablet/ Nonfasted
C_{max} (μg/mL)	22.0±6.42	16.9±8.84
T_{max} (hours)	1.69±0.69	1.08±0.46
$AUC_{0-\infty}$ (μg•hours/mL)	64.1±17.9	63.9±28.9
Terminal half-life, $t_{1/2}$ (hrs)	7.66±2.05	11.98±5.52

Pharmacokinetics in dogs with reduced kidney function: In a study involving four beagle dogs with induced acute renal failure, there was no observed change in drug bioavailability after administration of 200 mg single oral etodolac doses. In a study evaluating an additional four beagles, no changes in electrolyte, serum albumin/total protein and creatinine concentrations were observed after single 200 mg doses of etodolac. This was not unexpected since very little etodolac is cleared by the kidneys in normal animals. Most of etodolac and its metabolites are eliminated via the liver and feces. In addition, etodolac is believed to undergo enterohepatic recirculation[4].

EFFICACY

A placebo-controlled, double-blinded study demonstrated the anti-inflammatory and analgesic efficacy of EtoGesic (etodolac) tablets in various breeds of dogs. In this clinical field study, dogs diagnosed with osteoarthritis secondary to hip dysplasia showed objective improvement in mobility as measured by force plate parameters when given EtoGesic tablets at the label dosage for 8 days.

INDICATIONS

EtoGesic is recommended for the management of pain and inflammation associated with osteoarthritis in dogs.

DOSAGE AND ADMINISTRATION

The recommended dose of etodolac in dogs is 10 to 15 mg/kg body weight (4.5 to 6.8 mg/lb) administered once daily. Due to tablet sizes and scoring, dogs weighing less than 5 kg (11 lb) cannot be accurately dosed. The effective dose and duration should be based on clinical judgment of disease condition and patient tolerance of drug treatment. The initial dose level should be adjusted until a satisfactory clinical response is obtained, but should not exceed 15 mg/kg once daily. When a satisfactory clinical response is obtained, the daily dose level should be reduced to the minimum effective dose for longer term administration.

CONTRAINDICATIONS

EtoGesic is contraindicated in animals previously found to be hypersensitive to etodolac.

PRECAUTIONS

Treatment with EtoGesic tablets should be terminated if signs such as inappetence, emesis, fecal abnormalities, or anemia are observed. Dogs treated with nonsteroidal anti-inflammatory drugs, including etodolac, should be evaluated periodically to ensure that the drug is still necessary and well tolerated.

EtoGesic, as with other nonsteroidal anti-inflammatory drugs, may exacerbate clinical signs in dogs with pre-existing or occult gastrointestinal, hepatic or cardiovascular abnormalities, blood dyscrasias, or bleeding disorders.

As a class, cyclooxygenase inhibitory NSAIDs may be associated with gastrointestinal and renal toxicity. Sensitivity to drug-associated adverse effects varies with the individual patient. Patients at greatest risk for renal toxicity are those that are dehydrated, on concomitant diuretic therapy, or those with renal, cardiovascular, and/or hepatic dysfunction. Since many NSAIDs possess the potential to induce gastrointestinal ulceration, concomitant use of etodolac with other anti-inflammatory drugs, such as other NSAIDs and corticosteroids, should be avoided or closely monitored.

Studies to determine the activity of EtoGesic tablets when administered concomitantly with other protein-bound drugs have not been conducted in dogs. Drug compatibility should be monitored closely in patients requiring adjunctive therapy.

The safety of EtoGesic has not been investigated in breeding, pregnant or lactating dogs or in dogs under 12 months of age.

INFORMATION FOR DOG OWNERS

EtoGesic, like other drugs of its class, is not free from adverse reactions. Owners should be advised of the potential for adverse reactions and be informed of the clinical signs associated with drug intolerance. Adverse reactions may include decreased appetite, vomiting, diarrhea, dark or tarry stools, increased water consumption, increased urination, pale gums due to anemia, yellowing of gums, skin or white of the eye due to jaundice, lethargy, incoordination, seizure, or behavioral changes. **Serious adverse reactions associated with this drug class can occur without warning and in rare situations result in death (see Adverse Reactions). Owners should be advised to discontinue EtoGesic therapy and contact their veterinarian immediately if signs of intolerance are observed.** The vast majority of patients with drug related adverse reactions have recovered when the signs are recognized, the drug is withdrawn, and veterinary care, if appropriate, is initiated. Owners should be advised of the importance of periodic follow-up for all dogs during administration of any NSAID.

WARNINGS

Keep out of reach of children. Not for human use. Consult a physician in cases of accidental ingestion by humans. **For use in dogs only.** Do not use in cats.

All dogs should undergo a thorough history and physical examination before initiation of NSAID therapy. Appropriate laboratory tests to establish hematological and serum biochemical baseline data prior to, and periodically during, administration of any NSAID should be considered. **Owners should be advised to observe for signs of potential drug toxicity (see Information for Dog Owners and Adverse Reactions).**

ADVERSE REACTIONS

In a placebo-controlled clinical field trial involving 116 dogs, where treatment was administered for 8 days, the following adverse reactions were noted:

Adverse Reaction	EtoGesic Tablets % of Dogs	Placebo % of Dogs
vomiting	4.3%	1.7%
regurgitation	0.9%	2.6%
lethargy	3.4%	2.6%
diarrhea/loose stool	2.6%	1.7%
hypoproteinemia	2.6%	0
urticaria	0.9%	0
behavioral change, urinating in house	0.9%	0
inappetence	0.9%	1.7%

Following completion of the clinical field trial, 92 dogs continued to receive etodolac. One dog developed diarrhea following 2-1/2 weeks of treatment. Etodolac was discontinued until resolution of clinical signs was observed. When treatment was resumed, the diarrhea returned within 24 hours. One dog experienced vomiting which was attributed to treatment, and etodolac was discontinued. Hypoproteinemia was identified in one dog following 11 months of etodolac therapy. Treatment was discontinued, and serum protein levels subsequently returned to normal.

Post-Approval Experience:

As with other drugs in the NSAID class, adverse responses to EtoGesic tablets may occur. The adverse drug reactions listed below are based on voluntary post-approval reporting. The categories of adverse event reports are listed below in decreasing order of frequency by body system.

Gastrointestinal: Vomiting, diarrhea, inappetence, gastroenteritis, gastrointestinal bleeding, melena, gastrointestinal ulceration, hypoproteinemia, elevated pancreatic enzymes.
Hepatic: Abnormal liver function test(s), elevated hepatic enzymes, icterus, acute hepatitis.
Hematological: Anemia, hemolytic anemia, thrombocytopenia, prolonged bleeding time.
Neurological/Behavioral/Special Senses: Ataxia, paresis, aggression, sedation, hyperactivity, disorientation, hyperesthesia, seizures, vestibular signs, keratoconjunctivitis sicca.
Renal: Polydipsia, polyuria, urinary incontinence, azotemia, acute renal failure, proteinuria, hematuria.
Dermatological/Immunological: Pruritus, dermatitis, edema, alopecia, urticaria.
Cardiovascular/Respiratory: Tachycardia, dyspnea.

In rare situations, death has been reported as an outcome of some of the adverse responses listed above. To report suspected adverse reactions, or to obtain technical assistance, call (800) 477-1365.

SAFETY

In target animal safety studies, etodolac was well tolerated clinically when given at the label dosage for periods as long as one year (see Precautions).

Oral administration of etodolac at a daily dosage of 10 mg/kg (4.5 mg/lb) for twelve months or 15 mg/kg (6.8 mg/lb) for six months, resulted in some dogs showing a mild weight loss, fecal abnormalities (loose, mucoid, mucosanguineous feces or diarrhea), and hypoproteinemia. Erosions in the small intestine were observed in one of the eight dogs receiving 15 mg/kg following six months of daily dosing.

Elevated dose levels of EtoGesic (etodolac), i.e≥40 mg/kg/day (18 mg/lb/day, 2.7X the maximum daily dose), caused gastrointestinal ulceration, emesis, fecal occult blood, and weight loss. At a dose of ≥80 mg/kg/day (36 mg/lb/day, 5.3X the maximum daily dose), 6 of 8 treated dogs died or became moribund as a result of gastrointestinal ulceration. One dog died within 3 weeks of treatment initiation while the other 5 died after 3-9 months of daily treatment. Deaths were preceded by clinical signs of emesis, fecal abnormalities, decreased food intake, weight loss, and pale mucous membranes. Renal tubular nephrosis was also found in 1 dog treated with 80 mg/kg for 12 months. Other common abnormalities observed at elevated doses included reductions in red blood cell count, hematocrit, hemoglobin, total protein and albumin concentrations; and increases in fibrinogen concentration and reticulocyte, leukocyte, and platelet counts.

In an additional study which evaluated the effects of etodolac administered to 6 dogs at the labeled dose for approximately 9.5 weeks, the incidence of stool abnormalities (diarrhea, loose stools) was unchanged for dogs in the weeks prior to initiation of etodolac treatment, and during the course of etodolac therapy. Five of the dogs receiving etodolac, versus 2 of the placebo-treated dogs, exhibited excessive bleeding during an experimental surgery. No significant evidence of drug-related toxicity was noted on necropsy.

STORAGE CONDITIONS

Store at controlled room temperature, 15-30°C (59-86°F).

HOW SUPPLIED

EtoGesic (etodolac) is available in 150 and 300 mg single-scored tablets and supplied in bottles containing 7, 30 and 90 tablets.

NDC 0856-5520-03 – 150 mg – bottles of 7
NDC 0856-5520-04 – 150 mg – bottles of 30
NDC 0856-5520-05 – 150 mg – bottles of 90
NDC 0856-5530-03 – 300 mg – bottles of 7
NDC 0856-5530-04 – 300 mg – bottles of 30
NDC 0856-5530-05 – 300 mg – bottles of 90

REFERENCES

1. Vane, JR, RM Botting. Overview - mechanisms of action of anti-inflammatory drugs. In Improved Non-steroid Anti-inflammatory Drugs - COX-2 Enzyme Inhibitors. J Vane, J Botting, R Botting (ed.) 1996. Kluwer Academic Publishers. Dordrecht, The Netherlands.
2. Glaser, KB. Cyclooxygenase selectivity and NSAIDs: cyclooxygenase-2 selectivity of etodolac (Lodine®). *Inflammopharmacol* (1995) 3:335-345.
3. Gervais, F, RR Martel, E Skamene. The effect of the non-steroidal anti-inflammatory drug etodolac on macrophage migration *in vitro* and *in vivo*. *J. Immunopharmacol* (1984) 6:205-214.
4. Cayen, MN, M Kraml, ES Ferdinandi, EL Greselin, D Dvornik. The metabolic disposition of etodolac in rats, dogs, and man. *Drug Metab. Revs.* (1981) 12:339-362.

Fort Dodge Animal Health
Fort Dodge, Iowa 50501 USA

5530D Revised March 2001 00781

FIGURE 5-1 An example of a package insert (Copyright © FDAH and reprinted with permission)

TORBUGESIC®-SA CIV

Fort Dodge **Analgesic**

Butorphanol Tartrate, USP Injection

NADA No.: 141-047

Active Ingredient(s): Each mL of TORBUGESIC®-SA contains 2 mg butorphanol base (as butorphanol tartrate, USP); 3.3 mg citric acid, USP; 6.4 mg sodium citrate, USP; 4.7 mg sodium chloride, USP; and 0.1 mg benzethonium chloride, USP; q.s. with water for injection, USP.

Indications: TORBUGESIC®-SA (butorphanol tartrate, USP) is indicated for the relief of pain in cats caused by major or minor trauma, or pain associated with surgical procedures.

Pharmacology:

Description: Butorphanol tartrate, USP is a synthetic, centrally acting, narcotic agonist-antagonist analgesic with potent antitussive activity. The results from laboratory and clinical studies suggest the existence of several distinct types of receptors that are responsible for the activity of opioid and opioid-like drugs. When activated, the μ(mu)-receptors are involved in analgesia, respiratory depression, miosis, physical dependence and feelings of well-being (euphoria). When activated, the κ(kappa)-receptors are involved in analgesia, as well as less intense (as compared to μ-receptors) miosis and respiratory depression. Butorphanol is considered to be a weak antagonist at the μ-receptor, but a strong agonist at the κ-receptor. Thus, butorphanol provides analgesia with a lower incidence and/or intensity of adverse reactions (e.g., miosis and respiratory depression) than traditional opioids.

Butorphanol tartrate is a member of the phenanthrene series. The chemical name is Morphinan-3, 14-diol, 17-(cyclobutylmethyl)-, (-)-, (S- (R*, R*))-2,3-dihydroxybutanedioate (1:1) (salt). It is a white, crystalline, water soluble substance having a molecular weight of 477.55; its molecular formula is $C_{21}H_{29}NO_2 \bullet C_4H_6O_6$.

Chemical Structure:

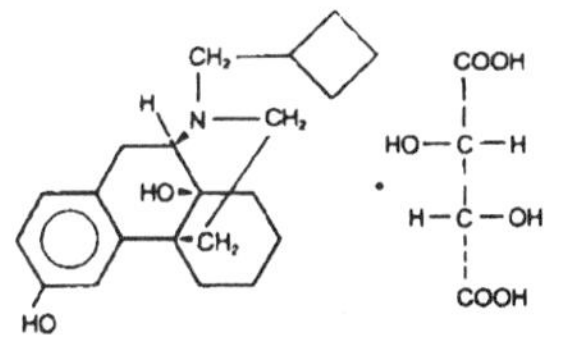

Clinical Pharmacology:

Feline Pharmacology: The magnitude and duration of analgesic activity of butorphanol were studied in cats under controlled laboratory conditions using both a visceral pain model and a somatic pain model.[1,2] Subcutaneous butorphanol dosages of 0.4 mg/kg produced analgesia significantly ($p<0.05$) greater than the placebo for up to two hours in the somatic pain model. At the label dose (0.4 mg/kg), cardiopulmonary depressant effects were minimal after treatment with butorphanol as demonstrated in cats.[1,2]

Clinical studies confirmed the analgesic effect of butorphanol administered subcutaneously in the cat. In field trials the overall analgesic effect was rated as satisfactory in approximately 75% of butorphanol treated cats. The duration of activity in cats responding to butorphanol ranged from 15 minutes to 8 hours. However, in 70% of responding cats the duration of activity was 3 to 6 hours following subcutaneous administration.

Safety Studies in Cats: Daily subcutaneous injections of butorphanol in cats, beginning at a dosage of 2 mg/kg the first week and doubling each week to a final dosage of 16 mg/kg on the fourth week, resulted in no deaths. No evidence of toxicity was observed during the first three weeks of the experiment, other than pain on injection. During the fourth week, transient incoordination, salivation, or mild seizures were observed within the first hour in the cats following the 16 mg/kg dosage (40 times the recommended clinical dosage). No other clinical, serum chemistry, or gross necropsy evidence of drug toxicity was encountered in any of the cats.

In subacute safety studies, butorphanol was injected subcutaneously to each of six cats at dosages of 0 (saline), 0.4, 1.2 or 2.0 mg/kg, every six hours for six days and continued once daily for a total of 21 days. The only adverse clinical effect observed was pain on injection. Histopathologic changes indicative of minimal to slight irritation were noted at the injection sites in 3 of 6 cats in the low dose group, 4 of 6 cats in the middle dose group and 6 of 6 cats in the high dose group. Histopathologic changes of focal renal tubular dilation were noted in half of the cats in the high dose group.

Dosage and Administration: The recommended dosage in cats is 0.4 mg of butorphanol per kilogram body weight (0.2 mg/lb) given by subcutaneous injection. This is equivalent to 1.0 mL of TORBUGESIC®-SA per 10 lbs of body weight.

Pre-clinical model studies and clinical field trials in cats demonstrated that the analgesic effects of TORBUGESIC®-SA are seen within 20 minutes and persist in the majority of responding cats for 3 to 6 hours following subcutaneous injection (see Feline Pharmacology). The dose may be repeated up to 4 times per day for up to 2 days.

Precaution(s): Store at controlled room temperature 15° to 30°C (59° to 86°F).

Caution(s): Federal law restricts this drug to use by or on the order of a licensed veterinarian.

TORBUGESIC®-SA, a potent analgesic, should be used with caution with other sedative or analgesic drugs as these are likely to produce additive effects.

Safety for use in pregnant female cats, breeding male cats or kittens less than 4 months of age has not been determined. Use of TORBUGESIC®-SA can therefore not be recommended in these groups.

Warning(s): Not for human use.

Adverse Reactions: In clinical trials in cats, pain on injection, mydriasis, disorientation, swallowing/licking and sedation were reported.

References: Available upon request.

Presentation: 10 mL vials TORBUGESIC®-SA (butorphanol tartrate, USP) Veterinary Injection, 2 mg base activity per mL.

NDC 0856-4531-01 — 10 mL — vials

4530C

FIGURE 5-2 An example of a controlled substance package insert (Copyright © FDAH and reprinted with permission)

Drug Reference Books

What's in the *PDR*?

- Manufacturer's Index (white pages): Alphabetical listing of pharmaceutical manufacturers included in the *PDR*
- Brand & Generic Name Index (pink pages): Gives the page number of each product by brand and generic name. The drugs listed in the *PDR* are chosen by the drug manufacturer because they pay a fee to have their drugs in this references. Therefore, not all drugs manufactured in the United States are included in the *PDR*.
- Product Category Index (blue pages): Lists all fully described products by prescribing category
- Product Identification Guide (gray pages): Full-color, actual-size photos of tablets, capsules, and other dosage forms, arranged alphabetically by manufacturer

What's in the *VPB* and *CVP*?

- Manufacturer/Distributor Index (white pages): Alphabetical listing of pharmaceutical manufacturers included in the *VPB* and *CVP*
- Brand Name and Ingredient Index (green pages): Gives the page number of each product by brand name and ingredients found in these products
- Prescription drugs may be dispensed by pharmacists (human-labeled drugs) or trained veterinary staff. Therapeutic Index (pink pages): Lists all fully described products by prescribing category
- Biological Charts (blue pages): List groups of vaccines and bacterins by appropriate species, delineated by their antigens
- Anthelmintic and Parasiticide Charts (buff pages): List groups of internal parasiticides and endectocides by appropriate species, delineated by the corresponding parasite
- Withdrawal Time Charts (yellow pages): Alphabetically summarize information on withdrawal times for food products according to the species and route of administration

TIP

The **expiration date** is the date before which a drug meets all specifications and after which the drug can no longer be used. Expiration dates are assigned based on the stability of or experience with the drug. Expiration dates for drugs that are mixed in the clinic vary depending on reconstitution and refrigeration status. Check the package insert for information regarding expiration dates on these products.

DISPENSING VERSUS PRESCRIBING

Veterinary drugs are those approved only for use in animals. Veterinary drugs are not subject to regulatory action by the FDA; however, FDA regulations require that the drug label clearly describe the approved use of the drug in these ways: for use in certain indications, for use in certain species, for use by a certain route of administration, for use at a certain dose, and for use over a certain length of time. Any use of an approved label drug outside of its "approved use" is subject to regulatory action.

Some drugs, such as antibiotics, are utilized in both human and veterinary medicine. The FDA specifies in its Compliance Policy Guide that "food-producing animals" (e.g., cattle, sheep, pigs, chickens) generally should not receive drugs that are utilized in human and veterinary practice. Illegal drug residues may be present in milk, meat, and/or eggs from these animals and may present an unknown health risk to the consumer of these products. Most diseases in food-producing animals are treatable with approved veterinary-use-only drugs that can be selected and employed as appropriate by the veterinarian. Any manufacturer, distributor, or pharmacy that suggests the use of human-labeled drugs for animals must comply with FDA regulations on the administration of these drugs.

TABLE 5-2 Veterinarian/Client/Patient Relationship Guidelines

A veterinarian/client/patient relationship (VCPR) exists when all of the following conditions have been met:
• The veterinarian has assumed responsibility for making clinical judgments regarding the health of the animal(s) and the need for medical treatment, and the client has agreed to follow the veterinarian's instructions.
• The veterinarian has sufficient knowledge of the animal(s) to initiate at least a general or preliminary diagnosis of the medical condition of the animal(s). This means that the veterinarian has recently seen and is personally acquainted with the keeping and care of the animal(s) by virtue of an examination of the animal(s) or by medically appropriate and timely visits to the premises where the animal(s) are kept.
• The veterinarian is readily available for follow-up evaluation, or has arranged for emergency coverage, in the event of adverse reactions or failure of the treatment regimen.

Adapted from the AVMA Guidelines for Veterinary Prescription Drugs in *VPB* 12th ed.

In the case of non-food-producing animals, the Compliance Policy Guide makes an exception. Generally, veterinarians can prescribe human-labeled drugs for any non-food-producing animal without violating any FDA regulations. If such human-labeled drugs harm the animal's health, however, the Compliance Policy Guide warns of FDA regulatory action or referral to the state veterinary licensing authority for investigation.

Veterinarians must establish a veterinarian/client/patient relationship prior to prescribing any medication for an animal (Table 5-2). If an animal requires a drug, the veterinarian can either write a prescription to be filled at a human pharmacy, or dispense the drug from the veterinary clinic. Instructions for drug use for the patient are part of the medical record and are often written in abbreviated Latin. Veterinarians abbreviate instructions because Latin terminology is universally used in medical professions and is concise, convenient, and timesaving. Table 5-3 on page 68 shows common Latin abbreviations.

Some veterinarians order drugs by prescription because keeping an extensive pharmaceutical inventory ties up working capital and runs the risk of outdating, spoilage, and obsolescence. A **prescription** is an order to a pharmacist, written by a licensed veterinarian (or physician), to prepare the prescribed medicine, to affix the directions, and to sell the preparation to the client. The prescription is a legally recognized document and the writer is held responsible for its accuracy. A pharmacist may dispense human-labeled drugs and the veterinary staff may dispense veterinary-use-only drugs.

Through a written prescription, the veterinarian gives the pharmacist all the information necessary to dispense safe and effective drugs for the animal. Prescriptions may be written on standard forms or telephoned in to the pharmacy. Figure 5-3 shows the seven parts of a prescription.

1. The veterinarian's name, address, and telephone number, all usually preprinted at the top of the form. (If the prescription is for a controlled substance, the DEA number of the prescriber must be given to the pharmacist. DEA numbers may be listed on the preprinted form, but may also be kept confidential and excluded from the preprinted prescription pad.)

TIP

Veterinary medicine is generally regulated according to a standard of care; that is, a veterinarian may be liable on charges of substandard practice if the care provided is below that which is considered standard veterinary practice.

TIP

Latin abbreviations for pharmacology terms and directions make medical-record and prescription writing concise and efficient.

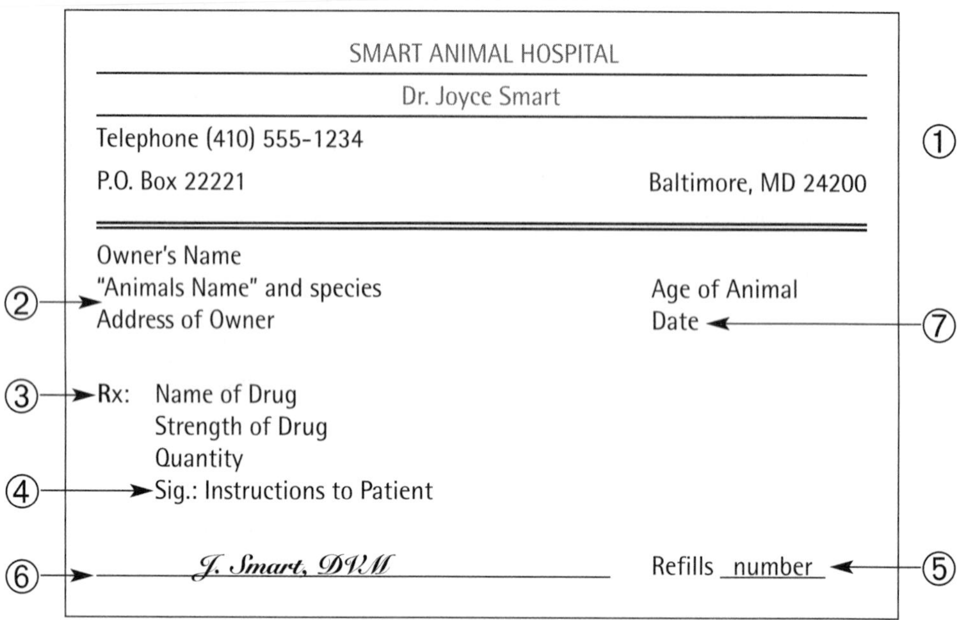

SMART ANIMAL HOSPITAL

Dr. Joyce Smart

Telephone (410) 555-1234 ①

P.O. Box 22221 Baltimore, MD 24200

Owner's Name

② "Animals Name" and species Age of Animal

Address of Owner Date ⑦

③ Rx: Name of Drug

Strength of Drug

Quantity

④ Sig.: Instructions to Patient

⑥ J. Smart, DVM Refills number ⑤

FIGURE 5-3 The seven parts of a prescription

2. The client's name and address, as well as the species and name of the patient
3. The name of the drug, the strength of the drug, and the quantity to be given to the patient
4. The instructions for giving the drug to the patient; these must include the amount to be given, the route of administration, the frequency of administration, and the duration of administration
5. The number of refills permitted
6. The veterinarian's signature
7. The date of the prescription

Other things that may be on the prescription include cautionary statements to be included on the label ("keep out of reach of children," "use gloves when applying," etc.) and withdrawal times. Warning labels can be purchased and placed on the prescription vial as well. Sample warning labels are illustrated in Figure 5-4.

Figure 5-5 shows three typical prescriptions with parts 1 through 7 indicated. Prescription A is an order for a cream to be applied to the skin to treat an infection;

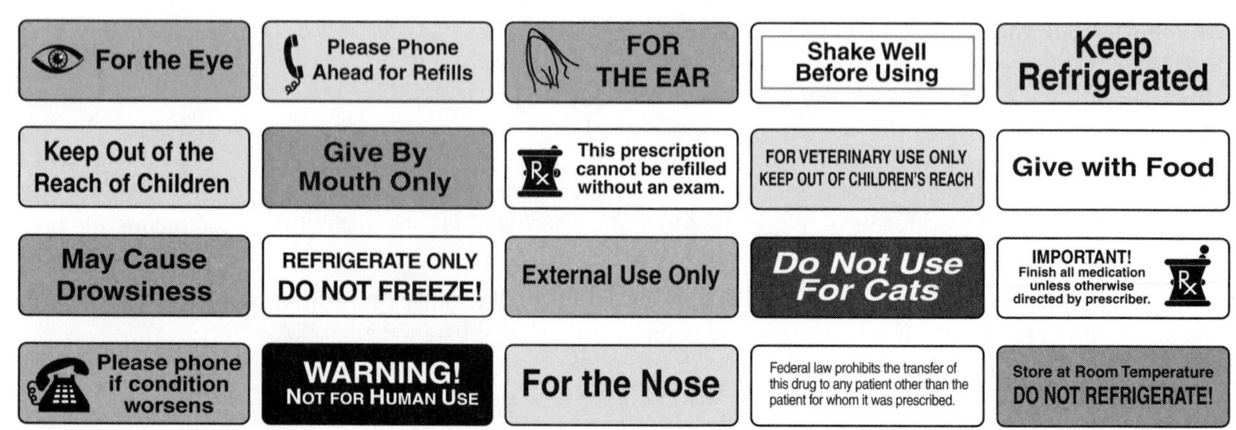

FIGURE 5-4 Warning labels are placed on prescription medication containers to emphasize key facts about the drug.

prescription B is for insulin injection; prescription C is for vitamin C (ascorbic acid) tablets. Figure 5-6 shows three other prescriptions written by veterinarians.

For additional practice in interpreting drug orders and understanding drug labels, use Chapters 5 and 6 on the Accu-Calc CD.

All veterinary prescription drugs should be properly labeled when dispensed. The label should contain the information that appears on the prescription, with the exception of the veterinarian's signature. Remember that veterinary staff members cannot refill or dispense medications without veterinarian approval. All medications should be dispensed in childproof containers. Labels with cautionary statements may also be used on the prescription bottle to help clients remember key facts about the drugs they are giving their animals.

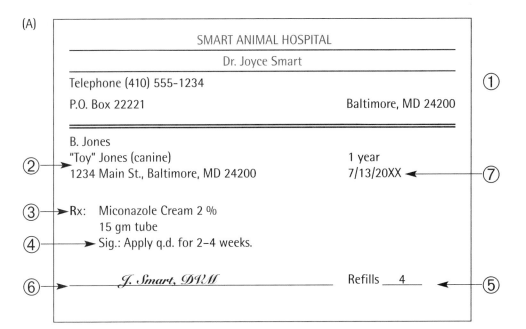

TIP

Ways to avoid errors in prescription writing include writing legibly, using a zero in front of any numbers with a decimal point (for example, 0.02 instead of .02), and specifying both dose and strength of drug required.

(A)

SMART ANIMAL HOSPITAL
Dr. Joyce Smart
Telephone (410) 555-1234 ①
P.O. Box 22221 Baltimore, MD 24200

② → B. Jones
"Toy" Jones (canine) 1 year
1234 Main St., Baltimore, MD 24200 7/13/20XX ← ⑦

③ → Rx: Miconazole Cream 2 %
15 gm tube
④ → Sig.: Apply q.d. for 2–4 weeks.

⑥ → *J. Smart, DVM* Refills 4 ← ⑤

(B)

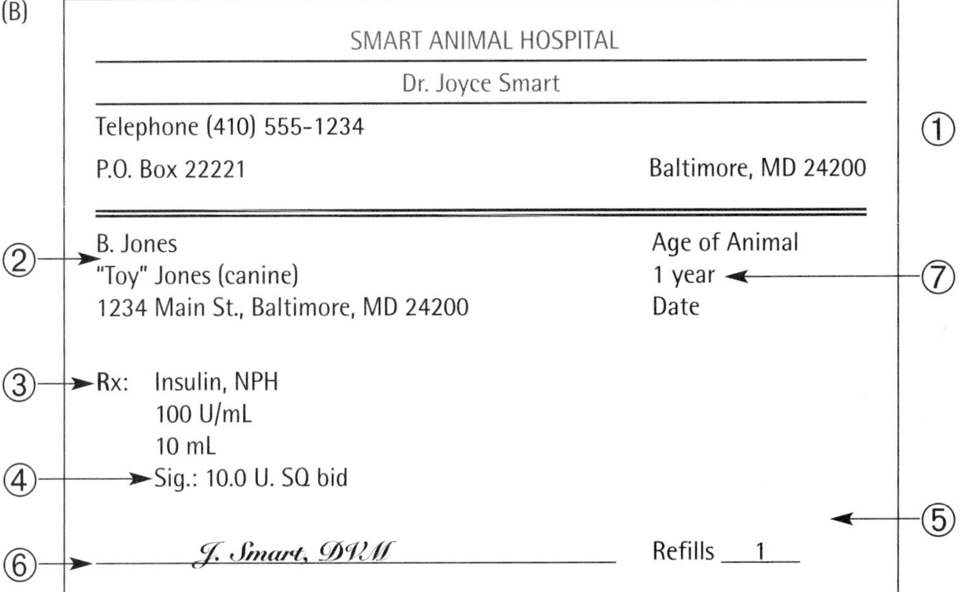

SMART ANIMAL HOSPITAL
Dr. Joyce Smart
Telephone (410) 555-1234 ①
P.O. Box 22221 Baltimore, MD 24200

② → B. Jones Age of Animal
"Toy" Jones (canine) 1 year ← ⑦
1234 Main St., Baltimore, MD 24200 Date

③ → Rx: Insulin, NPH
100 U/mL
10 mL
④ → Sig.: 10.0 U. SQ bid
← ⑤
⑥ → *J. Smart, DVM* Refills 1

FIGURE 5-5 Three example prescriptions *(continues)*

(C)

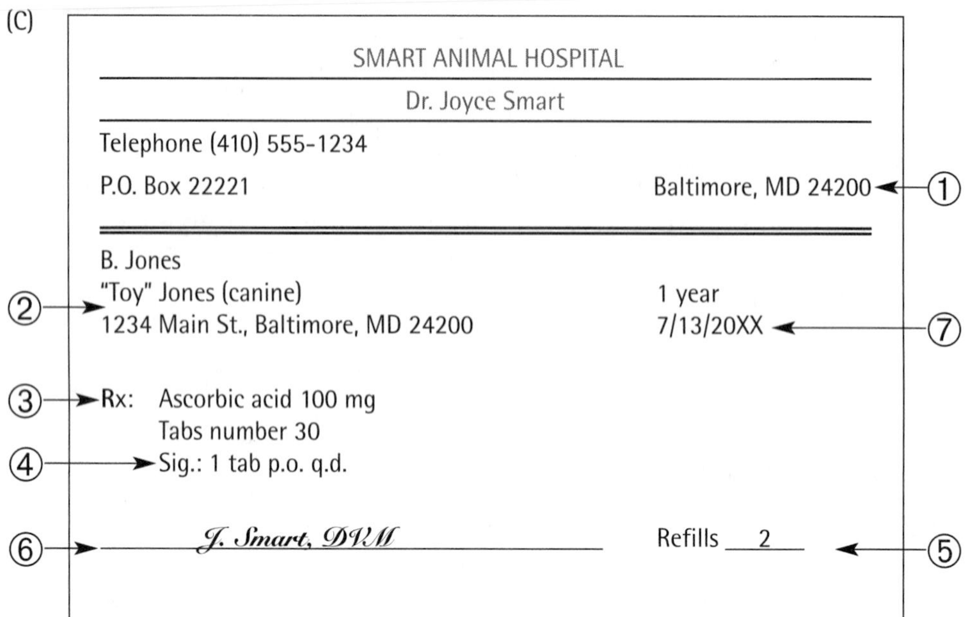
SMART ANIMAL HOSPITAL
Dr. Joyce Smart
Telephone (410) 555-1234
P.O. Box 22221
Baltimore, MD 24200
B. Jones
"Toy" Jones (canine)
1234 Main St., Baltimore, MD 24200
1 year
7/13/20XX
Rx: Ascorbic acid 100 mg
Tabs number 30
Sig.: 1 tab p.o. q.d.
J. Smart, DVM
Refills 2

FIGURE 5-5 *Continued*

1

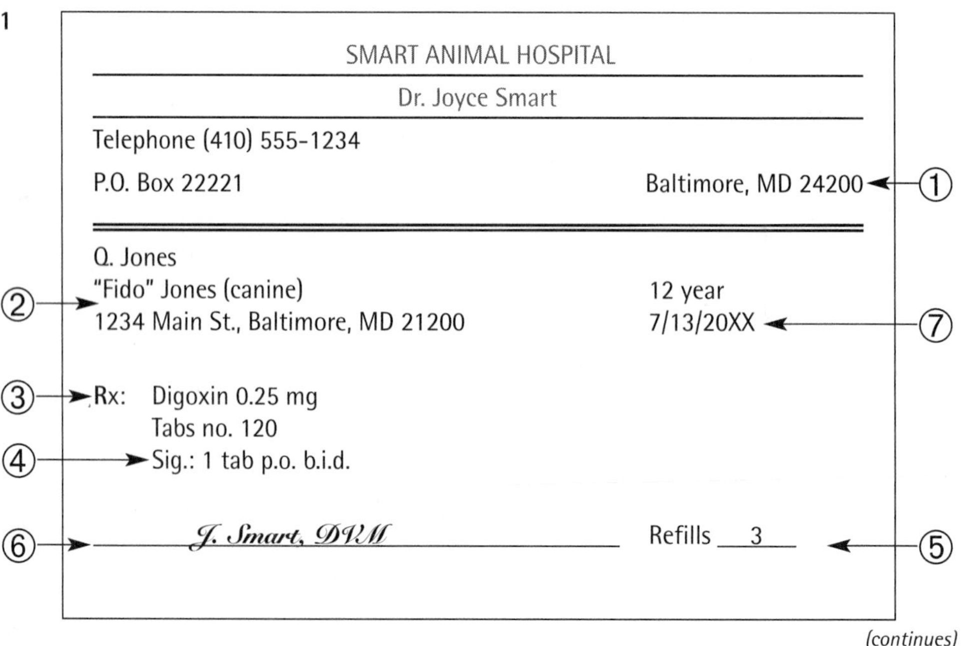
SMART ANIMAL HOSPITAL
Dr. Joyce Smart
Telephone (410) 555-1234
P.O. Box 22221
Baltimore, MD 24200
Q. Jones
"Fido" Jones (canine)
1234 Main St., Baltimore, MD 21200
12 year
7/13/20XX
Rx: Digoxin 0.25 mg
Tabs no. 120
Sig.: 1 tab p.o. b.i.d.
J. Smart, DVM
Refills 3

(continues)

FIGURE 5-6 Prescription samples

2

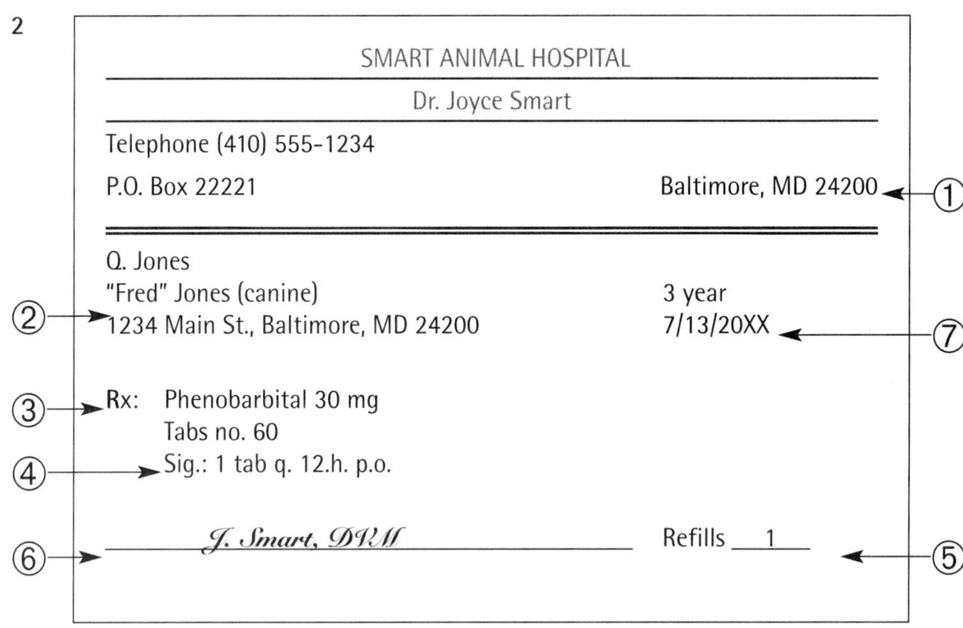

SMART ANIMAL HOSPITAL

Dr. Joyce Smart

Telephone (410) 555-1234

P.O. Box 22221 Baltimore, MD 24200 ①

Q. Jones
"Fred" Jones (canine) 3 year
② 1234 Main St., Baltimore, MD 24200 7/13/20XX ⑦

③ Rx: Phenobarbital 30 mg
Tabs no. 60
④ Sig.: 1 tab q. 12.h. p.o.

⑥ J. Smart, DVM Refills 1 ⑤

3

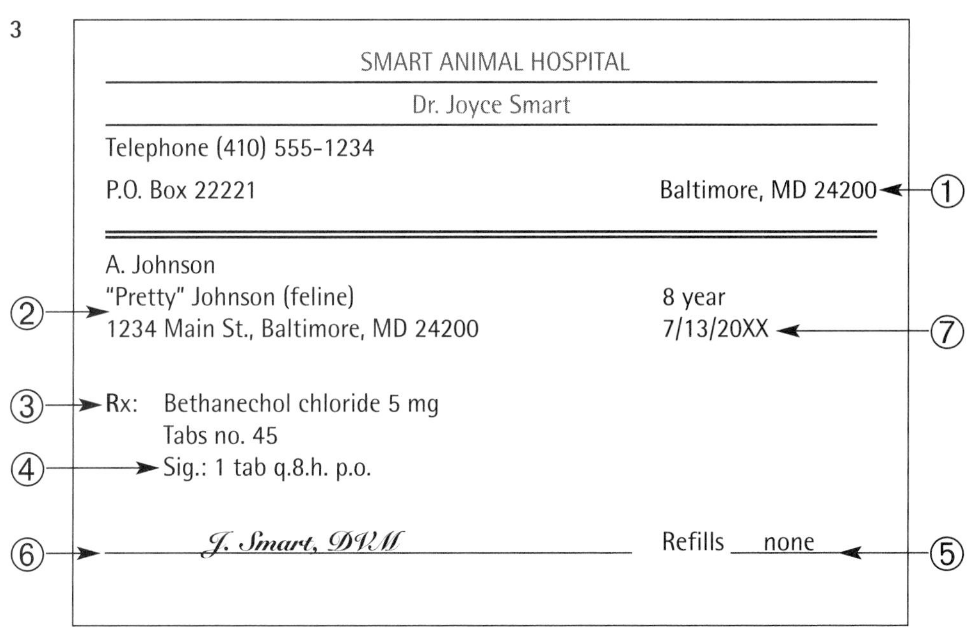

SMART ANIMAL HOSPITAL

Dr. Joyce Smart

Telephone (410) 555-1234

P.O. Box 22221 Baltimore, MD 24200 ①

A. Johnson
② "Pretty" Johnson (feline) 8 year
1234 Main St., Baltimore, MD 24200 7/13/20XX ⑦

③ Rx: Bethanechol chloride 5 mg
Tabs no. 45
④ Sig.: 1 tab q.8.h. p.o.

⑥ J. Smart, DVM Refills none ⑤

FIGURE 5-6 *Continued*

TABLE 5-3 Pharmacology Abbreviations

Abbreviation	Meaning
%	percent
®	registered trade name when superscript by drug name
ac	before meals (ante cibum)
ad lib	as much as desired (ad libitum)
bid	twice daily (bis in die)
BSA	body surface area
c̄	with
cal	calorie
cap	capsule
cc	cubic centimeter (same as ml)
cm	centimeter
conc	concentration
D_5W	5% dextrose in water
dr	dram; equal to 1/8 oz or 4 ml
ED	effective dose
ED_{50}	median effective dose
fl oz	fluid ounce
g or gm	gram
gal	gallon
gr	grain; unit of weight approximately 65 mg
gt	drop (gutta)
gtt	drops (guttae)
hr	hour
IA	intra-arterial
IC	intracardiac
ID	intradermal
IM	intramuscular
IP	intraperitoneal
IT	intratracheal
IU	international units
IV	intravenous
kg	kilogram
km	kilometer
L	liter
lb or # behind a number	pound(s)
LD	lethal dose
LD_{50}	median lethal dose
LRS	lactated Ringer's solution
m	meter
mcg or μg	microgram
mED	minimal effective dose
mEq	milliequivalent
mg	milligram

TABLE 5-3 *Continued*

Abbreviation	Meaning
MIC	minimum inhibitory concentration
MID	minimum infective dose
ml or mL	milliliter (same as cc)
MLD	minimum lethal dose
mm	millimeter (also used to mean muscles)
npo	nothing by mouth (non per os)
NS	normal saline
OTC	over the counter
oz	ounce
p̄	after
pc	after meals (post cibum)
PDR	*Physician's Desk Reference*
pH	hydrogen ion concentration (acidity and alkalinity measurement)
po	orally (per os)
ppm	parts per million
PR	per rectum
prn	as needed (pro re nata)
pt	pint
q	every
q12h	every twelve hours
q4h	every four hours
q6h	every six hours
q8h	every eight hours
qd	every day
qh	every hour
qid	four times daily (quarter in die)
qn	every night
qod or eod	every other day
qp	as much as desired
qt	quart
Rx	prescription
s̄	without
sid	once daily
sig	let it be written as (used when writing prescription)
sol'n or soln	solution
SQ, SC, subQ, or subc	subcutaneous
T	tablespoon or tablet (or temperature)
tab	tablet
tid	three times daily (ter in die)
tsp	teaspoon
vol	volume
VPB	*Veterinary Pharmaceuticals and Biologicals*

The label on the prescription must be complete and contain directions that clients can understand and follow. The complete label must have:

- the name and address of the dispenser (animal hospital and veterinarian's names are usually preprinted on the label)
- the client's name (address is optional)
- the animal's name and species
- the drug name, drug strength, and drug quantity to be given
- the date of the order
- the directions for use
- any refill information (may be included if warranted)

Figure 5-7 shows the components of a drug label. Table 5-4 summarizes information required for prescriptions, labels, and medical records.

SUMMARY

Drugs are marketed under generic and brand names. The generic name is the official name of the drug and may be used by anyone to identify that drug; the brand name can be used for identification purposes only by the corporation that produces the drug.

Drug information is available through many resources: the *United States Pharmacopoeia (USP)*, drug bottles, package inserts, the *Physicians' Desk Reference (PDR)*, *Compendium of Veterinary Products (CVP)*, *Veterinary Pharmaceuticals and Biologicals (VPB)*, and other references.

A veterinarian or physician must order prescription drugs. Prescription drugs may be dispensed by pharmacists (human-labeled drugs) or trained veterinary staff. Specific information is required on prescriptions and on medication labels.

SMART ANIMAL HOSPITAL/Dr. J. Smart
P.O. Box 22221
Baltimore, MD 24200

K. Ross-Barnes
818 Division St., Baltimore, MD 24200
"Silver" Ross-Barnes (horse) Date 7/13/20XX

Directions: Apply 4 drops daily to both eyes

J. J. Smith, D. V. M.

Gentocin Ophthalmic Drops 3 mg/ml
Refills 0 5 mL

For veterinary use only—keep out of reach of children

FIGURE 5-7 Sample prescription drug label

TABLE 5-4 Basic Information for Records, Prescriptions, and Labels

Information	Records	Prescriptions	Labels
• Name, address, and telephone number of the veterinarian	+	+	+
• Name, address, and telephone number of the client	+	+	Name only
• Identification of the animal(s) treated, species, and numbers of animals treated, when possible	+	+	+
• Date of treatment, prescribing, or dispensing of drug	+	+	+
• Name and quantity of the drug to be prescribed or dispensed	+	+	+
• Dose and duration directions for use	+	+	+
• Number of refills authorized	+	+	May be included

CHAPTER REFERENCES

Reiss, B., & Evans, M. (2002). *Pharmacological aspects of nursing care* (6th ed. revised by Bonita E. Broyles, pp. 2–31). Clifton Park, NY: Thomson Delmar Learning.

Rice, J. (1999). *Principles of pharmacology for medical assisting* (3rd ed.). Clifton Park, NY: Thomson Delmar Learning.

Wynn, R. (1999). *Veterinary pharmacology.* Cambridge, MA: Harcourt Learning Direct.

CHAPTER REVIEW

Matching

Match the English words to the pharmacological abbreviation.

1. ______ one tablet by mouth twice a day
2. ______ one tablet every 12 hours by mouth
3. ______ one tablet every eight hours by mouth
4. ______ two units subcutaneously after meals
5. ______ one milliliter by mouth every 12 hours
6. ______ one tablet by mouth every day for five days
7. ______ one tablet by mouth once a day
8. ______ one tablet by mouth three times a day
9. ______ apply every day for two to four weeks
10. ______ one tablet by mouth every day

a. Apply qd 2-4 weeks
b. 1 ml q12h po
c. 1 T tid po
d. 1 T q12h po
e. 1 T qd
f. 1T qd po X 5d
g. 1 T q8h po
h. 1 T sid po
i. 2U SQ pc
j. 1 T bid po

Multiple Choice

Choose the one best answer.

11. What term describes the date before which a drug meets all specifications and after which the drug can no longer be used?
    a. dispensing date
    b. prescribing date
    c. expiration date
    d. termination date

12. What is the order given or written to a pharmacist by a licensed veterinarian to prepare the medicine, to affix the directions, and to sell the preparation to the client?
    a. dispensatory note
    b. prescription
    c. drug order
    d. package insert

13. On a drug label, which part is usually in capital letters with a superscript R by it?
    a. generic name
    b. drug concentration
    c. manufacturer's name
    d. trade name

Fill in the Blank

14. A drug is usually marketed under both its ______ name and its ______ name.

Write the English meanings for the following prescription abbreviations:

15. bid ____________________
16. npo ____________________
17. q4h ____________________
18. tid ____________________
19. prn ____________________
20. po ____________________

Case Study

21. A client brings in a puppy for its first examination at your clinic. The routine physical exam is normal; however, the puppy has intestinal parasites. The puppy is prescribed an oral liquid dewormer that you prepare and dispense in a syringe to the client. You advise the client to bring in a follow-up stool sample two to three weeks after the last dose of dewormer is given to the puppy.

    When the client gets home, he attempts to give the puppy its dewormer. He is having trouble restraining the puppy to insert the syringe rectally. The client believes that the medication has to be given rectally because the worms were present in the puppy's stool. How could this confusion have been avoided?

CHAPTER 6

Systems of Measurement In Veterinary Pharmacology

OBJECTIVES

Upon completion of this chapter, the learner should be able to

- review basic math concepts.
- differentiate between the household, apothecary, and metric systems of measurement.
- describe the base units in the metric system.
- describe the commonly used metric prefixes.
- convert between and within the household, apothecary, and metric systems.
- perform dose calculations.
- perform solution calculations.
- perform reconstitution calculations.

KEY TERMS

metric system
apothecary system
household system
micro-
milli-
centi-
kilo-
meter
liter
gram
minim
dram
grain
solution
solvent
solute
miscible
immiscible

INTRODUCTION

On a bright sunny day, you decide to take a brisk walk on your break. As you walk you notice that you are sweating quite a bit and wonder if you drank enough water today to prevent you from becoming dehydrated. You know you should drink about eight (eight-ounce) glasses of water per day. You brought your liter bottle of water to work today and drank one bottle of water. Did you drink enough water today to replace normal water loss? Can you mathematically determine if you did or did not?

THE BASICS

Veterinary professionals use mathematical concepts on a daily basis to determine drug doses, fluid rates, and dilution of chemicals. The more comfortable you feel with math, the better you will be able to perform these calculations.

Basic mathematical concepts include the use of percents, decimals, and ratios. Reducing fractions, interpretation of values, and conversion of fractions, decimals, percents, and ratios are things you should already feel comfortable with. You should be able to perform simple algebra to solve for an unknown and use proportions to solve for unknowns. If you need to review these concepts, refer to Chapter 1: Math Review on the Accu-Calc CD.

HOW DO WE MEASURE?

Measurement is the use of standard units to determine the weight, length, or volume of substances. Without standard units, pharmacology would be an inaccurate science; everything flows from our ability to standardize weight, length, and volume.

The units of measurement used in pharmacology are the **metric system**, the **apothecary system**, and the **household system**. The household system uses household measures when an approximate dose is acceptable. The household method uses a system of weights and measures based on 1 pound containing 16 ounces. Most household measuring devices lack standardization, but are calibrated in units that most people are familiar with. The apothecary system uses weights and measures based on 480 grains equal to 1 ounce and 12 ounces equal to 1 pound (this is different from the household system where 16 fluid ounces equal 1 pound). Only a few apothecary units are used today. Pharmacology mainly uses the metric system, and you must become an expert at metric conversion to avoid harmful or fatal mistakes in preparation and dosing of drugs. The metric system is based on a standard to which other factor-of-ten units are compared. You must also be able to convert among the three systems, as some drugs and dosages vary in their representation in the veterinary community.

The Household System

Household measures are approximate measurements that lack standardization and are not accurate for measuring medicines. For example, there may be three or four teaspoons per tablespoon depending on the reference source or size of each utensil. Though inaccurate, the household system may be used in some circumstances in which standardization is not crucial.

The inch and foot are common standard units for measuring length in the household system. In the United States, a standard piece of paper is 8-1/2 inches by 11 inches. This figure communicates a length we can understand and reproduce because of standardization.

The pound is a common standard unit used to measure weight in the household system. If a box of sugar has a weight of five pounds, then we say it's a five-pound box of sugar. If a box of table salt contains one of these pounds, then we say it's a one-pound box of table salt. Anyone familiar with a pound can purchase sugar or salt and know the amount being purchased, because of standardization.

The fluid ounce is a common standard unit used to measure liquid volume in the household system. If a water glass has a volume of eight fluid ounces, then we say it's an eight-ounce water glass. Eight ounces is its final measurement. If a bottle of soda holds 12 of these fluid ounces, then we say it's a 12-ounce bottle of soda. Twelve ounces is its final measurement. These figures communicate a volume we can understand and reproduce because of standardization.

The Metric System

The term *metric* is borrowed from the Greek term *metron*, which means measure or standard. The metric system of measure was formulated by the French government in the late eighteenth century as a product of the French Revolution. The metric system was developed to standardize measures and weights for European countries. It has been used in the United States since 1890. Table 6-1 lists the fundamental units of the metric system. In the metric system, prefixes are used to denote the size of a metric unit. These metric units are all based on factors of 10. The metric prefixes used most commonly in the veterinary field are:

micro- = one millionth of a unit = 0.000001

milli- = one thousandth of a unit = 0.001

centi- = one hundredth of a unit = 0.01

kilo- = one thousand units = 1000

TIP

Abbreviations for the following metric prefixes:

micro- = mc or μ = 0.000001 × base unit

milli- = m = 0.001 × base unit

centi- = c = 0.01 × base unit

kilo- = k = 1000 × base unit

Each prefix in the metric system changes the value of the basic unit by a factor of 10.

The **meter** (m) is the metric standard unit used to measure length. The expression *centimeter* (cm) is the other important term in measuring length in the metric system. One meter consists of 100 centimeters. In other words, it takes one hundred centimeters of length to equal one meter.

Some pharmacological doses are based on surface area of an animal (m^2 of body surface area for calculation of chemotherapy drugs for cancer patients). Charts are available for conversion from weight to surface area. This information is covered in Chapter 20 on antineoplastic drugs.

The **liter** (l) is the metric standard unit used to measure liquid volume. The expression *milliliter* (ml) is the other important term in measuring liquid volume in the metric system. One liter consists of 1,000 milliliters. In other words, it takes one thousand milliliters of liquid to equal one liter.

If water were placed in a cube measuring one centimeter on every side, the amount in the cube would be a cubic centimeter of water. The term *cubic centimeter* is abbreviated as cc. A very important relationship exists between cubic centimeters and milliliters: the volume of a cubic centimeter (cc) of liquid is equivalent to one milliliter.

1 ml = 1 cc

The **gram** (g) is the metric standard unit of weight measurement. Refer back to the preceding explanation of the cubic centimeter, in which the weight of the water in the one cubic centimeter is 1 g. Thus, 1 cc = 1 ml = 1 g, which you read as the volume of one cubic centimeter of water equals one milliliter of water, which in turn weighs one gram.

The denominations most commonly used in measures of weights for medications and weights of powders used in preparing medications are milligrams and grams. When using metric terminology, it helps to remember prefixes: the prefix *milli-* is from the Latin meaning one one-thousandth of a unit. Therefore, one milligram equals one one-thousandth of a gram. Another common denomination is *kilogram*. Animal weights and drug dosages may be represented in kilogram units. The term *kilo* is from the Greek meaning one thousand times the unit, so one kilogram equals 1,000 grams.

Metric weight also includes the microgram, which is one-thousandth of a milligram or one-millionth of a gram. Microgram is abbreviated as mcg or μg. The relationships of metric weights are summarized in Table 6-2.

A fundamental operation using the metric system is converting to lower or higher denominations. When calculating weights or liquid volumes of medications, you'll often have to convert between grams, milligrams, micrograms, and kilograms, and between milliliters and liters. Remember that this knowledge can prevent disastrous errors! Metric system guidelines are summarized in Table 6-3.

TABLE 6-1 Fundamental Units of the Metric System

The fundamental unit of length = the meter
The fundamental unit of liquid volume = the liter
The fundamental unit of weight = the gram

TABLE 6-2 Important Metric Relationships

Mass and/or Weight	Gram (g) is the base unit
1 microgram (mcg, μg)	0.000001 of a gram
1 milligram (mg)	0.001 of a gram
1 gram (g)	*1 gram*
1 kilogram (kg)	1000 grams
Volume	**Liter (l) is the base unit**
1 milliliter (ml)	0.001 of a liter
1 liter (l)	*1 liter*
Length	**Meter (m) is the base unit**
1 micrometer (mcm, μm)	0.000001 of a meter
1 millimeter (mm)	0.001 of a meter
1 meter (m)	*1 meter*
1 kilometer (km)	1000 meters

TIP

Rounding numbers that will alter drug doses is ultimately up to the discretion of the veterinarian. When performing calculations it is best not to perform any rounding until the end product is achieved. A decision about rounding can be made at that time.

Converting Grams to Milligrams One method used to convert between metric units is by dimensional analysis (commonly referred to as unit cancellation). Let's use the example of converting 1.5 g to mg. Solve the equation 1.5 g = __________ mg

Step 1: Because the unknown factor in the given formula is the number of milligrams contained in 1.5 g, you have to know your metric equivalents. In this example you need to know 1000 mg = 1 g.

Step 2: You know that 1000 mg = 1 g, so you can create a conversion factor. Conversion factors are used to move between units and always have a value of one. Because they have a value of one, conversion factors do

TABLE 6-3 Metric System Guidelines

1. Arabic numbers are used to designate whole numbers: 1, 25, 2500, and so forth.
2. Decimal fractions are used for quantities less than one: 0.1, 0.01, 0.001, and so forth.
3. To insure accuracy, place a zero before the decimal point: 0.1, 0.01, and so forth.
4. The Arabic number precedes the metric unit of measurement: 1 g, 30 ml, and so forth.
5. Abbreviations are generally used.
6. If the term is used, the prefixes are written in lowercase letters: milli, centi, and so forth.
7. Capitalize the measurement and symbol when it is named after a person: Celsius (C)
8. Periods are not used with most abbreviations or symbols.
9. Abbreviations for units are the same for singular and plural forms. An "s" is not added to indicate the plural form.

not change the value of the end product. We would write this conversion factor as 1000 mg/1 g.

Step 3: The next step is to determine in what format to write the conversion factor. We have two choices: 1000 mg/1 g or 1 g/1000 mg. The choice is made based on the unit of measure we want to cancel out and the unit of measure we want to end up with. In this case we want to cancel out g and end up with mg. Therefore we want g on the bottom and mg on the top of the conversion factor.

Step 4: We can now set up our conversions in an equation. Start with what we have (1.5 g) and use the conversion factor to set up the equation.

(1.5 g) (1000 mg/1 g) = X

Step 5: We can now perform the calculation. Remember, values on the bottom of the conversion factors are divided into the numbers on the top. Values on the top of the conversion factors are multiplied by the numbers on the top. In this case we take 1.5 g, multiply it by 1000 mg, and then divide by 1 g.

(1.5 g × 1000 mg) ÷ 1 g = 1500 mg

Keep in mind that units should cancel as well to give us the unit we want at the end.

Step 6: To make sure you have the correct answer, prove your work by looking at the units involved. Smaller units (in this case mg) should have a larger number in front of them versus larger units (in this case g). In this case mg has 1500 and g has 1.5. Always double-check your answer to avoid errors in calculations.

A shortcut method used to convert from grams to milligrams is as follows: When given a value for grams, obtain the equivalent amount of milligrams by moving the decimal point three places to the right of that value (Figure 6-1). When moving

1 GRAM = HOW MANY MILLIGRAMS?

(1) Take the 1-gram expression and place the decimal point in the proper place.

1 gram is the same as 1.0 gram (decimal point in place)

(2) Move that decimal point three places to the right. Note that as the decimal point is moved, you must add zeros to mark each movement.

1.0 gram = 1.000. = 1,000 milligrams

(after converting, you remove the decimal point if no digits follow it.)

1.5 GRAMS = HOW MANY MILLIGRAMS?

(1) The 1.5-gram expression has the decimal point already in place.

(2) Move that decimal point three places to the right. Remember to add zeros to mark each movement.

1.5 grams = 1.500. = 1,500 milligrams

FIGURE 6-1 Converting from grams to milligrams

the decimal point three places to the right, you're essentially multiplying the value by 1000, which is the equivalent difference between grams and milligrams. There are 1000 milligrams in each gram of weight. Please note that from now on we will use the abbreviations mg and g for milligram and gram units.

Converting Milligrams to Grams Sometimes drug preparations will require you to convert milligrams to grams. Using the dimensional analysis method, convert 2500 mg to g. Solve the equation 2500 mg = _________ g

Step 1: Because the unknown factor in the given formula is the number of grams contained in 2500 mg, you have to know your metric equivalents. In this example, you need to know 1000 mg = 1 g.

Step 2: You know that 1000 mg = 1 g, so you can create a conversion factor. We would write this conversion factor as 1000 mg/1 g.

Step 3: The next step is to determine in what format to write the conversion factor. We have two choices: 1000 mg/1 g or 1 g/1000 mg. In this case we want to cancel out mg and end up with g. Therefore, we want mg on the bottom and g on the top of the conversion factor.

Step 4: We can now set up our conversions in an equation. Start with what we have (2500 mg) and use the conversion factor to set up the equation.

$$(2500 \text{ mg})\ (1 \text{ g}/1000 \text{ mg}) = X$$

Step 5: We can now perform the calculation. Remember, values on the bottom of the conversion factors are divided into the numbers on the top. Values on the top of the conversion factors are multiplied by the numbers on the top. In this case we take 2500 mg, multiply it by 1 g, and then divide by 1000 mg.

$$(2500 \text{ mg} \times 1 \text{ g}) \div 1000 \text{ mg} = 2.5 \text{ g}$$

Keep in mind that units should cancel as well to give us the unit we want at the end.

Step 6: To make sure you have the correct answer, prove your work by looking at the units involved. Smaller units (in this case mg) should have a larger number in front of them versus larger units (in this case g). In this case mg has 2500 and g has 2.5. Always double-check your answer to avoid errors in calculations.

To use the shortcut method of converting milligrams to grams, apply the following rule: When given the value of the milligram quantity, obtain the equivalent amount of grams by moving the decimal point three places to the left (Figure 6-2). By doing this, you're essentially dividing the milligram value by 1000. This is because 1000 milligrams is equivalent to 1 gram. This movement of the decimal point to the left is the exact opposite of what you just did to convert grams to milligrams.

Converting Kilograms to Grams Some drug dosages for small animals or laboratory species may be expressed in grams. Because some of these animals are weighed on a kilogram scale, you may need to convert between kg and g. Using the dimensional analysis method, convert 45 kg to g (Figure 6-3). Solve the equation 45 kg = _______ g

Step 1: Because the unknown factor in the given formula is the number of grams contained in 45 kg, you have to know your metric equivalents. In this example, you need to know 1000 g = 1 kg.

Step 2: You know that 1000 g = 1 kg, so you can create a conversion factor. Conversion factors are used to move between units and always have a value of one. Because they have a value of one, conversion factors do

1,000 mg = HOW MANY g?

(1) Place decimal point in proper place.

1,000 mg = 1000. mg

(2) Move that decimal point three places to the left.

1000. mg = 1.000. g or 1 g

2,500 mg = HOW MANY g?

(1) Place decimal point in proper place.

2,500 mg = 2500. mg

(2) Then move decimal point three places to left.

2500. mg = 2.500. g or 2.5 g

FIGURE 6-2 Converting from milligrams to grams

not change the value of the end product. We would write this conversion factor as 1000 g/1 kg.

Step 3: The next step is to determine in what format to write the conversion factor. We have two choices: 1000 g/1 kg or 1 kg/1000 g. The choice is made based on the unit of measure we want to cancel out and the unit of measure we want to end up with. In this case we want to cancel out kg and end up with g. Therefore, we want kg on the bottom and g on the top of the conversion factor.

Step 4: We can now set up our conversions in an equation. Start with what we have (45 kg) and use the conversion factor to set up the equation.

$$(45\ \text{kg})\ (1000\ \text{g}/1\ \text{kg}) = X$$

Step 5: We can now perform the calculation. Remember, values on the bottom of the conversion factors are divided into the numbers on the top. Values on the top of the conversion factors are multiplied by the numbers on the top. In this case we take 45 kg, multiply it by 1000 g, and then divide by 1 kg.

$$(45\ \text{kg} \times 1000\ \text{g}) \div 1\ \text{kg} = 45{,}000\ \text{g}$$

Keep in mind that units should cancel as well to give us the unit we want at the end.

Step 6: To make sure you have the correct answer, prove your work by looking at the units involved. Smaller units (in this case g) should have a larger number in front of them versus larger units (in this case kg). In this case g has 45,000 and kg has 45. Always double-check your answer to avoid errors in calculations.

If using the shortcut method, apply the following rule: When given a value for kilograms, obtain the equivalent amount of grams by moving the decimal point three places to the right of that value (Figure 6-3). When moving the decimal point three places to the right, you're essentially multiplying the value by 1000, which is the equivalent difference between kilograms and grams. There are 1000 grams in each kilogram of weight.

45 kg = HOW MANY g?

(1) Place decimal point in proper place.

45 kg = 45. kg

(2) Move that decimal point three places to the right.

45. kg = 45.000. g or 45,000 g

FIGURE 6-3 Converting from kilograms to grams

Converting Grams to Kilograms Some drug dosages for small animals or laboratory species may be expressed in kilograms. Because some of these animals are weighed on a gram scale, you may need to convert between g and kg. Using the dimensional analysis method, convert 25 g to kg. Solve the equation 25 g = ________ kg

Step 1: Because the unknown factor in the given formula is the number of kilograms contained in 25 g, you have to know your metric equivalents. In this example, you need to know 1000 g = 1 kg.

Step 2: You know that 1000 g = 1 kg, so you can create a conversion factor. We would write this conversion factor as 1000 g/1 kg.

Step 3: The next step is to determine in what format to write the conversion factor. We have two choices: 1000 g/1 kg or 1 kg/1000 g. In this case we want to cancel out g and end up with kg. Therefore, we want g on the bottom and kg on the top of the conversion factor.

Step 4: We can now set up our conversions in an equation. Start with what we have (25 g) and use the conversion factor to set up the equation.

$$(25 \text{ g})\ (1 \text{ kg}/1000 \text{ g}) = X$$

Step 5: We can now perform the calculation. Remember, values on the bottom of the conversion factors are divided into the numbers on the top. Values on the top of the conversion factors are multiplied by the numbers on the top. In this case we take 25 g, multiply it by 1 kg, and then divide by 1000 g.

$$(25 \text{ g} \times 1 \text{ kg}) \div 1000 \text{ g} = 0.025 \text{ kg}$$

Keep in mind that units should cancel as well to give us the unit we want at the end.

Step 6: To make sure you have the correct answer, prove your work by looking at the units involved. Smaller units (in this case g) should have a larger number in front of them versus larger units (in this case kg). In this case g has 25 and kg has 0.025. Always double-check your answer to avoid errors in calculations.

TIP

When you convert from larger units to smaller units, the quantity gets larger. When you convert from smaller units to larger units, the quantity gets smaller.

If using the shortcut method, apply the following rule: When given the value of the gram quantity, obtain the equivalent amount of kilograms by moving the decimal point three places to the left (Figure 6-4). By doing this, you're essentially dividing the gram value by 1000. This is because 1000 grams is equivalent to 1 kilogram. This movement of the decimal point to the left is the exact opposite of what you just did to convert kilograms to grams.

25 g = HOW MANY kg?

(1) Place decimal point in proper place.

25 g = 25. g

(2) Then move decimal point three places to left.

25. g = 0.025. kg or 0.025 kg

FIGURE 6-4 Converting from grams to kilograms

Converting Liters to Milliliters Use of liquids and liquid medications may sometimes require you to convert liters to milliliters. Using the dimensional analysis method, convert 2.525 l to ml. Solve the equation 2.525 l = __________ ml

Step 1: Because the unknown factor in the given formula is the number of milliliters contained in 2.525 l, you have to know your metric equivalents. In this example, you need to know 1000 ml = 1 l.

Step 2: You know that 1000 ml = 1 l, so you can create a conversion factor. We would write this conversion factor as 1000 ml/1 l.

Step 3: The next step is to determine in what format to write the conversion factor. We have two choices: 1000 ml/1 l or 1 l/1000 ml. In this case we want to cancel out l and end up with ml. Therefore, we want l on the bottom and ml on the top of the conversion factor.

Step 4: We can now set up our conversions in an equation. Start with what we have (2.525 l) and use the conversion factor to set up the equation.

(2.525 l) (1000 ml/1 l) = X

Step 5: We can now perform the calculation. Remember, values on the bottom of the conversion factors are divided into the numbers on the top. Values on the top of the conversion factors are multiplied by the numbers on the top. In this case we take 2.525 l, multiply it by 1000 ml and then divide by 1 l.

(2.525 l × 1000 ml) ÷ 1 l = 2525 ml

Keep in mind that units should cancel as well to give us the unit we want at the end.

Step 6: To make sure you have the correct answer, prove your work by looking at the units involved. Smaller units (in this case ml) should have a larger number in front of them versus larger units (in this case l). In this case ml has 2525 and l has 2.525. Always double-check your answer to avoid errors in calculations.

If using the shortcut method, apply the following rule: When given a value for liters, to obtain the equivalent amount of milliliters simply move the decimal point three places to the right of that value (Figure 6-5). When you move the decimal point to the right, you're multiplying the value by 1000, which is equivalent to the difference between liters and milliliters. There are 1000 milliliters in each liter of a liquid.

2.525 l = HOW MANY ml?

(1) 2.525 l: move decimal point three places to right

(2) Then 2.525 l = 2.525 ml = 2525 ml.

FIGURE 6-5 Converting from liters to milliliters

Converting Milliliters to Liters Sometimes drug preparation will require you to convert milliliters to liters. Using the dimensional analysis method, convert 500 ml to l. Solve the equation 500 ml = ________ l

Step 1: Because the unknown factor in the given formula is the number of l contained in 500 ml, you have to know your metric equivalents. In this example, you need to know 1000 ml = 1 l.

Step 2: You know that 1000 ml = 1 l, so you can create a conversion factor. We would write this conversion factor as 1000 ml/1 l.

Step 3: The next step is to determine in what format to write the conversion factor. We have two choices: 1000 ml/1 l or 1 l/1000 ml. In this case we want to cancel out ml and end up with l. Therefore, we want ml on the bottom and l on the top of the conversion factor.

Step 4: We can now set up our conversions in an equation. Start with what we have (500 ml) and use the conversion factor to set up the equation.

$$(500 \text{ ml}) (1 \text{ l}/1000 \text{ ml}) = X$$

Step 5: We can now perform the calculation. Remember, values on the bottom of the conversion factors are divided into the numbers on the top. Values on the top of the conversion factors are multiplied by the numbers on the top. In this case we take 500 ml, multiply it by 1 l, and then divide by 1000 ml.

$$(500 \text{ ml} \times 1 \text{ l}) \div 1000 \text{ ml} = 0.500 \text{ l}$$

Keep in mind that units should cancel as well to give us the unit we want at the end.

Step 6: To make sure you have the correct answer, prove your work by looking at the units involved. Smaller units (in this case ml) should have a larger number in front of them versus larger units (in this case l). In this case ml has 500 and l has 0.500. Always double-check your answer to avoid errors in calculations.

If using the shortcut method, apply the following rule: When given a value for milliliters, to obtain the equivalent amount of liters simply move the decimal point three places to the left of that value (Figure 6-6). When you move the decimal point to the left, you're dividing the value by 1000, which is equivalent to the difference between liters and milliliters. There are 1000 milliliters in each liter of a liquid.

The Apothecary System

The apothecary system (also called the *common system*) is derived from the British apothecary system of measures, and is a system of liquid units of measure used chiefly by pharmacists. *Apothecary* comes from the Greek word for storehouse, which can be loosely translated into "drugstore." In seventeenth century England "druggists"

500 ml = HOW MANY l?

(1) Place decimal point in proper place.

500 ml = 500. ml

(2) Then move decimal point three places to left.

500. ml = 0.500. l or 0.500 l

FIGURE 6-6 Converting from milliliters to liters

agreed that they would stock only drugs and that grocers would only sell food and not drugs.

The fluid ounce is the basic unit of liquid measure in the apothecary system. Table 6-4 shows other units of liquid measure and their relationship to the fluid ounce.

A **minim** is essentially the liquid volume of a drop of water from a standard medicine dropper (Figure 6-7). Sixty minims (or drops) make up a fluid **dram**. Eight of these fluid dram units, the equivalent of 480 minims (or drops), make up one fluid ounce. Many medicines are dispensed in quantities of fluid ounces, and liquid medicine bottles are available in standard sizes ranging from 1-ounce bottles to 16-ounce bottles. The most frequently dispensed fluid-ounce quantities are four ounces and six ounces.

The **grain** (gr) is the basic unit of weight measurement in the apothecary system. Other units of weight and their relationship to the grain are shown in Table 6-5. Prescription vials to dispense oral medications are available in a variety of dram weight sizes. Eight drams are equal to one ounce.

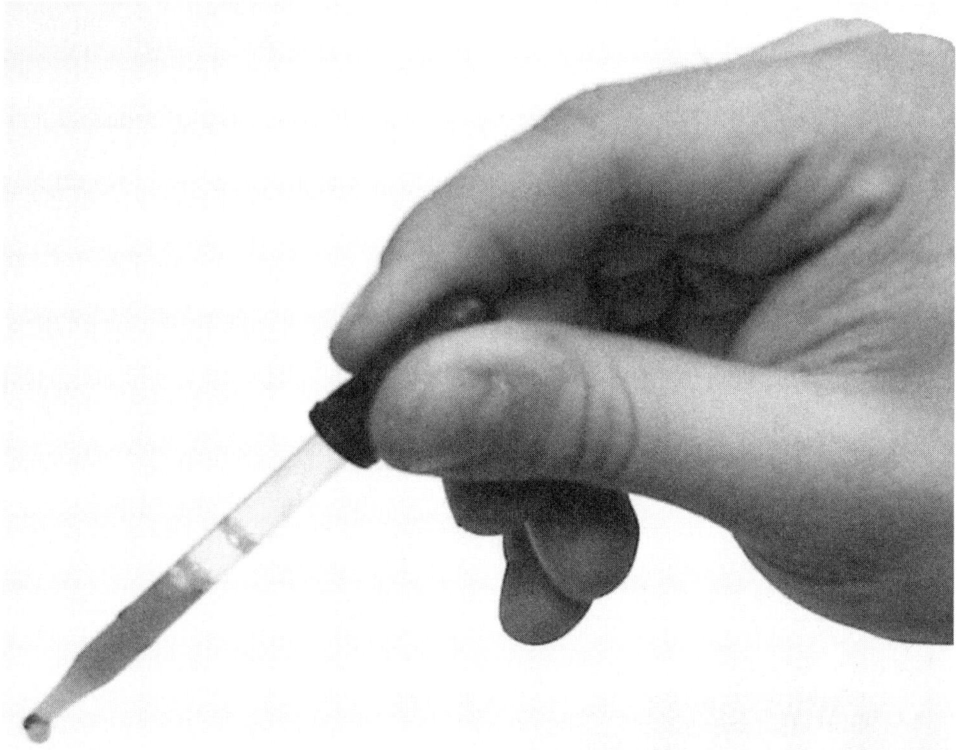

FIGURE 6-7 One minim equals the liquid volume of a drop of water from a standard medicine dropper.

TABLE 6-4 Apothecary System Liquid Measures

60 minims = 1 fluid dram
8 fluid drams (480 minims) = 1 fluid ounce
16 fluid ounces = 1 pint
2 pints = 1 quart
4 quarts = 1 gallon

TABLE 6-5 Apothecary System Weight Measures

60 grains (gr) = 1 dram (dr)
8 drams (480 grains) = 1 ounce (oz)
1.2 ounces (oz) = 1 pound (lb)

Conversions Between Metric and Apothecary Systems of Measure

Sometimes pharmacology calls for conversions between the metric and apothecary systems. Pounds may not be added to grams, nor milliliters subtracted from fluid ounces. The differences must be bridged with conversion to a single system.

A given relationship can serve as this bridge. The relationship between metric and apothecary weight is:

1 g = 15.432 gr (rounded to 15.4 gr)

TIP

Technically, 1 grain = 64.9 mg (rounded to 65 mg); however, both 1 gr = 65 mg and 1 gr = 60 mg conversion factors are still used.

This bridge converts grams to grains. It reads as follows: 1 gram is equivalent to 15.4 grains. You may also interpret it as: a substance that weighs 1 gram in the metric system weighs 15.4 grains in the apothecary system.

A more useful bridge is the expression:

1 gr = 0.065 g or 65 mg

This bridge converts grains to milligrams. It reads as follows: 1 grain is equivalent to 65 one-thousandths of a gram, which is equivalent to 65 milligrams. Therefore, 1 grain is equivalent to 65 milligrams. Table 6-6 lists the most useful bridges among the metric, household, and apothecary (common) systems of measure. Table 6-7 shows metric-to-metric conversions. If you need extra practice with systems of measurement and unit conversions, refer to Chapters 2 and 3 on the Accu-Calc CD.

Temperature Conversions

In the veterinary field, you will need to take and record an animal's body temperature accurately. The type of thermometer you use will determine whether the reading will be in the Fahrenheit or Celsius scale. You need to understand both these scales.

Gabriel Fahrenheit developed the mercury thermometer. He used a mixture of salt and ice to experiment with temperature. The coldest mixture he could make he called "zero." He noted that water froze at 32° and it boiled at 212° on this scale. The Fahrenheit system is used in the United States and uses the references of water freezing at 32°F and boiling at 212°F.

TABLE 6-6 Eqivalents for the Metric, Household, and Apothecary Systems of Measure

Length
Metric base unit = meter
1 meter = 1.0936 yards
1 centimeter = 0.39370 inch
1 inch = 2.54 centimeters
1 kilometer = 0.62137 mile
1 mile = 5280 feet or 1.6093 kilometers
1 foot = 0.3048 meter
Weight
Metric base unit = gram
1 kilogram = 2.2 pounds
1 pound = 453.59 grams
1 pound = 16 ounces
1 grain = 65 milligrams or 1 grain = 60 milligrams
1 dram = 3.888 grams
1 ounce = 28.35 grams
1 ton = 2000 pounds
1 gram = 0.035274 ounces
Volume
Metric base unit = liter
1 liter = 1.0567 quarts
1 gallon = 4 quarts
1 gallon = 8 pints
1 pint = 2 cups = 16 fluid ounces
1 cup = 8 fluid ounces
1 gallon = 3.7854 liters
1 quart = 32 fluid ounces
1 quart = 0.94633 liter
1 minim = 0.06 milliliter
1 fluid dram = 3.7 milliliter
1 ounce = approximately 30 milliliters
1 milliliter = 1 cubic centimeter

Anders Celsius suggested a temperature scale based on the freezing and boiling points of water. He felt that the point at which water freezes should be 0 and the point at which water boils should be 100. The Celsius temperature scale, which is used in the metric system, states that water freezes at 0°C and boils at 100°C.

Conversion between the Fahrenheit and Celsius scales is based on the relationship between scales as to the freezing and boiling points of water. In the Celsius scale, there is a 100-degree difference between the freezing and boiling points of water (100°C – 0°C = 100). In the Fahrenheit scale, there is a 180-degree difference between the freezing and boiling points of water (212°F – 32°F = 180). The ratio between these two differences is 1.8 (180 ÷ 100 = 1.8). Using this ratio and the variation in freezing point, conversions between Celsius and Fahrenheit can be calculated.

To convert a temperature reading from Fahrenheit to Celsius, subtract 32 from the Fahrenheit reading and divide by 1.8.

$$°C = \frac{(°F - 32)}{1.8}$$

TABLE 6-7 Review of Metric-to-Metric Conversions

Linear Measure: base unit is meters (m)
1 m = 100 centimeters (cm) 1 m = 1000 millimeters (mm) 1 m = 1,000,000 micrometers or microns (mcm, μm or μ) 1000 m = 1 kilometer (km) or 0.001 m = 1 km
Volume Measure: base unit is liter (l)
1 l = 100 centiliters (cl) 1 l = 1000 milliliters (ml) 1000 l = 1 kiloliter (kl)
Weight Measure: base unit is gram (g)
1 g = 100 centigrams (cg) 1 g = 1000 milligrams (mg) 1 g = 1,000,000 micrograms or 0.000001 g = 1 microgram (mcg or μg) 0.001 mg = 1 microgram (mcg or μg) 1 mg = 1000 micrograms (mcg or μg)

For example, convert 98.6°F to Celsius.

Step 1: Take the Fahrenheit reading (98.6) and subtract 32. This gives 66.6.
Step 2: Divide the result (66.6) by 1.8. This gives 37.
You have converted 98.6°F to 37°C.

To convert a temperature reading from Celsius to Fahrenheit, multiply the Celsius reading by 1.8 and add 32.

$$°F = 1.8°C + 32$$

For example, convert 100°C to Fahrenheit.

Step 1: Multiply the Celsius reading (100) by 1.8. This gives 180.
Step 2: Take the result (180) and add 32. This gives 212.
You have converted 100°C to 212°F.

TIP

For converting temperatures, use the following equations

$$°F = 1.8°C + 32$$

$$°C = \frac{°F - 32}{1.8}$$

DOSE CALCULATIONS

Dose calculations are performed daily in veterinary practice. Let's look at how some of the dose calculations performed in a veterinary setting are done.

Determining the Dose and the Amount of Drug Dispensed

Let's say you know the amount of a drug needed per kilogram of the animal's weight (the drug dosage), and you need to calculate the total dose of drug. The total dose is the total amount of drug given to the animal in one administration.

Dose in mg An example of calculating a dose in mg is as follows: The dosage of a drug is 2 mg per kilogram body weight of the animal. How many total mg should be given to an animal weighing 22 lb?

First you must convert lb to kg. From Table 6-6 you see that 1 kg = 2.2 lb. The animal weighs 22 lb. Therefore, use the conversion factor 2.2 lb = 1 kg to get the weight into kg units. You do this by dividing 22 lb by 2.2 lb to get kilograms.

22 lb ÷ 2.2 = 10 kilograms

In other words, you are taking the original weight and setting up a unit conversion to switch from one system of measure to the other.

$$22\text{lb}\,\frac{(1\text{ kg})}{(2.2\text{lb})} = 10\text{ kg}$$

If the animal needs 2 mg of drug for every 1 kilogram of body weight, and the animal weighs 10 kilograms, then multiply 10 × 2 mg = 20 mg total dose. In other words, you are taking the weight and setting up a cancellation method to incorporate the dosage.

$$10\text{ kg}\,\frac{(2\text{ mg})}{(1\text{ kg})} = 20\text{ mg total dose}$$

The formula just used is animal weight (in lb) divided by 2.2 (to convert to kg), then multiplied by amount of drug per kilogram.

Wt (lb) / 2.2 × dose = total dose to animal

Remember, the total amount of medicine and the size of dose must be measured in the same units.

You now know the size of each dose to be given to the animal and the total amount of medicine to prescribe. To measure the dose, we need to have the dose in a measurable unit. For example, tablets can be dispensed as a unit and milliliters can be dispensed as a unit.

Dose in Tablets If we use the example of the 22-lb animal needing 20 mg of drug, how many tablets will the owner give? To find this, you need to know what size tablets the drug comes in. If you determine from the clinic inventory that the tablets come in 40 mg, 80 mg, and 100 mg tablets, how many tablets per dose would you have this owner give the animal?

You know the animal needs 20 mg per dose, so you would divide the 20 mg dose by 40 mg (the tablet size closest to the dose).

(20 mg) (1 tablet/40 mg) = ½ tablet per dose (assuming that the tablets are scored)

TIP

Sometimes the same dose of a drug may be cheaper when a scored tablet of a higher dose is divided in half rather than using a full tablet of the lower dose. For example, splitting a 40 mg tablet in half to get a 20 mg dose may be cheaper than using 1 20 mg tablet.

Dose in ml If we use the example of the 22-lb animal needing 20 mg of drug again, how many ml of drug will the owner give? If you determine from the drug vial that the concentration of drug is 10 mg/ml, how many ml would you have the owner give?

You know the animal needs 20 mg per dose, so you would divide 20 mg by the concentration of 10 mg/ml.

(20 mg) (1 ml/10 mg) = 2 ml

Dose in Units Some liquid medications, including insulin, heparin, and penicillin, are measured in units (U) or international units (IU). These medications are standardized in units based on their strengths. The strength varies from one medicine to another, depending on their source, their condition, and the method by which they are obtained.

If you needed to give a 1000-lb cow 50,000 U/kg of penicillin G IM, how many ml would you give based on a concentration of 300,000 U/ml?

First, you would convert 1000 lb to kg, using the conversion factor 2.2 lb = 1 kg

$$(1000 \text{ lb}) \frac{(1 \text{ kg})}{(2.2 \text{ lb})} = 454.54 \text{ kg}$$

Next, you would take the weight in kg and multiply it by the dosage

$$(454.54 \text{ kg})\ (50{,}000 \text{ U/kg}) = 22727272.73 \text{ U}$$

Then you would take the units and divide by the concentration to give you a dose in ml

$$(22727272.73 \text{ U})\ (1 \text{ ml}/300{,}000 \text{ U}) = 75.8 \text{ ml}$$

Calculating Total Dose

If we use the example of the animal needing to be treated with 20 mg per dose, what would we need to dispense if this animal needed the drug bid for seven days? How many tablets would you dispense? Previously we determined that each dose in tablet form was ½ tablet. Because the animal needs the drug bid (twice daily), the animal will get 1 tablet per day (1/2 tablet × 2 doses per day = 1 tablet per day). The animal needs to take the medication for one week. So you take the 1 tablet per day and multiply it by 7 days: 1 tablet × 7 days = 7 tablets total to be dispensed.

Calculating Number of Doses

To find the number of doses needed for a total dose, use this formula

number of doses = total amount of medicine divided by size of each dose

Let's look at the following examples.

Example 1: Let's say that the veterinarian prescribes 200 mg of medicine to an ill animal. Knowing the size of each dose (20 mg), then

200 mg/20 mg = 10

Therefore, the animal would need to take a total of 10 doses.

Example 2: Let's say a single dose of the drug to be given is 1 g. How many total doses are contained in 10 g?

We know that the total amount of medicine is 10 g. The size of each dose is 1 g. You already have similar units (grams). You would calculate the problem as follows:

number of doses = 10 g/1 g = 10 doses

Sometimes you'll need to convert between units of measure within the same dose calculation.

Example 3: Let's say the dose of a drug given to an ill sow is 200 mg. How many doses are in 10 g?

We know that the total amount of medicine is 10 g. The size of each dose is 200 mg. Now you have to convert to similar units. Convert 10 g to mg so that the common unit is mg. From Table 6-2, we know that 1 g = 1000 mg. Therefore, 10 g = 10,000 mg.

number of doses = 10,000 mg/200 mg = 50 doses

Therefore, there are 50 doses in 10 g.

Determining the Amount in Each Dose

Now you know the total amount of medicine prescribed and the number of doses in the total amount. What's the quantity of each individual dose? Use the following formula:

quantity in each dose = quantity in total amount/the number of doses

The following example demonstrates a determination of the amount in each dose. If the total amount of drug is 100 mg in a solution of 100 ml, which represents 20 doses, how much drug is in one dose?

quantity in each dose = 100 mg/20 doses = 5 mg in each dose

For additional problems on calculating doses and equipment used to dispense these doses refer to chapters 4, 7, 8, and 9 on the Accu-Calc CD.

SOLUTIONS

Solutions are mixtures of substances not chemically combined with each other. The dissolving substance of a solution is referred to as the **solvent** and is usually a liquid. The dissolved substance of a solution is referred to as the **solute** and is usually the solid or particulate part of the mixture. Substances that form solutions are called **miscible** and those that do not are called **immiscible**.

When working with solutions, we want to know the amount of solute (particles) in the solvent. The amount of solute dissolved in solvent is known as the *concentration*. Concentrations may be expressed as parts (per some amount), weight per volume, volume per volume, and weight per weight. They are usually then reported out as percents or percent solution. Remember that a percent is the parts per the total times 100. Here are some rules of thumb for working with concentrations.

- *Parts:* Parts per million means 1 mg of solute in a kg (or l) of solvent. Ratios or fractions must be translated into percents of solution to perform many of the necessary calculations. For example, 1:1000 epinephrine is equal to what percent concentration? To change 1:1000 to a percent, divide the 1 by 1000 and multiply by 100

  $$(1 \div 1000) \times 100 = 0.1\%$$

- *Liquid in liquid:* The percent concentration is the volume per 100 volumes of the total mixture. The two volumes may be expressed in any unit as long as they are the same unit within the percent. For example, 1 ml/100 ml, 5 oz/100 oz, 15 l/100 l
- *Solids in solids:* The percent concentration is the weight per 100 weights of total mixture. The two weights may be expressed in any units as long as they are the same unit within the percent. For example, 60 mg/100 mg, 55 oz/100 oz, 4.5 g/100 g
- *Solids in liquid:* The percent concentration is the weight in grams per 100 volume parts in milliliters. The weight must be in grams and the volume must be in milliliters. For example, dextrose 5% = 5 g/100 ml. Dextrose 5% can also be expressed as 5000 mg/100 ml or 50 mg/ml, as long as the number is based originally on grams.

TIP

When using ratios it is important to understand what the notation represents. A 1:2 ratio means of the total amount (2 parts) 1 part is the medication. A 1:1000 ratio means of the total amount (1000 parts) 1 part is medication. Therefore, a 1:1 ratio means of the total amount (1 part) the entire 1 part is medication. In other words a 1:1 ratio contains all medication.

Percent Concentration Calculations

Occasionally you may have to prepare a drug solution from a pure drug or stock solution. *Pure drugs* are substances, in solid or liquid form, that are 100 percent pure. A *stock solution* is a relatively concentrated solution from which more dilute (weaker) solutions are made.

One method of determining the amount of pure drug needed to make a solution is the ratio-proportion method. The formula used to determine the amount of pure drug needed to make the solution is

$$\frac{\text{Amount of drug}}{\text{Amount of finished solution}} = \frac{\text{\% of finished solution}}{\text{100 \% (based on a pure drug)}}$$

For example, if we wanted to determine how much sodium chloride is needed to make 500 ml of a 0.9% solution, we would perform the following:

$$\frac{X\text{ g}}{500\text{ ml}} = \frac{0.9\%}{100\%}$$

To solve for X, we cross-multiply to get

$$100(\%)\ X(g) = 450(\%)(g)$$

then divide each side by 100% to isolate X

$$X = 4.5\text{ g of sodium chloride}$$

The amount of drug used to prepare a solution adds to the total volume of the solvent. The amount of volume contributed by a dry drug varies by its individual structure and is usually not accounted for. However, it is fairly easy to determine the volume a liquid drug will add to the solvent volume. To determine the amount of solvent needed to make the finished solution, subtract the volume of liquid drug from the amount of finished solution.

For example, you need to prepare a liter of 4% formaldehyde fixative solution from a 37% stock solution

$$\frac{X\text{ ml}}{1000\text{ ml}} = \frac{4\%}{37\%}$$

$$37(\%)\ X(ml) = 4000(\%)(ml)$$
$$X = 108\text{ ml of stock solution}$$
$$1000\text{ ml} - 108\text{ ml} = 892\text{ ml of solvent should be added}$$

Another way to determine volume for a desired final volume is via the volume concentration method. The equation for this is:

$$V_S \times C_S = V_d \times C_d$$

V_S = volume of the beginning or stock solution
C_S = concentration of the beginning or stock solution
V_d = volume of the final solution
C_d = concentration of the final solution

Here's an example of this equation in action. How much water must be added to a liter of 90% alcohol to change it to a 40% solution?

To solve this, consider what you have. You have 1 liter of 90% alcohol. Now consider what you want. You want a 40% solution and you will use whatever amount of water you need to make this happen.

$$1000\text{ ml} \times 90\% = V_d \times 40\%$$
$$90{,}000(ml)(\%) = 40V_d(\%)$$
$$2250\text{ ml} = V_d$$

You can also convert the percents to decimals before solving the preceding equation.

$$1000\text{ ml} \times 0.90 = V_d \times 0.40$$
$$900\text{ ml} = 0.40\ V_d$$
$$2250\text{ ml} = V_d$$

Remember that 2250 ml is your final volume. Therefore, you must subtract the original amount of 90% alcohol.

$$2250\text{ ml} - 1000\text{ ml} = 1250\text{ ml}$$

Here's another example: How much of a 1:25 solution of NaCl is needed to make 3 liters of 1:50 solution? First you must convert 1:25 and 1:50 to percents

$(1 \div 25) \times 100 = 4\%$
$(1 \div 50) \times 100 = 2\%$

Next, use the volume concentration equation to complete the problem.

$V_S \times 4\% = 3\,l \times 2\%$
$V_S\, 4(\%) = 6(l)(\%)$
$V_S = 1.5\,l$

Therefore, you need 1.5 liters of 4% NaCl and 1.5 liters of solvent to make a 2% NaCl solution.

Here's one last example: How much water do you need to add to a 1% solution to make a 10% solution? The answer is that it cannot be done. You cannot make a more dilute solution more concentrated simply by adding water. You should always check to make sure your answer makes sense before mixing or dispensing a product.

Sometimes drug concentrations are listed in percents. The percent concentrations are usually found on the front of the drug vial or bottle. Occasionally, these containers also have the concentration listed in mg/ml. Here is an example.

Lidocaine is a drug used as a topical anesthetic and as an anti-arrhythmic drug. The dosage for a dog is 3 mg/kg. Calculate the dose of lidocaine for a 15-lb dog.

First, we change lb to kg by using the 2.2 lb = 1 kg conversion factor

(15 lb) (1 kg/2.2 lb) = 6.81 kg

Next we calculate the dose needed in mg

6.81 kg (3 mg/kg) = 20.45 mg

The front of the vial says the concentration of lidocaine is 2%. Because percents are parts per the total, 2% is 2 parts per 100. If you remember that percents are g/100 ml, then 2% equals 2 g/100 ml. In this case we need to calculate a dose in ml, so the units of weight (mg and g in this case) need to be the same.

Let's change 20.45 mg to g

20.45 mg (1 g/1000 mg) = 0.02045 g

Then calculate the dose based on 2% = 2 g/100 ml

0.02045 g (100 ml/2 g) = 1.02 ml

This dog would get 1.02 ml per dose.

Additional problems concerning concentration calculations can be found in Chapters 9 and 10 on the Accu-Calc CD.

Reconstitution Problems

Some parenteral medications are not stable when suspended in solution. Such drugs are usually stored in a powder form because in time (hours, days, or weeks) the drug begins to deteriorate in solution. For this powdered drug to be administered parenterally, it must first be dissolved or reconstituted in sterile water, saline, or dextrose solution. Usually a vial label or package insert indicates the amount of solution necessary to dissolve the powder in the vial, as well as the solvent that should be used when dissolving the powder.

In multiple-dose vials, the powder to be dissolved often adds to the total final volume of the liquid being reconstituted. The label in these cases indicates the amount of solvent to use to reconstitute the powder and the total volume this reconstitution will achieve. This total volume will include the volume added by both the powder and the liquid. Consider the following example.

TIP

When there is a choice as to the amount of diluent to use for reconstitution of a drug, the volume used and the final concentration should be circled or written on the vial to avoid any dose calculation errors. The date and time of the reconstitution or expiration date, as well as the initials of the person who reconstituted the drug, should also be written on the vial.

Cefazolin sodium comes in a 1-g vial size. The label instructions state that when 2.5 ml of sterile saline is added to the vial, the total final volume will be 3 ml. The label also states that the approximate average concentration when cefazolin sodium is reconstituted as directed will yield a concentration of 330 mg/ml. The dose calculation is based on this concentration.

Some labels or package inserts allow a choice of dilution amounts, which will provide different concentrations of the drug based on which dilution is chosen. The amount of diluent (liquid) to use is based on the strength most appropriate for the dose ordered (to yield the smallest possible volume of drug to be given) and personal preference. Consider the following example.

A brand name of ampicillin comes in a vial containing 25 g of ampicillin. It can be diluted with sterile water for injection in volumes of 104.5 ml (final concentration 200 mg/ml), 79.0 ml (final concentration 250 mg/ml), or 41.0 ml (final concentration 400 mg/ml). If you diluted the 25 g ampicillin vial with 41.0 ml, how many ml would be given to an animal needing an 800 mg dose of ampicillin?

The concentration of ampillicin in this case is 400 mg/ml. If this animal needs 800 mg per dose, the following calculation is made

800 mg (1 ml/400 mg) = 2 ml

Some drugs that require reconstitution can be prepared for various parenteral routes. The dilutions for these routes of administration may vary. Use caution to ensure that the proper reconstitution for the desired route of administration is performed!

SUMMARY

The ability to perform mathematical calculations in a veterinary setting is critical for patient care. Three systems of measurement are used in determining drug doses and solution concentrations: household, apothecary, and metric. The metric system is the most commonly used. The base units of the metric system are meter (for length), liter (for volume), and gram (for weight). Prefixes applied to the base units denote the size of the metric unit. Common prefixes include micro- (0.000001), milli- (0.001), centi- (0.01), and kilo- (1000). Conversions can be done within a system or between systems using conversion factors. Appendix A contains conversion tables for reference when doing mathematical conversions.

Solutions are mixtures of substances not chemically combined with each other. The dissolving substance of a solution is referred to as the solvent and is usually a liquid. The dissolved substance of a solution is referred to as the solute and is usually the solid or particulate part of the mixture. Substances that form solutions are called miscible and those that do not are called immiscible.

When working with solutions, we want to know the amount of solute (particles) in the solvent. The amount of solute dissolved in solvent is known as the concentration. Concentrations may be expressed as parts (per some amount), weight per volume, volume per volume, and weight per weight. Solution calculations are used to determine amounts of solute or solvent to add to dilute or concentrate a solution.

CHAPTER REFERENCES

Rice, J. (2002). *Medications and mathematics for the nurse* (9th ed.). Clifton Park, NY: Thomson Delmar Learning.

Rice, J. (1999). *Principles of pharmacology for medical assisting* (3rd ed.). Clifton Park, NY: Thomson Delmar Learning.

Wynn, R. (1999) *Veterinary pharmacology*. Cambridge, MA: Harcourt Learning Direct.

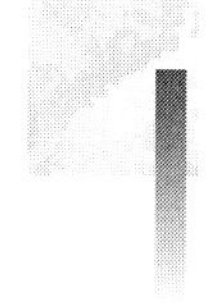

CHAPTER REVIEW

Matching

Match the abbreviation or term to its definition.

1. ______ centi-
2. ______ milli-
3. ______ kilo-
4. ______ micro-
5. ______ pt
6. ______ qt
7. ______ cc
8. ______ g
9. ______ l
10. ______ m

a. liter
b. gram
c. ml
d. 2 pints
e. 1000 times the base unit
f. meter
g. 1/100th of the base unit
h. 1/1000th of the base unit
i. 1/1,000,000th of the base unit
j. 16 fl. oz or 16 fluid ounces

Multiple Choice

Choose the one best answer. You may refer to the unit conversion chart in Appendix A.

11. In the metric system, the fundamental unit of liquid volume is the
    a. meter.
    b. gram.
    c. liter.
    d. inch.
12. In the metric system, the fundamental unit of weight is the
    a. meter.
    b. gram.
    c. liter.
    d. inch.
13. One gram is equivalent to how many milligrams?
    a. 1 mg
    b. 10 mg
    c. 100 mg
    d. 1000 mg
14. 38 fluid ounces are equal to how many pints?
    a. 0.421 pint
    b. 2.375 pints
    c. 38 pints
    d. 608 pints
15. 7.6 quarts are equal to how many pints?
    a. 14 pints
    b. 15.2 pints
    c. 28.8 pints
    d. 30.4 pints
16. How many quarts are equal to 17 gallons of liquid?
    a. 4.25 quarts
    b. 34 quarts
    c. 8.5 quarts
    d. 68 quarts
17. 3.5 fluid ounces equal how many milliliters?
    a. 7 ml
    b. 105 ml
    c. 56 ml
    d. 1655.5 ml
18. If the dose of a drug is 5 mg, how many doses are contained in 10 mg?
    a. 1
    b. 10
    c. 2
    d. 20

19. If the dose of a drug is 16 g, how many doses (in g) are contained in 50,000 mg?

    a. 0.0032
    b. 31.25
    c. 3.125
    d. 312.5

20. If you have a total of 86 g of medicine, and each dose of medicine is 400 mg, how many doses do you have?

    a. 0.215
    b. 34.4
    c. 4.6
    d. 215

21. The dose of a drug is 1 mg per kilogram of body weight. How many mg should be given to a rabbit weighing 10 pounds?

    a. 4.5 mg
    b. 10 mg
    c. 5.5 mg
    d. 45 mg

22. The dose of a drug is 0.05 mg per kg of body weight. How many mcg should be given to a cat weighing 17.6 kg?

    a. 0.008 mcg
    b. 880 mcg
    c. 0.88 mcg
    d. 8880 mcg

23. You have a total of 18 l of medicine, which represents 240 doses. How many ml are in each dose?

    a. 0.75 ml
    b. 75 ml
    c. 1.33 ml
    d. 432 ml

24. If one dose is 0.005 g, how many doses (in mg) are in 100 mg of medicine?

    a. 0.05
    b. 20
    c. 20,000
    d. 200

25. Translate 1:70 into a percent.

    a. 10%
    b. 1.4%
    c. 1%
    d. 2.5%

26. How much water would you add to a 5 g vial of drug powder to make a 2.5% solution?

    a. 200 ml
    b. 50 ml
    c. 100 ml
    d. 150 ml

27. How much of a 1:20 solution of NaCl is needed to make 1 liter of 1:50 solution of NaCl?

    a. 2000 ml
    b. 1000 ml
    c. 500 ml
    d. 400 ml

Fast Conversions

Convert the following.

28. 1.550 g = ______ mg
29. 674 mg = ______ g
30. 2.55 l = ______ ml
31. 600 cc= ______ ml = ______ l
32. 15 mg = ______ g
33. 25 mcg = ______ mg
34. 400 cc = ______ l
35. 0.003 g = ______ mg
36. 0.015 mg = ______ mcg
37. 176 lb = ______ kg
38. 1000 g = ______ kg
39. ¼ l = ______ ml
40. 51 cm = ______ inches
41. Which is larger, 0.125 mg or 0.25 mg?
42. 10 lb = ______ kg
43. 13 mm = ______ m
44. 125 ml + 3 l = ______ ml
45. How many 250 mg tablets are needed to equal 1 g? ______
46. Hou would you dispense 5 mg of a drug that comes in 20 mg tablets (assuming the tablet is properly scored)?

    ______

47. 32 oz = ______ ml
48. 2.5 cm = ______ inches
49. 32 kg = ______ lb
50. 100 mm = ______ cm

Temperature Conversions

51. 98°F = ______ °C
52. 102°F = ______ °C
53. 212°F = ______ °C
54. 25°C = ______ °F
55. –40°C = ______ °F

Case Studies

56. A 13-year-old, M/N domestic short hair (DSH) named Buttons has a bite wound in the right side of the mandible. The owner does not know when the cat got the bite wound because Buttons is an outside cat. He has not been eating or drinking the past few days, is lethargic, and on physical examination (PE) has temperature (T) = 103.5°F, heart rate (HR) = 180 bpm, and respiration rate (RR) = 45 breaths/min. Other than the mandibular wound, he is healthy. He is current on vaccinations. The vet decides to sedate Buttons so that the wound can be clipped, cleaned, and debrided. You must calculate the dose of injectable anesthetic for Buttons, who weighs 12.5 lb. The dosage of ketamine anesthetic is 22 mg/kg IM. The concentration listed on the vial of ketamine is 100 mg/ml. What volume of ketamine do you give this cat?
    a. What is his weight in kg?
    b. What dose of ketamine should he get in mg?
    c. What dose of ketamine should he get in ml?
57. Eliza, a 3-yr.-old, female spayed (F/S) German Shepherd, was presented for signs of weakness and joint pain of three months' duration. On PE, vital signs were within normal limits (WNL). The hips were palpated because of this breed's predilection for developing hip dysplasia. Pelvic radiographs showed no signs of hip dysplasia. Because Eliza is a young, active dog that spends a lot of time outside, a Lyme titer was drawn. Her blood test returned positive and she was started on doxycycline antibiotic. The dosage is 5 mg/kg po sid for 30 days. Eliza weighs 100 lb. Doxycycline comes in 50 mg and 100 mg tablets.
    a. What is Eliza's weight in kg?
    b. What dose of doxycycline should she get in mg?
    c. How many 50 mg tablets would she get per dose?
    d. How many 100 mg tablets would she get per dose?
58. A 13-yr.-old, M/N Golden Retriever named Chaska (45 kg) has not been able to bear weight on his rear limbs for the past 2½ days. On PE, moderate pain was elicited on palpation of the pelvic joints. Temperature, pulse, and respiration (TPR) were WNL; the rest of the PE was normal. Pelvic radiographs revealed a shallow acetabulum on both right (R) and left (L) sides and narrowing of the joint space. Chaska was diagnosed (dx'd) as having hip dysplasia with degenerative joint disease. The owner was given both surgical and medical options, but because of the dog's age the owner opted for medical treatment. This dog was put on carprofen 1 mg/lb bid. Before treatment was started, a blood sample was collected to assess liver enzymes. Carprofen comes in 25 mg, 75 mg, and 100 mg tablets. You are to dispense enough medication for 14 days.
    a. How many tablets do you dispense to this owner?
59. Kyra, a 7-yr.-old, F/S mixed-breed dog (107 lb) is presented to the clinic with signs of increased vocalization and urine leaking. She does not appear to be PU/PD (exhibiting increased urination/increased drinking) or having accidents due to lack of training. She usually leaks urine after she has lain down (her bed is damp). A urinalysis (UA) is WNL, as is the PE. Scout radiographs of the urinary bladder are unremarkable. Based on her history and pattern of urine leaking, the vet determines that she might have estrogen-responsive incontinence (she was spayed at an early age). She is prescribed DES (diethylstilbestrol) at a dose of 1 mg po sid for 3 days, followed by maintenance therapy of 1 mg po per week. DES comes in 1 mg and 5 mg tablets.
    a. How many mg does this dog receive per dose?
    b. How many mg does this dog receive for the first three days?

(continues)

c. How many tablets does this dog receive for the first three days?
d. How many mg does this dog receive for three weeks of treatment?
e. How many tablets does this dog receive for three weeks of treatment?

60. Fenbendazole is an antiparasitic drug used in the treatment of roundworms, hookworms, whipworms, and some species of tapeworms. For dogs the dosage is 50 mg/kg po for 3 days.

a. How many mg would you give a 50 lb dog?
b. Fenbendazole comes in a 10% suspension. How many mg/ml is that?
c. How many ml would you give this dog per dose?
d. How many ml would you give this dog per treatment regimen?

CHAPTER 7

Drugs Affecting the Nervous System

KEY TERMS

anticonvulsant
tranquilizer
sedative
anti-anxiety drug
analgesic
opioid antagonist
narcotic antagonist
neuroleptanalgesic
anesthetic
CNS stimulant
euthanasia solution
cholinergic drug
parasympathomimetic
anticholinergic drug
parasympatholytic
adrenergic drug
sympathomimetic
adrenergic blocking agent
sympatholytic

OBJECTIVES

Upon completion of this chapter, the learner should be able to

- explain the basic anatomy and physiology of the nervous system.
- explain the role of neurotransmitters at a synapse.
- describe the role of the autonomic nervous system (ANS).
- describe the branches of the ANS and how they affect body function.
- differentiate between the types of sympathetic receptors.
- differentiate between the types of parasympathetic receptors.
- describe the types of CNS drugs, including anticonvulsants, tranquilizers/sedatives, analgesics, opioid antagonists, neuroleptanalgesics, anesthetics, CNS stimulants, and euthanasia solutions.
- describe the types of ANS drugs, including cholinergic drugs, anticholinergic drugs, adrenergic drugs, and adrenergic blocking agents.

INTRODUCTION

A six-month-old F puppy is brought into the clinic for spaying. While the owner is filling out the paperwork, she asks what kind of anesthetic is used on puppies for routine surgeries. You tell her the name of the anesthetic used in your clinic and she asks you a series of questions: Is it the safest kind available? Does it provide pain relief after surgery? What delivery method is the safest? What drugs provide pain relief, and can they be used with anesthetic agents? What about pain relief for the puppy after she goes home? Do all anesthetics provide postsurgical pain relief?

BASIC NERVOUS SYSTEM ANATOMY AND PHYSIOLOGY

The nervous system is the main regulatory and communication system in the animal's body. The function of the nervous system is to receive stimuli and transmit information to nerve centers to initiate an appropriate response.

The basic unit of the nervous system is the *neuron* (Figure 7-1). There are three types of neurons: sensory (carry impulses toward the central nervous system), associative (carry impulses from one neuron to another), and motor (carry impulses away from the central nervous system). The parts of the neuron include the cell body, one or more dendrites, one axon, and terminal end fibers. The cell body has a nucleus and is responsible for maintaining the life of the neuron. The *dendrites* are root-like structures that receive impulses and conduct them toward the cell body. The *axon* is a single process that extends away from the cell body and conducts impulses away from the cell body. The *terminal end fibers* are the branching fibers that lead the impulse away from the axon and toward the synapse.

The space between two neurons or between a neuron and its receptor is the *synapse* (Figure 7-2). A chemical substance called a *neurotransmitter* allows the impulse to move across the synapse from one neuron to another. There are different neurotransmitters for different functions.

Figure 7-3 shows the two divisions of the nervous system: the central and the peripheral. The brain and spinal cord constitute the portion of the nervous system known as the central nervous system (CNS). The function of the CNS is to interpret information sent by impulses from the peripheral nervous system and return instructions through the peripheral nervous system for appropriate cellular actions. CNS stimulation may increase nerve cell activity or may block nerve cell activity. The CNS is encased in a multilayered connective tissue membrane called the *meninges*. The three layers of the meninges are the dura mater, the arachnoid membrane, and the pia mater. The CNS is cushioned by and nourished by the cerebrospinal fluid (CSF). The CSF is clear, colorless fluid produced by special capillaries within the ventricles of the brain.

The peripheral nervous system (PNS) consists of cranial nerves, spinal nerves, and the autonomic nervous system. Twelve *cranial nerves* originate from the undersurface of the brain. They provide a variety of functions depending on the different areas they serve. The *spinal nerves* arise from the spinal cord and are paired. The dorsal roots of spinal nerves carry sensory impulses from the periphery to the spinal cord. The ventral root carries motor impulses from the spinal cord to muscle fibers or glands.

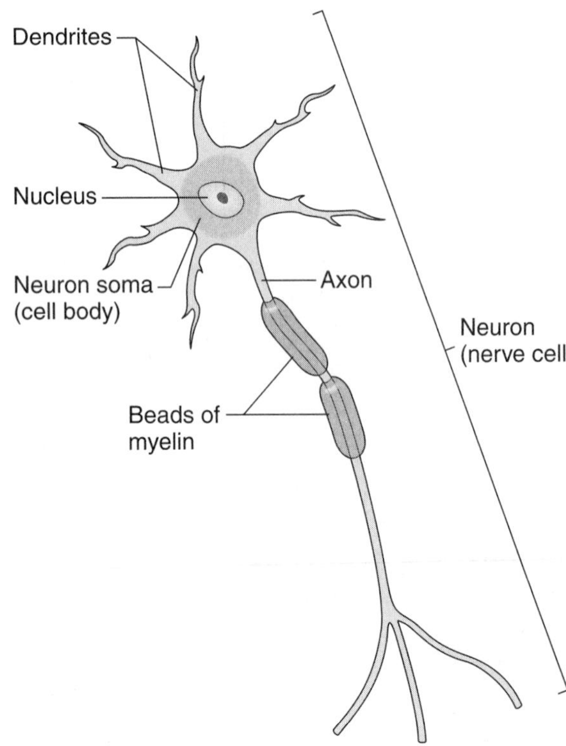

FIGURE 7-1 Structures of a neuron. Dendrites conduct impulses toward the neuron cell body; axons conduct impulses away from the cell body.

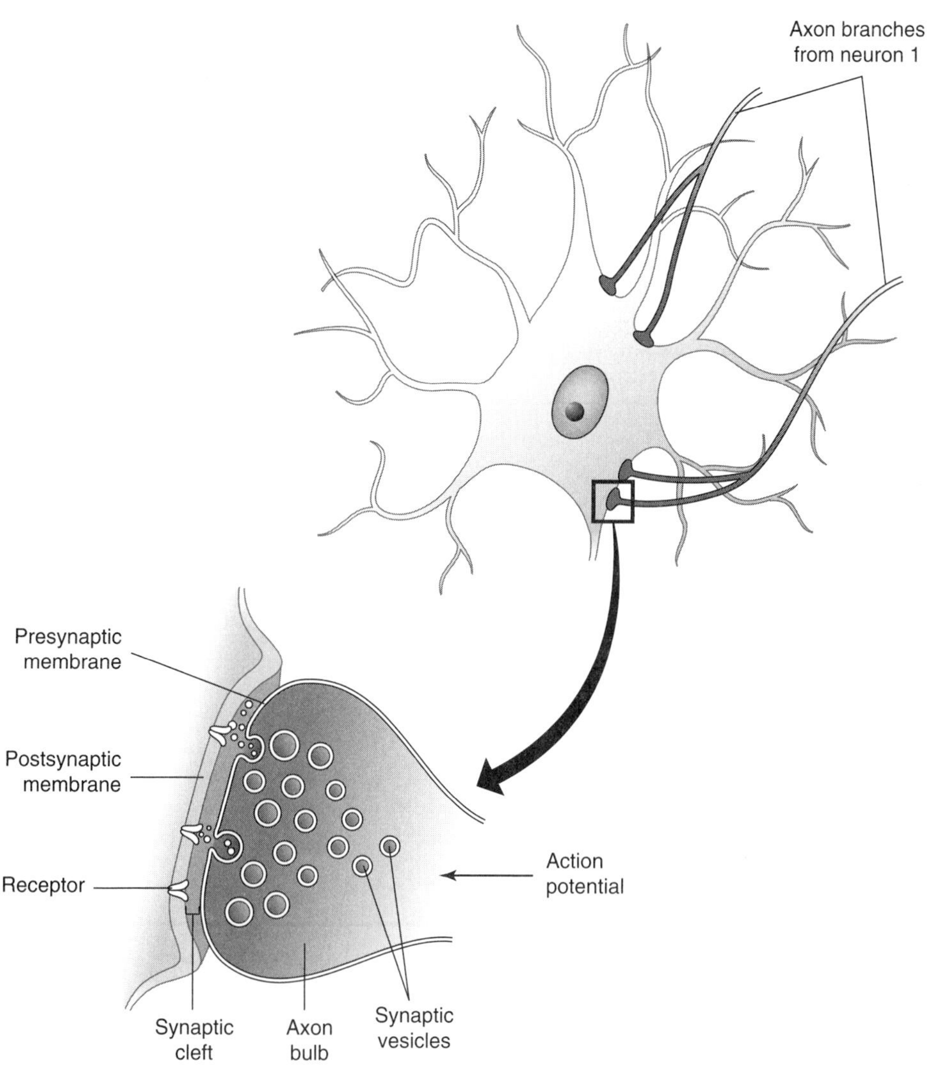

FIGURE 7-2 Synapse structure and function. Transmission across a synapse occurs when a neurotransmitter is released by the presynaptic neuron, diffuses across the synaptic cleft, and binds to a receptor in the postsynaptic membrane.

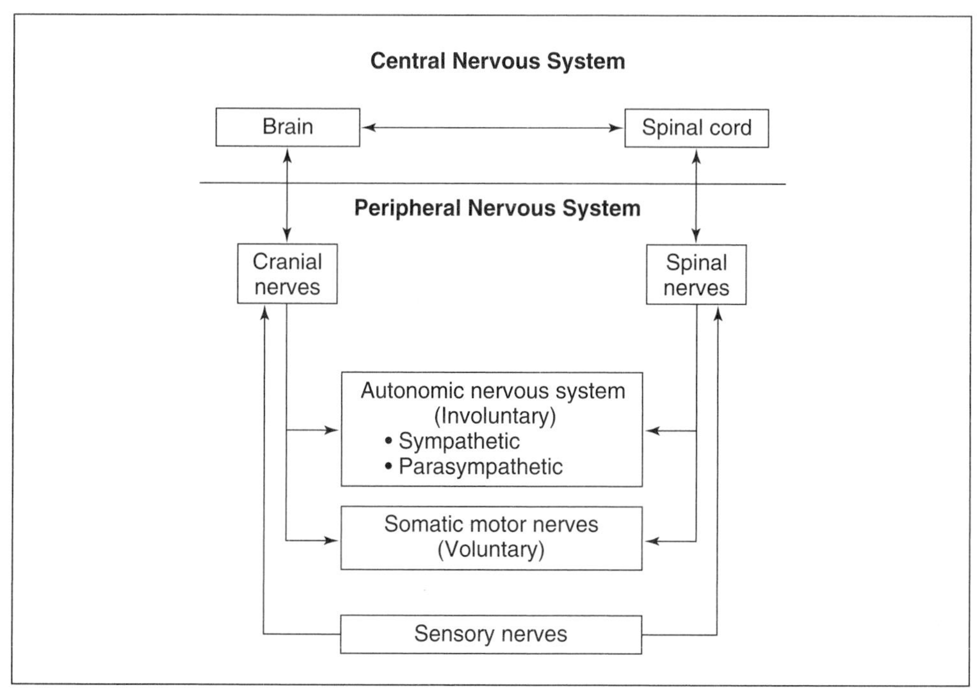

FIGURE 7-3 Divisions of the nervous system.

The autonomic nervous system (ANS) is an involuntary response system that innervates smooth muscle, cardiac muscle, and glands. The two divisions of the ANS are the sympathetic nervous system and the parasympathetic nervous system.

Sympathetic Nervous System or Adrenergic System. This branch is known as the *fight-or-flight system*. It is responsible for increasing heart rate, respiratory rate, and blood flow to muscles. It also decreases gastrointestinal function and causes pupillary dilation. The sympathetic nervous system is found in the thoracic and lumbar regions between T1 and L3. It has short preganglionic fibers and long postganglionic fibers. *Acetylcholine* is the neurotransmitter released at the preganglionic synapse and *epinephrine* or *norepinephrine* is the neurotransmitter released at the postganglionic synapse (Figure 7-4).

Sympathetic receptors include:

- alpha-1, found in the smooth muscles of blood vessels. Stimulation of alpha-1 receptors causes constriction of the arterioles (except in the gastrointestinal tract), increasing the blood pressure.

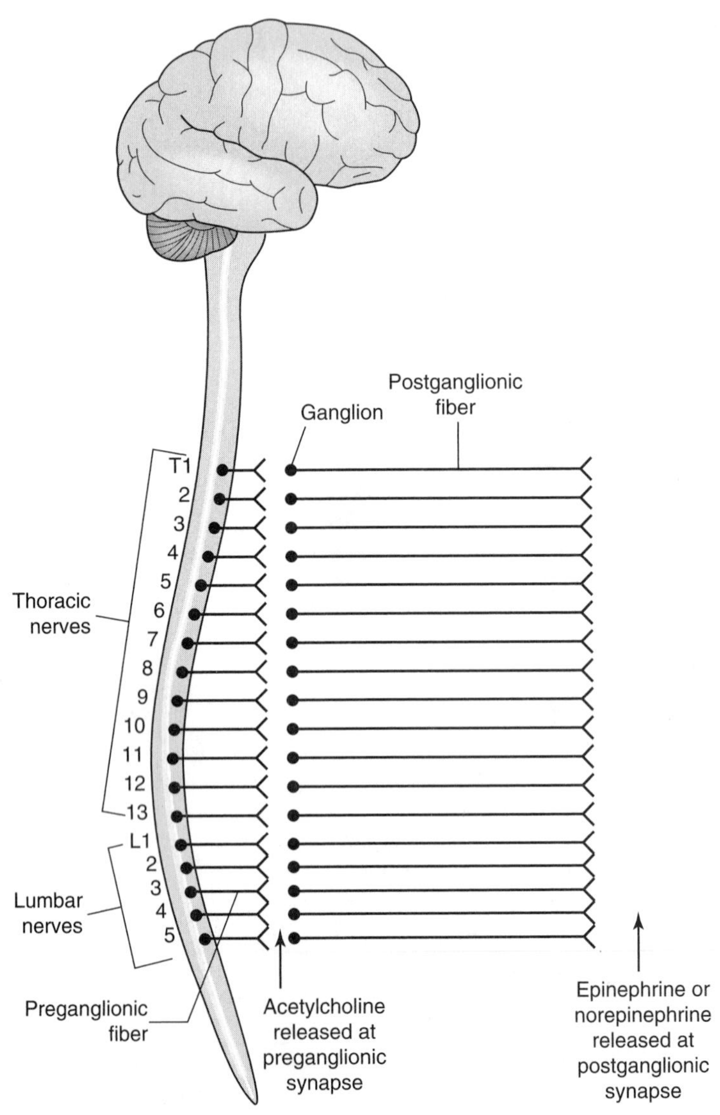

FIGURE 7-4 Sympathetic nervous system. The sympathetic nervous system, also known as the fight-or-flight system, has acetylcholine released at the preganglionic synapse and epinephrine or norepinephrine released at the postganglionic synapse.

- alpha-2, found in the postganglionic sympathetic nerve endings. Stimulation of alpha-2 receptors causes inhibition of norepinephrine release, resulting in decreased blood pressure and constriction of skeletal muscle.
- beta-1, located in the heart. These cause increased heart rate, conduction, and contractility.
- beta-2, found mainly in smooth muscles of the lung. Stimulation of these receptors causes bronchodilation and dilation of skeletal blood vessels.
- dopaminergic, located in the renal, mesenteric, and cerebral arteries. Stimulation of dopaminergic receptors causes dilation of the coronary vessels, dilation of the blood vessels of the kidney, and dilation of mesenteric blood vessels.

Parasympathetic Nervous System or Cholinergic System. This branch of the ANS is known as the *homeostatic system*. Parasympathetic nervous system effects are generally opposite to sympathetic nervous system effects. It is responsible for returning heart rate, respiratory rate, and blood flow to normal levels. It also returns gastrointestinal function to normal and causes the pupils to constrict to normal size. The parasympathetic nervous system is found in the brainstem region and sacral segments. It has long preganglionic fibers and short postganglionic fibers. Acetylcholine (ACh) is released at both the pre- and postganglionic synapses (Figure 7-5).

Parasympathetic receptors include muscarinic receptors, which stimulate smooth muscles and slow the heart rate; and nicotinic receptors, which affect skeletal muscles.

TIP

Because acetylcholine is the only neurotransmitter present in the parasympathetic nervous system, it makes sense that this system is referred to as the cholinergic system.

CENTRAL NERVOUS SYSTEM DRUGS

Central nervous system drugs include anticonvulsants, tranquilizers/sedatives, barbiturates, dissociatives, opioids, opioid antagonists, neuroleptanalgesics, stimulants, and analgesics.

Seizure-Stopping (Anticonvulsants)

Seizures are periods of altered brain function due to recurrent abnormal electrical impulses; they are characterized by loss of consciousness, increased muscle tone and movement, and altered sensations. Seizures occur in animals for a variety of reasons, such as traumatic, idiopathic (unknown), infectious, toxic, and metabolic factors. Seizures result from abnormal electric discharges by the cerebral neurons. **Anticonvulsants** are drugs that help prevent seizures. Ongoing seizures, known as *status epilepticus*, are treated as an emergency situation. Periodic, recurring seizures are treated with long-term preventative therapy using oral drugs. Many types of anticonvulsants work by suppressing the abnormal electric impulses from the seizure focus to other areas of the cerebral cortex, thus preventing the seizure but not eliminating the cause of the seizure. The goal of anticonvulsant therapy is to obtain the greatest degree of control over the seizures without causing severe side effects. All anticonvulsants are classified as CNS depressants and may cause ataxia, drowsiness, and hepatotoxicity (especially phenobarbital and primidone). Anticonvulsants include the following.

Phenobarbital. Phenobarbital is a long-acting barbiturate that depresses the motor centers of the cerebral cortex (Table 7-1). Barbiturates work by impairing chemical transmission of impulses across synapses in the brainstem. Phenobarbital is a C-IV controlled substance that is used orally (tablet and elixir) and parenterally (IV) to control seizures. Examples include Solfoton®, Luminal®, Barbita®, and a variety of generic forms. Induction of liver enzymes may increase the rate of phenobarbital (the animal develops tolerance) and other drug metabolism. An increase in liver enzymes, especially alkaline phosphatase and alanine aminotransferase (ALT), is commonly seen in

TIP

Most anticonvulsant drugs are taken prophylactically to prevent the occurrence of seizures.

TIP

Anticonvulsant medication should never be stopped suddenly (even when switching between anticonvulsants). The dosage should be reduced gradually to avoid inducing seizure activity.

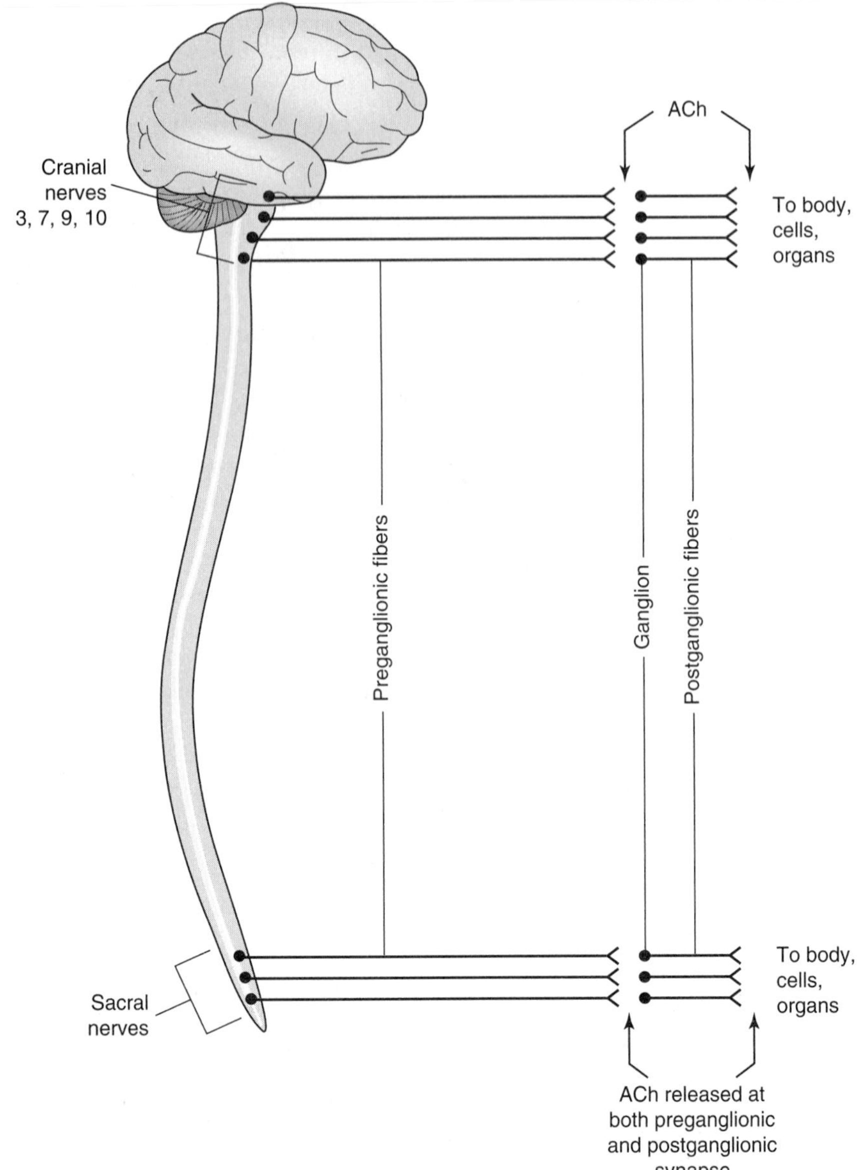

FIGURE 7-5
Parasympathetic nervous system. The parasympathetic nervous system, also known as the homeostatic system, has acetylcholine released at both the preganglionic and postganglionic synapse.

patients on phenobarbital. Side effects include ataxia, drowsiness, liver damage, PU/PD, and polyphagia. Blood samples can be taken to determine phenobarbital levels and doses altered accordingly.

Pentobarbital. Pentobarbital is a short-acting barbiturate that is administered IV to control seizures. Pentobarbital is especially useful when controlling seizures caused by toxins (such as strychnine). It is a C-II controlled substance that lasts about one to three hours. It can also be given orally, but this administration route is less common. Examples include Nembutal® and generic products. Side effects include respiratory depression and hypothermia. It is very irritating when given SQ or perivascularly, so these administration routes are avoided. Pentobarbital is also used as an euthanasia solution (covered later in this chapter).

TIP

When diazepam is given IV, it is given slowly to avoid vascular and cardiac problems.

Primidone. Primidone is structurally similar to phenobarbital and is broken down to phenobarbital and phenylethymalondiamide. It is given orally; examples include Mylepsin®, Neurosyn®, and generic brands. It can induce liver enzymes that increase its own and other drug metabolism. Side effects include ataxia, PU/PD, and polyphagia.

TABLE 7-1 Barbiturate Classifications (by duration of action)

Classification	Duration of action	Examples
Long-acting	6–8 hours	phenobarbital
Short-acting	1 hour	pentobarbital
Ultrashort-acting	10–15 minutes	thiopental methohexital

Diazepam. Diazepam (Valium®) is a C-IV controlled substance used intravenously for the treatment of status epilepticus (rapid succession of epileptic seizures). Once the seizures are brought under control with IV diazepam, oral anticonvulsant therapy is initiated. Diazepam can also be given orally as an adjunct to seizure treatment. It has a shorter duration of action (three to four hours) than phenobarbital. Diazepam works by increasing gamma-aminobutyric acid (GABA), a substance that inhibits impulse transmission in nerve cells. Side effects include CNS excitement (a paradoxical or contradictory response) and weakness.

Clorazepate. Clorazepate is a C-IV controlled substance in the same drug classification as diazepam. It is used orally as an adjunct anticonvulsant (usually used with phenobarbital) and for behavioral phobias. Trade names include Cloraze®Caps, Tranxene®-SD, and GenENE®. Side effects include sedation and ataxia.

Potassium Bromide. KBr is used as an adjunct to anticonvulsant therapy when seizures cannot be controlled by phenobarbital or primidone alone. It is either sprinkled on the food or squirted in liquid form into the mouth. KBr has a long half-life, and a narrow therapeutic range; it takes several days to reach a steady state. This chemical must be compounded and is ordered USP grade from chemical supply houses. The powder form is usually mixed with Karo® syrup. A side effect of potassium bromide use is electrolyte imbalance.

Valproic acid. Valproic acid (Depakene®) is an oral anticonvulsant (in tablet or syrup forms) that appears to increase brain levels of gamma-aminobutyric acid (GABA). Its use is limited in veterinary practice because of side effects such as hepatotoxicity, vomiting, and diarrhea. It may be used in animals that are unresponsive to other anticonvulsants. Complete blood counts (CBCs) and liver enzyme panels are routinely performed at least every six months as a precaution in animals taking valproic acid.

Phenytoin. The anticonvulsant phenytoin (Dilantin®) is used in human medicine, but has undesirable side effects (including rapid blood pressure decreases) that have severely limited its use in animals. Anticonvulsant activity occurs only after accumulation of several doses of phenytoin, and this presumed lack of efficacy has also limited its use.

TIP

Anticonvulsants help prevent seizures but do not eliminate the cause of the seizures.

Calming (Tranquilizers, Sedatives, and Anti-Anxiety Agents)

Tranquilizers are drugs that calm animals and are used to reduce anxiety and aggression in animals. **Sedatives** are drugs that decrease irritability and excitement in animals and are used to quiet excited animals. **Anti-anxiety drugs** lessen anxiousness but do not make the animal drowsy. Drugs that fall into one or more of these categories are phenothiazine derivatives, butyrophenones, benzodiazepines, and alpha-2 agonists (Table 7-2).

Phenothiazine derivatives work via an unknown mechanism, but are believed to block dopamine and the alpha-1 receptors found in the smooth muscle cells of peripheral blood vessels. This group of drugs causes sedation and relieves fear and anxiety, but does not produce analgesia. The primary uses of phenothiazines include sedation prior to minor procedures or as preanesthetic agents. Phenothiazines depress the chemoreceptor trigger zone in the brain and therefore are also used as anti-emetics, usually to prevent motion sickness in animals. Clinically, giving more phenothiazine

TIP

Phenothiazine drugs do not produce analgesia.

TABLE 7-2 Properities of Sedatives, Tranquilizers, and Anti-anxiety Drugs

	Phenothiazines	Benzodiazepines	Alpha-2 agonists
Classification:			
Sedation	X		X
Anti-anxiety	X	X	
Antiemetic	X		
Analgesic			X (some of short duration)
Antiarrhythmic	X		
Antihistimine effect	X		
Peripheral vasodilation	X		
Seizure threshold reducer	X		
Muscle relaxant		X	X
Emetic			X

drug to an animal does not necessarily mean greater effectiveness. The main side effects of this drug group are the development of hypotension, the lowering of the seizure threshold in animals with a seizure history, protrusion of the nictitating membrane or third eyelid, and the potential development of paraphimosis (retraction of the prepuce, causing swelling of the penis that prevents it from being retracted) in horses. This group does not cause respiratory depression or pronounced adverse effects on the heart. Examples in this group include acepromazine (Promace® and generic), chlorpromazine (Thorazine®), prochlorperazine/isopropamide (Darbazine®, Compazine®), and promazine (Sparine®).

Benzodiazepines are anti-anxiety drugs; this group includes diazepam (Valium®), zolazepam (found in Telazol®), midazolam (Versed®), and clonazepam (Klonopin®). These drugs have anticonvulsant activity, produce muscle relaxation, and reduce anxiousness in the animal while it remains alert.They work by increasing GABA, an inhibitory neurotransmitter in the brain. They are sometimes used as appetite stimulants in cats and in combination with ketamine for short-term anesthesia. These drugs have the benefit of causing minimal cardiovascular and respiratory depression. These drugs are C-IV controlled substances and do not provide any analgesic effects. Side effects include CNS excitement and weakness. The reversal agent for benzodiazepines is flumazenil (Romazicon®).

Alpha-2 agonists are drugs that bind to alpha-2 receptors on neurons that normally release the neurotransmitter norepinephrine. When molecules of this group of drug bind to the alpha-2 receptor, norepinephrine production is decreased. Because norepinephrine maintains alertness, its absence produces sedation. Alpha-2 agonists produce a calming effect, some analgesia, and muscle relaxation, and decrease the animal's ability to respond to stimuli. Cattle are very sensitive to this group of drugs and need to be dosed with caution. Side effects include bradycardia and heart block, so premedication with anticholinergics like atropine is recommended. The following are xamples of alpha-2 agonists.

TIP

Because xylazine causes vomiting, it is contraindicated in animals with bloat, gastric torsion, or other conditions in which vomiting is dangerous.

Xylazine (Rompun®, AnaSed®, and Gemini®). Xylazine may be combined with ketamine for short-term procedures such as castration in horses and cats and surgical wound repair. It also causes vomiting in cats and some dogs, so it is used as an emetic especially in cats. It produces analgesia and is used in horses for pain associated with colic. (Note that horses may appear sedate, but can still respond to stimuli by kicking!) It is used extra-label in cattle for surgical procedures such as cesarean (C-)

sections. Small-animal (20 mg/ml) and large-animal (100 mg/ml) concentrations are available. A main side effect is profound cardiovascular effects, especially when given IV or without an accompanying anticholinergic drug. Respiratory side effects include hypoventilation and cyanosis. Xylazine can also slow insulin secretion by the pancreas, resulting in transient hyperglycemia. This should be kept in mind when interpreting blood sample results of animals given xylazine.

Reversal agents for xylazine include yohimbine (Yobine®, Antagonil®) and tolazoline (Tolazine®), which are alpha-adrenergic blocking agents.

detomidine (Dormosedan®). Detomidine produces better analgesia than xylazine. It is labeled for use in horses for sedation, but the sedated animal can still respond to stimuli. It also causes severe cardiovascular and respiratory side effects.

medetomidine (Domitor®). Medetomidine is labeled for use in dogs more than 12 weeks of age as a sedative and analgesic. It is used for minor surgical procedures, and as a restraint for diagnostic and dental procedures. Muscle twitching can be seen in dogs sedated with medetomidine. Blood pressure is initially increased, followed by decreased heart rate due to a vagal response in the animal; therefore, anticholinergic use is recommended with this drug. Side effects include bradycardia and decreased respiration rates.

Medetomidine has a reversal agent called atipamezole (Antisedan®). Atipamezole is an alpha-2 antagonist. Dosing charts for both atipamezole and medetomidine are provided in the package insert. The volumes given per animal weight are the same for both drugs. Care must be taken to accurately read the IM versus IV doses in these charts.

TIP

When using xylazine, make sure you know the concentration per milliliter (100 mg/ml versus 20 mg/ml), to prevent overdosing or underdosing a patient: 10% solution is 100 mg/ml and 2% solution is 20 mg/ml.

Pain-Relieving (Analgesics)

Analgesics are prescribed for pain relief and are categorized as either nonnarcotic or narcotic. The choice of analgesic prescribed depends on the severity of pain. Mild to moderate pain of skeletal muscles and joints is frequently relieved with the use of nonnarcotic analgesics. Nonnarcotic analgesics are not addictive and are less potent than narcotic analgesics. Nonnarcotic analgesics act on peripheral nervous system receptor sites, whereas narcotic analgesics act mostly on the CNS. Nonnarcotic analgesics are covered in Chapter 16 on anti-inflammatory and pain-reducing drugs.

Narcotic analgesics are usually used for moderate to severe pain in smooth muscles, organs, and bones. Narcosis is a reversible state of drug-induced CNS depression. *Narcotic* refers to opioid (natural) or opioid-like (synthetic) products that were required to become prescription drugs in 1914 (under the Harrison Narcotic Act of 1914).

Opioids are naturally obtained from opium poppy plants and are also available synthetically. They produce analgesia and sedation and relieve anxiety. They do not produce anesthesia; animals still respond to sound and sensation when taking opioids. Side effects of opioids may include respiratory depression and excitement if given rapidly. They affect the regulatory centers in the brain for body temperature control and may cause panting in animals. They cross the placenta very slowly and are used in C-sections. They have central parasympathetic effects, producing salivation, defecation, and vomiting associated with the GI tract and slowed heart rate and hypotension associated with the cardiovascular system. Opioids are used as preanesthetics and postanesthetics for analgesia and sedation, in combination with other drugs for surgical procedures, for restraint, and as antitussives and antidiarrheals. Cats and horses are extremely sensitive to opioids.

Opioids produce their effect by the action of opioid receptors, which are located in high concentrations in the nervous tissue but are also found in the gastrointestinal tract, urinary tract, and smooth muscle. Five major types of opioid receptors have been identified: mu (μ), kappa (κ), sigma (σ), delta (δ), and epsilon (ε). The action of the opioid analgesics seems to be centered at the mu, kappa, and sigma receptors. Mu receptors are found in the pain areas of the brain. They cause analgesia, euphoria, respiratory depression, and physical dependence. If a drug affects the mu receptor, it is classified as a controlled substance. Kappa receptors are found in the cerebral cortex and spinal cord. They produce spinal analgesia, sedation, and miosis but not respiratory depression. Sigma receptors are located in the brain and control whining, hallucination, and struggling effects. They also cause mydriasis (pupillary dilation).

Morphine is the opioid to which all others in this category are compared, with regard to activity. The following are examples of opioids.

Opium is a naturally occurring opioid that is known as paregoric when it is the camphorated tincture of opium. It is a C-III controlled substance and is used as an antidiarrheal in calves and foals.

Morphine sulfate (Duramorph® and Astramorph® PF are injectables; Roxanol®, MS Contin®, and generic tablets are oral forms) is a naturally occurring opium derivative that affects mu receptors and is a C-II controlled substance. It is used to treat severe pain, as a preanesthetic, and as an anesthetic. The dosage used for dogs can cause mania in cats, so a lower dosage rate is used in cats. Side effects include gastrointestinal stimulation (vomiting) and severe respiratory depression.

Meperidine (Demerol®, Pethidine®) is a synthetic opioid used extra-label in animals and is a C-II controlled substance. It is a mu agonist. It has a shorter duration of action and is one-eighth as potent as morphine. It is used for acute pain relief, as a preanesthetic, and as a neuroleptanalgesic when combined with other drugs (see below). It can be administered orally, IM, and IV. Meperidine has fewer gastrointestinal and respiratory side effects than morphine.

Hydromorphone (Dilaudid®) is a semisynthetic opioid that is five to seven times more potent than morphine. It is a mu agonist and a C-II controlled substance. It is used IV, IM, or SQ preoperatively because it produces more sedation and causes less vomiting in animals than morphine. Hydromorphone is also used to offset moderate to severe postoperative pain. Pain relief with hydromorphone typically lasts for four hours. Side effects include respiratory depression and bradycardia.

Butorphanol (Torbugesic®, Torbutrol®) is a synthetic opioid with kappa- and sigma-receptor activity. It is a C-IV controlled substance that provides two to five times more analgesia than morphine. It is a potent antitussive and is labeled as an antitussive agent in dogs. It is also used as an analgesic and preanesthetic in dogs, cats, and horses. Side effects include sedation and anorexia; however, anxiety and excitation have been noted in some animals. It produces less respiratory depression than other opioids. Butorphanol is available in two concentrations: 2 mg/ml (Torbutrol®, Torbugesic®-SA), which is approved for use as an antitussive and analgesic in dogs, and 10 mg/ml (Torbugesic®), which is approved as an analgesic for horses.

Hydrocodone (Hycodan®, Tussigon®) is a synthetic opioid and C-III controlled substance. It is used as an antitussive in dogs. Side effects include sedation and vomiting.

Fentanyl (transdermal fentanyl patches are marketed under the name Duragesic®) is a synthetic opioid and C-II controlled substance. It is about 200 times more potent than morphine and has been used IV, IM, and SQ as an analgesic/tranquilizer for minor surgical and dental procedures, and for chemical restraint in dogs, as a combination fentanyl/droperidol product under the trade name Innovar-Vet®. This product is no longer available. Fentanyl is now available in a transdermal patch delivery system. Transdermal fentanyl patches must be applied with gloves. The fur is clipped and the skin is cleaned and dried prior to application of the patch over the dorsal neck area. The animal should not be allowed to lick or to eat the patch. Patches should not be cut or torn as this may allow fentanyl to pass into the skin too quickly. Side effects include defecation, respiratory depression, and pain after injection. Most side effects are associated with higher doses. It is not approved for use in food-producing animals.

Etorphine (M-99®) is a synthetic opioid and C-II controlled substance. It has analgesic effects 1000 times more potent than morphine and is used in zoo and exotic animals for immobilization. Diprenorphine is its antagonist if people accidentally inject themselves with this potent—and potentially fatal—drug.

Buprenorphine (Buprenex®, Lepetan®, Temgesic®) is an opioid agent that provides long-term analgesia (8 to 10 hours) and is used postsurgically in veterinary medicine. It is considered 30 times more potent than morphine and has a high affinity for mu receptors. Side effects are rare but include respiratory depression. In the United States it is available only in low-concentration doses and injectable form. In October 2002 the DEA rescheduled buprenorphine from a schedule V to a schedule III narcotic under the Controlled Substances Act.

Pentazocine (Talwin®, Talwin®-V) is a synthetic opioid and C-IV controlled substance. It acts as an antagonist at the mu receptor. It has a short duration of action that has limited its use in animals. It is used for analgesia in horses and dogs.

Methadone (Dolophine®) is a synthetic opioid and C-II controlled substance. It is used for treatment of colic in horses. Side effects include respiratory depression and sedation.

Codeine (generic) is a synthetic opioid and C-III controlled substance. Its analgesic effects are less than morphine. It is used as an antitussive in dogs. It may be combined with acetaminophen for pain relief (especially in humans). Side effects are rare at low to moderate dosages; sedation is seen with higher dosages.

Diphenoxylate (Lomotil®) is a synthetic opioid and C-V controlled substance. It is combined with atropine and used as an antidiarrheal. *Loperamide* (Immodium®) is a synthetic opioid that is sold over the counter as an antidiarrheal for dogs weighing more than 10 kg. *Apomorphine* (generic) is a synthetic opioid that stimulates the chemoreceptor trigger zone in the brainstem to induce vomiting. These drugs are covered in Chapter 11 on GI drugs.

Opioid-Blocking (Opioid Antagonits)

Opioid antagonists, also known as **narcotic antagonists**, block the binding of opioids to their receptors. The opioid antagonist has a higher affinity for the opioid receptor site than the narcotic does. These drugs both displace bound molecules and prevent binding of new molecules. Opioid antagonists are used to treat the respiratory and CNS depression of opioid use. These drugs are given IV and effects are rapid. Examples include:

- *naloxone* (Narcan®, Naloxone® Injection), an opioid antagonist given IV or IM that has a high affinity for mu receptors. It does not reverse the analgesic effects of the original drug. Naloxone reverses meperidine, oxymorphone, and morphine. It is relatively free of side effects.
- *naltrexone* (Trexan®), an opioid antagonist given SQ or po. It is more often used for behavior disorders and excessive licking of pruritic dermatitis than it is as a reversal agent. It has rare side effects.

Pain-Relieving and Anxiety-Calming (Neuroleptanalgesics)

Neuroleptanalgesics are a combination of an opioid and a tranquilizer or sedative. These drugs cause a state of CNS depression and analgesia and may or may not produce unconsciousness. The opioid antagonists can reverse the opioid portion of a neuroleptanalgesic. Examples include fentanyl and droperidol (Innovar-Vet®), formerly the only commercially available neuroleptanalgesic. It is no longer available as a veterinary or human (Innovar®) product. Combinations prepared by veterinarians include xylazine and butorphanol, and acepromazine and morphine.

TIP

Anesthesia means lacking or without sensation; it does not mean unconsciousness.

No Pain (Anesthetics)

Anesthesia means without sensation. **Anesthetics** are drugs that interfere with the conduction of nerve impulses and are used to produce loss of sensation, muscle

TIP

There are no safe anesthetics, just safe anesthetists.

relaxation, and/or loss of consciousness. General anesthetics affect the CNS and produce loss of sensation with partial or complete loss of consciousness; local anesthetics block nerve transmission in the area of application, causing loss of sensation without loss of consciousness.

Local Anesthetics Local anesthetics block pain at the site of administration or application in the PNS and spinal cord. They can be used as nerve blocks in equine lameness exams to localize lesions, as nerve blocks in cattle to allow surgical and medical procedures, as an aid to help endotracheal tube placement in cats by preventing laryngeal spasm, and to ease skin irritation. Local anesthetics work by preventing the conduction of nerve impulses in the peripheral nerves. They may be applied topically to mucous membranes and the cornea, by infiltration of a wound or joint, by IV, and around nervous tissue. Local anesthesia usually lasts from 5 to 30 minutes. Their duration of action can be lengthened by use of epinephrine, which causes vasoconstriction and thus prolongs absorption time. Side effects include restlessness and hypotension from the added epinephrine. Local anesthetics can usually be recognized by their *-caine* ending. Types of local anesthetics are listed in Table 7-3.

- *Lidocaine* (Xylocaine®) is available in 0.5%–2% solution for injection and 2%–4% for topical use. It provides immediate onset and if mixed with epinephrine has a two-hour duration.
- *Proparacaine* (Ophthaine®, Ophthetic®) is a rapid-acting topical anesthetic used for ophthalmic procedures. It comes in 0.5% solution that is applied in -drop doses every 5 to 10 minutes for a maximum of 7 doses. It provides 5 to 10 minutes of anesthesia to the cornea with limited drug penetration to the conjunctiva.
- *Tetracaine* (LiquiChlor®, Neo-Predef® with tetracaine, Pontocaine®) is used topically on the skin and as an ophthalmic or otic solution.
- *Mepivacaine* (Carbocaine®) is available in 1–2% injectable form. It has an immediate onset of action and lasts 90 to 180 minutes.
- *Bupivacaine* (Marcaine®) is available in 0.25%–0.5% for injection and 0.75% for epidural use. It provides immediate onset and lasts for four to six hours.

General Anesthetics General anesthetics covered in the following sections are either used strictly as anesthetics or as induction agents to produce general anesthesia. Induction agents are used to provide enough sedation for animals to allow inhalant anesthetics to be given. Induction agents are given by injection and tend to have a rapid onset of action. There are two major categories of general anesthetics: injectable and inhalant. Figure 7-6 summarizes agents used in general anesthesia.

Injectable Anesthetics Injectable anesthetics include the barbiturates, dissociatives, and miscellaneous.

Barbiturates Barbiturates are CNS depressants that are derived from barbituric acid. Barbiturates are used mainly as anticonvulsants, anesthetics, and euthanasia solutions. They are easy and inexpensive to administer; however, they can cause potent cardiovascular and respiratory depression. They are highly protein-bound drugs, and plasma proteins can serve as reservoirs for these drugs. Acidotic animals (for example, animals in shock, with diabetes mellitus, and having other systemic diseases) show less binding of barbiturates to plasma proteins. In animals with hypoproteinemia and acidosis, barbiturate doses must be decreased to avoid side effects associated with overdosing.

Barbiturates are classified according to their duration of action: long-acting, short-acting, or ultrashort-acting (refer to Table 7-1). They may also be classified according to the side chain on the barbituric acid: either *oxybarbiturate* (a side chain connected by

TABLE 7-3 Types of Local Anesthesia

Type	Description	Benefit	Use	Drug example
Infiltration anesthesia	Small amounts of anesthetic solution are injected into the tissue surrounding the site to be worked on (surgical site, wound repair site, etc.)	Because small amounts of anesthetic are used, there is reduced danger of systemic side effects	• Wound suturing • Wound debriding • Skin biopsies	• lidocaine • mepivacaine • tetracaine • bupivacaine
Topical anesthesia	Anesthetic agent is applied directly onto the surface of the skin or eye; also used to aid in diagnostic procedures and intubation in cats	Systemic absorption is limited from these sites	• Eye examinations • Minor skin irritation • Catheter passing (gel is applied to the catheter tip) • Larynx is sprayed to liquid applied to prevent spasming during intubation in cats	• tetracaine • proparacaine
Nerve block anesthesia	Anesthetic solution is injected along the course of a nerve so that the area it innervates is desensitized	Localization of pain relief Ability to determine the source of pain	• Helps locate areas of injury • Provides local desensitization	• lidocaine • mepivacaine • bupivacaine
Line block anesthesia	A continuous line of local anesthetic is given SQ in the tissues proximal to the targeted area	Allows for larger area of desensitization, but limits systemic effects	• Surgical procedures	• lidocaine • mepivacaine • bupivacaine
Regional (epidural)	Anesthetic agent is injected into a nerve plexus or area of the spinal cord (subarachnoid space)	Provides adequate restraint and may prevent movement (can affect respiratory muscles if given in the cranial parts of the spinal cord)	• Surgeries like C-sections, tail amputations, anal sac removal, and surgery of the rear limb	• lidocaine • mepivacaine • bupivacaine

oxygen) or *thiobarbiturate* (a side chain connected by sulfur). Oxybarbiturates are usually long-acting or short-acting, whereas thiobarbiturates are usually ultrashort-acting. Thiobarbiturates have an ultrashort duration of activity because they are very fat-soluble and move out of the CNS rapidly to fat stores within the body. The following are examples of barbiturates.

Phenobarbital (generic) is a long-acting oxybarbiturate that lasts 8 to 12 hours. It is a C-IV controlled substance and is used as an anticonvulsant (discussed previously).

Pentobarbital (generic, Nembutal®) is a short-acting oxybarbiturate that lasts one to two hours. It is a C-II controlled substance that used to be used as a general anesthetic, but is now used mainly as an anticonvulsant and euthanasia solution.

Thiopental (Pentothal®) is an ultrashort-acting thiobarbiturate that lasts 5 to 30 minutes. It is available in a vial as a sterile powder that must be reconstituted with sterile water for use. Thiopental is a C-III controlled substance that is used as an anesthetic

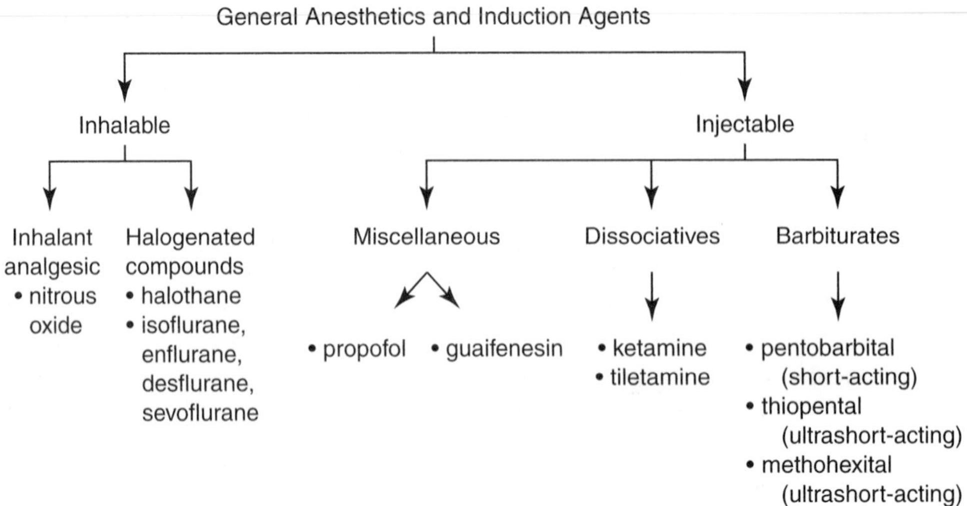

FIGURE 7-6 General anesthetics and induction agents.

induction agent and can only be given IV because it is very alkaline. If thiopental accidentally is injected perivascularly, severe inflammation, tissue swelling, and tissue necrosis may result. Care must be taken when administering thiobarbiturates to thin animals (like sighthounds), because they lack fat stores. Following IV injection, thiobarbiturates rapidly enter the CNS (highly perfused) and then redistribute to fat (poorly perfused). The duration of action of thiobarbiturate is short because of this rapid redistribution of drug out of the CNS. However, the thiobarbiturate does not leave the body: it just leaves the CNS and goes to fat. The animal may not appear anesthetized, so another dose may be given. This second dose may not be able to be redistributed to the fat (as the fat already has drug in it) and thus stays in the CNS. Because the thiobarbiturate has nowhere to go, the amount available in the CNS increases and brain concentrations of drug may go extremely high. Likewise, obese animals may accumulate excessive stores of thiobarbiturates, which leave poorly perfused tissue like fat very slowly and result in prolonged anesthesia. Thiobarbiturates used in obese animals must be dosed based on lean body weight. Thiobarbiturates can also cause apnea if given too fast (resulting in the need for artificial respirations to be given) and CNS excitement if given too slowly. Typically animals are given one-third to one-half of the calculated dose rapidly and then the rest is given to effect. Side effects include cardiac arrhythmias and transient apnea (no breathing).

Methohexital (Brevane®) is a methylated oxybarbiturate similar in structure to the ultrashort-acting thiobarbiturates. It is used in sighthounds because of its rapid redistribution and metabolism by the liver. Its duration of action is short (5 to 10 minutes). It is a C-III controlled substance. Methohexital can cause profound respiratory depression.

> **TIP**
> Unlike other general anesthetics, dissociatives cause CNS stimulation.

Dissociatives Dissociatives belong to the cyclohexamine family (which includes the street drug PCP or angel dust) and cause muscle rigidity (catalepsy), amnesia, and mild analgesia by altering neurotransmitter activity. Dissociatives are used only for restraint, diagnostic procedures, and minor surgical procedures because dissociatives do not relieve deep pain. They are usually used in combination with other agents for surgical procedures. They cause minor cardiac stimulation, respiratory depression, and exaggerated reflexes. During induction and recovery, tremors, spasticity, and convulsions may occur. Examples of dissociatives include the following.

> **TIP**
> Apneustic breathing is an inspiration followed by a long pause and short expiration. It seems like the animal is holding its breath.

Ketamine (Ketaset®, Ketalar®, Vetalar®) is a C-III controlled substance usually used in combination with acepromazine, xylazine, and/or diazepam to provide muscle relaxation and deepen anesthesia. It is given IM or IV. It is approved for cats and primates and is used extra-label in other species. Animals keep their eyes open when given ketamine; therefore, ocular lubricants are applied to the eye when using this drug. Spastic muscle jerking and increased salivation may also be seen with this drug. Pain at the injection site is frequently noted with this drug, due to its low pH. Apneustic breathing is seen with ketamine use. Anesthetic depth in animals given ketamine may be difficult to assess because palpebral reflexes are altered.

Tiletamine (found in combination with zolazepam in the product Telazol®) is a C-III controlled substance manufactured as an injectable anesthetic approved for dogs and cats. Tiletamine provides better analgesia than ketamine. In cats tiletamine is metabolized first, whereas in dogs zolazepam is metabolized first. Therefore, in dogs a premedication or postsurgical sedative is usually recommended. Ocular lubricants are also needed for animals given this drug. Pain at the injection site is frequently noted with this drug, due to its low pH.

Ketamine-diazepam mixtures are prepared in clinics and used for IV induction. This mixture is made by combining equal volumes of diazepam (5 mg/ml) and ketamine (100 mg/ml). It may begin to precipitate if stored for more than one week. It may be mixed in a syringe as needed to prevent problems with precipitation.

Miscellaneous Guaifenesin (Guailaxin®, Gecolate®) is a skeletal muscle relaxant used in combination with an anesthetic drug to induce general anesthesia in the horse. It comes in a 5% or 10% solution. Generally, large volumes are given to horses to induce general anesthesia, with the addition of small increments to maintain or extend anesthesia.

Propofol (Rapinovet®, PropoFlo®) is a short-acting injectable anesthetic agent that produces rapid and smooth induction when given slowly IV. A single bolus of propofol lasts two to five minutes. It provides sedation and minimal analgesic activity, so other agents may be given to the animal as well. It is a white emulsion made of egg lecithin, soybean oil, and glycerol without preservatives; this allows bacteria to grow if the emulsion is contaminated. Unused portions of propofol should be discarded because of the potential for contamination. It should not be used if there is evidence of separation of the phases of the emulsion. Side effects include cardiac arrhythmias and apnea, but these are less severe than with other agents.

Inhalant Agents Inhalant anesthetics are brought into the body via the lungs and are distributed by the blood into different tissues. The main target of inhalation anesthetics, also called *volatile anesthetics*, is the brain. Inhalant anesthetics are volatile liquids that are purchased in a liquid form (they are liquid at room temperature) and are administered by inhalation in combination with air or oxygen. Advantages of inhalant anesthetics include the ability to alter the depth of anesthesia, because anesthetic continually enters and leaves the body; the constant delivery of oxygen; and the emergency access for ventilation in intubated animals. Methods of administering anesthetics by inhalation are summarized in Table 7-4.

How do inhalant anesthetics get into the blood? The liquid form of the drug is vaporized through a vaporizer as oxygen passes over it in the anesthetic machine. The gas travels in the respiratory system until it gets to the alveoli of the lung, where it diffuses across the alveolar membrane (gas particles move from an area of high concentration in the alveoli to an area of low concentration in the blood capillary). The concentration of gas is high in the alveoli and low in the blood when we induce anesthesia; therefore, there is rapid diffusion of anesthetic into the blood. The anesthetic goes from the blood to the areas of the body that are well perfused (like the brain). Diffusion occurs in the opposite direction when we turn off the vaporizer after a procedure. The concentration of gas in the alveoli goes down when we turn the vaporizer off, although the concentration of anesthetic is high in the blood; therefore, there is diffusion of particles from the blood into the alveoli. The gas is expired from the alveoli. If we provide the animal with pure oxygen when we turn off the vaporizer, this diffusion will happen more rapidly because of the greater difference in concentration.

TIP

Volatile means that the substance evaporates quickly. Anesthetic agents that have a high vapor pressure will evaporate easily.

TABLE 7-4 Methods of Administering Anesthetics by Inhalation

Method	Description	Description/Example
Open-drop	Liquid anesthetic is dropped onto a cloth and extended over the animal's nose and mouth	• This method is not used anymore because there is lack of control over the amount of anesthetic delivered, and there is no respiratory assistance. It is also not currently used because these agents (such as ether) are flammable.
Semi-closed	Anesthetic is provided through a mask connected to a reservior (usually a gas anesthetic machine)	• Need to use a rapid-acting anesthetic like halothane or isoflurane and related anesthetics • Provides greater control of anesthesia delivered to the patient as compared to the open-drop method • Exhaled gases still leak into the environment
Closed	Anesthetic is delivered by anesthesia machine after the liquid has vaporized to the inhalant (gas) form	• Animal is intubated with an endotracheal tube and inhalant anesthetic is delivered directly to the respiratory system • Examples include isoflurance and related anesthetics; halothane

TIP

A single vaporizer should not be used for multiple drugs, even if those drugs have similar vapor pressures. Vaporizers should be temperature-, flow-, and backpressure-compensated.

TIP

↓ MAC = more potent anesthetic = less anesthetic needed to produce the effect.

↑ MAC = less potent anesthetic = more anesthetic needed to produce the effect.

Inhalant anesthetics are compared on a value called *minimum alveolar concentration* (MAC). MAC is the lowest concentration of an anesthetic that produces no response to painful stimuli in 50 percent of patients. It is often referred to as the "strength of anesthetic." Inhalant anesthetics with a lower MAC are more potent and those with a higher MAC are less potent. Halothane has a lower MAC than isoflurane; therefore, you need to provide the patient a higher concentration or percent of isoflurane to get the animal under anesthesia.

Another factor used in comparing inhalant anesthetics is *blood-to-gas solubility*, the measure of the inhalation anesthetic to distribute between the blood and gas phases in the body. It is a measure of the tendency of an inhalation anesthetic to exist as a gas or to dissolve in blood. Anesthetic gases that are highly soluble must saturate the blood before molecules entering the blood are "available" to distribute to the tissues. Because it takes time to saturate the blood, it takes longer for highly soluble anesthetics to "fill" the blood so that "leftover" molecules can then distribute to the tissue. Less soluble anesthetics need not saturate the blood, and therefore pass into the blood from the alveolus (absorption of drug) and then more readily move to the tissues (distribution of drug).

An analogy of sugar and tea can be used to represent anesthetic gas and its absorption and distribution in blood. Hot tea represents a solution that is highly soluble;

cold tea represents a solution with low solubility. If the same volume of sugar (anesthetic gas) is dissolved in both hot and cold tea, the sugar will remain in solution in the hot tea and will precipitate out of solution in the cold tea. The precipitated (leftover) sugar at the bottom of the cup of cold tea can be used for other purposes. Anesthetic gas that is poorly soluble quickly saturates the blood so that additional gas is "leftover" for distribution to tissues.

An inhalant anesthetic with high blood-to-gas solubility will have a longer induction time and longer recovery time than an inhalant anesthetic with low blood-to-gas solubility. Inhalant anesthetics with low blood-to-gas solubility stay in the alveoli (gas phase) and a large concentration difference between the agent in blood and in alveoli builds. This large concentration difference between the agent in blood and in alveoli allows rapid entry of the inhalant anesthetic into the circulation and then distribution into the tissues (like the brain). In contrast, an inhalant anesthetic with high blood-to-gas solubility is rapidly absorbed into the blood and tissues and thus no large concentration difference between the agent in blood and in alveoli occurs. This small concentration difference between the agent in blood and in alveoli allows wider and more even distribution of the agent throughout the body. This results in slower induction and recovery rates. Examples of inhalant agents include the following.

TIP

The higher the blood-to-gas solubility of an inhalant anesthetic, the longer the induction and recovery times.

Inhalant Analgesics Nitrous oxide, or laughing gas, is an inhalant analgesic that is stored in blue cylinders in the United States. It diffuses rapidly throughout the body and can enter gas-filled body compartments (like the stomach and bowels), thus increasing the pressure in these compartments. Therefore, nitrous oxide is contraindicated in cases of gastric dilatation, pneumothorax, and twisted intestines. At the end of surgery, it is recommended to leave the animal on 100% oxygen for about 10 minutes and keep the endotracheal tube in place. This is to prevent diffusion hypoxia, which occurs when nitrous oxide rapidly diffuses out of the tissues back to the blood and then back to the alveoli. The alveoli are flooded with nitrous oxide, which dilutes the oxygen in the lung and causes hypoxia (low levels of oxygen).

Inhalant Anesthetics The volatile anesthetic agents are halogenated hydrocarbons (carbon- and hydrogen-based molecules that have fluorine, chlorine, bromine, and/or iodine attached; Figure 7-7). In general, increasing the halogenation of the molecule increases its potency and reduces its flammability. The addition of fluoride atoms increases the agent's stability.

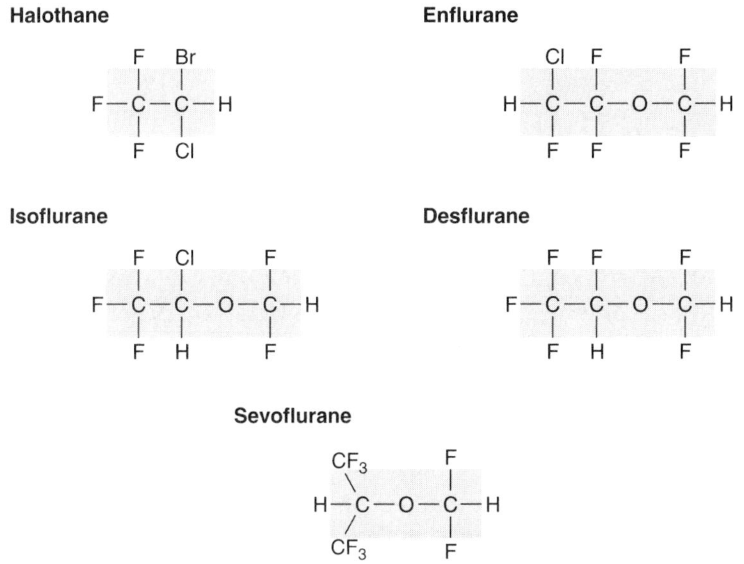

```
Halothane                    Enflurane

   F  Br                        Cl  F      F
   |  |                         |   |      |
F—C—C—H                     H—C—C—O—C—H
   |  |                         |   |      |
   F  Cl                        F   F      F

Isoflurane                   Desflurane

   F  Cl     F                  F   F      F
   |  |      |                  |   |      |
F—C—C—O—C—H                 F—C—C—O—C—H
   |  |      |                  |   |      |
   F  H      F                  F   H      F

              Sevoflurane

              CF3     F
                \     |
               H—C—O—C—H
                /     |
              CF3     F
```

FIGURE 7-7 Chemical structures of volatile anesthetic agents.

TIP

Advantages of inhalant anesthesia include easily and rapidly controlled anesthetic depth, increased inspired oxygen to the patient with the delivery of inhalation agents, provision of respiratory support via endotracheal tube placement, and minimal metabolism requirements for recovery. Disadvantages of inhalant anesthesia include waste gas exposure and dose-dependent cardiopulmonary depression.

Halothane (Fluothane®) is a nonflammable inhalant anesthetic administered via a precision vaporizer, because of its high vapor pressure (precision vaporizers limit the concentration of vaporization). Halothane is susceptible to decomposition; therefore, it is stored in amber-colored bottles and thymol is added as a preservative. Hepatic problems often occur with halothane use because about 25 percent of the halothane that is delivered to the patient is metabolized by the liver (the metabolites are hepatotoxic). Other problems that may occur with halothane are malignant hyperthermia, cardiac arrhythmias, bradycardia, and tachypnea (that will eventually increase the amount of anesthetic that the animal receives). Animals on halothane need close monitoring beause they can change planes of anesthesia more quickly.

Isoflurane (Forane®, Aerrane®) is a nonflammable inhalant anesthetic that causes rapid induction of anesthesia and short recoveries following anesthetic procedures. These drugs must be administered via a precision vaporizer. They do not cause the cardiac arrhythmia problems that halothane does. However, vigilant monitoring is needed because of the animal's ability to change planes of anesthesia quickly and the very short recovery period. Using a tranquilizer or sedative prior to induction can smooth recovery. Masking an animal with isoflurane and related anesthetics may be difficult due to the ability of these drugs to irritate the respiratory system. A side effect of isoflurane is respiratory depression (greater than that of halothane) and, like halothane, it can trigger malignant hyperthermia.

Isomers of isoflurane include enflurane (Ethrane®), desflurane (Suprane®), and sevoflurane (Ultane®, SevoFlo®). These compounds vary in the amount and type of chemicals that are attached to the base molecule. This group of drugs is nonflammable and has fewer cardiovascular side effects than the other inhalant anesthetics. These isomers quickly enter the bloodstream and escape to the brain, making them good for mask inductions. Enflurane metabolism releases fluoride ions, but apparently not to a level that causes nephrotoxicity. Enflurane has been found to increase intracranial pressure and should not be used in animals with a history of seizure activity. Desflurane cannot be delivered by standard vaporizers and requires the use of electrically heated vaporizers. In high doses, desflurane can reduce blood pressure. Sevoflurane is a profound respiratory depressant and close monitoring of animals receiving sevoflurane is needed. Sevoflurane undergoes temperature-dependent degradation by the soda lime and barium lime crystals used in carbon dioxide absorber canisters; therefore, sevoflurane cannot be used in low-flow or closed-system anesthesia. Both desflurane and sevoflurane have low tissue solubility, resulting in rapid elimination of the drug by the body and rapid awakening. Because this group of drugs provides limited analgesia, pre- and postoperative analgesics must be administered to alleviate pain in surgical patients.

CNS Stimulants (↑ CNS)

CNS stimulants are used to reverse CNS depression caused by CNS depressants. *Doxapram* (Dopram-V®,) stimulates the brainstem to increase respiration in animals with apnea or bradypnea. It is commonly used when animals have C-sections. It is given sublingually or via the umbilical cord to the neonates who received CNS depressants in the form of anesthetics from the dam through the placenta. It should be used with caution in animals with a seizure history.

Methylxanthines, a group of drugs that inhibit an enzyme that normally breaks down cyclic AMP, include substances such as caffeine, theophylline, and aminophylline. Methylxanthines are typically used as bronchodilators; however, one of their adverse effects is CNS stimulation. Theophylline (generic, Theo-dur®) and aminophylline (generic) are very similar compounds that are available in oral and injectable forms. Aminophylline contains about 80% theophylline and is better tolerated by the gastrointestinal tract; however, theophylline comes in sustained-release forms that can be given less frequently. Methylxanthines cause CNS stimulation, gastrointestinal irritation, and bronchodilation.

Euthanasia Solutions

Euthanasia solutions are used to humanely end an animal's life. They usually contain pentobarbital. When pentobarbital is the only narcotic agent present, it is a C-II controlled substance (Sleep Away®), it is a C-III controlled substance when in combination with other agents (Beuthanasia®). T-61 is a nonnarcotic, nonbarbiturate, noncontrolled-substance, general anesthetic euthanasia solution that causes muscle paralysis.

AUTONOMIC NERVOUS SYSTEM DRUGS

Autonomic nervous system drugs work either by acting like neurotransmitters or by interfering with neurotransmitter release. They may affect either the parasympathetic or the sympathetic nervous systems.

Parasympathetic Nervous System Drugs

Two groups of drugs affect the parasympathetic nervous system: the cholinergics and the anticholinergics.

Cholinergic Drugs/Parasympathomimetics **Cholinergic drugs**, also known as **parasympathomimetics**, mimic the action of the parasympathetic nervous system. They work either by mimicking the action of acetylcholine (direct-acting) or by inhibiting acetylcholine breakdown (indirect-acting). Table 7-5 summarizes the effects of cholinergic drugs. Direct-acting cholinergics are selective for muscarinic receptors and affect the smooth muscles of the urinary and gastrointestinal tracts. Side effects of cholinergic drugs include bradycardia, diarrhea and vomiting, and increased secretions (intestinal, bronchial, ocular). Examples include:

- *bethanechol* (Urecholine®), a direct acting cholinergic used to treat gastrointestinal and urinary atony.
- *metoclopramide* (Reglan®), a direct-acting cholinergic used to control vomiting and aid in gastric emptying.
- *pilocarpine* (Akarpine®, Pilocar®, IsoptoCarpine®), a direct-acting cholinergic used as an ophthalmic solution to decrease the intraocular pressure seen in glaucoma. Local irritation is an additional side effect.
- *edrophonium* (Tensilon®), an indirect-acting cholinergic used to diagnose myasthenia gravis.
- *neostigmine* (Prostigmine®, Stiglyn®) and physostigmine (Antilirium®, Eserine®), indirect-acting cholinergics used to treat rumen atony, intestinal atony, and urine retention.
- *demecarium* (Humorsol®) and isoflurophate (Floropryl®), indirect-acting cholinergics used to manage glaucoma.
- *organophosphates,* indirect-acting cholinergics used in antiparasitic products. If used improperly or in debilitated animals, toxicity may be seen that can be reversed with 2-PAM (Protopam®). 2-PAM reactivates cholinesterase.

Anticholinergic Drugs/Parasympatholytics **Anticholinergic drugs** inhibit the actions of acetylcholine by occupying the acetylcholine receptors. These drugs are also referred to as **parasympatholytics**, antimuscarinic agents, or antispasmodics. The major body

TABLE 7-5 Effects of Cholinergic Drugs

Body Tissue	Effect of Cholinergic Drugs
Cardiovascular	Decreases heart rate, causes vasodilation (lowers blood pressure), and slows conduction of the AV node
Lung (bronchi)	Stimulates bronchial smooth muscle contraction and increases bronchial secretions
Gastrointestinal	Increases motility of the smooth muscles of the stomach, increases peristalsis, and relaxes sphincter muscles.
Urinary	Contracts urinary bladder muscles, relaxes sphincter muscles of the urinary bladder, and stimulates urination
Ocular	Causes miosis (pupillary constriction)
Skeletal muscle	Maintains muscle strength and tone
Glandular	Increases salivation, perspiration, and tear production

tissues affected by the anticholinergic drugs are the heart, respiratory tract, gastrointestinal tract, urinary bladder, eye, and exocrine glands. By blocking the parasympathetic nerves, the sympathetic or adrenergic nervous system dominates. Anticholinergic and cholinergic drugs have opposite effects. Table 7-6 summarizes the effects of anticholinergic drugs. Side effects of anticholinergics may include tachycardia, constipation, dry mouth, dry eye, and drowsiness. Examples of anticholinergics include:

- *atropine* (generic), used as a preanesthetic agent to prevent bradycardia and to decrease salivation, to dilate pupils for ophthalmic examination, to control ciliary spasm of the eye, to decrease gastrointestinal motility, to treat bradycardia, and as an antidote for organophosphate poisoning.
- *glycopyrrolate* (Robinul-V®), similar to atropine but with a longer duration of action. It is mainly used as a preanesthetic agent.
- *aminopentamide* (Centrine®), used to control diarrhea and vomiting.
- *propantheline* (Pro-Banthine®), which decreases gastric secretions and gastrointestinal spasms, treats urinary incontinence, and decreases colonic peristalsis in horses.

TABLE 7-6 Effects of Anticholinergic Drugs

Body Tissue	Effect of Anticholinergic Drugs
Cardiovascular	Increases heart rate
Lung (bronchi)	Dilates bronchi and decreases bronchial secretions
Gastrointestinal	Relaxes smooth muscle tone of the gastrointestinal tract, decreases gastrointestinal motility and peristalsis, and decreases gastric and intestinal secretions
Urinary	Relaxes urinary bladder muscles, increases constriction of the internal sphincter muscle of the urinary bladder, and causes urine retention
Ocular	Causes mydriasis (pupillary dilation) and paralyzes the ciliary muscle
CNS/Muscular system	Decreases muscle rigidity and can cause drowsiness and disorientation
Glandular	Decreases salivation, perspiration, and tear production

Sympathetic Nervous System Drugs

Two groups of drugs affect the sympathetic nervous system: the adrenergics and the adrenergic blocking agents.

Adrenergic Drugs/Sympathomimetics Drugs that simulate the action of the sympathetic nervous system are called **adrenergic drugs** or **sympathomimetics**. They act on one or more adrenergic receptors located on the cells of smooth muscles. Catecholamines are chemicals that can cause a sympathomimetic response. Table 7-7 summarizes the effects of adrenergic (and adrenergic blocking) agents. Side effects include tachycardia, hypertension, and cardiac arrhythmias. Examples of naturally occurring catecholamines are epinephrine, norepinephrine, and dopamine. Synthetic catecholamines include:

- *epinephrine* (generic, Adrenalin®), an injectable adrenergic drug that affects alpha-1, beta-1, and beta-2 receptors. It increases heart rate, cardiac output, constriction of blood vessels to the skin, dilation of bronchioles, and dilation of blood vessels to muscles. It is used in emergency situations for cardiac resuscitation and treatment of anaphylaxis.
- *norepinephrine* (Noradrenalin®, Levophed®, Levarterenol®), an injectable adrenergic drug that affects alpha-1 and beta-1 receptors and is used mainly to increase blood pressure.
- *isoproterenol* (Isuprel®), an adrenergic drug that affects beta-1 and beta-2 receptors and is used mainly to cause bronchodilation.
- *dopamine* (Intropin®), an adrenergic drug that affects beta-1 receptors and is used to treat shock and congestive heart failure.
- *dobutamine* (Dobutrex®), an adrenergic drug that affects beta-1 receptors and is used to treat heart failure.
- *phenylpropanolamine* (generic, Ornade®, Prolamine®, Dexatrim®, Dimetapp®), an adrenergic drug that affects alpha-1 and beta-1 receptors and is used to treat urinary incontinence in dogs.
- *isoetharine* (Bronkosol®), *albuterol* (Proventil®) and *terbutaline* (Brethine®), adrenergic drugs that affect beta-2 receptors and are used to treat bronchospasm by producing bronchodilation.
- *ephedrine* (Vatronol®), an adrenergic drug that affects alpha-1, beta-1, and beta-2 receptors and is used to produce bronchodilation.
- *xylazine* (Rompun®, AnaSed®), an adrenergic drug that affects alpha-2 receptors and was covered previously in this chapter under tranquilizers/sedatives.

TABLE 7-7 Effects of Adrenergics and Adrenergic Blocking Agents

Receptor	Adrenergic Drug Effect	Adrenergic Blocking Drug Effect
alpha-1	Increases force of heart contraction, increases blood pressure, and causes mydriasis	Vasodilation and miosis
alpha-2	Inhibits release of norepinephrine and dilates blood vessels, producing hypotension	None
beta-1	Increases heart rate and force of heart contraction	Decreases heart rate
beta-2	Dilates bronchioles and relaxes gastrointestinal tract	Constricts bronchioles

Adrenergic Blocking Agents/Sympatholytics Drugs that block the effects of the adrenergic neurotransmitters are called **adrenergic blocking agents** or **sympatholytics**. They act as antagonists to the adrenergic agonists by blocking the alpha- and beta-receptor sites. They can block the receptor site either by occupying the receptor or by inhibiting the release of the neurotransmitter.

Alpha blockers usually promote vasodilation and a decrease in blood pressure. Side effects of alpha blockers include tachycardia and hypotension. Examples of drugs used in veterinary practice include:

- *phenoxybenzamine* (Dibenzyline®), used in small animals to decrease urethral sphincter tone and in horses to prevent or treat laminitis. It should not be used in horses with colic.
- *prazosin* (Minipress®), used to treat animals with heart failure and hypertension.
- *yohimbine* (Yobine®) and *atipamezole* (Antisedan®), reversal agents already discussed in this chapter.

Beta blockers decrease heart rate and blood pressure. Side effects of beta blockers include bradycardia and hypotension. Examples of beta blockers used in veterinary practice include:

- *propranolol* (Inderal®), which blocks beta-1 and beta-2 receptors and is used to treat cardiac arrhythmias and cardiac disease in animals.
- *metoprolol* (Lopressor®) and *atenolol* (Tenormin®), which blocks beta-2 receptors and are used to treat hypertension.
- *timolol* (Timoptic®), which blocks beta-1 and beta-2 receptors and is used as an ophthalmic preparation to treat glaucoma.

SUMMARY

The nervous system has a basic unit called a neuron, which consists of a cell body, dendrites, and an axon. The space between two neurons or between a neuron and its receptor is the synapse. Neurotransmitters move across the synapse to cause action.

There are two branches of the nervous system: the CNS and the PNS. The PNS consists of cranial nerves, spinal nerves, and the autonomic nervous system. The ANS is divided into two branches: the sympathetic (fight-or-flight pathway) and the parasympathetic (the homeostatic pathway).

CNS drugs include anticonvulsants that control seizures, tranquilizers/sedatives/anti-anxiety drugs that reduce anxiety and excitement in animals, analgesics that minimize pain, opioid antagonists that treat respiratory and CNS depression, neuroleptanalgesics that cause CNS depression and analgesia, anesthetics that remove pain sensation, CNS stimulants that counteract CNS depression, and euthanasia solutions that humanely end an animal's life.

Drugs that work on the ANS include cholinergics, anticholinergics, adrenergics, and adrenergic blocking agents. ANS drugs affect the cardiovascular system, bronchi, gastrointestinal tract, urinary tract, CNS/muscular system, glands, and ocular system. Table 7-8 summarizes the types of drugs covered in this chapter.

TABLE 7-8 Types of Drugs Covered in this Chapter

Drug Category	Examples
Anticonvulsants	• phenobarbital • pentobarbital • primidone • diazepam • clorazepate • potassium bromide • valproic acid • phenytoin
Tranquilizers/sedatives/anti-anxiety	• phenothiazine derivatives: acepromazine, chlorpromazine, prochlorperazine/isopropamide, promazine • benzodiazepines: diazepam, zolazepam, midazolam, clonazepam • alpha-2-agonists: xylazine, detomidine, medetomidine
Narcotic analgesics	• opium • morphine • meperidine • hydromorphone • butorphanol • hydrocodone • fentanyl • etorphine • buprenorphine • pentazocine • diphenoxylate • apomorphine • loperamide • methadone • codeine
Opioid antagonists	• naloxone • naltrexone
Neuroleptanalgesics	• xylazine & butorphanol • acepromazine & morphine
Local anesthetics	• lidocaine • proparacaine • mepivacaine • tetracaine • bupivacaine
General anesthetics (injectable)	• barbiturates: phenobarbital, pentobarbital, thiopental, methohexital • dissociatives: ketamine, tiletamine • miscellaneous: guaifenesin, propofol
Analgesic (inhalant)	• nitrous oxide
General anesthetic (inhalant)	• halothane • isoflurane, enflurane, desflurane, sevoflurane

(continues)

TABLE 7-8 *Continued*

Drug Category	Examples
CNS stimulants	• doxapram • methylxanthines
Euthanasia solutions	• pentobarbital combinations
Cholinergics	• bethanechol • metoclopramide • pilocarpine • edrophonium • neostigmine • demecarium • organophosphates
Anticholinergics	• atropine • glycopyrrolate • aminopentamide • propantheline
Adrenergics	• epinephrine • norepinephrine • isoproterenol • dopamine • dobutamine • phenylpropanolamine • isoetharine • albuterol • terbutaline • ephedrine • xylazine
Adrenergic blocking agents	• alpha-blockers: phenoxybenzamine, prazosin, yohimbine • beta-blockers: propranolol, metoprolol, timolol

CHAPTER REFERENCES

Amundson Romich, J. (2000). *An illustrated guide to veterinary medical terminology*. Clifton Park, NY: Thomson Delmar Learning.

Delmar's A-Z NDR-97 nurses drug reference. Clifton Park, NY: Thomson Delmar Learning.

Hendrix, P. (May 2001). Perioperative pain management in cats and dogs. *Practice Builder, DVM: The Newsletter of Veterinary Medicine*, http://www.dvmnewsmagazine.com/dvm/content/printContentPopup.jsp?id=5556

Reiss, B., & Evans, M. (2002), *Pharmacological aspects of nursing care* (sixth edition, revised by Bonita E. Broyles, pp. 358–400). Clifton Park, NY: Thomson Delmar Learning.

Veterinary pharmaceuticals and biologicals (12th ed.). (2001). Lenexa, KS: Veterinary Healthcare Communications.

Veterinary values (5th ed.). (1998). Lenexa, KS: Veterinary Healthcare Communications.

Wynn, R. (1999). *Veterinary pharmacology*. Cambridge, MA: Harcourt Learning Direct.

CHAPTER REVIEW

Matching

Match the drug name with its action.

1. ______ phenobarbital
2. ______ xylazine
3. ______ acepromazine
4. ______ phenylpropanolamine
5. ______ diazepam
6. ______ morphine
7. ______ ketamine
8. ______ glycopyrrolate
9. ______ bethanecol
10. ______ methohexital

a. analgesic drug
b. anticonvulsant drug
c. phenothiazine derivative tranquilizer
d. adrenergic drug
e. cholinergic drug
f. anticholinergic drug
g. thiobarbiturate
h. dissociative anesthetic
i. alpha-2 agonist
j. anticonvulsant and tranquilizer that causes muscle relaxation

Multiple Choice

Choose the one best answer.

11. Which neurotransmitter is released at the postganglionic synapse of parasympathetic nerves?
    a. acetylcholine
    b. epinephrine
    c. dopamine
    d. mu
12. Which neurotransmitter is released at the postganglionic synapse of sympathetic nerves?
    a. acetylcholine
    b. epinephrine
    c. dopamine
    d. mu
13. Why do some animals become "resistant" to their dose of phenobarbital?
    a. They excrete this drug through the kidneys.
    b. They cannot biotransform this drug.
    c. Liver enzymes are induced causing the animal to develop tolerance.
    d. The half-life of the drug is shortened due to the high level of protein binding.
14. Which of the following sedatives produces analgesia in the horse, but still allows the horse to respond to stimuli by kicking?
    a. xylazine
    b. ketamine
    c. diazepam
    d. yohimbine
15. Obese animals that are anesthetized with thiobarbiturate would
    a. only be anesthetized for a short time.
    b. be anesthetized for a long time.
    c. would not go under anesthesia, because of their fat stores.
    d. have severe perivascular inflammation.
16. Which group of anesthetics causes muscle rigidity and mild analgesia?
    a. barbiturates
    b. dissociatives
    c. inhalants
    d. propofol
17. For which inhalant analgesic is it recommended to leave the animal on 100% oxygen for about 10 minutes following the procedure to prevent diffusion hypoxia?
    a. nitrous oxide
    b. halothane
    c. isoflurane
    d. sevoflurane
18. Which anesthetic is a white emulsion, the effects of which last about two to five minutes, when given slow IV?
    a. barbiturate
    b. dissociative
    c. propofol
    d. guaifenesin

True/False

Circle a. for true or b. for false.

19. All euthanasia solutions are C-II controlled substances.

a. true
b. false

20. Cholinergic drugs tend to be used to improve the cardiovascular system.

a. true
b. false

Case Studies

21. A 13-yr.-old F/S Poodle (Figure 7-8) with a 5-year history of epilepsy has been treated with phenobarbital. You check her liver enzymes every six months and this time notice that her liver enzymes are increasing.

a. Is an increase in liver enzymes normally seen with phenobarbital treatment?
b. This poodle's brother is also being treated with phenobarbital for seizures. He has been on 1/4 grain of phenobarbital bid but is now having seizures again. What phenomenon is occurring?
c. These dogs are going on vacation and the owner would like to receive an anti-anxiety drug for each of them. What anti-anxiety drug would *not* be used?
d. These dogs need refills on their phenobarbital. Before dispensing the medication that has been approved by the veterinarian, what do you need to do?

22. A 4-yr.-old, 85#, M/N Brittany Spaniel (Figure 7-9) is presented to the clinic for a laceration he received on his morning walk. His physical examination is normal except for his obesity and the laceration that must be sutured. The veterinarian intends to induce this dog with thiobarbiturate and wants you to calculate a dose.

a. What should you consider in determining a dose for this dog?
b. Why would this dog have an anesthetic overdose if the dose were calculated based on his 85-lb weight?
c. When giving thiobarbiturate IV, how can you avoid overdosing the patient?

FIGURE 7-8 Poodle. Photo by Isabelle Francais

FIGURE 7-9 Brittany Spaniel. Photo by Isabelle Francais

CHAPTER 8

Cardiovascular Drugs

OBJECTIVES

Upon completion of this chapter, the learner should be able to

- describe the anatomy and physiology of the cardiovascular system, including blood flow through the heart and the electrical conduction system.
- describe the components of the ECG.
- differentiate between preload and afterload.
- outline ways the heart can improve its workload in heart disease.
- describe the clinical usages of the following drugs in the treatment of heart disease: positive inotropes, antiarrhythmics, vasodilators, diuretics, anticoagulants, hemostatic drugs, blood-enhancing drugs.

KEY TERMS

inotropy
arrhythmia
preload
afterload
cardiac glycoside
catecholamine
antiarrhythmic drug
renin
angiotensin I
angiotensin II
vasodilator
diuretic
anticoagulant
hemostatic drug
blood-enhancing drug
erythropoietin

INTRODUCTION

A 14-year-old F/S Cocker Spaniel is brought into the clinic because her owner feels that the dog is not as energetic as usual. The Spaniel is up to date on her vaccinations and has been heartworm-tested annually. On physical exam, the vet notes that her mucous membrane color is not as pink as it has been in the past, hears a murmur upon auscultation of her heart, and finds that some fluid appears to be present in the abdominal cavity. These signs indicate possible heart disease, so the appropriate diagnostic tests are performed on the dog. Chest radiographs show an enlarged heart. The dog's ECG is normal and her blood values are normal. The dog is diagnosed with congestive heart failure and the vet prescribes a diuretic and a mixed vasodilator. The veterinarian also prescribes a sodium-restricted diet without any treats. The owner asks you how all these drugs are going to help her dog. Why can't she keep her dog on the same dog food that she has been feeding for many years? What do you tell her? Can you explain the rationale behind low-sodium diets for animals with congestive heart failure? How can we assess success or failure of cardiac treatment? Is there any way to determine if the drugs and diet restriction are working? When should the owner bring the dog in again for reexamination?

BASIC CARDIAC ANATOMY AND PHYSIOLOGY

The functions of the cardiovascular system include delivery of oxygen, nutrients, and hormones to the various body tissues and the transport of waste products to the appropriate waste removal system. There are three major parts of the cardiovascular system: the heart, the blood vessels, and the blood.

The heart is a hollow, muscular organ that provides the power to move blood through the body. The heart consists of a thick muscular wall called the *myocardium*; an inner membrane that lines the chambers and valves called the *endocardium*; and the outer membrane of the heart called the *epicardium*. A double-walled membrane called the *pericardium* surrounds the heart. The heart consists of chambers: the *atria* are the upper chambers and the *ventricles* are the lower chambers. In animals with four-chambered hearts (mammals and birds), the right atrium receives deoxygenated blood from the venous circulation. Blood is pumped to the right ventricle, which in turn pumps blood to the lungs to receive oxygen. The left atrium receives oxygenated blood and pumps it to the left ventricle, which in turn pumps it to the aorta. Separating these chambers are four valves that help control the direction of blood flow in the heart. There are two atrioventricular valves (tricuspid and mitral) and two semilunar valves (pulmonic and aortic). Figures 8-1 and 8-2 illustrate the structures of the heart. Table 8-1 describes blood flow through the heart.

The primary function of the heart is twofold: to pump fresh, oxygenated blood to the tissues of the body and to take waste products such as carbon dioxide away from the tissues. Blood carries oxygen to tissues via arteries and arterioles and takes wastes away from tissues via venules and veins. Diffusion of oxygen and carbon dioxide occurs in the single-cell-layer-thick capillaries.

The conduction of electrical impulses of the heart originates in the sinoatrial (SA) node. The SA node is located in the wall of the right atrium and regulates heart rate. When the cells in the SA node depolarize, a wavelike sequence of depolarization passes through the right and then the left atrium. *Depolarization* occurs when sodium

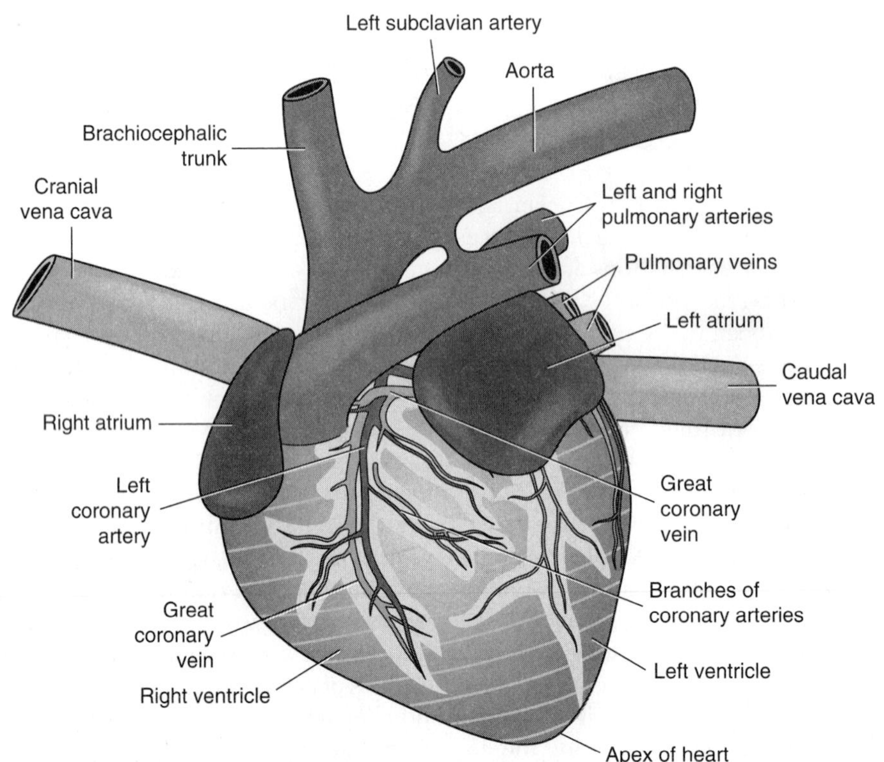

FIGURE 8-1 External heart structures

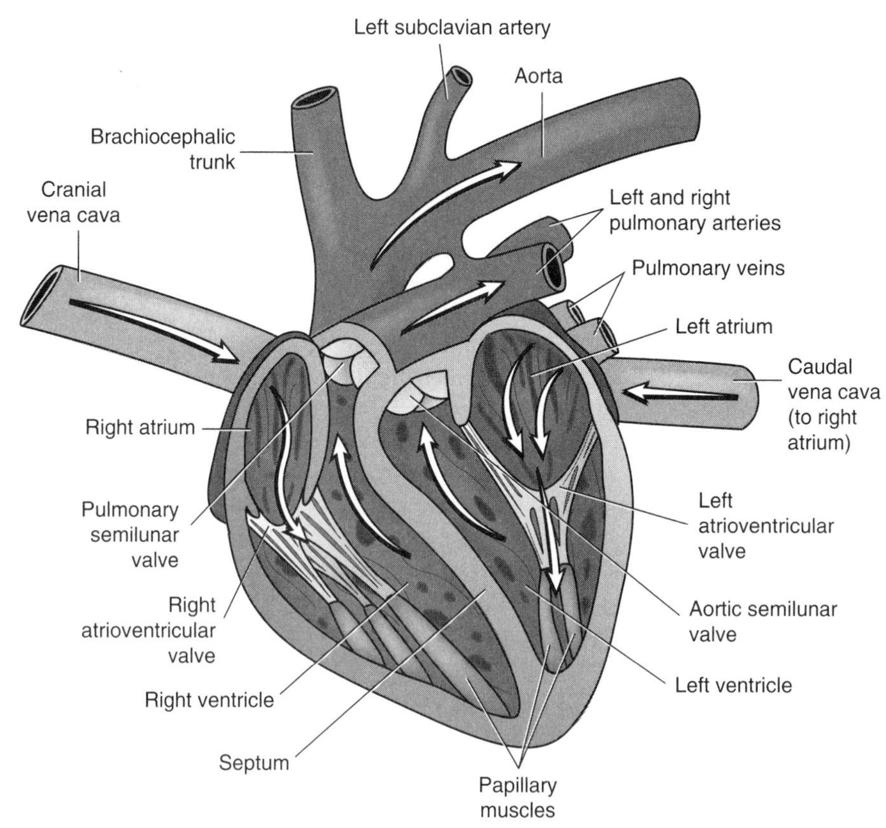

FIGURE 8-2 Internal heart structures

channels open, allowing sodium ions into the cell, which then allows the cells of the myocardium to contract. Depolarization continues cell to cell very quickly, so that it seems as though both the right atrium and the left atrium are contracting at the same time. As the atria contract, blood is pushed into the ventricles. The wave of depolarization spreads from the atria, but is prevented from entering the ventricles by a cellular barrier. The depolarization wave can enter the ventricles only through the atrioventricular (AV) node. The AV node is located in the interatrial septum. As the depolarization wave enters the AV node, it is delayed for a fraction of a second; then it enters

TABLE 8-1 Blood Flow Through the Heart

↓ The right atrium receives blood from all tissues, except the lungs, through the cranial and caudal venae cavae. Blood flows from here through the tricuspid valve into the right ventricle. (This is systemic circulation.)
↓ The right ventricle pumps the blood through the pulmonary semilunar valve and into the pulmonary artery, which carries it to the lungs. (This is pulmonary circulation.)
↓ The left atrium receives oxygenated blood from the lungs through the four pulmonary veins. The blood flows through the mitral valve into the left ventricle. (This is pulmonary circulation.)
↓ The left ventricle receives blood from the left atrium. From the left ventricle, blood goes out through the aortic semilunar valve and into the aorta and is pumped to all parts of the body, except the lungs. (This is systemic circulation.)
↓ Blood is returned by the venae cavae to the right atrium and the cycle continues.

the ventricles. The delay allows the atria to complete contraction before the ventricles contract. After the depolarization wave passes through the AV node, it goes through the interventricular septum, along a conduction system known as the *bundle branches*. Bundle branches conduct electrical impulses to the Purkinje fibers at the apex of the heart. From the Purkinje fibers, an impulse travels rapidly through the ventricular muscle cells, causing contraction from the apex toward the heart valves. At this point blood is pushed from the ventricles through the pulmonic and aortic semilunar valves to the lung and body. Figure 8-3 summarizes the conduction system of the heart.

Heart rate is controlled primarily by the autonomic nervous system. Parasympathetic (cholinergic) nerve endings (vagal fibers) are located close to the SA node. When parasympathetic nerves are stimulated, the neurotransmitter acetylcholine is released at the junction of the nerve and the cardiac muscle. This acts to slow the heart rate by inhibiting impulse formation and electrical conduction in the heart.

Sympathetic (adrenergic) nerve fibers also innervate part of the heart. When sympathetic nerves are stimulated, the neurotransmitter norepinephrine is released. This acts to increase heart rate by promoting impulse formation and electrical conduction in the heart. Sympathetic stimulation also decreases the time between impulses, thus reducing the duration of the refractory period (time between consecutive muscle contractions).

Conduction of electrical impulses can be seen on an electrocardiogram (ECG). The P wave of the ECG corresponds to the depolarization of the atria (and the contraction of the atrial cells). A short delay occurs between the time when the depolarization wave enters the AV node and the time when the ventricles contract. This delay is represented on the ECG by the flat line after the P wave known as the *P-R interval*. Depolarization of the ventricles (and the contraction of the ventricular cells) and repolarization of the atria are represented as the large Q-R-S complex. After the ventricles contract, the ventricular muscle cells relax and repolarize (sodium channels close, potassium channels open, and potassium moves into the cell). This repolarization of the ventricles is represented by the T-wave. Figure 8-4 shows the appearance of the ECG.

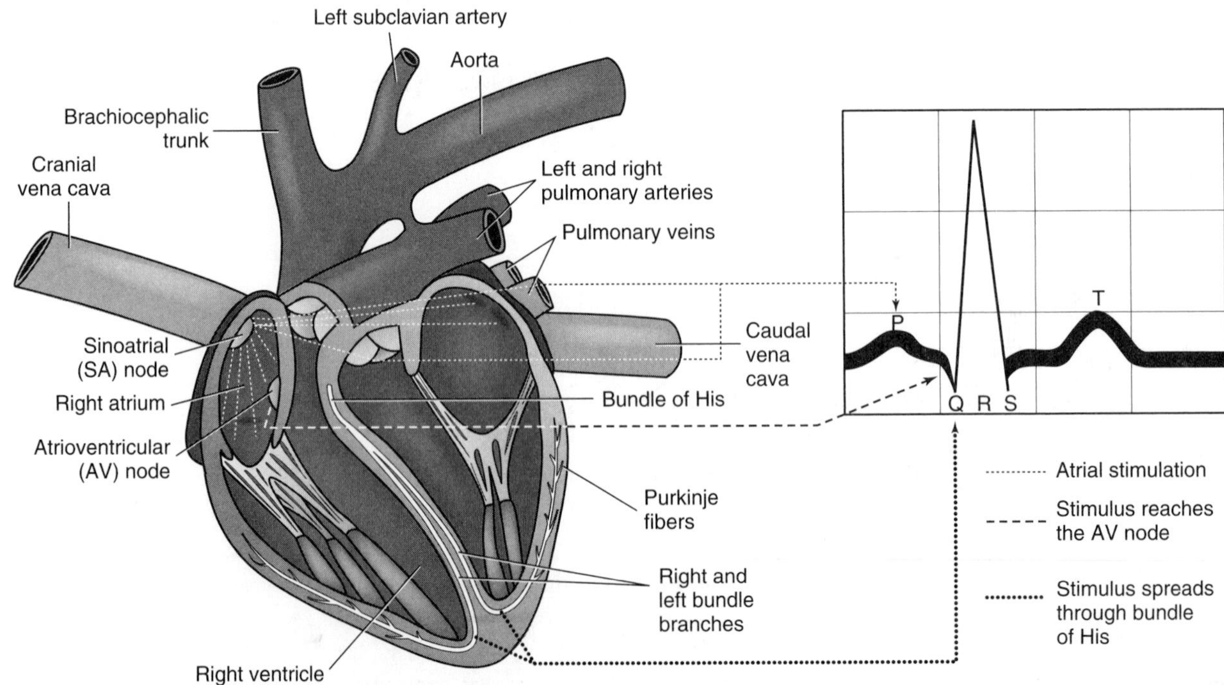

FIGURE 8-3 Conduction systems of the heart

Additional assessment of the heart is accomplished by *auscultation*, listening to the heart. Auscultation allows for determination of heart rate and rhythm. Cardiac contractions are divided into the states of systole and diastole. *Systole* refers to contraction of the heart chambers; diastole refers to relaxation of the chambers, during which time the chambers are filling with blood. The force of contraction is known as **inotropy**.

A *rhythm* is a recurrence of an audible action at regular intervals. The heart's contractions are supposed to be rhythmic. The rate and regularity of the heart rhythm, or *heartbeat*, is innate to the myocardial cells as directed by the SA node. Normal heart rhythm is called the *sinus rhythm* because it starts in the SA node. If the SA node does not function properly and is unable to send the impulse to the rest of the heart, other areas of the conduction system can take over and initiate a heartbeat. The resulting abnormal rhythm is called an **arrhythmia**. Depending on the type of arrhythmia present, drugs may be used to control heartbeat irregularities.

The workload of the heart is divided into preload and afterload. **Preload** is the volume of blood entering the right side of the heart, or the ventricular end-diastolic volume. **Afterload** is the force needed to push blood out of the ventricles, or the impedance to ventricular emptying presented by aortic pressure. Preload problems are usually associated with right-sided heart disease, whereas afterload is associated with left-sided heart disease. Preload and afterload, along with contractility (the force of ventricular contraction), make up *stroke volume*, the amount of blood ejected from the left ventricle with each heartbeat. Stroke volume multiplied by heart rate is the cardiac output. *Cardiac output* is the volume of blood expelled from the heart in one minute.

If the heart is not working properly, compensatory mechanisms to improve the workload of the heart may take over. Some examples of compensatory mechanisms of the heart include:

- *increasing heart rate* to increase cardiac output, so long as the heart rate is not so fast as to limit the chambers' filling with blood.
- *increasing stroke volume*, because increased force of contraction should result in more blood being pumped.

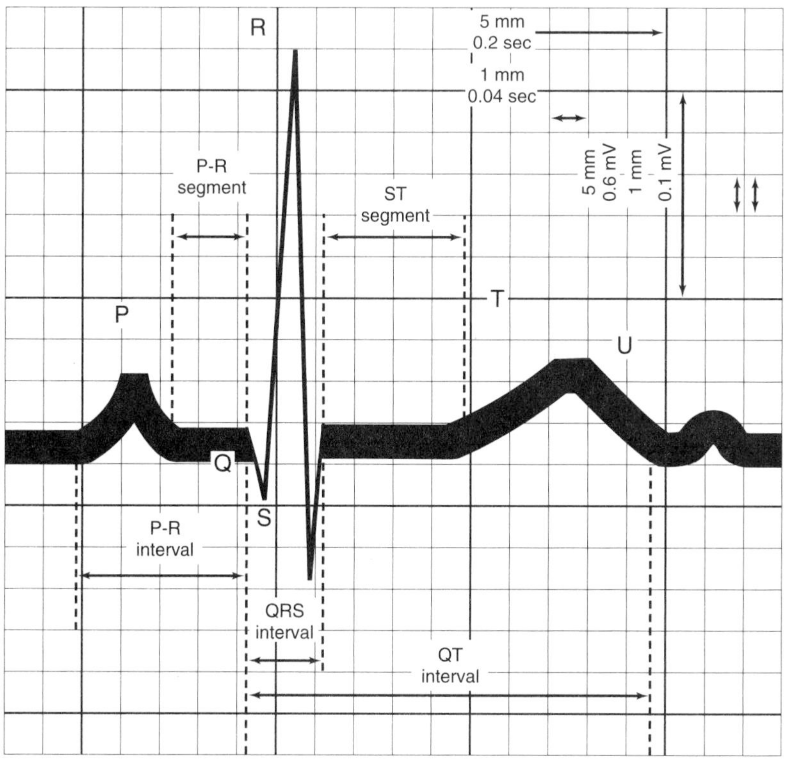

FIGURE 8-4 Anatomy of an electrocardiogram. *Source:* From Glazek, Basic electrocardiography—Part I, *Veterinary Technician 17*(10): 721, 1996.

- *increasing efficiency* of the heart muscle.
- *enlarging the heart* in an attempt to pump out more blood. This can be achieved either by dilating the chambers or by thickening the myocardium.

CARDIOVASCULAR DRUGS

Cardiovascular drugs can influence aspects of cardiac performance in the following ways:

- Increasing or decreasing the force of myocardial contraction. Drugs that increase the force of contraction are called *positive inotropic drugs*. Drugs that reduce the force of contraction are called *negative inotropic drugs*.
- Increasing or decreasing heart rate by altering the rate of impulse formation at the SA node. Drugs that increase heart rate are called *positive chronotropic drugs*. Drugs that decrease heart rate are called *negative chronotropic drugs*.
- Increasing or decreasing the conduction of electrical impulses through the myocardium. Drugs that increase the rate of electrical conduction are called *positive dromotropic drugs*. Drugs that decrease the rate of electrical conduction are called *negative dromotropic drugs*.

TIP

Sympathetic response through beta-1 receptors produces positive inotropic and chronotropic effects on the heart. Parasympathetic response through cholinergic receptors produces negative inotropic and chronotropic effects on the heart.

Increasing Force

Inotropy is the force of contraction. Positive inotropic drugs aim to improve the strength of myocardial contraction. Although these drugs may increase the heart's demand for oxygen, they are commonly used to increase the strength of contractions by failing heart muscle. Examples of positive inotropic drugs are cardiac glycosides and catecholamines.

TIP

All cardiac glycosides have a low therapeutic index; that is, the therapeutic dose is very close to the toxic dose. Careful patient monitoring for signs of toxicity (vomiting, lethargy, and arrhythmias) is a very important part of the animal care protocol for patients prescribed any cardiac glycoside.

Cardiac glycosides, or digitalis drugs, are derived from natural sources (the foxglove plant) and have been in use for hundreds of years to treat heart problems in humans. Digitalis drugs increase the strength of cardiac contractions, decrease heart rate, have an antiarrhythmic effect, and decrease signs of dyspnea. Digitalis increases the strength of contraction by inhibiting the sodium-potassium pump, which increases intracellular calcium concentrations. This increase in intracellular calcium concentration causes the myocardial fibers to contract more efficiently. Cardiac glycosides are used in animals to treat such heart problems as congestive heart failure (CHF), atrial fibrillation, and supraventricular tachycardia. Side effects include anorexia, vomiting, diarrhea, and cardiac arrhythmias. Blood levels must be monitored to prevent toxicity. Digitalis compounds have interactions with many other drugs, so multiple drug dosing must be monitored very closely. Examples of digitalis drugs include digoxin (Cardoxin® LS, Cardoxin®, Lanoxin®, Lanoxicaps®), which comes in elixir, injectable, tablet, and capsule formulations, and digitoxin (Crystodigin®), which comes in tablet form. Digoxin has a shorter duration of action than digitoxin and thus it is less likely to result in cumulative toxic effects. Because of toxicity issues associated with cardiac glycoside use, blood must be routinely collected to determine digoxin levels.

TIP

Digoxin and digitoxin are positive inotropes and negative chronotropes. They cause the heart to pump harder and the heart rate to slow.

Catecholamines are sympathomimetic drugs that have the following influences: they increase the force and rate of myocardial contraction, which increases cardiac output; they constrict peripheral blood vessels, which increases blood pressure; and they increase blood glucose levels. Receptors in the sympathetic nervous system are referred to as *adrenergic receptors* and are classified as alpha-1 (α_1), alpha-2 (α_2), beta-1 (β_1) and beta-2 (β_2). Alpha-1 receptors are found primarily in smooth muscle tissue of peripheral blood vessels and in the sphincters of the gastrointestinal system and urinary tract. When α_1 receptors are stimulated, contraction occurs in the smooth muscles associated with them. Alpha-2 receptors seem to function as controllers of neurotransmitters in the synaptic space. Stimulation of α_2 receptors results in reduction of neurotransmitter release. Beta-1 receptors are located primarily in heart muscle and fatty tissue. Stimulation of β_1 receptors in the heart increases heart rate and causes more forceful heart contractions. Stimulation of β_1 receptors in fatty tissue promotes the breakdown of stored fat to fatty acids, which can be used by the body as energy sources. Beta-2 receptors are located primarily in bronchial smooth muscle and in the walls of blood vessels of skeletal muscle, the brain, and the heart. Stimulation of β_2 receptors in bronchial smooth muscle produces muscle relaxation, thereby causing bronchodilation. Table 8-2 summarizes the effects of stimulation of catecholamine receptors. The following are the primary catecholamines (sympathomimetic drugs).

Epinephrine. Epinephrine has both alpha and beta activity. It causes smooth muscle relaxation in the bronchi (bronchodilation), raises blood glucose levels, and increases heart rate and contractility. It is also used as a cardiac stimulant following cardiac arrest. It may be given intracardiac, intratracheal, intramuscular, or subcutaneous. It is manufactured in two concentrations (1:1000 solution and 1:10,000 solution), so care should be taken when ordering, performing dose calculations, or measuring drug doses to make sure that the proper concentration is chosen. The 1:10,000 solution is the preferred concentration for use in veterinary medicine. Side effects include the possible development of arrhythmias and hypertension.

Dopamine. Dopamine (Inotropin®, Dopastat®, Dopamine® HCl, Dopamine® HCl in 5% dextrose) is a precursor to norepinephrine and works on both alpha and beta-1 receptors. Dopamine causes increased heart contractility, increased heart rate, and increased blood pressure. It is used in treating acute heart failure, severe shock, and oliguric (scanty amount of urine) renal failure. Side effects include tachycardia, dyspnea, and vomiting.

Dobutamine. Dobutamine (Dobutrex®) is a direct beta-1 agonist, with slight beta-2 and alpha activity. It produces increased cardiac output without the dilation of blood vessels seen with dopamine. It is given as a constant-rate IV infusion.

Isoproterenol. Isoproterenol (Isuprel®) has beta activity and is used in the treatment of cardiac arrhythmias and bronchial constriction. Side effects include tachycardia, weakness, and tremors; therefore, it is no longer frequently used in veterinary practice.

TABLE 8-2 Catecholamine Receptors

Receptor	Effect when Stimulated	Example of Clinical Use Affecting Receptor
α_1	Constriction of peripheral blood vessels	Constricts blood vessels within the area and prevents rapid diffusion of drug away from the injection site (for example, epinephrine can be injected with a local anesthetic)
α_2	Reduces blood pressure	Reduces blood pressure, but has CNS effect that can cause drowsiness and sedation
β_1	Increases heart rate and force of heart contractions; increases blood glucose	Treats animals with decreased cardiac function
β_2	Causes bronchodilation	Treats diseases where bronchodilation is needed

Fixing the Rhythm

An *arrhythmia* is a variation from the normal beating of the heart, which often results in reduced cardiac output due to the abnormal pumping activity. The rhythm of the heart is generally controlled by the SA node; however, spontaneous electrical discharge (automaticity) may occur anywhere in the heart. Any electrical activity initiated by spontaneous discharge is considered an arrhythmia. **Antiarrhythmic drugs** are divided into categories according to the types of arrhythmia they treat and the actions they have on the cells of the heart. Antiarrhythmic drugs can have the following influences on the heart: decreasing automaticity, altering the rate of electrical impulse conduction, or altering the refractory period of the heart muscle between consecutive contractions. Table 8-3 summarizes antiarrhythmic drugs.

The following points highlight special concerns about antiarrhythmic drugs.

- Digoxin levels may increase in animals taking quinidine; the dosage of digoxin may have to be lowered.
- Reactions to procainamide are likely in animals that are sensitive to procaine and other "-caine" local anesthetics.
- Be certain not to use the lidocaine product with epinephrine when giving lidocaine IV. Lidocaine is not effective when given orally (there is a high first-pass effect). Cats are sensitive to lidocaine, so dosing must be carefully monitored.
- Because propranolol blocks β_1 and β_2 activity, it affects the heart (lowering heart rate and blood pressure) and bronchi (causing bronchoconstriction).

TIP

Supraventricular arrhythmias originate above the ventricles (for example, the SA node or AV node). Ventricular arrhythmias originate in the ventricles (for example, premature ventricular complexes or PVCs).

TABLE 8-3 Antiarrhythmic Drugs

Class of Antiarrhythmic	Drug Name	Action
Class IA: Local Anesthetics This group works as local anesthetics to the nerves and myocardial membrane	A. quinidine (Cin-Quin®, Cardioquin®, Quinidex®)	A. Suppresses myocardial excitability and increases conduction times. Used orally to treat atrial and ventricular arrhythmias. Side effects include vomiting, diarrhea, and weakness.
	B. procainamide (Pronestyl®, Procamide® SR, Procan® SR)	B. Suppresses myocardial excitability and increases conduction times. Used orally to treat premature ventricular contractions (PVC), ventricular tachycardia, and some atrial tachycardias. Side effects are vomiting, diarrhea, and weakness.
Class 1B: Membrane Stabilization This group works by blocking the influx of sodium into the cell, thus stabilizing the myocardium and preventing depolarization	A. lidocaine (Xylocaine®, Lidocaine® for Injection 1% and 2%)	A. Depresses myocardial excitability. Used IV to control PVCs and treat ventricular tachycardia. Side effects are rare.
	B. tocainide (Tonocard®)	B. Action similar to lidocaine, but is only given orally. Used orally to treat ventricular arrhythmias (ventricular tachycardia and PVCs). Side effects include ataxia, vomiting, and hypotension.

TABLE 8-3 *Continued*

Class of Antiarrhythmic	Drug Name	Action
Class 1B: Membrane Stabilization (cont.)	C. mexiletine (Mexitil®)	C. Action is similar to lidocaine, but is used orally. Used to treat ventricular arrhythmias (ventricular tachycardia and PVCs). Side effects include vomiting and unsteadiness.
Class II: Beta-Adrenergic Blockers This group works by blocking beta-adrenergic receptors or by preventing release of norepinephrine from the adrenergic neuron	propranolol (Inderal®, Intensol®)	Reduces automaticity by blocking β_1 and β_2 receptors, thus reducing oxygen demand by the myocardium. Used to treat hypertrophic cardiomyopathy and some ventricular arrhythmias. Side effects include bradycardia, lethargy, and depression.
Class III: Action Potential Prolongation This group works by lengthening the time between action potentials, which decreases the sinus rate	bretylium (Bretylol®) amiodarone (Cordarone®)	Increases the duration of the action potential. Both are used for emergency treatments of ventricular tachycardia and fibrillation. These drugs are used in animals that are resistant to other drugs. Side effects include hypotension and vomiting.
Class IV: Calcium-Channel Blockers This group works by blocking channels that allow calcium to enter the myocardial cell. The entry of calcium into the cell facilitates muscle contractility	A. verapamil (Isoptin®, Calan®) B. nifedipine (Procardia®, Adalat®) C. diltiazem (Cardizem®)	A. Blocks calcium passage, relaxing vascular smooth muscle to lower blood pressure and inhibit the cardiac conduction system. Given orally and IV to treat supraventricular tachycardia and atrial flutter. B., C. Blocks calcium passage, dilating coronary and peripheral blood vessels which results in the reduction of cardiac workload. Both are used orally to treat atrial fibrillation and supraventricular tachycardia. Side effects of calcium-channel blockers include hypotension and edema.

Correcting Constriction

As heart failure occurs, the animal's body tries to compensate for the loss of cardiac function. The first reaction to a failing heart is to increase heart rate. Next, blood vessels are constricted because of a nervous system reaction and a substance called **renin**, an enzyme released by the kidney. Renin causes angiotensinogen to be converted to **angiotensin I**. Angiotensin I is converted by angiotensin-converting enzyme (ACE) to **angiotensin II**. Angiotensin II causes further vasoconstriction and stimulates aldosterone secretion by the adrenal cortex. Aldosterone influences the tubules of the nephron to reabsorb sodium ions and thus retain water (where sodium goes, water follows). The retained water is physiologically utilized to expand the circulating blood volume, which improves tissue perfusion.

Vasodilators are drugs used to dilate arteries and/or veins, which alleviates vessel constriction and improves cardiac output. The actions of vasodilators often greatly improve the condition of an animal with CHF. Types of vasodilators include the following.

Angiotensin-converting enzyme inhibitors (ACE inhibitors). ACE inhibitors are combined vasodilators (both venous and arterial) used in the treatment of heart failure and hypertension. They prevent the conversion of angiotensin I to angiotensin II (Figure 8-5). Examples include enalapril (Enacard®, Vasotec®), lisinopril (Zestril®), benazepril (Lotensin®), and captopril (Capoten®). Side effects include hypotension and gastrointestinal problems.

Arteriole dilators. Hydralazine (Apresoline®) is an arteriole dilator used to reduce the afterload associated with congestive heart failure (CHF). Side effects include hypotension and gastrointestinal signs.

Venodilators. Nitroglycerin ointment (Nitro-Bid®, Nitrol®) is a venodilator used to improve cardiac output and reduce pulmonary edema. It is applied as a dose in inches to the skin (wear gloves when applying nitroglycerin). It is also available in transdermal system form (patch). Care must be taken so that the animal does not lick or chew the patch if this method of delivery is used. Side effects include rashes and irritation at the application site.

Combined vasodilators. Prazosin (Minipress®) is a combined vasodilator (both venous and arterial). It is used to treat CHF, dilated cardiomyopathy, hypertension, and pulmonary hypertension. Side effects include hypotension and gastrointestinal problems.

TIP

How does the patch work? The transdermal system of drug delivery uses a small adhesive patch that is applied to the skin (fur is clipped/shaved). The transdermal system usually has four layers: an impermeable back to keep the drug from leaking, a reservoir containing the drug, a membrane with tiny holes to control the rate of drug release, and an adhesive layer or gel to keep it in place.

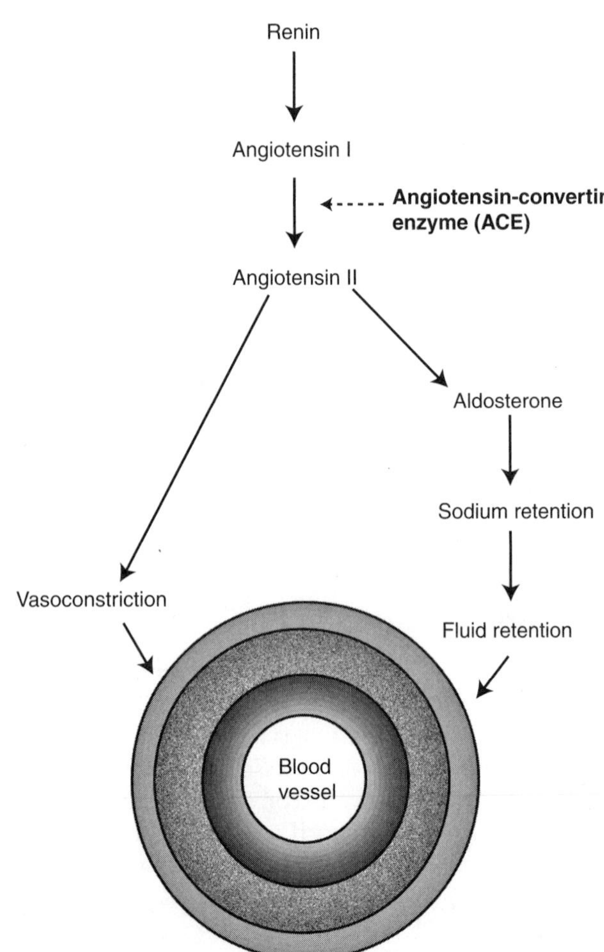

FIGURE 8-5 Renin-angiotensin system. ACE inhibitors act as antagonists of the renin-angiotensin system by interfering with the conversion of angiotensin I to angiotensin II. This ultimately results in vasodilation.

Losing Fluid

Diuretics are drugs that increase the volume of urine excreted by the kidneys and thus promote the release of water from the tissues. This process, called *diuresis*, lowers the fluid volume in tissues (extracellular fluid volume). Diuretics are used in the treatment of hypertension because they promote fluid loss and thus reduce blood pressure. These drugs are covered in Chapter 12 on the urinary system. Table 8-4 summarizes diuretic drugs.

TABLE 8-4 Diuretic Drugs

Diuretic Category	Action	Examples	Side Effects
Thiazide diuretics	Inhibit sodium and chloride reabsorption from the distal convoluted tubule of the nephron	• hydrochlorothiazide (HydroDIURIL®) • chlorothiazide (Diurel®) • hydroflumethiazide (Saluron®) • bendroflumethiazide (Corzide®)	Hypokalemia (low blood potassium levels)
Loop diuretics	Inhibit reabsorption of sodium and chloride in the loop of Henle, reducing the ability of the kidneys to concentrate urine; more potent than thiazides in promoting sodium and water excretion	• furosemide (Lasix®, Disal®, Diuride®) • ethacrynic acid (Edecrin®)	Hypokalemia
Potassium-sparing diuretics	Work in a variety of ways 1. Inhibit aldosterone (mineralocorticoid that normally increases sodium retention and thus water in the kidney) 2. Block sodium reabsorption in the distal convoluted tubule 3. Inhibit sodium reabsorption in the distal convoluted tubule	1. spironolactone (Aldactone®) 2. trimterene (Dyazide®) 3. amiloride (Midamor®)	Hyperkalemia
Osmotic diuretics	Large molecules that can be filtered by the glomerulus but have limited capability of being reabsorbed into the blood. High concentration of osmotic agent is left in the kidney tubule, which carries large amounts of fluid with it	• mannitol (Osmitrol®) • glycerin (Osmoglyn®)	Vomiting, electrolyte imbalance
Carbonic anhydrase inhibitors	Block the action of carbonic anhydrase, an enzyme that normally is involved in the reabsorption of sodium, potassium, bicarbonate, and water	• acetazolamide (Diamox®) • dichlorphenamide (Daranide®)	Gastrointestinal disturbances, CNS signs (depressed CNS activity)

TIP

Anticoagulants do not alter the size of an existing thrombus or clot.

DRUGS AFFECTING BLOOD

Drugs affecting blood are categorized by their effects and goals for use: to stop clots from forming, to stop bleeding, or to enhance blood and its products.

Clot Stopping

Anticoagulants are used to inhibit clot formation by inactivating one or more clotting factors (Figure 8-6). They are used clinically to inhibit clotting in catheters, to prevent

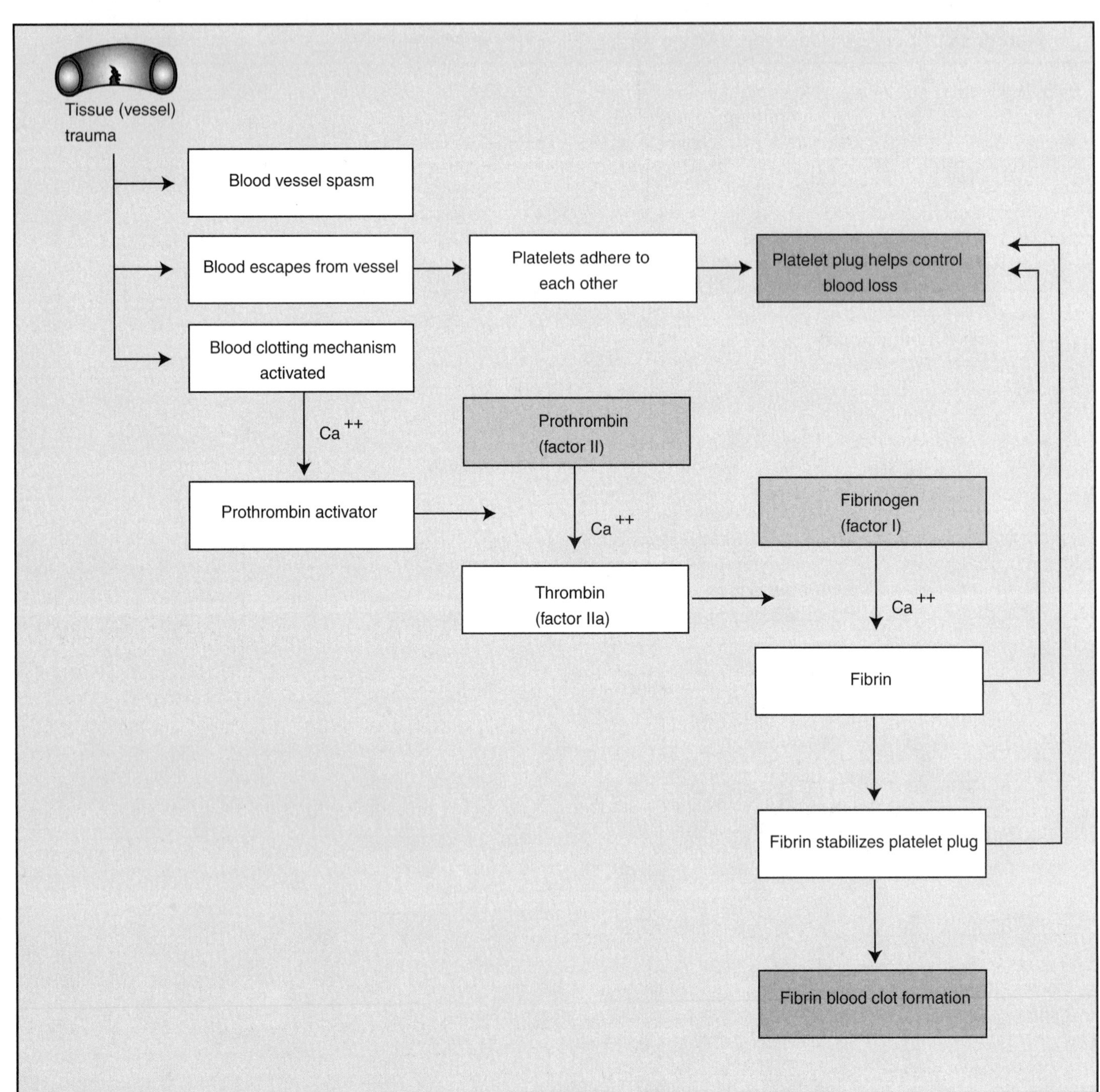

FIGURE 8-6 The clotting pathway

blood samples from clotting, to preserve blood transfusions, and to treat emboli that may occur. Examples of anticoagulants include the following.

Heparin. Heparin prevents the conversion of prothrombin to thrombin (Figure 8-7). Mainly used in the process of testing blood and in transfusion, heparin is used to treat heart disorders such as thromboembolism, where a clot, or thrombus, is constricting an artery of the heart. Heparin does not break down clots, but prevents them from getting bigger. It is also used to treat disseminated intravascular coagulation (DIC) and laminitis in horses. Side effects include bleeding and thrombocytopenia (decrease in blood clotting cells). Heparin overdose is treated with protamine sulfate. The strongly basic protamine combines with the strongly acidic heparin to form a stable complex with no anticoagulant activity.

TIP

Heparin is not effective when administered orally because it is rapidly inactivated by HCl in the stomach.

Ethylenediamine tetraacetic acid (EDTA). EDTA (Meta-Dose®) chelates calcium, which prevents clots from forming. It is used in lavender-topped blood tubes to prevent clotting of blood samples and as an injectable chelating agent in the treatment of lead poisoning. Side effects include vomiting, diarrhea, and renal toxicity.

Coumarin derivatives. This type of anticoagulant binds vitamin K_1, thereby inhibiting the formation of prothrombin. Coumarin derivatives are used for the long-term treatment of thromboembolisms and include the trade-named drugs Coumadin®, Dicumarol®, and Sofarin®. Overdose of coumarin can be treated with vitamin K_1. Side effects include bleeding and weakness.

Aspirin. Aspirin has antiplatelet activity because it inhibits the stickiness of platelets through the inhibition of thromboxane. It is used to prevent thromboembolism associated with heartworm disease in small animals and cardiomyopathies in cats. Side effects include bleeding, gastrointestinal bleeding, and vomiting. Caution should be exercised when using aspirin in cats (see Chapter 16).

Blood transfusion anticoagulants. Acid citrate dextrose (ACD®) and citrate phosphate dextrose adenine (CPDA-1®) are chemicals found in preservative bags used to collect blood for preservation. Both work by chelating calcium.

Bleeding Stopping

Hemostatic drugs help promote the clotting of blood. This category of drugs can be divided into parenteral and topical.

Parenteral hemostatic drugs include:

- *vitamin K_1* or *phytonadione,* a synthetic version of vitamin K_1. Vitamin K is involved in the clotting cascade; if it is not available, formation of some clotting factors is hindered. It is used in veterinary medicine for the treatment of rodenticide poisoning and for bleeding disorders. Brand-name formulations include AquaMEPHYTON®, Mephyton®, and Konakion®. Side effects include anaphylactic reactions and bleeding from the injection site.
- *protamine sulfate,* a basic protein used to treat heparin overdose and bracken fern poisoning in cattle. It is given slowly IV and is marketed generically. Side effects include hypotension and bradycardia.

TIP

The synthesis of new clotting factors after giving phytonadione takes 6 to 12 hours; therefore, emergency needs for clotting factors must be met by administering blood products.

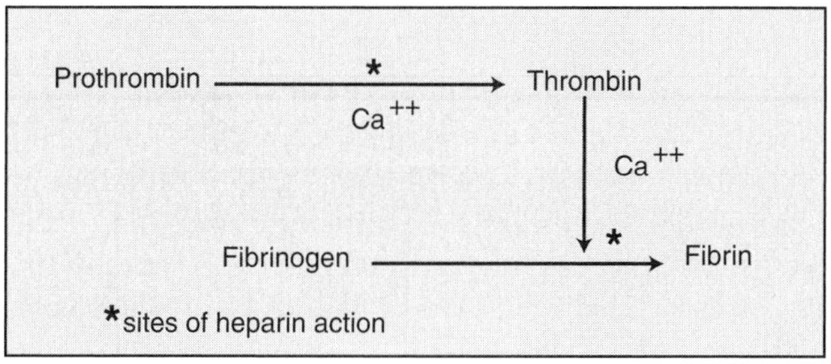

FIGURE 8-7 Heparin exerts its anticoagulant activity by interfering with the conversion of prothrombin to thrombin.

Topical hemostatic drugs are used to control capillary bleeding. They include:

- *silver nitrate sticks,* which act as an astringent (agent that constricts tissue) to stop capillary bleeding.
- *hemostat powder* (ferrous sulfate), to control capillary bleeding from superficial cuts and wounds and after dehorning of cattle, sheep, or goats. Irritation, redness, and swelling may be seen with this product, and if seen use of this product should be discontinued.
- *gelfoam gelatin sponge,* which provides hemostasis following surgical procedures. This sponge, available in a variety of sizes and shapes, is applied onto the bleeding tissue to quickly stop the flow of blood. The gelatin material is gradually absorbed by the body without inducing excessive scar formation.
- *thrombogen topical thrombin solution,* which acts directly to promote the conversion of fibrinogen to fibrin. Topical thrombin is derived from animal sources and is available as a sterile powder, which is usually reconstituted with sterile distilled water or isotonic saline before use. The speed with which the drug promotes clotting of blood is proportional to its concentration.

Blood-Enhancing Drugs

Blood-enhancing drugs include drugs that affect red blood cells. Erythrocytes or red blood cells (RBCs) are made in the bone marrow in response to a substance produced in the kidneys called **erythropoietin**. Red blood cells carry oxygen to tissues. Hemoglobin is the portion of the RBC that transports oxygen. Hemoglobin is made up of a heme protein and an iron-containing compound. Adequate amounts of iron and other vitamins and minerals are essential to the formation of RBCs. Drugs that affect the production or quality of RBCs include the following.

Iron. Iron compounds include iron dextran, ferrous sulfate, ferric hydroxide, and others. Iron compounds are used clinically in the treatment of baby pig anemia and as a nutritional supplement. Iron comes in both oral and injectable forms; examples of brand-name forms are Iron Dextran Complex Injection (Armedexan®, Ferrextran®, Pigdex 100®, Imposil®), Iron-Gard 100®, and Purina Oral Pigemia®. Side effects include discoloration at the injection site and muscle weakness.

Erythropoietin. This protein, made by the kidneys, stimulates the differentiation of bone marrow stem cells to form RBCs. It is used to treat anemia in animals with chronic renal failure. It is marketed as a genetically engineered hemopoietin (epoetin alpha) under the trade names Epogen® and Procrit®. Allergic reactions are sometimes seen with erythropoietin products.

SUMMARY

The functions of the cardiovascular system include the delivery of oxygen, nutrients, and hormones to body tissues and the transportation of wastes to waste removal systems of the body. Abnormalities that can affect the heart include decreased myocardial contraction, changes in heart rate, and changes in heart rhythm.

Positive inotropes are drugs that improve the strength of myocardial contraction. Examples of positive inotropes are cardiac glycosides and catecholamines. Antiarrhythmics are drugs that correct abnormal heart rhythms and include local anesthetics, membrane stabilizers, beta-adrenergic blockers, drugs that prolong the action potential, and calcium-channel blockers. Vasodilators counteract the vasoconstriction associated with heart disease and may dilate veins only, arterioles only, or both veins and arterioles (called combined or mixed vasodilators). Examples of vasodilators are ACE inhibitors, arteriole dilators, venodilators, and combined vasodilators. Diuretics decrease the fluid volume that a diseased heart is required to pump, thereby decreasing the stress on the heart.

Anticoagulants inhibit clot formation; they include heparin, EDTA, coumarin derivatives, aspirin, and chemicals used in blood preservation for transfusion. Hemostatic drugs are substances that help promote clotting of blood. This category of drugs includes vitamin K_1 and protamine sulfate (parenteral) and silver nitrate sticks, hemostat powder (ferrous sulfate), gelfoam gelatin sponge, and thrombogen topical thrombin solution (topical). Blood-enhancing drugs affect red blood cells. Drugs that affect the production or quality of RBCs include iron and erythropoietin.

CHAPTER REFERENCES

Amundson Romich, J. (2000). *An illustrated guide to veterinary medical terminology*. Clifton Park, NY: Thomson Delmar Learning.

Delmar's A-Z NDR-97 nurses drug reference. Clifton Park, NY: Thomson Delmar Learning.

Reiss, B., & Evans, M. (2002), *Pharmacological aspects of nursing care* (sixth edition, revised by Bonita E. Broyles, pp. 534–629). Clifton Park, NY: Thomson Delmar Learning.

Veterinary pharmaceuticals and biologicals (12th ed.). (2001). Lenexa, KS: Veterinary Healthcare Communications.

Veterinary values (5th ed.). (1998). Lenexa, KS: Veterinary Healthcare Communications.

Wynn, R. (1999). *Veterinary pharmacology*. Cambridge, MA: Harcourt Learning Direct.

CHAPTER REVIEW

Matching

Match the drug name with its action.

1. ______ digoxin
2. ______ propranolol
3. ______ prazosin
4. ______ coumarin
5. ______ epinephrine
6. ______ protamine sulfate
7. ______ hydralazine
8. ______ silver nitrate
9. ______ heparin
10. ______ lidocaine

a. cardiac glycoside that increases the strength of contractions and decreases heart rate
b. catecholamine that increases force of myocardial contraction
c. membrane-stabilizing antiarrhythmic used to control PVCs and ventricular tachycardia
d. beta-adrenergic blocker used to treat hypertrophic cardiomyopathy
e. arteriole dilator
f. combined vasodilator
g. anticoagulant that prevents the conversion of prothrombin to thrombin
h. hemostatic drug used to treat heparin overdose
i. topical hemostatic drug
j. anticoagulant that binds vitamin K_1

Multiple Choice

Choose the one best answer.

11. Coumarin toxicity, found when animals ingest rat poisoning, is treated with
   a. heparin.
   b. vitamin K_1.
   c. EDTA.
   d. silver nitrate sticks.

12. Erythropoietin is used to
   a. increase WBC count in infectious disease.
   b. increase urine loss to decrease fluid retention.
   c. stop bleeding.
   d. increase RBC production.

13. Which group of drugs prevents the conversion of angiotensin I to angiotensin II?
   a. arteriole dilators
   b. venodilators
   c. ACE inhibitors
   d. combined vasodilators

14. Which antiarrhythmic drug is used to treat ventricular fibrillation?
   a. propranolol
   b. bretylium
   c. verapamil
   d. lidocaine

15. Catecholamines function as what autonomic nervous system response?
   a. sympathetic
   b. parasympathetic
   c. cholinergic
   d. sympathetic-blocking

16. What drug is manufactured in both 1:1000 and 1:10,000 concentrations?
   a. epinephrine
   b. isoproterenol
   c. dopamine
   d. dobutamine

17. Which cardiovascular drug should be monitored for toxicity levels through blood testing?
   a. digoxin
   b. epinephrine
   c. dopamine
   d. lidocaine

18. The use of diuretics in treating heart disease is believed to
   a. increase preload.
   b. decrease preload.
   c. increase afterload.
   d. decrease afterload.

19. The use of vasodilators in treating heart disease is believed to
   a. increase preload.
   b. decrease preload.
   c. increase afterload.
   d. decrease afterload.

20. Which of the following are false regarding the ECG?
   a. The P wave represents atrial depolarization.
   b. The Q-R-S complex represents ventricular depolarization.
   c. The T wave represents ventricular repolarization.
   d. None of the above are false.

Case Studies

21. A 12-yr.-old M/N Boston Terrier (14#) (Figure 8-8) comes into the clinic for exercise intolerance and coughing at night. On physical examination, his TPR are normal; however, he has developed ascites and is very lethargic. The veterinarian requests chest X-rays and an ECG. The tests reveal that the dog is in heart failure.
   a. What medication group could be given to this dog to reduce preload (think of reducing fluid volume that the heart must pump around)?
   b. What medication group could be given to this dog to reduce afterload (think of reducing the tension against which the heart has to pump)?

22. A 7-yr.-old F Poodle (10#) (Figure 8-9) presents to the clinic with weakness. On physical examination, the veterinarian notes an arrhythmia, so she hooks the dog up to an ECG. The ECG reveals that the dog has a ventricular arrhythmia.
   a. Considering that this is an emergency, what drug could be given IV?
   b. What drug could this dog be sent home on once she is stabilized?

FIGURE 8-8 Boston Terrier. Photo by Isabelle Francais

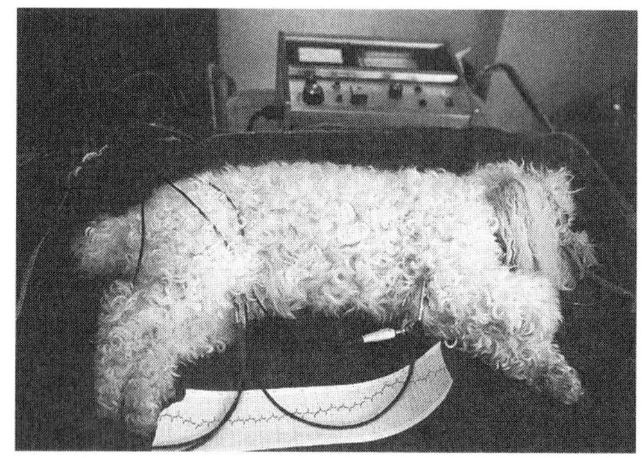

FIGURE 8-9 Poodle undergoing an ECG. *Source:* Lodi Veterinary Hospital, S.C.

CHAPTER 9

Respiratory System Drugs

OBJECTIVES

Upon completion of this chapter, the learner should be able to

- describe the basic anatomy and physiology of the respiratory tract.
- differentiate between ventilation and respiration.
- describe the clinical usages of the following drugs in the treatment of respiratory disease: expectorants, mucolytic drugs, antitussives, decongestants, bronchodilators, antihistamines, respiratory stimulants

KEY TERMS

expectorant
mucolytic
antitussive
decongestant
bronchodilator
cholinergic blocking agent
beta-2-adrenergic agonist
antihistamine
respiratory stimulant

INTRODUCTION

A four-year-old male (M) Beagle is brought into the clinic with a two-day history of a harsh, honking cough. When questioned, the owner says that the dog was boarded last week while the family was on vacation. A physical exam shows that the dog's TPR are normal, and he does not appear dehydrated. While in the clinic, he coughs quite a bit, and the veterinarian determines that this dog probably has kennel cough (a disease that can be caused by the bacterium *Bordetella bronchiseptica* and several viruses, including canine parainfluenza virus). Because this disease is usually self-limiting, no treatment is initiated. The owner, however, tells you that the dog is keeping the family awake all night with the incessant coughing. He wants to know if there is anything he can give the dog to control its cough. Is this a wise thing to do? Should all coughs be suppressed? The desire to suppress or not suppress a cough is based on the type of cough it is. Can you explain the difference to the owner of this beagle? Can you explain the rationale behind the treatment chosen for this dog? What drugs are available for this dog, and what are their potential side effects?

BASIC RESPIRATORY ANATOMY AND PHYSIOLOGY

The *respiratory system* is the body system that brings oxygen from the air into the body for delivery via the blood to the cells. *Respiration* is the exchange of gases (oxygen and carbon dioxide) between the atmosphere and the cells of the body. *Ventilation* is the term used to describe the bringing in of fresh air.

The respiratory system is divided into two parts: the upper respiratory tract, consisting of the nostrils, nose, nasal cavities, pharynx, and larynx; and the lower respiratory tract, consisting of the trachea, bronchi, bronchioles, and alveoli (Figures 9-1 and 9-2). Several important anatomical features play a role in respiratory therapies.

- The upper respiratory tract is lined with epithelial cells that contain cilia (microscopic hairs) and goblet cells. Goblet cells secrete mucus, which can trap foreign particles that are then moved out of the respiratory tract via the movement of cilia.
- The bronchioles are lined with smooth muscle that can be dilated with sympathetic nervous stimulation and constricted with parasympathetic nervous stimulation.
- The alveoli (small, terminal, saclike structures where oxygen and carbon dioxide exchange) produce a chemical called *surfactant*. Surfactant keeps the alveoli open by reducing surface tension of these small sacs.
- Irritation to the respiratory tract is sensed by cough receptors in the respiratory tract that send a signal to the cough center of the brainstem. The brainstem in turn sends impulses to the respiratory muscles to produce a forceful expiration or cough. Nonproductive or dry coughs are not accompanied by the expulsion of fluid or material from the respiratory tract. Productive coughs help expel mucus and foreign material from the respiratory tract. It is usually recommended that only nonproductive coughs be suppressed, because productive coughs help clear the respiratory tract.

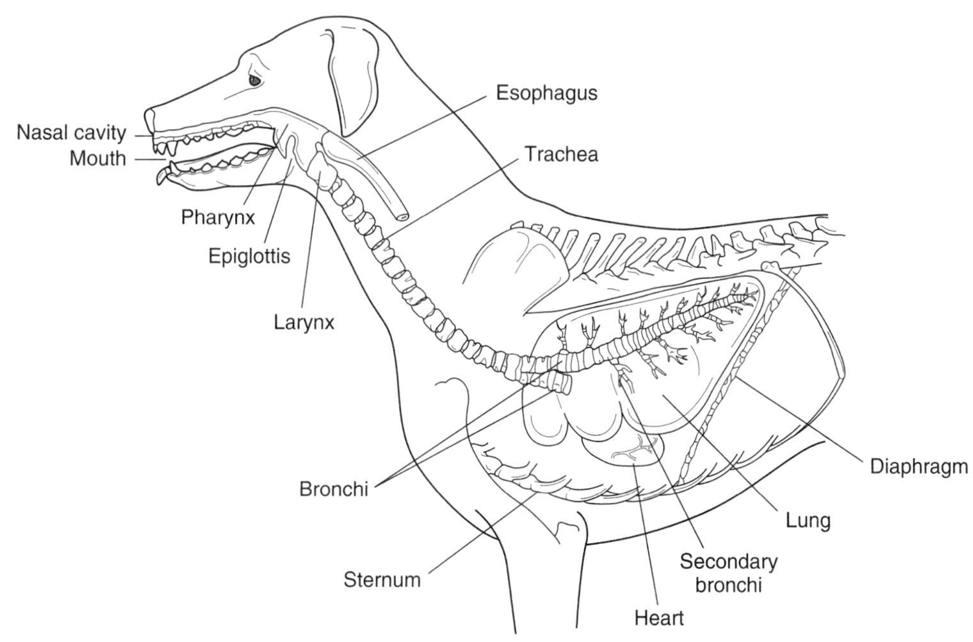

FIGURE 9-1 Structures of the respiratory system

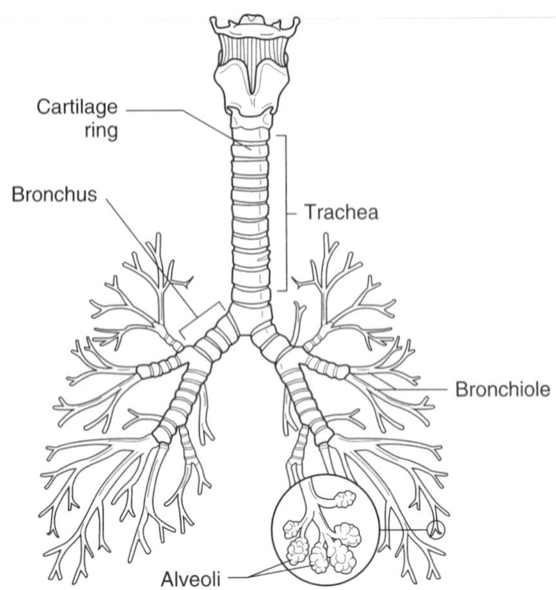

FIGURE 9-2 Lower respiratory tract structures

RESPIRATORY DRUGS

Respiratory drugs include expectorants, mucolytics, antitussives, decongestants, bronchodilators, and antihistamines. The drugs in these categories affect the respiratory tract in different ways.

Cough-Causing (Expectorants)

Expectorants increase the flow of respiratory secretions to allow material to be coughed up from the lungs. Expectorants do this by increasing the fluidity of mucus, which is more effectively coughed up than thicker mucus. Expectorants are believed to work by acting on the goblet cells or by reducing the stickiness of the mucus. Increased secretory activity also helps keep dry, irritated tissue moist, thus protecting it from further trauma. Glyceryl guaiacolate (guaifenesin) is a secretory expectorant used orally in humans to increase secretions from the airway and clear it of fluid. Guaifenesin is also used in horses as a muscle relaxant, as part of a general anesthesia protocol. Trade names of guaifenesin products include Guailazin® and Gecolate®.

TIP

A *nebulizer* is a device used to change liquid medication to an aerosol form. The aerosolized medication is beneficial when inhaled and deposited into the lung. The effects of nebulized medication are almost immediate.

Mucus-Breaking (Mucolytics)

Mucolytic drugs decrease the viscosity or thickness of respiratory secretions. Acetylcysteine (Mucomyst®, Mucosol®) is a mucolytic expectorant used to break up thick mucoid secretions in the airway to promote better respiration. Acetylcysteine is administered orally, intravenously, or as an aerosol through nebulization. Acetylcysteine is also used to treat acetaminophen toxicity in cats; it helps metabolize the acetaminophen and decreases its toxicity in the liver.

TIP

Antitussives are used to suppress nonproductive coughs. A productive cough is usually not suppressed.

Cough-Controlling (Antitussives)

Antitussives suppress coughing. Antitussives can be centrally acting (working on the cough centers of the brainstem) or locally acting (soothing irritation to the mucosal lining of the respiratory tract that initiates coughing). Locally acting antitussives are usually syrups or lozenges and are not used commonly in veterinary practice.

Butorphanol (Torbugesic®, Torbutrol®) is a centrally acting antitussive that is available by injection (with differing concentrations per ml) and oral tablet form. It is now classified as a C-IV controlled substance; therefore, careful documentation of its use is required. Butorphanol is also used as a preanesthetic and analgesic, because its main side effects are sedation and ataxia.

Hydrocodone (Hycodan®, Tussigon®) is a C-III controlled substance that is a centrally acting narcotic. It comes in tablet and syrup forms, and is used primarily for harsh, nonproductive coughs. Side effects include sedation and slowing of gastrointestinal motility that may result in constipation.

Codeine is a C-IV, centrally acting antitussive that may or may not be mixed with aspirin or acetaminophen. Codeine is available under many generic and trade names in injectable, syrup, and tablet forms. Its side effects also include sedation and decreased gastrointestinal motility.

Dextromethorphan is a non-narcotic, centrally acting antitussive that is chemically similar to codeine. Dextromethorphan is not addictive like codeine and is sold over the counter. It is available in syrups, but owners should be advised that products containing other "cold products" in the syrup might be harmful to animals. Dextromethorphan is generally not very effective in nonhuman species.

Trimeprazine (is a centrally acting antitussive) and prednisolone (a corticosteroid) are found in the combination drug Temaril-P®. This drug has antitussive and antipruritic effects and is available in spansules and tablets. Side effects include sedation, depression, and hypotension.

TIP

Spansules are medicine-containing capsules that are coated to slow the dissolution rate, so that the medicine is delivered more slowly.

Congestion-Reducing (Decongestants)

Decongestants decrease the congestion of nasal passages by reducing swelling. Decongestants may be given by spray or orally as a liquid or tablet. These products have limited use in veterinary practice, but have been used to help treat feline upper respiratory tract disease. Decongestants include phenylephrine (Neo-Synephrine®) and pseudoephedrine (Sudafed®). Phenylephrine has cardio-stimulatory properties and should not be used in animals with hypertension or tachycardia (increased heart rate).

Bronchi-Widening (Bronchodilators)

TIP

Bronchodilators ↓ airway resistance and ↑ airflow.

Bronchodilators widen the lumen of the bronchi and counteract bronchoconstriction. Bronchoconstriction occurs when acetylcholine is released by the parasympathetic nerves, causing increased respiratory secretions; histamine is released from mast cells through allergic or inflammatory reactions, causing it to bind to receptors on bronchiole smooth muscle; or beta-2-adrenergic receptors are blocked as a result of other drug treatments. **Cholinergic blocking agents**, more commonly known as *anticholinergics*, produce bronchodilation by counteracting the first mechanism: they bind to acetylcholine receptors and prevent bronchoconstriction. Cholinergic blockers include aminopentamide (Centrine®), atropine (Atropine Sulfate for Injection), and glycopyrrolate (Robinul-V®). Side effects include dry mouth, dry eyes, and tachycardia. Atropine has more side effects, including vomiting, constipation, and urinary retention.

Another group of bronchodilators works by stimulating beta receptors and hence are referred to as **beta-2-adrenergic agonists**. Beta-2 receptors are involved in bronchodilation and stabilization of mast cells (reducing histamine release). Bronchodilators in this category are epinephrine (which is usually reserved for life-threatening situations and has beta-1 and alpha-1 activity), isoproteronol (which also causes beta-1 stimulation and thus affects the heart), and terbutaline and albuterol (which have primarily beta-2 activity and thus little effect on the heart). These drugs go by the following generic and trade names: epinephrine (Adrenalin Chloride® and generically labeled forms), isoproteronol (Isuprel®), terbutaline (Brethine®), and albuterol (Proventil®, Ventolin®). Side effects include tachycardia (increased heart rate), and CNS excitement and weakness.

TIP

Some theophylline products are sustained release. The reported release rates for humans may or may not correlate to actual release rates in veterinary patients.

A last group of bronchodilators are the methylxanthines. Methylxanthines work by inhibiting an enzyme in smooth muscle cells. The enzyme normally promotes bronchoconstriction, but when it is inhibited bronchodilation occurs. Aminophylline

(Aminophyllin®), theophylline (Slo-BID Gyrocaps®, Theo-dur®), caffeine, and theobromine (found in chocolate) are examples of methylxanthines. Side effects include CNS stimulation and gastrointestinal irritation.

TIP

Some over-the-counter antihistamines may contain other active ingredients. Take care to counsel your clients to choose OTC products that contain only the desired antihistamine compound.

Histamine-Blocking (Antihistamines)

Antihistamines block the effects of histamine, a chemical released from mast cells that combines with H-1 receptors on bronchiole smooth muscle to cause bronchoconstriction (histamine also works on other body systems) (Figure 9-3). Antihistamines are usually used in prevention of respiratory problems such as heaves (in horses) and feline asthma. Generic names for antihistamines usually end with *-amine*. Examples in this category include diphenhydramine (Benadryl®) and chlorpheniramine (ChloroTrimeton®). Side effects include CNS depression and anticholinergic effects (dry mouth, urinary retention).

Respiratory Stimulants

Respiratory stimulants are drugs that stimulate the animal to increase its respirations. *Doxapram* is a central nervous stimulant that is usually used in neonates to stimulate respiration after Cesarean section or dystocia. Doxapram is also used to restore reflexes after anesthesia. It is available for injection as Dopram-V®. Side effects include hypertension, arrhythmias, and seizures.

Naloxone (Narcan®) is used to stimulate respiration following narcotic overdose and is available in injectable forms for IV, IM, and SQ use. Side effects are rare. The duration of action of naloxone may be shorter than that of the narcotic being reversed, so the patient should be closely monitored, in case additional doses are needed.

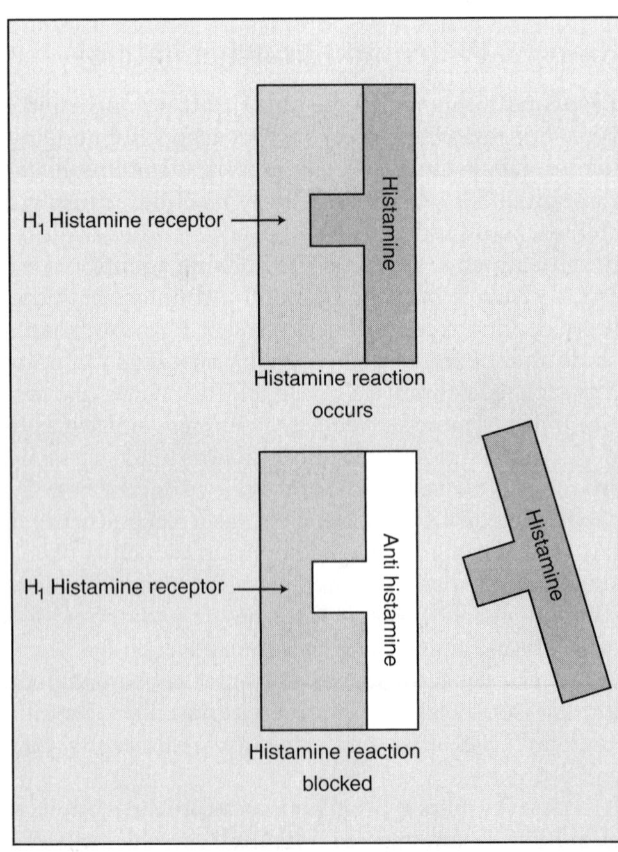

FIGURE 9-3 Mechanism of antihistamine action

Yohimbine is a respiratory stimulant used to reverse xylazine administration or overdose. It is available as Yobine® and Antagonil® in an injectable form for IV and IM use. Side effects include CNS stimulation and muscle tremors.

Other Drugs Used in the Respiratory System

Other drugs used in the respiratory system include corticosteroids (Chapter 16), antimicrobials (Chapter 14), and diuretics (Chapter 12). These drugs are addressed under different chapters and information regarding these drug categories is found in those chapters. The respiratory drugs covered in this chapter are summarized in Table 9-1.

TABLE 9-1 Respiratory Drugs Covered in this Chapter

Drug Category	Action	Examples
Expectorants	Increase the flow of respiratory secretions to allow the coughing up of material from lungs.	• guaifenesin (Guailaxin®, Gecolate®, Robitussin Cough Syrup®, Triaminic®)
Mucolytics	Decrease viscosity or thickness of respiratory secretions.	• acetylcysteine (Mucomyst®)
Antitussives	Work centrally by affecting cough centers of brainstem. Work locally by soothing irritation to mucous lining of respiratory tract (limited use in veterinary practice).	• butorphanol (Torbugesic®, Torbutrol®) • hydrocodone (Hycodan®) • codeine • dextromethorphan (Dimetapp®) • trimeprazine and prednisolone (Temaril-P®)
Decongestants	Decrease the congestion of nasal passages by reducing swelling (limited veterinary use).	• phenylephrine (Neo-Synephrine®) • pseudoephedrine (Sudafed®)
Bronchodilators	Either cholinergic blockers (work by counteracting the action of acetylcholine by binding to acetylcholine receptors), beta-2-adrenergic agonists (stimulate $beta_2$ receptors that cause bronchodilation), or methylxanthines (inhibit an enzyme in smooth muscle cells that normally causes bronchoconstriction)	Cholinergic blockers: • aminopentamide (Centrine®) • atropine (Atropine Sulfate for Injection) • glycopyrrolate (Robinul-V®) Beta-2-adrenergic agonists: • epinephrine (Adrenalin Chloride®) • isoproteronol (Ventoline®) • albuterol (Alupent®, Proventil®) Methylxanthines: • aminophylline (Aminophyllin®) • theophylline (Theo-Dur®)

SUMMARY

Respiratory drugs include expectorants, mucolytics, antitussives, decongestants, bronchodilators, antihistamines, and respiratory stimulants. Expectorants increase the flow of respiratory secretions to allow material to be coughed up from the lungs. Guaifenesin is an expectorant. Mucolytic drugs decrease the viscosity or thickness of respiratory secretions. Acetylcysteine is a mucolytic drug that is also used for acetaminophen toxicity in cats. Antitussives suppress coughing. Antitussives can be centrally acting (working on the cough centers of the brainstem) or locally acting (soothing irritation to the mucosal lining of the respiratory tract that initiates coughing). Examples of centrally acting antitussives are butorphanol, hydrocodone, codeine, trimeprazine, and dextromethorphan. Decongestants decrease the congestion of nasal passages by reducing swelling; they have limited use in veterinary practice. Bronchodilators expand the bronchi and counteract bronchoconstriction. Bronchodilators are cholinergic blockers, beta-2-adrenergic agonists, or methylxanthines. Cholinergic blockers include atropine, beta-2-adrenergic agonists include terbutaline, and methylxanthines include aminophylline and theophylline. Antihistamines block the effects of histamine, a chemical released from mast cells that combines with H-1 receptors on bronchiole smooth muscle to cause bronchoconstriction. The generic names of antihistamines usually end with *-amine*. Respiratory stimulants include doxapram, naloxone, and yohimbine.

CHAPTER REFERENCES

Amundson Romich, J. (2000). *An illustrated guide to veterinary medical terminology*. Clifton Park, NY: Thomson Delmar Learning.

Delmar's A-Z NDR-97 nurses drug reference. Clifton Park, NY: Thomson Delmar Learning.

Reiss, B., & Evans, M. (2002), *Pharmacological aspects of nursing care* (6th ed., revised by Bonita E. Broyles, pp. 462–503). Clifton Park, NY: Thomson Delmar Learning.

Veterinary pharmaceuticals and biologicals (12th ed.). (2001). Lenexa, KS: Veterinary Healthcare Communications.

Veterinary values (5th ed.). (1998). Lenexa, KS: Veterinary Healthcare Communications.

Wynn, R. (1999). *Veterinary pharmacology*. Cambridge, MA: Harcourt Learning Direct.

CHAPTER REVIEW

Matching

Match the drug name with its action.

1. ______ guiafenesin
2. ______ theophylline
3. ______ terbutaline
4. ______ acetylcysteine
5. ______ doxapram
6. ______ butorphanol
7. ______ phenylephrine
8. ______ atropine
9. ______ dextromethorphan
10. ______ diphenhydramine

a. respiratory stimulant
b. methylxanthine
c. beta-2-adrenergic agonist
d. expectorant
e. mucolytic
f. centrally acting antitussive
g. decongestant
h. anticholinergic (cholinergic blocker)
i. non-narcotic centrally acting antitussive
j. antihistamine

Multiple Choice

Choose the one best answer.

11. What type of respiratory drug inhibits cough production?
    a. antihistamine
    b. antitussive
    c. decongestant
    d. expectorant
12. What drug is a mucolytic and also is used to treat acetaminophen toxicity?
    a. guaifenesin
    b. hydrocodone
    c. theophylline
    d. acetylcysteine
13. Which of the following are controlled substances?
    a. theophylline
    b. aminophylline
    c. butorphanol
    d. guaifenesin
14. Which respiratory drug can be used to prevent respiratory problems?
    a. expectorant
    b. mucolytic
    c. antihistamine
    d. antitussive
15. Which of the following is an antihistamine?
    a. diphenhydramine
    b. hydrocodone
    c. Temaril-P®
    d. butorphanol
16. Which group of bronchodilators works by inhibiting an enzyme in smooth muscle cells that normally causes vasoconstriction?
    a. anticholinergics
    b. beta-2-adrenergic agonists
    c. methylxanthines
    d. all of the above work in a similar fashion
17. Which of the following drugs are used to stimulate respiration in neonates after a cesarean section?
    a. naloxone
    b. doxapram
    c. yohimbine
    d. theophylline
18. Which respiratory drug category decreases the viscosity of respiratory secretions?
    a. expectorant
    b. mucolytic
    c. antitussive
    d. decongestant
19. The cough center is located in what part of the brain?
    a. cerebrum
    b. cerebellum
    c. brainstem
    d. meninges

20. Which antitussive combination drug also has a corticosteroid in it?
    a. acetaminophen with codeine
    b. hydrocodone
    c. Temaril-P®
    d. butorphanol

Case Study

A 2-yr.-old F/S Chihuahua (10#) (Figure 9-4) presents to the clinic with a dry, harsh cough. On physical examination, T=104°F, HR = 120 bpm, and RR = 24 breaths/min. Upon auscultation of the lung fields, harsh referred sounds are heard. The trachea is palpated to determine if the dog has a collapsing trachea, but she does not. The owner states that this dog was boarded while the owner was on vacation. The veterinarian suspects that this dog has kennel cough because she was not vaccinated before she was kenneled.

21. What category of drug might the veterinarian prescribe for this dog?
22. The veterinarian knows that most cases of kennel cough resolve in three to seven days. She decides to prescribe butorphanol for this dog, as well as antibiotics to treat a secondary bacterial infection she believes the dog has. What type of antitussive is butorphanol?
23. Using your answer to question 22, what side effect of butorphanol do you think this owner should be warned about?
24. Is butorphanol a controlled substance?

FIGURE 9-4 Long-haired Chihuahua. Photo by Isabelle Francais

CHAPTER 10

Hormonal and Reproductive Drugs

OBJECTIVES

Upon completion of this chapter, the learner should be able to

- describe the basic anatomy and physiology of the endocrine system.
- describe how endocrine function is regulated.
- describe the functions of pituitary gland hormones and how they influence other glands and organs.
- describe the usual disease mechanisms for endocrine disease.
- describe the pharmaceutical management of endocrine disease.
- describe the role of insulin and glucagon in regulating blood glucose levels.
- describe the role of the thyroid gland in regulating metabolism.
- list the stages of estrous and describe the function of the hormones in each stage.
- describe the functions of FSH, LH, GnRH, estrogen, progesterone, and prostaglandins in the estrous cycle.
- differentiate between the follicular and luteal phases of the estrous cycle.
- describe the role of testosterone as a growth promotant.
- outline the role of endocrine drugs in regulating the reproductive status of animals, including estrus cycle, abortion, and pregnancy maintenance.
- describe uses of reproductive drugs other than influencing reproductive status.

KEY TERMS

hormone
insulin
glucagon
diabetes mellitus
thyroxine
tri-iodothyronine
hypothyroidism
hyperthyroidism
mineralocorticoid
glucocorticoid
adrenocortical insufficiency
hyperadrenocorticism
proestrus
estrus
metestrus
diestrus
anestrus
follicular phase
luteal phase
androgen
testosterone
estrogen
progesterone
prostaglandin
gonadotropin
growth promotants
anabolic steroids

INTRODUCTION

A five-year-old M/N Dachshund comes into the clinic for an examination. He has been lethargic lately and seems to be gaining weight. When you weigh him, you discover that he has not gained weight since his last appointment, but that his abdomen hangs lower than it used to. He is still a bit overweight, but has been for many years. The owner tells you that he has been drinking a lot and urinating a lot. On physical exam, you find that TPR are normal, but note areas of alopecia (abnormal hair loss). The veterinarian orders some blood tests to determine what is going on with this dog. What tests do you think will be run on this dog's blood? What are some examples of endocrine diseases that this dog might have? Are the treatments for these endocrine diseases safe? What do you tell the owner? Are you able to adequately explain endocrine diseases to clients? Can you explain to the client with the Dachshund the disease process occurring in this dog and how the treatment proposed can help this pet?

BASIC ENDOCRINE SYSTEM ANATOMY AND PHYSIOLOGY

The endocrine system is composed of ductless glands that secrete chemical messengers called *hormones* into the bloodstream. Hormones enter the bloodstream and are carried throughout the body to affect a variety of tissues and organs. Tissues and organs that the hormones act upon are called *target organs*. The glands of the endocrine system (Figure 10-1) include:

- one pituitary gland (with two lobes)
- one thyroid gland (right and left lobes fused ventrally)
- four parathyroid glands (in most species)
- two adrenal glands
- one pancreas
- one thymus
- one pineal gland
- two gonads (ovaries in females, testes in males)

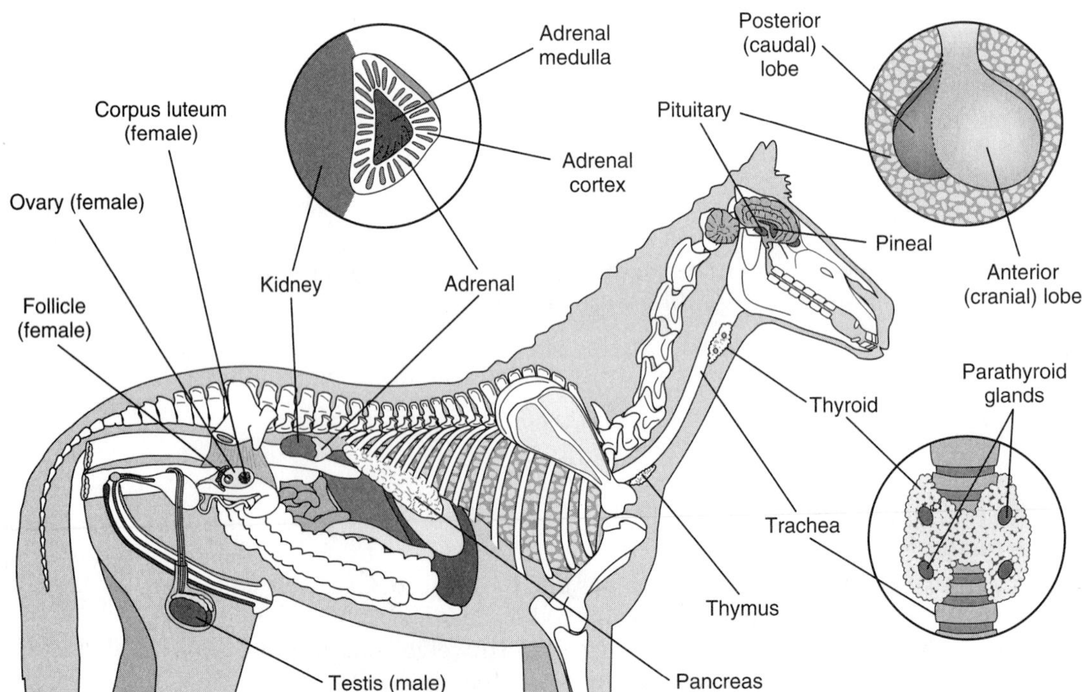

FIGURE 10-1
Endocrine gland locations

REGULATION OF THE ENDOCRINE SYSTEM

The endocrine system is controlled by a feedback mechanism that includes the hypothalamus, pituitary gland, and the other endocrine glands (Figure 10-2). The pituitary gland is located at the base of the brain below the hypothalamus. The hypothalamus secretes releasing and inhibiting factors that affect the release of substances from the pituitary gland. When signaled by releasing factors from the hypothalamus, the pituitary gland secretes hormones that control other endocrine glands.

Two types of feedback loops are involved in regulating the endocrine system. Feedback loops are described as either negative or positive (Figure 10-3). An example of the negative feedback loop in action is the release of corticosteroid from the cortex of the adrenal gland. If there are low blood levels of corticosteroid, a signal is sent to the hypothalamus. The hypothalamus then secretes a releasing factor that signals the anterior pituitary gland to secrete adrenocorticotropic hormone (ACTH). ACTH is delivered via the bloodstream to the adrenal gland and causes the adrenal cortex to produce more corticosteroid. If too much corticosteroid is produced, a signal is sent to the hypothalamus to secrete inhibiting factor to the anterior pituitary gland. The anterior pituitary gland will then decrease its production and release of ACTH, thus lowering the amount of corticosteroid ultimately produced by the adrenal cortex.

Positive feedback loops occur less frequently in the body. An example is the release of oxytocin from the posterior pituitary gland during parturition. The presence of the fetus in the pelvic canal causes oxytocin to be released from the posterior pituitary. Oxytocin signals the uterus to contract. The uterine contraction signals the hypothalamus to produce more releasing factor. The releasing factor signals the posterior pituitary gland to secrete more stored oxytocin (oxytocin is actually produced in the hypothalamus and stored in the posterior pituitary). Continued uterine contractions signal more releasing factor to be released from the hypothalamus, resulting in the release of more oxytocin from the posterior pituitary in an effort to complete labor.

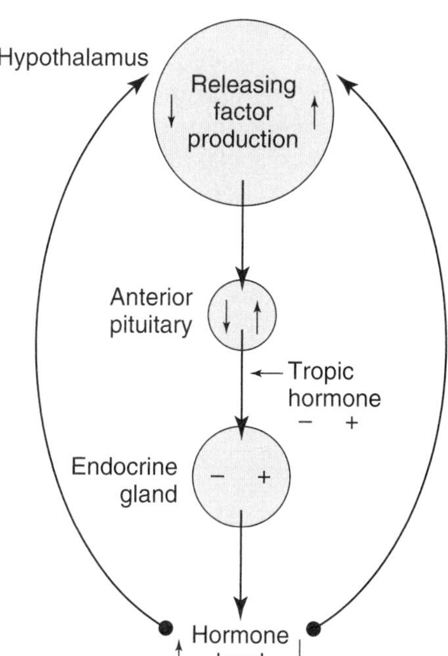

FIGURE 10-2 Secretions of the pituitary gland

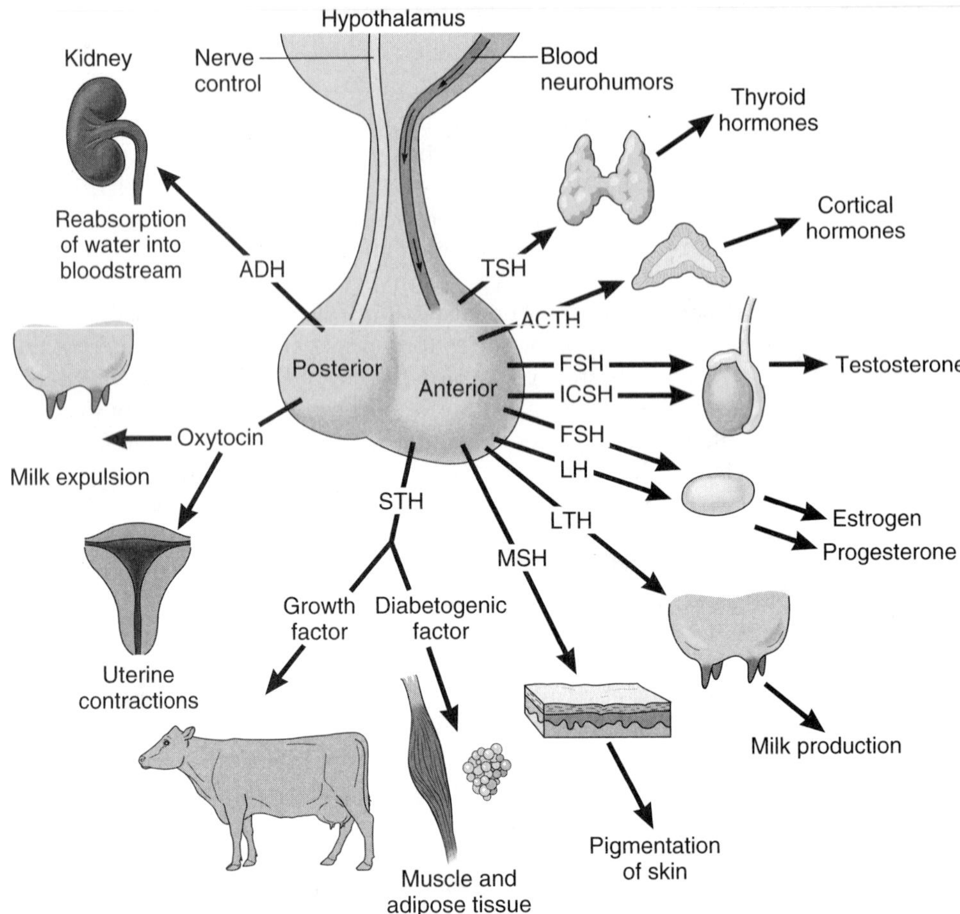

FIGURE 10-3 Feedback control mechanism. Positive and negative feedback mechanisms control the levels of a particular hormone in the blood by secreting releasing factors or inhibiting factors that affect hormone release.

HORMONAL DRUGS

A **hormone** is a chemical substance produced by cells in one part of the body and transported to another part of the body where it influences and regulates cellular activity and organ function. Veterinarians use hormones and hormone-like preparations for two purposes: to replace hormones diminished by disease, or to block excess hormones created by disease.

Master Gland Control

The pituitary gland, also known as the master gland, has two divisions: the anterior (front) and the posterior (rear). Anterior pituitary hormones regulate an animal's growth and the proper functioning of its thyroid, gonads (testes and ovaries), and various endocrine glands. The posterior pituitary secretes antidiuretic hormone (ADH, vasopressin) and oxytocin (labor-producing hormone) that have been produced by the hypothalamus. Table 10-1 summarizes the pituitary gland hormones and their functions.

Anterior pituitary hormones are used in veterinary practice for the following:

- *Thyroid stimulating hormone* (TSH) is used in the diagnosis of primary hypothyroidism.

TABLE 10-1 Pituitary Gland Hormones and Functions

Portion of Pituitary Gland	Hormone Produced	Hormone Action
Anterior Pituitary	thyroid stimulating hormone (TSH)	Augments growth and secretions of the thyroid gland
	adrenocorticotropic hormone (ACTH)	Augments the growth and secretions of the adrenal cortex (corticosteroids and mineralocorticoids)
	follicle stimulating hormone (FSH); FSH is a type of gonadotropic hormone	Augments secretion of estrogen and growth of eggs in the ovaries (female) and the production of sperm (male)
	luteinizing hormone (LH); LH is a type of gonadotropic hormone	Augments ovulation and aids in maintenance of pregnancy (females)
	interstitial cell-stimulating hormone (ICSH); ICSH is now considered to be LH and is a type of gonadotropic hormone	Stimulates testosterone secretion (males)
	prolactin (also known as lactogenic hormone or luteotropin)	Augments milk secretion and influences maternal behavior
	growth hormone (GH); also known as somatotropin	Accelerates body growth
	melanocyte stimulating hormone (MSH)	Augments skin pigmentation
Posterior Pituitary	antidiuretic hormone (ADH); also known as vasopressin	Maintains water balance in the body by augmenting water reabsorption in the kidneys
	oxytocin	Stimulates uterine contractions during parturition and milk letdown from the mammary ducts

- *Adrenocorticotropic hormone* (ACTH) is used to stimulate the adrenal cortex to secrete corticosteroids in the diagnosis of adrenal cortex disease.
- *Follicle stimulating hormone* (FSH) and *luteinizing hormone* (LH) (and its signal from the hypothalamus, GnRH) are discussed in the reproductive section.
- *Growth hormone* (GH), also known as somatotropin, is used to increase growth rate and feed efficiency in livestock, and is believed to increase milk production in dairy cows. In healthy animals, it is released throughout the animal's life to promote tissue building by increasing protein synthesis. The FDA approved a

genetically produced bovine somatotropin (BST) for use in 1993; the trade name of BST is Posilac®. There is sharp debate about its use because of potential milk residues and potential overproduction of milk that would result in low milk prices for farmers.

The posterior pituitary secretes antidiuretic hormone (ADH, vasopressin) and oxytocic (labor-producing) hormone. Antidiuretic hormone is available commercially and has been used in dogs and cats to treat diabetes insipidus. Diabetes insipidus is a disease characterized by the inability to concentrate urine, caused by the posterior pituitary's failure to release sufficient antidiuretic hormone. This results in increased thirst and urination (PD/PU). *Vasopressin* (Pitressin®) injection has been used to reverse this condition. Vasopressin is available from both natural and synthetic sources.

TIP

Oxytocin increases both the frequency and force of contractions of uterine smooth muscle.

Oxytocin is available as a synthetic commercial injection whose primary effects are in stimulating the smooth muscle of the uterus and mammary gland. Veterinarians use it to contract the uterus and induce labor in animals at term. This is especially helpful in the queen, bitch, and sow, where the possibility of manual manipulation is limited. Oxytocin also helps expel the placenta and uterine debris after Cesarian section. Oxytocin can also induce milk letdown by stimulating the smooth muscle that surrounds the milk-secreting cells of the mammary gland. This muscle contraction forces milk into the sinuses, making the milk readily available to the suckling young. Oxytocin is also used as adjunct therapy in animals with an open pyometra, so that the increase in uterine contractions can help expel pus from the uterus. If used in animals with a closed pyometra, the increase in uterine contractions and the inability to dispel pus through the undilated cervix may cause uterine tearing. Oxytocin is marketed generically and does not have a milk or meat withdrawal time.

Blood Glucose Regulation

To remain healthy, animals must maintain blood glucose levels within a narrow, minimally fluctuating range. The two hormones that maintain this range are insulin and glucagon. **Insulin**, which is formed in the pancreas, responds primarily to a rise in blood glucose; it promotes the uptake and utilization of glucose for energy in body cells, and the storage of glucose in the liver as glycogen (the chief source of carbohydrate storage in animals). **Glucagon**, in contrast, increases blood glucose levels by promoting the breakdown of liver glycogen into glucose, which exits the liver and enters the bloodstream.

Diabetes mellitus is a complex disease of carbohydrate, fat, and protein metabolism caused by lack of insulin or inefficient use of insulin in animals. Diabetic animals have elevated blood glucose levels, glucose in the urine, frequent thirst and urination, and alterations in fat metabolism that can lead to toxic effects and diabetic coma. Diabetes mellitus, though recorded in all species, shows up most frequently (about 1 in 1,000) in dogs, especially obese female dogs that are middle-aged or older. The exact cause of diabetes mellitus in domestic animals is not known. However, dogs with the condition often have a chronic pancreatic disorder affecting the islets of Langerhans, the pancreatic cells that secrete insulin.

TIP

Signs of an insulin overdose (hypoglycemia) include lethargy, weakness, ataxia, and seizures. Owners need to be aware of these signs and know that this is a medical emergency.

Veterinarians treating diabetic dogs or cats consider both dietary and medical management. Because a diabetic animal has difficulty using carbohydrates, its diet must be low in carbohydrates and high in protein. The diabetic diet should also be high in soluble fiber, which slows digestion and reduces postprandial (after eating) hyperglycemia. Along with a managed diet, the animal usually receives insulin injections; trial doses establish the final doses that restore the animal's former physical condition and activity. Owners should also be counseled about signs and treatment of insulin overdose, which include lethargy, weakness, ataxia, and potentially seizures.

Insulin preparations differ with respect to degree of purity, source (beef or pork origin), and onset and duration of action. Insulin purity is based on the concentration of proinsulin contamination in the insulin product. *Proinsulin* is a long chain of amino acids that are used by the beta cells of the pancreas to produce insulin. The greater the concentration of proinsulin, the greater the likelihood of local or systemic allergic reactions. Purified insulin products generally do not contain more than 10 parts per million (ppm) of proinsulin contamination.

Insulin has traditionally been extracted from bovine or porcine pancreas or a combination of both. Porcine (pork) insulin is similar in structure to naturally occurring canine and human insulin, whereas bovine (beef) insulin is similar to naturally occurring feline insulin. The more similar the commercial insulin is to the host insulin, the less likely it is that the recipient animal will have problematic immunological reactions. Recombinant DNA and synthetically processed insulin are now available, but have not been as successful in treatment of veterinary cases of diabetes mellitus. Manufacturers have discontinued bovine (beef) insulin production, which has decreased the options veterinarians have for choosing a type of insulin product to use.

The onset and duration of insulin activity in the body may be controlled by modification of regular insulin. Regular insulin has the most rapid onset and the shortest duration of action. By precipitating insulin with zinc, various modified insulins can be made. Another way of modifying insulin to achieve a longer onset and duration of action is to precipitate insulin with zinc and a large protein, protamine. These modifications result in NPH (isophane insulin suspension) and PZI (protamine zinc) insulin products.

The insulin categories, based on onset and duration of action, are as follows:

- Short-acting insulin (*regular crystalline insulin*, which is given IV, and *semilente*, which is given IM or SQ). Short-acting insulin is used initially to treat diabetic ketoacidosis and control blood glucose until the animal is stabilized. Animals are not usually kept on short-acting insulin for long-term glucose management. Short-acting insulin effects last from one-half to two hours. Examples include Regular Iletin® I and II, Humulin® R, Novolin® R, Regular Insulin, Velosulin®, and Purified Pork Insulin.
- Intermediate-acting insulin (*NPH* and *lente*, which are given SQ). Intermediate-acting insulin is used for control of blood glucose in uncomplicated diabetes mellitus in cats and dogs. This category of insulin has a longer duration of action than the short-acting insulins and a more rapid onset of action than the long-acting insulins. Its duration of action is between 6 to 24 hours in dogs and 4 to 10 hours in cats. Examples include NPH Iletin® I and II, Lente Iletin® II, Humulin® N, Novolin® N, and Lente Purified Pork.
- Long-acting insulin (*protamine zinc insulin* (PZI®) and *ultralente*, which are given SQ). Long-acting insulin is poorly absorbed from tissue, therefore, it maintains a long-lasting blood level. The duration of action for long-acting insulin is 6 to 28 hours in dogs and 6 to 24 hours in cats. PZI was the insulin of choice for cats until it was taken off the human market (manufacturing of PZI has begun again on a much smaller scale). At that time, veterinarians started using ultralente. Examples of long-acting insulin include Humulin® U, PZI Vet®, and PZI.

Insulin concentration is expressed in units of insulin per milliliter. Insulin is available in 40 units/ml (U-40), 100 units/ml (U-100), and 500 units/ml (U-500). U-40 insulin is used when administering small amounts of insulin to cats and small dogs. U-40 insulin is given with U-40 syringes (Figure 10-4). U-100 insulin is given with U-100 syringes. Drawing up small volumes on a U-40 syringe is easier and more accurate than drawing up small volumes on a U-100 syringe.

Although insulin products are generally stable at room temperature, they are usually stored in the refrigerator and not frozen. Exposure for even short periods to freezing or high temperatures can permanently degrade insulin products. If the product becomes discolored or a precipitate forms, it should be discarded. Insulin should

TIP

- Dosage adjustments may be necessary when changing insulin types.
- Insulin appears cloudy in the vial except for Regular Insulin, which is clear.
- Insulin should be gently rotated in the palms of the hand for mixing; it should never be shaken, to prevent damage to the molecular structure of the insulin.

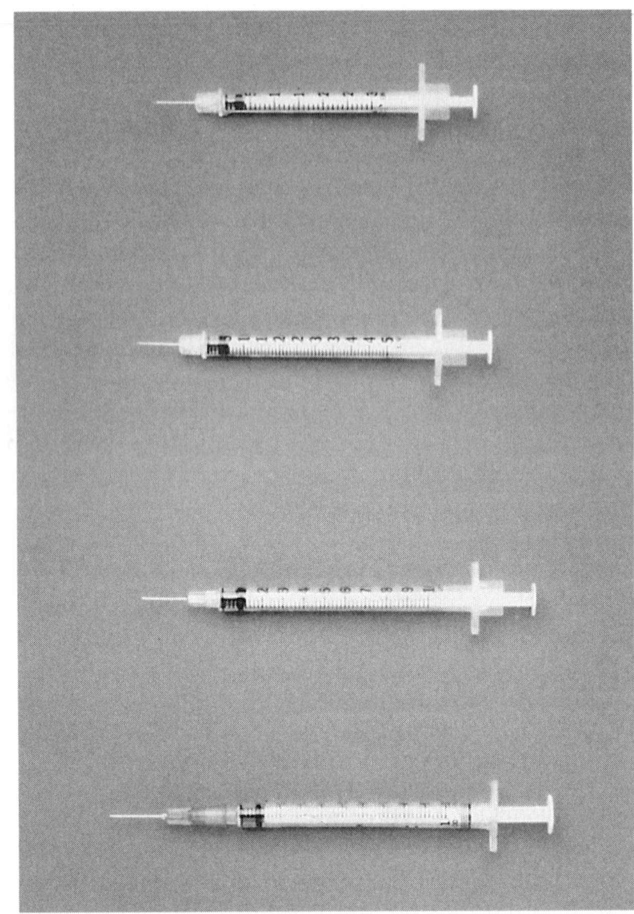

FIGURE 10-4 U-40 insulin syringes from 0.3 ml to 1 ml. Courtesy of Becton, Dickinson and Company.

be gently rolled between the palms to mix the product before it is withdrawn from the vial.

Oral hypoglycemic agents have been used with some success in animals. Oral hypoglycemic agents work by stimulating pancreatic beta cells to secrete insulin; therefore, some pancreatic function is required for these drugs to act. The oral hypoglycemics used in veterinary practice are in the chemical class known as the sulfonylureas. The representative drug in this class is *glipizide* (Glucotrol®). Glipizide, a second-generation sulfonylurea, is more potent than older agents and must be dosed less frequently. Because most dogs with diabetes mellitus tend to have the insulin-dependent form of the disease, glipizide has not been effective. Glipizide has been shown to be effective in treating about 25 percent of diabetic cats. Side effects of glipizide include gastrointestinal disturbances, hypoglycemia, and liver toxicity.

In a healthy animal, glucagon is used to regulate blood glucose levels when they become low. Low blood sugar or hypoglycemia is associated with insulinomas (tumors of the pancreas) that cause increased secretion of insulin. *Diazoxide* is a drug that is labeled for oral use in treating hyperglycemia by directly inhibiting pancreatic insulin secretion. Clients giving diazoxide to their pets should be instructed to monitor for signs of hyperglycemia, hypoglycemia, or gastrointestinal problems. It can also cause sodium and water retention and should be used cautiously in animals with heart or kidney disease. Trade names of diazoxide include Proglycem® suspension and Eudemine® tablets.

Other drugs used to increase blood glucose levels include corticosteroids, epinephrine, and progesterone. These drug categories have been covered under other sections.

TIP

The "T" designation in T_4 (thyroxine) and T_3 (tri-iodothyronine) refers to the iodine atoms in each hormone: thyroxine has four, tri-iodothyronine has three.

Regulation of Metabolic Rate

The thyroid gland is an organ located within the neck near the larynx. Hormones produced by the thyroid are necessary for an animal's normal growth and reproduction. *Thyroid hormone* is a collective term for two active hormones found in the thyroid gland: **thyroxine** (T_4) and **tri-iodothyronine** (T_3). The synthesis of these hormones takes place in a series of chemical steps. Iodides consumed in food and water are absorbed through the gastrointestinal tract and enter the blood stream. When blood passes through the thyroid gland, iodide is trapped and converted to iodine. Iodine combines with tyrosine (an amino acid) to form iodotyrosine. Iodotyrosine molecules combine to form T_4 and T_3, which are stored in the thyroid gland until they are released. When T_4 is released into the bloodstream, some of it is converted to T_3. Figure 10-5 summarizes the steps of T_4 and T_3 synthesis.

Thyroid hormones act as catalysts in the body and affect many metabolic, growth, reproductive, and immune functions. They help regulate lipid and carbohydrate metabolism. Thyroid hormones also affect heat production in the body.

Hypothyroidism, a disease characterized by a deficiency of thyroid hormone, may be caused by many things. Abnormalities of the thyroid gland can result from disorders of iodide trapping, conversion of iodide to iodine, and/or the release of thyroid hormone from its storage sites. Another cause of hypothyroidism is a disorder of the anterior pituitary gland in which inadequate amounts of thyroid stimulating hormone are released. Whatever the cause, the signs of hypothyroidism reflect thyroxine's necessity to virtually all the cells and organ systems of the animal. These signs include decreased coat luster or hair loss; weight gain without any increase in appetite; listlessness; intolerance to cold; reproductive failure; and skin that is more susceptible to mites, bacteria, and scales.

Hypothyroidism is a serious condition in dogs and is diagnosed in a variety of ways. One way to diagnose hypothyroidism is to measure serum total T_4 and T_3. However, hypothyroid dogs may test in the normal range when only serum levels of thyroid hormone are measured. Another diagnostic test is the thyroid stimulation test, in which thyroid hormones are measured before and after the administration of TSH. Blood is drawn prior to the test, and then TSH is given IV. Another blood sample is collected four to six hours after the TSH injection. If the level of thyroid hormone is low following TSH injection, then the dog is hypothyroid. *Thyrotropin* is a purified form of TSH that

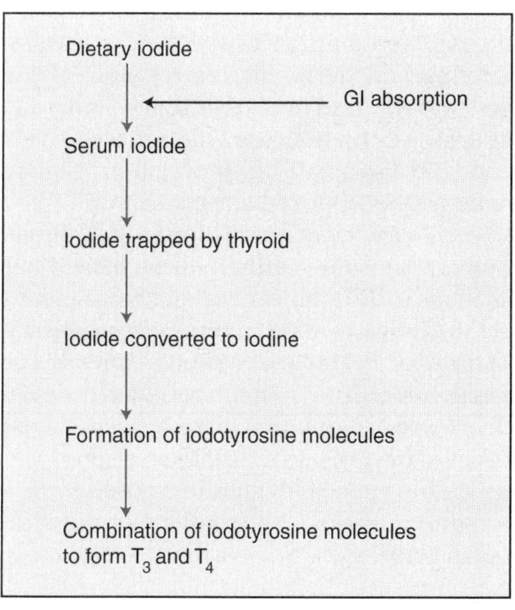

FIGURE 10-5 Summary of the biosynthesis of T_4 and T_3

is collected from bovine anterior pituitary glands and used to make commercial TSH. Trade names for TSH are Dermathycin® and Thyrotropar®. They are available in powder forms that must be reconstituted prior to use. Reconstituted thyrotropin can be stored for up to three weeks before losing its effectiveness.

Another test used to detect hypothyroidism is the thyrotropin releasing hormone (TRH) response test. The TRH response test is performed by obtaining blood samples before and six hours after IV administration of *TRH* (Relefact®). An increase in serum T_4 of 1.5 times baseline is normal; levels below that are considered abnormal.

The use of thyroid hormone to treat hypothyroidism is called thyroid replacement therapy. The goal in treating hypothyroid animals is to achieve an euthyroid state (normal thyroid state) by supplying the animal with appropriate concentrations of thyroid hormone. Drugs used in thyroid replacement therapy include:

- *levothyroxine sodium* (T_4): Levothyroxine is the synthetic isomer of T_4. It is the drug of choice for treating hypothyroidism in all animals. Its advantages are that it is chemically pure, is inexpensive, and has a long half-life, which helps in lowering dosing frequency. Side effects are rare. Examples are Soloxine®, Thyro-Form®, Thyro-Tab®, and Synthroid®.
- *liothyronine sodium* (T_3): Liothyronine is a synthetic form of T_3 that is chemically pure but has a shorter half-life than levothyroxine. It is used in animals that do not respond well to treatment with levothyroxine. Side effects are rare. Examples are Cytobin® and Cytomel®.

TIP

What is a stimulation test? Stimulation tests evaluate an endocrine gland's responsiveness to exogenous stimulating factors. Stimulation tests measure plasma levels of hormone after administration of a substance that should increase production of the hormone. If this increase does not occur, the endocrine gland is not functioning as it should.

Hyperthyroidism is excessive functional activity of the thyroid gland. Excessive secretion of thyroid hormone results in increased levels of metabolism, excessive heat production, and increased nervous system activity. Seen mainly in cats, hyperthyroidism is characterized by increased thirst, weight loss despite increased appetite, increased stool production, restlessness, and tachycardia (increased heart rate). Treatment includes destruction or removal of the dysfunctional thyroid gland or the administration of antithyroid drugs.

The thyroid can be surgically excised to remove the source of excess thyroid hormone. Prior to surgery, animals are brought to an euthyroid state with drugs that suppress thyroid function.

The thyroid gland can be destroyed by radioactive isotopes of iodine (sodium iodine I-131). Domestic animals require iodine for normal thyroid function, though no one knows the minimum amount of iodine that various species require. It is known that iodine (in the form of iodide salts), dosed substantially above dietary requirements, inhibits thyroid hyperfunction. Veterinarians use radioactive iodine, I-131, to evaluate thyroid gland function and to destroy hyperactive thyroid tissue. Radiation is emitted by the trapped isotope, which destroys thyroid cells without excessively damaging the surrounding tissue. Because I-131 has a half-life of only 8 days, more than 99 percent of the radiant energy will be gone in about 56 days.

TIP

An animal's response to stress is multidimensional. The sympathetic nervous system releases norepinephrine and/or epinephrine and works on target organs like the heart (to increase the heart rate). The adrenal medulla responds to sympathetic nervous stimulation by releasing the hormones epinephrine and norepinephrine.

Antithyroid drugs work by blocking excess hormone production in the thyroid. *Methimazole* (Tapazole®) is the drug of choice for treating hyperthyroidism. It interferes with the incorporation of iodine in the molecules of T_3 and T_4. Thyroid hormone existing in the blood is not altered. Side effects include vomiting, anorexia, and lethargy. *Propylthiouracil* (PTU®) has been used as a drug for treating hyperthyroidism and has the added effect of interfering with the conversion of T_4 to T_3 in the peripheral circulation. Side effects, including blood disorders (especially in cats), have limited the use of PTU in veterinary practice.

Propranolol (Inderal®), a beta-adrenergic blocking agent, has been used to suppress the tachycardia associated with hyperthyroidism. Propranolol is covered in Chapter 8 on cardiovascular drugs.

Regulation of the Adrenal Cortex

The adrenal glands are paired glands located near the cranial portion of the kidney. Each gland has an outer region (adrenal cortex) and an inner region (adrenal medulla). The hypothalamus regulates both regions of the adrenal gland. The medulla receives direct nervous stimulation that originates in the hypothalamus, travels to the brainstem, the spinal cord, and the sympathetic nerves. The adrenal medulla releases epinephrine and norepinephrine when stimulated by the sympathetic nervous system. The hypothalamus regulates the adrenal cortex by secreting releasing hormones for ACTH, which stimulates the anterior pituitary gland to secrete ACTH. ACTH stimulates the adrenal cortex.

TIP

All hormones produced from the adrenal cortex are steroids. Steroids have a parent molecule that consists of three six-carbon ring structures and one five-carbon ring structure.

The hormones produced by the adrenal cortex are produced in different regions, but all are steroids. **Mineralocorticoids** regulate blood volume and electrolyte concentration in the blood. The principal mineralocorticoid is aldosterone, which acts to conserve sodium ions and water in the body. Aldosterone is secreted in direct response to sodium and potassium ions.

Glucocorticoids, also secreted by the adrenal cortex, help regulate nutrient levels in the blood. They do this by increasing cellular utilization of energy sources (proteins and fats) and conserving glucose, causing blood glucose levels to increase. Glucocorticoids also stimulate liver cells to produce glucose from amino acids and fat. ACTH from the anterior pituitary gland controls glucocorticoid secretion. The main glucocorticoid is cortisol, which increases blood glucose levels.

Diseases of the adrenal gland include **adrenocortical insufficiency** (too little hormone produced) and **hyperadrenocorticism** (too much hormone produced). Adrenocortical insufficiency, also known as Addison's disease, is a progressive condition associated with adrenal atrophy, usually caused by immune-mediated inflammation. Adrenal atrophy results in deficient production of corticosteroids and mineralocorticoids. Signs of this disease include lethargy, weakness, anorexia, vomiting, diarrhea, and PU/PD (polyuria = increased urination and polydipsia = increased thirst/drinking). Adrenocortical insufficiency is diagnosed by the ACTH stimulation test. Exogenous ACTH is available as an animal extract ACTH® gel or as synthetic ACTH. A blood sample is obtained for determination of a resting plasma cortisol level. ACTH® gel is then given and another blood sample is collected one to two hours later. Low levels of cortisol indicate adrenocortical insufficiency. Adrenocortical insufficiency can also be diagnosed by determining ACTH blood levels, which are elevated in primary hypoadrenocorticism.

Treatment of adrenocortical insufficiency includes use of *desoxycorticosterone* (commonly referred to as DOCP) (Percorten-V®), which is a long-acting mineralocorticoid, in conjunction with prednisone or prednisolone, both of which are corticosteroids. DOCP is thought to act by controlling the rate of protein synthesis. DOCP is given every 25 days and has relatively few side effects. Those side effects, which are related to increased renal absorption of sodium, may include edema and electrolyte imbalance. DOCP should be used with caution in animals with cardiovascular and renal disease. Corticosteroids are covered in Chapter 16 on anti-inflammatory and pain-reducing drugs.

Another drug used to treat adrenocortical insufficiency is *fludrocortisone acetate* (Florinef). Fludrocortisone is a potent corticosteroid with both glucocorticoid and mineralocorticoid activity. It is given once or twice daily, but it has not been as effective as DOCP in the treatment of adrenocortical insufficiency. Side effects include edema and electrolyte imbalance.

Hyperadrenocorticism, also known as Cushing's disease, can be caused by prolonged administration of adrenocortical hormones, by adrenocortical tumors, or by pituitary disorders. Animals with hyperadrenocorticism have signs of PU/PD, hair loss, and a pendulous abdomen due to abnormal nutrient metabolism (Figure 10-6).

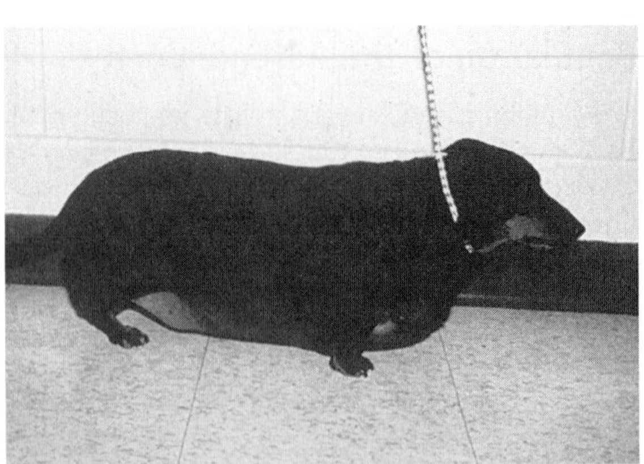

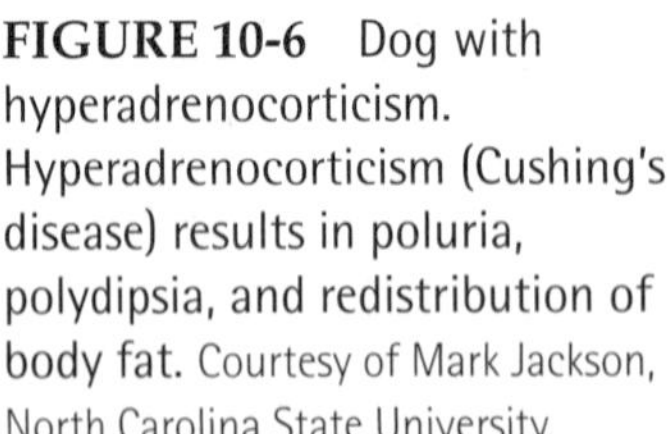

FIGURE 10-6 Dog with hyperadrenocorticism. Hyperadrenocorticism (Cushing's disease) results in poluria, polydipsia, and redistribution of body fat. Courtesy of Mark Jackson, North Carolina State University.

Hyperadrenocorticism is diagnosed by a low-dose dexamethasone suppression test and/or ACTH stimulation test. The low-dose dexamethasone suppression test has historically been used as the initial diagnostic test for confirming hyperadrenocorticism, but the ACTH stimulation test can provide information on the pituitary gland prior to initiation of treatment. The low-dose dexamethasone test measures plasma cortisol concentrations before and four and eight hours after IV administration of dexamethasone. If plasma cortisol levels are high eight hours after dexamethasone administration, hyperadrenocorticism is present. The ACTH stimulation test measures serum cortisol concentrations before and following the administration of ACTH. This test can differentiate between iatrogenic (caused by treatment) and naturally occurring hyperadrenocorticism. ACTH gel is commercially made from porcine pituitary glands and stimulates the production and release of glucocorticoids.

TIP

Iatrogenic Addison's disease occurs when animals are supplemented with glucocorticoids and not tapered off the dose gradually. Sudden withdrawal of exogenous glucocorticoid does not give the adrenal cortex-hypothalamus-anterior pituitary feedback loop time to adjust to the changes in glucocorticoid level. Iatrogenic Addison's disease is more likely to occur when supplementation with high doses of glucocorticoids is done for long periods of time.

Treatment of hyperadrenocorticism due to adrenocortical tumors is aimed at destroying part of the adrenal cortex using *mitotane* (Lysodren®), also known as *o,p'*-DDD in veterinary medicine. Side effects of mitotane use include neurologic signs, lethargy, vomiting, and diarrhea. Adverse effects are common if there is a very rapid decrease in plasma cortisol levels. To prevent gastrointestinal side effects, mitotane should be given with food.

Other treatment options for hyperadrenocorticism include *ketoconazole* (Nizoral®), an antifungal drug that also blocks the enzymes needed to produce steroid compounds; and *selegiline* (Anipryl®), a monoamine oxidase inhibitor. Ketoconazole's low toxicity and minimal effect on mineralocorticoid production make it a viable option in treatment of this disease. Selegiline is used for treating pituitary-dependent hyperadrenocorticism because it increases synthesis and release of dopamine. Hypothalamic dopamine deficiency plays a role in the pathology of pituitary-dependent hyperadrenocorticism in dogs. Response to selegiline treatment varies; it may take one to two months after initiation of treatment for signs of improvement to appear. Side effects of selegiline include vomiting, diarrhea, and restlessness.

TIP

Because mitotane destroys some of the adrenal cortex, clients who give mitotane to their animals need to know the signs of adrenocortical insufficiency: lethargy, weakness, PU/PD, and gastrointestinal problems. Due to the extreme toxicity associated with mitotane, clients should wear gloves for administration or wash their hands thoroughly after handling this drug.

The last set of hormones secreted by the adrenal cortex is the androgens (male sex hormones) and estrogens (female sex hormones). Because their effect is usually masked by the hormones of the testes and ovaries, abnormalities of androgens are usually not seen with adrenal cortex disease.

BASIC REPRODUCTIVE SYSTEM ANATOMY AND PHYSIOLOGY

The reproductive system is responsible for the process of producing offspring. The reproductive organs, whether male or female, are called the *genitals*. The structures that produce sex cells are called *gonads* and consist of the ovaries in females and testes in males. The male reproductive system consists of the testes (located in the scrotum), epididymis, ductus deferens, accessory sex glands, urethra, and penis. Sperm are produced in the seminiferous tubules of the testes. The Leydig's cells in the testes produce testosterone. Figure 10-7 shows the structures of the male reproductive system.

The female reproductive system consists of the ovaries, uterine tubes, uterus, cervix, vagina, and vulva (Figure 10-8). Ova are produced in the graafian follicle of the ovary. Estrogen and progesterone are produced in the ovary (progesterone is also produced other places in pregnant animals).

The estrous cycle in animals consists of

- **proestrus**: The period of the cycle before sexual receptivity. It involves the secretion of FSH by the anterior pituitary gland, which causes the follicles to develop in the ovary. FSH stimulates ovarian release of estrogen, which helps prepare the reproductive tract for pregnancy.
- **estrus**: The period of the cycle in which the female is receptive to the male. During estrus, FSH levels decrease and LH levels increase, causing the graafian follicle to rupture and release its egg (ovulation).
- **metestrus**: The period of the cycle after sexual receptivity. The corpus luteum (CL) forms in the ovary and produces progesterone if the animal is pregnant. If the animal is not pregnant, the CL decreases in size and become a corpus albicans.
- **diestrus**: The period of the cycle that is a short phase of inactivity in polyestrous animals.
- a**nestrus**: The period of the cycle when the animal is sexually quiet; this is a long phase in seasonally polyestrous and monestrous animals.

Another way of categorizing the estrous cycle is by the follicular and luteal phases. The **follicular phase** is the stage of the estrous cycle in which the graafian follicle is present. Estrogen is the predominant hormone during the follicular phase. The **luteal phase** is the stage of the estrous cycle in which the corpus luteum is present. Progesterone is the predominant hormone during the luteal phase.

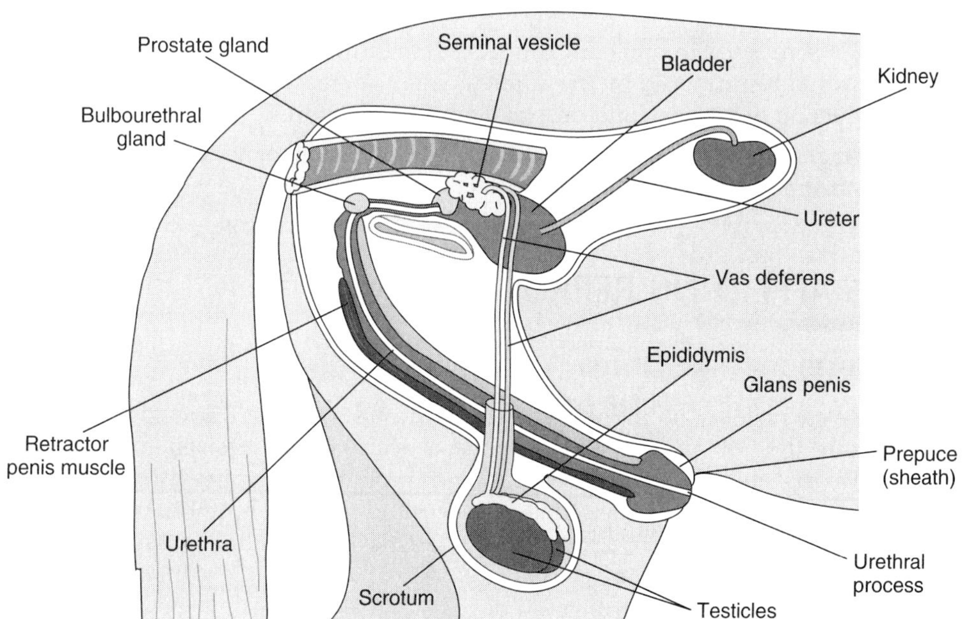

FIGURE 10-7 Reproduction tract of a stallion

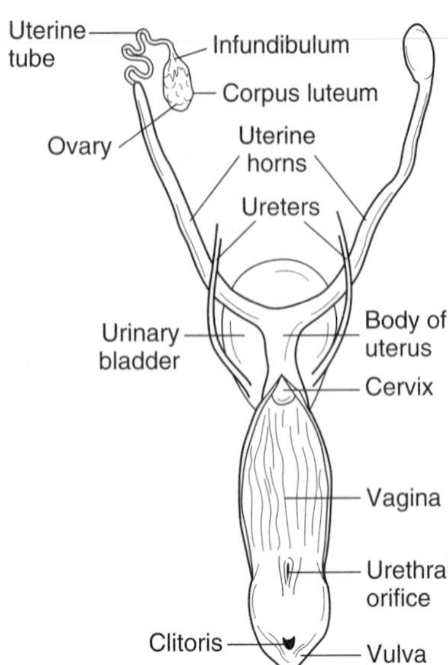

FIGURE 10-8 Reproduction tract of a bitch

Control of the reproductive system, like the endocrine system, occurs via the hypothalamus and pituitary glands. The hypothalamus makes gonadotropin-releasing hormone (gonadorelin or GnRH) in response to various stimuli such as daylight length and feedback mechanisms. The hormones regulating reproduction are briefly described here and in Figure 10-9.

- GnRH causes the release of FSH and LH from the anterior pituitary gland. GnRH release is controlled by levels of FSH and LH via a negative feedback loop.
- FSH causes the growth and maturation of ovarian follicles. Follicles produce estrogen. Increased estrogen levels signal the hypothalamus to produce less GnRH.
- LH causes ovulation of mature follicles and the formation of the CL. The CL produces progesterone. Increased progesterone levels signal the hypothalamus to produce less GnRH.
- ACTH is released in pregnant animals as parturition approaches. ACTH causes cortisol to be produced by the adrenal cortex. Increased cortisol levels increase production of estrogen and prostaglandin by the uterus.
- Prostaglandin breaks down the CL at the end of pregnancy and at the end of diestrus in nonpregnant animals.

DRUGS AFFECTING REPRODUCTION

Male Hormone-like Drugs

Androgens are male sex hormones produced in the testes, ovaries, and adrenal cortex. **Testosterone**, the primary male sex hormone, is primarily synthesized in the interstitial cells (Leydig's cells) of the testes. Testosterone production is initiated and controlled by follicle stimulating hormone and interstitial cell stimulating hormone. Testosterone has both androgenic (promoting male characteristics) and anabolic (tissue-building) effects. In addition to producing the male sex organs and secondary male sex characteristics, testosterone helps spermatozoa develop and also helps develop and maintain accessory sex organs (like the prostate gland).

TIP

Testosterone products that are derived naturally from animal testes are not effective when administered orally, because the liver rapidly inactivates them. Synthetic forms of testosterone are effective when administered orally, subcutaneously, or intramuscularly.

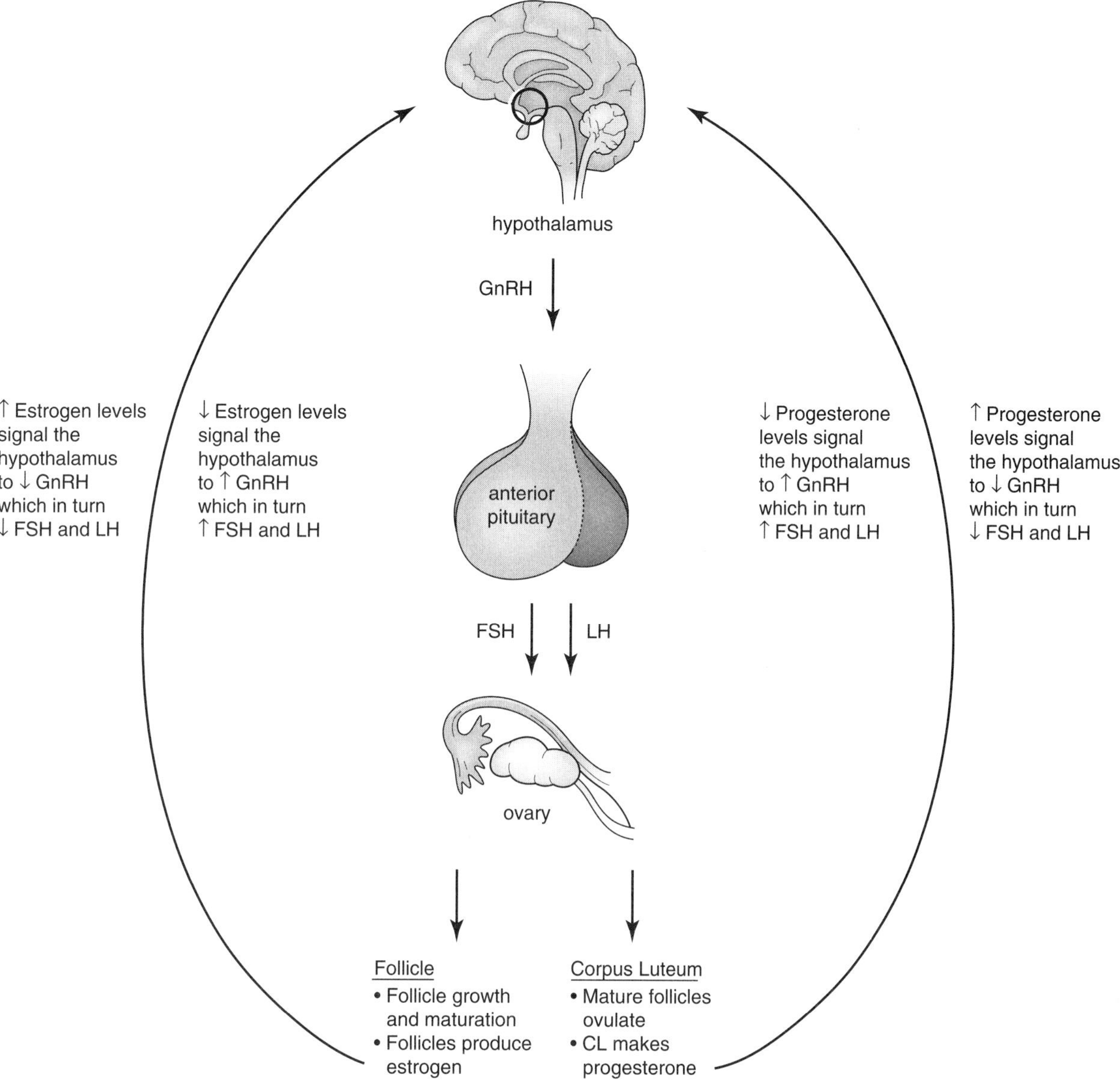

FIGURE 10-9 Control of the reproductive system through release of GnRH, LH, and FSH

Veterinarians use testosterone to treat conditions such as infertility and hypogonadism (the decreased function and retarded growth and sexual development of the gonads). Testosterone is also used to produce estrus detectors or "teaser animals" in cattle and for testosterone-responsive urinary incontinence in dogs. Testosterone propionate USP in oil can be given subcutaneously, intramuscularly, as subcutaneous implants, or as an oral tablet. Testosterone products are C-III controlled substances. Side effects of testosterone include the development of perianal tumors, prostatic disorders, and behavior changes. Examples of testosterone products include:

- testosterone cypionate in oil (generic and Depo-Testosterone®).
- testosterone enanthate in oil (generic).
- testosterone propionate in oil (generic).
- danazol (Danocrine®), a weak, synthetically produced androgen used primarily for immune-mediated hemolytic anemia in animals.

TIP

Teaser animals are animals that are used to detect females in heat, but are altered so that they can not impregnate the females.

Mibolerone is an androgenic compound that works by blocking the release of LH from the anterior pituitary gland by negative feedback. When LH release is blocked, the follicle does not fully develop, resulting in failure of ovulation and corpus luteum development. Mibolerone is used to prevent estrus in adult female dogs not intended for breeding and for treatment of false pregnancies. An example is Cheque® drops, which should not be used in cats because of its low margin of safety. Side effects include reproductive organ problems (such as vulvovaginitis) and behavior changes. Implants of mibolerone are discussed under growth promotants.

A nonsurgical neutering drug, *zinc gluconate neutralized by arginine* (Neutersol®), is a chemical sterilant approved for use in 3- to 10-month old male dogs. This drug is used as an intratesticular injection administered once in each testicle. It should not be used in cryptorchid animals (those with undescended testicles), animals with testicular disease, or animals with scrotal irritation. Most testicles atrophy to some degree following intratesticular injection of this product; however, variability in size between the left and right testicles may be noticed. Testosterone production is not completely eliminated, so diseases that occur as a result of testosterone production (such as prostatic disease and testicular tumors) may not be prevented. Clients should be advised that this product does not kill sperm present at the time of injection; therefore, treated dogs should be kept away from females in heat for at least 60 days. Side effects include vocalization upon injection, scrotal pain, vomiting, and anorexia.

Female Hormone-like Drugs

The ovaries, testicles, adrenal cortex, and placenta produce estrogens. The ovarian follicle (the ovum and its encasing cells) and placenta in the female are the main sources of **estrogen**, a hormone that promotes female sex characteristics and stimulates and maintains the reproductive tract and accessory reproductive organs, including duct growth of the mammary glands. Estrogen stimulates female reproductive function, and is necessary for the uterus to contract and respond to oxytocin. Estrogen also enhances such female secondary sex characteristics as mammary development, plumage, and beak color. Veterinarians use synthetic estrogens in lieu of the more expensive natural estrogens. Intramuscular injections of synthetic estrogen prevent implantation of fertilized ova in dogs; however, due to serious side effects (such as pyometra and aplastic anemia) the use of estrogen for mismating is not recommended. Estrogens can be helpful in correcting urinary incontinence, vaginitis, and dermatitis in dogs who have had their ovaries removed. In cattle, estrogens treat persistent corpus luteum, aid in expulsion of retained placentas and mummified feti, and promote weight gain. In horses, estrogen is used to induce estrus in the nonbreeding season. Side effects of estrogens include bone marrow suppression, endometrial hyperplasia, follicular cyst formation, and pyometra. Examples of estrogen include estradiol cypionate (ECP® injectable, Depo-Estradiol Cypionate®), and diethylstilbestrol (DES tablets and specialized preparations due to the unavailability of DES). Implants are discussed under growth promotants.

The "Gest"s (Progesterone Drugs)

Progesterone is a female sex hormone produced and secreted after ovulation by the corpus luteum (Latin for *yellow body*), a group of ovarian cells. Progesterone decreases uterine activity when a female is in estrus (heat) or pregnant; progesterone deficiency may cause embryonic death in some animals. Veterinarians use progesterone and progestins (a group of compounds similar in effect to progesterone) with some success in cows (for reducing embryonic death), mares, and other domestic animals, but recommend it selectively. Progestins also block estrus in the bitch. When estrus is blocked, the ovary does not grow follicles and ovulation ceases. Progestins also help cattle breeders by synchronizing the breeding and birth cycles so that newborn care is better programmed to use capital facilities efficiently. Progestins help breeders of other species by delaying estrus (heat) during racing seasons, travel, and livestock shows. Progestins can also be used to treat behavior disorders and some forms of dermatitis. Examples include:

- *mege**st**rol acetate* (Ovaban®, Megace®). These oral tablets are used in small animals to postpone estrus, to alleviate false pregnancies, to treat behavior problems (especially in cats), and to treat skin problems. Side effects include hyperglycemia (usually a transient diabetes mellitus in cats) and adrenal cortex suppression.
- *medroxypro**gest**erone* acetate (Depo-Provera®, Provera®, Cycrin®). These oral and injectable drugs are used in small animals similarly to megestrol acetate. Side effects include pyometra, mammary changes, and behavior changes.
- *altreno**gest*** (Regu-Mate®). This oral and injectable preparation is used in horses to suppress estrus in mares, for estrus synchronization, and in low levels to maintain pregnancy in mares. When the drug is stopped, GnRH release is stimulated and the mare will cycle again. Side effects are minimal and include electrolyte imbalance and changes in liver enzyme values. Altrenogest can be absorbed by the skin and should not be handled by pregnant women.
- *pro**gest**erone* (Eazi-Breed Cidr® cattle insert). Another type of estrus synchronization tool is the application of an intravaginal insert containing progesterone. This intravaginal insert has progesterone in elastic rubber molded over a nylon spine that is inserted with a special applicator into the vagina. The insert is placed in the cranial portion of the vagina for seven days and then removed. When the progesterone insert is removed, plasma progesterone rapidly decreases, triggering estrus to occur within three days. One day before the insert is removed, an injection of dinoprost (Lutalyse®) is given to assure estrus synchronization.
- *melen**gest**rol* acetate or MGA, a synthetic form of progesterone, used to suppress estrus in feedlot heifers intended for breeding. Examples of MGA are Steakmaker® and MGA 200®. MGA is a premix that is fed for 10 days to feedlot heifers; when it is removed from the feed, the heifers come into heat. MGA increases rate of weight gain, improves feed efficiency, and suppresses estrus.

TIP

Progestins or progesterone products can be recognized by the *gest* in their generic names.

TIP

When inserting implants, it is important to treat the area with an external antiparasitic spray or repellent to prevent fly infestation.

The "Prost"s (Prostaglandin Drugs)

Many groups of **prostaglandins** occur naturally in the body (groups include A, B, C, D, E, and F). In the reproductive system, prostaglandin $F_{2\alpha}$ causes lysis of the corpus luteum, resulting in lowered progesterone levels in plasma and initiation of a new estrous cycle. Prostaglandin $F_{2\alpha}$ also causes contraction of uterine muscle, facilitating either expulsion of pus (in pyometra) or the mummified fetus (in fetal death), or an abortion. In cattle, prostaglandin $F_{2\alpha}$ is used in estrus synchronization (if given to cattle in diestrus, it will lyse any corpus luteum and all of the cattle will cycle at the same time, in about three to five days), treatment of silent heat, and treatment of pyometra. In small animals it is used to treat pyometra, cause abortion, and induce parturition. In mares it is used for estrus synchronization. The corpus luteum of mares is resistant to prostaglandins for 5 days after ovulation; therefore, 2 injections (13 days apart) are usually given to synchronize estrus. Pregnant women should not handle this medication, as it is absorbed through the skin and can cause uterine contractions. Side effects include bronchoconstriction and elevated blood pressure in animals and humans (through skin absorption). Examples include:

- *dino**prost*** tromethamine (Lutalyse®).
- *flu**prost**enol* (Equimate®) (labeled for mares only).
- *clo**prost**enol* sodium (Estrumate®) (labeled for cows only).

TIP

Prostaglandins can be recognized by the *prost* in their generic names.

TIP

Typical hormone uses in females include the following:

- prostaglandins, progestins, and gonadotropins control the estrus cycle.
- estrogens and prostaglandins prevent or terminate pregnancy.
- progesterones maintain pregnancy.

Gonad Stimulators (Gonadotropins)

Gonadotropins are hormones that stimulate the gonads. Gonadotropin drugs cause release of LH and FSH or simulate their activity. The three substances that play a role in this category are LH, FSH, and GnRH. LH activity is simulated with the use of hCG, FSH activity is simulated by PMSG or by FSH processed from the pituitary gland, and GnRH is synthetically prepared.

Gonadorelin or GnRH is produced in healthy animals by the hypothalamus. GnRH causes release of FSH and LH by the anterior pituitary gland. GnRH is used IM or IV to treat follicular cysts in cattle, for estrus synchronization in cattle, and to induce estrus in small animals. Side effects are rare. Examples of GnRH are Cystorelin® and Factrel®.

Pregnant mare serum gonadotropin (PMSG) and FSH obtained from the pituitary gland (FSH-P) act like naturally occurring FSH. PMSG is a FSH-like hormone that comes from a pregnant mare's endometrium. Its function in the pregnant mare is to cause follicular development and formation of multiple corpus lutea as a progesterone source during pregnancy. Inactive ovaries that are dormant for various reasons can be effectively stimulated into activity by PMSG. A single subcutaneous injection produces estrus and ovulation within two to five days in cows and within five to eight days in horses. It has been used as a follicle stimulant in horses, cattle, sheep, swine, dogs, and cats. FSH-P causes growth and maturation of the ovarian follicle.

Human chorionic gonadotropin (hCG) is a compound that has gonadotropic activity similar to LH. The source of hCG is the human placenta. hCG appears in the urine a few weeks after conception, reaches a peak 50 days into pregnancy, then decreases. The rabbit pregnancy test (in which a pregnant woman's urine causes ovarian changes in the rabbit) employs hCG. Veterinarians use hCG obtained from the urine of pregnant women to treat cystic ovaries in nymphomanic cattle, to detect cryptorchidism in dogs, to get infertile bitches to cycle, and to make breeding mares ovulate. Examples are Follutein®, Chorulon®, A.P.L.®, and LyphoMed®. Veterinarians use intravenous or intramuscular injections of hCG, because hCG is destroyed in the GI tract after oral administration. Because hCG is a protein, the main side effect is anaphylactic reaction. If an anaphylactic reaction is observed, antihistamines are given.

TIP

Pheromones are substances secreted to the outside of an animal. They are perceived through smell by other members of the same species. Pheromones cause specific behavior, usually sexual or territorial in nature. Pheromones produced synthetically can prevent some of these behaviors, such as urine marking. An example of a commercially produced pheromone product is Feliway®, an analogue of feline facial pheromones. It is labeled to prevent urine marking by cats.

PROMOTING GROWTH

Improved feed conversion efficiency (rate of conversion from food to tissue) promotes growth. A number of chemical compounds, called anabolic agents or **growth promotants**, can improve growth in animals. Antibiotics, typically considered agents to treat bacterial infections, can improve the efficiency of a healthy animal's gastrointestinal tract by changing its microbial population. Antibiotics used in this fashion are classified as growth promotants (Table 10-2).

TABLE 10-2 Some Antimicrobial Growth Promotants Used in Livestock

Compound	Effects
bacitracin	Promotes growth in poultry
flavomycin	Increases feed efficiency, promotes growth in poultry and cattle
virginiamycin	Promotes growth in poultry
avoparcin	Increases feed efficiency, promotes growth in poultry, pigs, and cattle
lasalocid	Increases feed efficiency, promotes growth in cattle
monensin	Increases feed efficiency, promotes growth in cattle

HORMONAL IMPLANTS

Growth-promoting implants have been used since the 1940s to increase feed efficiency and weight gain in beef cattle. *Implants* are small pellets or devices placed under the skin at the back of the ear (to avoid the possibility of hormone residue in food products, implants inserted in the ear are removed prior to slaughter). Figure 10-10 shows a diagram of an ear implant. Each pellet contains a growth-promoting hormone that is slowly released into the blood and is subsequently carried to tissues. When growth-promoting implants are placed in the ear, there is a rapid release of hormone from the implant. The level of growth-promoting hormone will begin to fall a few days after implantation, but it remains above the animal's normal level for a varying length of time, depending on the type of implant and the chemical in the implant. Growth promotants must be officially approved and are subject to local regulation. Many implants contain small amounts of antibiotic for local protection.

Diethylstilbestrol (DES) was one of the first growth-promoting compounds; however, it was taken off the market many years ago because of its potential danger for human consumers (increased chance of some forms of cancer and infertility). Veterinarians currently use implants of natural hormones (such as testosterone, progesterone, and estradiol) and synthetic hormones to provide a slow, sustained release of the growth promotant. Examples of growth promotants include the following.

Estradiol is a potent anabolic agent in ruminants and is administered either as a compressed-tablet implant or a silastic rubber implant. Estradiol given to steers increases growth rate by 10 to 20 percent, lean meat content by 1 to 3 percent, and feed efficiency by 5 to 8 percent. Estradiol is not an effective anabolic agent in pigs. Examples of estradiol products are Compudose® and Encore®.

Testosterone is used in combination with estradiol as a component in compressed-tablet implants to slow the release rate of estradiol. Testosterone is not used on its own as an anabolic agent in farm animals. Examples include Synovex H® (heifers), Synovex S® (steers), and Implus-H®.

Progesterone, like testosterone, is used in combination with estradiol to slow the release rate from implanted compressed tablets. There is no evidence to suggest that progesterone itself is anabolic in livestock. Examples include Synovex C® (calves more than 45 days old), Implus-S®, Component®, and CALF-oid Implant®.

Synthetics are synthetic hormones that generally have more potency and fewer adverse effects than naturally occurring hormones. The synthetic hormones used in veterinary practice include *trenbolone acetate* (TBA), *melengestrol acetate* (MGA), and *zeranol.*

TIP

Cattle must be given adequate nutrition in order for implants to positively affect feed efficiency and weight gain.

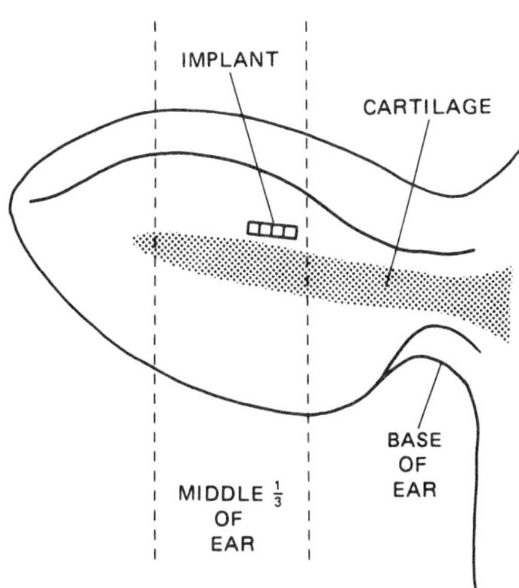

FIGURE 10-10 Example of ear implant placement

Veterinarians use TBA to promote growth in feedlot heifers and (to a lesser extent) sheep. It works like testosterone, but with greater activity. Unlike natural testosterone, TBA works by itself and with estradiol as a pellet-type implant in heifers and cull cows. It is not intended for use in breeding or dairy animals. Examples of TBA are Finaplix-H®, Finaplix-S®, and Revalor-S®.

Veterinarians use MGA, a synthetic form of progesterone, as a feed supplement to improve rate of weight gain and feed utilization. Examples of MGA are Steakmaker® and MGA 200® (these products were discussed previously with methods to suppress estrus in feedlot heifers). Heifers are fed this supplement daily to improve weight gain and feed efficiency (estrus is also suppressed). This supplement is not effective in steers.

Zeranol is an analog of a naturally occurring plant estrogen and is used as a subcutaneous ear implant to enhance weight gain and feed efficiency in cattle and sheep. Examples of zeranol are Ralgro® beef cattle implant and Ralgro® feedlot lamb implant.

Side effects of growth promotants that are synthetic hormones include mounting behavior, rectal prolapse, ventral edema, and udder development. Most growth promotants should not be given to dairy cattle or breeding animals.

TIP

All removed implants or inserts should be placed in a sealed container until they can be properly disposed of in accordance with local, state, and federal regulations.

TISSUE BUILDING

Anabolic steroids are tissue-building substances. Anabolic steroids cause a positive nitrogen balance and reverse tissue breakdown. These products are labeled for use in dogs, cats, and horses for anorexia, weight loss, debilitation, and to promote red blood cell formation. Anabolic steroids are C-III controlled substances. They should not be used in pregnant animals or animals intended for food. Testosterone is a natural anabolic steroid. Other examples include *stanozolol* (Winstrol-V®), which comes in injectable and tablet form; and *boldenone undecylenate* (Equipoise®), which comes in injectable form.

SUMMARY

The endocrine system consists of ductless glands that secrete hormones into the bloodstream. Disease of the endocrine system is usually a result of over- or underproduction of these hormones. Drugs used in the management of the endocrine system include substances to regulate pituitary hormones, blood glucose, metabolism via the thyroid gland, and the reproductive system.

The anterior pituitary gland produces and secretes chemicals that affect other endocrine glands and tends to be used in the diagnosis of endocrine diseases. The posterior pituitary gland secretes ADH, which affects urine production; and oxytocin, which affects uterine contractions.

Insulin and glucagon regulate blood glucose levels. Insulin is used to treat the hyperglycemia seen in patients with diabetes mellitus. Insulin is categorized as short-, intermediate-, or long-acting. Insulin concentration is expressed in units of insulin per milliliter. Insulin should be refrigerated and not frozen when stored. It is important never to shake insulin, as this results in degradation of the insulin molecule.

Hypothyroidism is a disease in which animals do not produce enough thyroid hormone. The drug of choice for treating hypothyroidism is levothyroxin, a synthetic form of thyroxine. Hyperthyroidism is a disease in which animals produce too much thyroid hormone. Hyperthyroidism is treated by surgically removing the dysfunctional thyroid gland, by destroying the dysfunctional thyroid gland with radioactive iodine, or by blocking hormone function with methimazole.

Corticosteroids are produced by the adrenal cortex. Adrenocortical insufficiency results in deficient production of corticosteroids (glucocorticoids and mineralocorticoids). Treatment of adrenocortical insufficiency includes use of desoxycorticosterone (DOCP), a long-acting mineralocorticoid; and prednisone, a glucocorticoid. Hyperadrenocorticism, a disease that results in overproduction/oversupplementation of glucocorticoids, is treated with mitotane. Mitotane destroys part of the adrenal cortex. Other treatment options for hyperadrenocorticism include ketoconazole and selegiline.

Drugs affecting reproduction in animals include testosterone, estrogen, progesterone, prostaglandin, and gonadotropins. Veterinarians use testosterone to treat infertility, hypogonadism, and urinary incontinence, and to produce teaser animals in cattle. Estrogen is used in animals to prevent conception and treat urinary incontinence in dogs; to treat persistent corpus lutea, expel retained placentas and mummified feti, and to promote weight gain in cattle; and to induce estrus in mares. Progesterone and progestins prevent estrus in dogs, synchronize breeding in cattle, and treat behavior problems and dermatitis in small animals. Progesterones and progestins can be recognized by "gest" in their generic names. Prostaglandins, mainly prostaglandin $F_{2\alpha}$, are used for estrus synchronization in cattle and horses, and for treating pyometra, causing abortion, and inducing parturition in small animals. Prostaglandins can be recognized by "prost" in their generic names. Gonadotropins cause release of LH and FSH or simulate their activity. Gonadotropins include GnRH (used to treat follicular cysts in cattle and induce estrus in small animals), PMSG and FSH-P (used to induce estrus or as a follicle stimulant), and hCG (used to treat cystic ovaries in cattle, to detect cryptorchidism and induce cycling in dogs, and to cause ovulation in mares).

Growth promotants, used to improve growth in animals, include antimicrobials, estradiol, testosterone, progesterone, synthetic hormones, and anabolic steroids. Side effects include mounting behavior, rectal prolapse, ventral edema, and udder development.

Table 10-3 summarizes the drugs covered in this chapter.

CHAPTER REFERENCES

Amundson Romich, J. (2000). *An illustrated guide to veterinary medical terminology*. Clifton Park, NY: Thomson Delmar Learning.

Britt, V. (2000). Glipizide. *Compendium of Continuing Education for the Practicing Veterinarian* 22(7): 678–679.

Delmar's A-Z NDR-97 nurses drug reference. Clifton Park, NY: Thomson Delmar Learning.

Reiss, B., & Evans, M. (2002), *Pharmacological aspects of nursing care* (6th ed, revised by Bonita E. Broyles, pp. 672–759). Clifton Park, NY: Thomson Delmar Learning.

Veterinary pharmaceuticals and biologicals (12th ed.). (2001). Lenexa, KS: Veterinary Healthcare Communications.

Veterinary values (5th ed.). (1998). Lenexa, KS: Veterinary Healthcare Communications.

Wynn, R. (1999). *Veterinary pharmacology*. Cambridge, MA: Harcourt Learning Direct.

TABLE 10-3 Drugs Covered in This Chapter

Endocrine System Condition or Action	Drug(s) Used to Treat Condition
Blood glucose regulation	• insulin • glipizide
Hypothyroid	• levothyroxine • liothryonine
Hyperthyroid	• methimazole • radioactive iodine I-131
Adrenocortical insufficiency	• desoxycorticosterone (DOCP) • fludrocortisone • prednisolone
Hyperadrenocorticism	• mitotane • ketoconazole • selegiline
Testosterone or testosterone-like products	• testosterone cypionate • testosterone enanthate • testosterone propionate • danazol • mibolerone
Estrogen or estrogen-like products	• estradiol cypionate • diethylstilbesterol
Progesterone or progesterone-like products	• megestrol acetate • medroxyprogesterone acetate • altrenogest • progesterone
Prostaglandins	• dinoprost • fluprostenol • cloprostenol
Gonadotropins	• gonadorelin • PMSG • FSH-P • hCG
Growth promotants (antimicrobial)	• bacitracin • flavomycin • virginiamycin • avoparcin • lasalocid • monensin
Growth promotants (hormonal)	• estradiol • testosterone • progesterone
Anabolic steroids	• stanozolol • boldenone undecylenate

CHAPTER REVIEW

Matching

Match the drug name with its action.

1. ______ levothyroxine
2. ______ vasopressin
3. ______ gonadorelin
4. ______ insulin
5. ______ oxytocin
6. ______ TBA (trenbolone acetate)
7. ______ estradiol
8. ______ methimazole
9. ______ megestrol
10. ______ dinoprost

a. a synthetic hormone used to induce estrus in horses during the nonbreeding season
b. a natural hormone used to promote growth in steers
c. drug of choice for treating hypothyroidism
d. chemical that has antidiuretic effect
e. GnRH used to treat follicular cysts in cattle
f. drug of choice for treating hyperthyroidism
g. drug used to treat diabetes mellitus in dogs
h. injectable drug used to stimulate uterine contractions
i. progestin used to treat behavior problems in cats
j. progesterone used to synchronize estrus in cattle

Multiple Choice

Choose the one best answer.

11. Which group of growth promotants causes tissue building and promotes red blood cell formation?
    a. estradiol
    b. progesterones
    c. anabolic steroids
    d. antimicrobials
12. Which estrus synchronization drug, if given to a herd of cattle, will lyse any CLs and cause the cattle to cycle all at the same time?
    a. prostaglandins
    b. GnRH
    c. estrogens
    d. testosterone
13. Adrenal cortex dysfunction, including adrenocortical insufficiency and hyperadrenocorticism, can be diagnosed by
    a. supplementing with desoxycorticosterone.
    b. supplementing with *o,p′*-DDD.
    c. clinical signs.
    d. measuring cortisol levels before and after administration of ACTH.
14. Decreased coat luster, hair loss, decreased appetite, listlessness, intolerance to cold, and reproductive failure are all signs of
    a. diabetes mellitus.
    b. hyperthyroidism.
    c. Addison's disease.
    d. hypothyroidism.
15. Insulin concentration is measured in
    a. units/mm.
    b. units/ml.
    c. cm/lb.
    d. g/ml.
16. Which insulin form is used initially to treat diabetic ketoacidosis and to stabilize glucose levels in newly diagnosed diabetic animals?
    a. Short-acting
    b. Intermediate-acting
    c. Long-acting
    d. Ultralong-acting

17. The link between the hypothalamus, pituitary gland, and some endocrine glands—in which low levels of hormone signal the hypothalamus to secrete more releasing factor, which in turn signals the pituitary gland to secrete stimulating hormones, resulting in increased hormone levels—is called a
    a. feedback stimulation mechanism.
    b. negative feedback loop.
    c. positive feedback loop.
    d. connected feedback loop.

True/False

Circle a. for true or b. for false.

18. Oxytocin is beneficial in treating both open and closed pyometra.
    a. true
    b. false
19. Bovine somatotropin (BST), is believed to cause increased milk production in dairy cattle.
    a. true
    b. false
20. TSH, ACTH, LH, and FSH are hormones controlled by the posterior pituitary gland.
    a. true
    b. false

Case Studies

21. A rancher would like his replacement heifers to calve three weeks before his mature cows. He would like to artificially inseminate the heifers with proven "calving ease bulls." Assume that all heifers are cycling females.
    a. What treatment options does this rancher have?
    b. The rancher decides to use 25 mg dinoprost IM. This drug comes in 5 mg/ml concentration. How much does each heifer get?
    c. Another rancher tells him to use a newer protocol for ovulation synchronization called Ovsynch. It uses GnRH (Cystorelin®) and dinoprost (Lutalyse®). Here is the protocol:
        - Day 1: Give 2 cc Cystorelin® (50 mcg/ml GnRH) IM.
        - Day 8: Give 5 cc Lutalyse® (5 mg/ml dinoprost) IM at 8 A.M.
        - Day 10: Give 2 cc Cystorelin® IM at 5 P.M.
        - Day 11: Breed all animals by artificial insemination at 8 A.M.
    d. On day 1 and day 10, what does the Cystorelin® do?
    e. On day 1 and day 10, how many total mcg does the rancher give each heifer?
    f. On day 8, what does the Lutalyse® do?
    g. On day 8, how many total mg does the rancher give each heifer?
    h. What should you tell this rancher (and remember yourself) about handling dinoprost?
22. A Jersey cow (Figure 10-11), 120 days in milk, is examined during a normal herd health visit at the farm for being anestrus. Physical exam: The cow is in good body flesh. Rectal palpation reveals a 40-mm soft circular structure on the left ovary. There are no other structures on either ovary. The uterus is normal size for the number of days postpartum. This cow is diagnosed with a cystic follicle (follicular cyst).
    a. What drug do you think the veterinarian will prescribe for this cow?
    b. This cow needs 100 mcg Cystorelin® IM. It comes in 50 mcg/ml concentration. How much should this cow receive?

FIGURE 10-11 Jersey cow. Courtesy of American Jersey Cattle Club.

23. A 3-yr.-old M/N Golden Retriever weighing 90# (Figure 10-12) presents to the clinic with hair loss and weight gain. He has been less active than normal. Physical exam reveals normal TPR, alopecia over the shoulders and hindquarters, and listlessness. Blood tests showed that the dog was hypothyroid.
    a. Explain to the owner what hypothyroidism is.
    b. How do we treat hypothyroidism? What form of hormone is this?
    c. The veterinarian prescribes Soloxine® at a dosage of 0.02 mg/kg daily. How much should this dog get?
    d. Soloxine® comes in 0.1 mg, 0.2 mg, 0.3 mg, 0.4 mg, 0.5 mg, 0.6 mg, 0.7 mg, and 0.8 mg pills. How many pills should this dog get per dose?

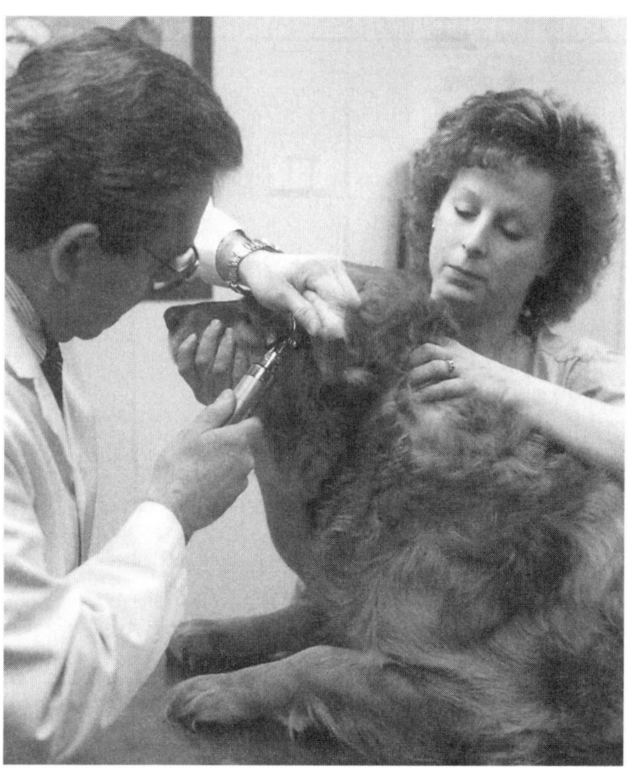

FIGURE 10-12 Golden Retriever. Courtesy of Brian Yacur and Guilderland Animal Hospital.

CHAPTER 11

Gastrointestinal Drugs

OBJECTIVES

Upon completion of this chapter, the learner should be able to

- describe the basic structure and function of the gastrointestinal tract.
- describe various signs of gastrointestinal disease.
- describe the use of the following drugs in veterinary medicine: antisialogues, antidiarrheals, laxatives, anti-emetics, systemic and nonsystemic antacids, emetics, antiulcer compounds, antifoaming agents, prokinetic agents, and digestive enzymes.
- outline the use of dental prophylaxis and treatment aids.

KEY TERMS

cholinergic drugs
anticholinergic drugs
sympathetic drugs
antisialogue
antidiarrheal
protectants
adsorbents
probiotics
laxative
cathartic
anti-emetic
emetic
antiulcer drugs
antacid
antifoaming agent
prokinetic agent
digestive enzyme

INTRODUCTION

A teenager calls the clinic and says that she has been babysitting a child and his dog. The dog got into the cupboard and spilled cleaning supplies on the floor. When the babysitter found the dog, he was lapping up some of the chemicals from the floor. She wants to know what she can give the dog to make him vomit up the chemicals. What do you tell her? Is vomiting warranted in this case? What information do you need before making that decision? What home options does the babysitter have, or is it better to bring the dog into the clinic?

BASIC DIGESTIVE SYSTEM ANATOMY AND PHYSIOLOGY

Gastrointestinal tract, *alimentary system*, *GI system*, and *digestive tract* are all terms that basically describe a long, muscular tube that begins at the mouth and ends at the anus. The anatomic structures therein include the oral cavity, pharynx, esophagus, stomach, small intestine, and large intestine. These structures vary from monogastric animals with simple stomachs to ruminant animals with multichambered forestomachs (Figure 11-1). In either type of animal, the digestive tract plays a role of bringing life-sustaining

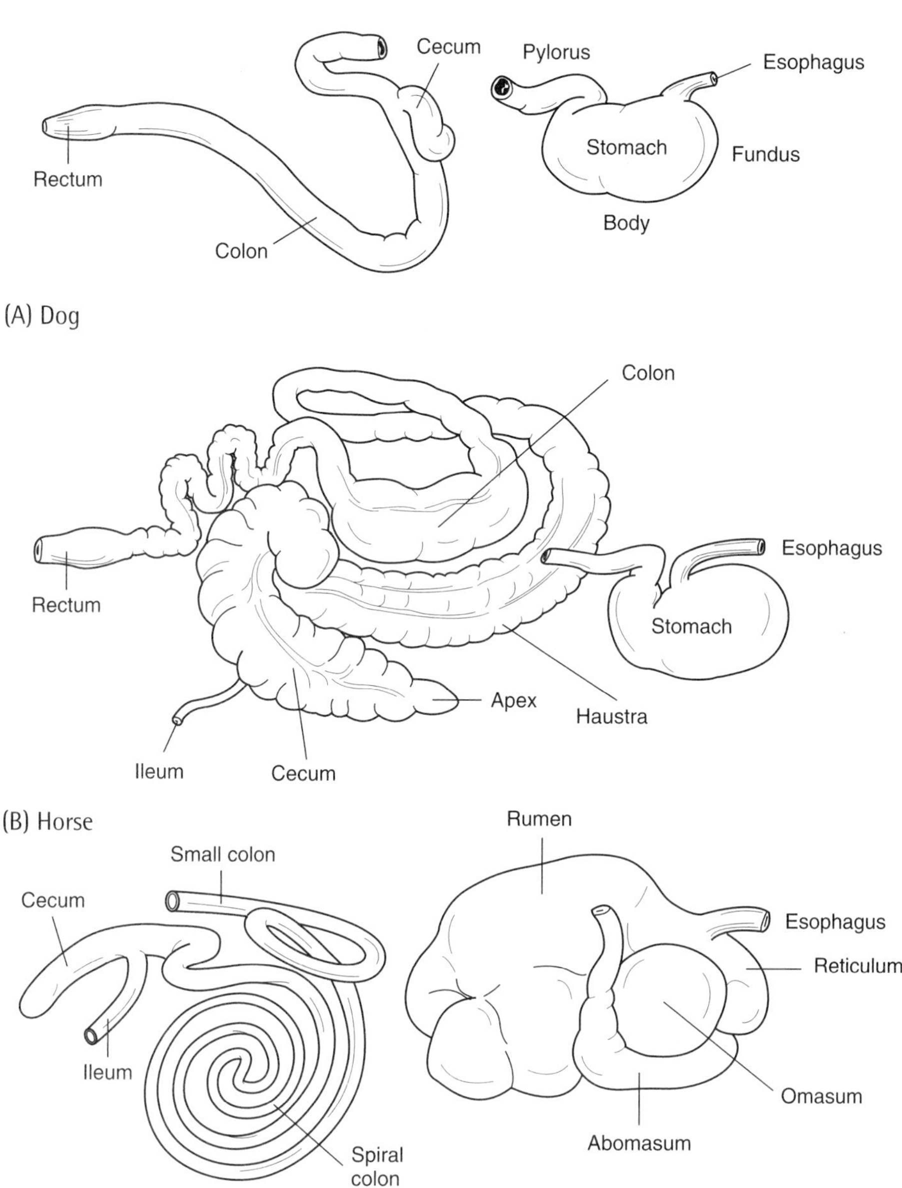

FIGURE 11-1 Gastrointestinal tracts (small intestinal segments omitted for clarity): (A) Dog (B) Horse (C) Ruminant

elements into the body, and taking waste products out of it. Proper functioning of this system depends on an unobstructed and regulated flow of these elements. In addition to the structures of this tube, there are accessory organs (liver, gall bladder, salivary glands, and pancreas) that aid the digestive tract through the production or secretion of enzymes that help in the digestion of food products.

Flow of material along the gastrointestinal tract occurs by *peristalsis*, a wavelike contraction of longitudinal and circular muscle fibers that moves food through the gut. Another process occurring in the gastrointestinal tract is *segmentation*: periodic, repeating intestinal constrictions that cause churning of the GI contents (Figure 11-2). Regulation of these actions is controlled by many mechanisms. One control mechanism of the GI tract is the autonomic nervous system (ANS), which consists of the sympathetic branch (fight-or-flight response) and the parasympathetic branch (homeostatic

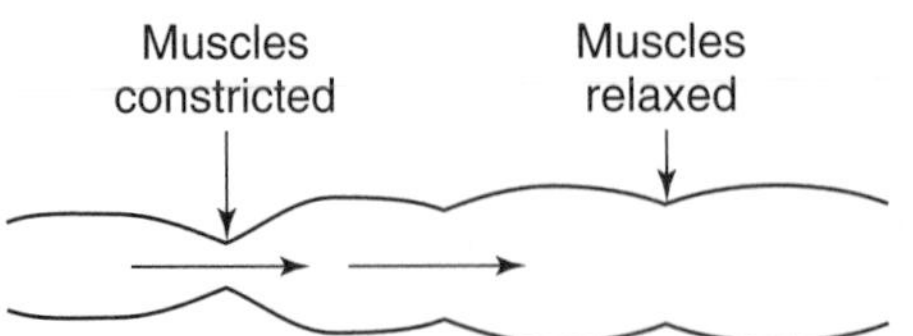

FIGURE 11-2 Peristalsis and segmentation: (A) Peristaltic contraction moves food through the digestive tract. (B) Segmentation helps break down and mix food through cement-mixer-type action.

response). Parasympathetic stimulation increases intestinal motility, increases GI secretions, and relaxes sphincters. **Cholinergic drugs** simulate these actions. **Anticholinergic drugs** inhibit these actions. Sympathetic stimulation decreases intestinal motility, decreases intestinal secretions, and inhibits the action of sphincters. **Sympathetic drugs** simulate these actions. Stretch receptors in the GI tract also increase peristalsis.

Chemical secretions also control the GI tract. Hormones released from the intestinal cells help control actions such as gall bladder emptying. Chemical substances such as histamine can attach to the H_2 receptors of the gastric cells, causing increased HCl production.

GASTROINTESTINAL DRUGS

TIP

Atropine comes in two concentrations: Atropine Injectable-SA (0.5 mg/ml) and Atropine Injectable-LA (2 mg/ml). Care should be taken when using atropine to ensure that the proper concentration is chosen.

Gastrointestinal drugs help maintain the unobstructed and regulated flow of food into the body and waste products out of the body. Infections and disease can alter the functioning of the GI tract. Bacterial infections resulting in endotoxin release from gram-negative bacterial cell walls can increase intestinal blood vessel permeability, resulting in increased fluid loss. Disease conditions such as liver disease can affect the production of bile and ultimately fat digestion. Although many diseases can affect gastrointestinal tract function, this chapter covers only drugs that directly affect the gastrointestinal tract.

Veterinary gastrointestinal drugs can be administered to encourage peristalsis, to suppress it, or to reduce its undesirable byproducts. Gastrointestinal drugs do this in a variety of ways resulting in a variety of outcomes.

Saliva Stopping

TIP

Glycopyrrolate does not appreciably cross into the CNS or placenta, making it more desirable for use in pregnant animals than atropine, which does cross into the CNS and placenta.

Antisialogues are drugs that decrease salivary flow. Veterinarians administer antisialogues intravenously, intramuscularly, or subcutaneously to limit this excess saliva production, which often occurs secondary to anesthetic drug use. Glycopyrrolate (Robinul®) and atropine (Atropine Injectable-SA®, Atropine Injectable-LA®) are antisialogues via their anticholinergic mechanism of action. Anticholinergic drugs block the effects of acetylcholine at parasympathetic nerve endings. The effect of anticholinergics is to reduce gastrointestinal motility and secretions (including saliva). Antisialogues, although used to decrease salivation, can also affect peristalsis. In addition to decreasing saliva production, anticholinergic drugs are used in the gastrointestinal tract to treat vomiting, diarrhea, and excess gastric secretion by blocking parasympathetic nerve impulses (see section on antidiarrheals and anti-emetics). Side effects of anticholinergics include dry mouth, constipation, CNS stimulation, tachycardia, and pupillary dilation.

Diarrhea Stopping

Diarrhea is abnormal frequency and liquidity of fecal material due to failure of the intestinal tract to adequately absorb fluids from the intestinal contents. Diarrhea is not a disease but rather a sign of underlying disease. There are many causes of diarrhea, including parasitic disease, bacterial infection, viral infection, dietary indiscretion, or a systemic nongastrointestinal disease. Diarrhea is a concern because it can cause excessive fluid loss that results in dehydration and electrolyte imbalance. Diarrhea also decreases the uptake of nutrients from the intestine if the contents are moving too quickly through it. The longer diarrhea is allowed to continue, the more likely it is that the animal will experience side effects. Side effects of diarrhea include electrolyte imbalances, muscle weakness, acid-base disturbances, and anorexia.

Antidiarrheals are drugs that decrease peristalsis of the gastrointestinal tract, thereby allowing fluid absorption from the intestinal contents. This helps to reverse diarrhea by decreasing the liquidity of stool. Antidiarrheals include anticholinergics, protectants, adsorbents, and narcotic analgesics.

Anticholinergics. Anticholinergics are used to treat tenesmus (straining to defecate) associated with colitis and vomiting related to colonic irritation. This group of drugs blocks acetylcholine release from the parasympathetic nerve endings, thus decreasing gastrointestinal motility and secretions. It should be kept in mind that in some cases decreased gastrointestinal motility is already associated with diarrhea; therefore, these drugs should be used with caution. Side effects of anticholinergics include dry mucous membranes, urine retention, tachycardia, and constipation. Drugs in this category include:

- atropine (generic, Atropine Injectable-SA®, Atropine Injectable-LA®).
- aminopentamide (Centrine®).
- isopropamide (Darbazine®).
- propantheline (Pro-Banthine®).
- methscopolamine (Pamine®).

Protectants/Adsorbents. This category of antidiarrheal drugs works either by coating inflamed intestinal mucosa with a protective layer (**protectants**), or by binding bacteria and/or digestive enzymes and/or toxins to protect intestinal mucosa from their damaging effects (**adsorbents** bind substances). Side effects from these drugs are uncommon except for the possible development of constipation. Drugs in these categories include:

- bismuth subsalicylate (Corrective Mixture®, Pepto-Bismol®). The bismuth portion of this drug coats the intestinal mucosa and has anti-endotoxic and weak antibacterial effects. The subsalicylate portion has anti-inflammatory effects because it reduces the production of prostaglandins. Subsalicylate is an aspirin-like product; therefore, this drug should not be used in cats. Corrective Mixture with Paregoric® has an opium tincture in it and is a C-V controlled substance. These products can blacken stool and can also cause opacities on radiographs.
- kaolin/pectin (Kao-Forte®, Kaopectolin®, K-P-Sol®). This combination drug has both adsorbent and protective qualities. Bacteria and toxins are adsorbed in the gut and the coating action protects inflamed intestinal mucosa.
- activated charcoal (Liqui-Char®, Superchar®, Actidose-Aqua®). Activated charcoal is a fine, black, tasteless powder used to adsorb many chemicals and drugs in the upper gastrointestinal tract. It is used primarily to treat ingestions of certain toxins and is typically administered via stomach tube.

Opiate-Related Agents. Opiates or narcotic analgesics control diarrhea by decreasing both intestinal secretions and the flow of feces, and increasing segmental contractions, thereby resulting in increased intestinal absorption. Side effects of these drugs

TIP

An *adsorbent* is a substance that binds other material to its surface. A gastrointestinal adsorbent adsorbs gases, toxins, bacteria, drugs, and digestive enzymes in the stomach and intestines. Adsorbents lack specificity and may prevent absorption of other drugs the animal was given orally.

TIP

Most adsorbents are clay-like material given in tablet or liquid form.

include CNS depression (excitement in horses and cats), ileus, urinary retention, bloat, and constipation with prolonged use. Examples include:

- diphenoxylate (Lomotil®, Lonox®, Diphenatol®), a C-V controlled substance with atropine added. Diphenoxylate is not an opium derivative, but is structurally similar to meperidine.
- loperamide (Imodium®, Imodium A-D®). This drug causes less CNS depression than other drugs in this category and can be purchased over the counter. Like diphenoxylate, it is structurally similar to meperidine.
- paregoric. This drug is a C-III controlled substance and may be combined with kaolin/pectin. Paregoric is camphorated tincture of opium.

Probiotics. Another approach in treating diarrhea is the use of **probiotics** to seed the gastrointestinal tract with beneficial bacteria such as *Lactobacillus spp.* and *Enterococcus faecium*. This approach is based on the theory that some forms of diarrhea (like that secondary to antibiotic use) are caused by disruption of the normal bacterial flora of the gastrointestinal tract. Plain yogurt with active cultures is often used to try to repopulate the gastrointestinal tract with beneficial bacteria. Trade names of probiotic products used in veterinary medicine include Fastrack® gel, Probiocin® oral gel for pets, and Probiocin® oral gel for ruminants. These products may have to be refrigerated to maintain the viability of the bacterial culture.

Metronidazole. Metronidazole (Flagyl®) is an antibiotic that is effective against anaerobic bacteria. It is sometimes used as an antidiarrheal on the theory that disruption of the normal-flora environment may increase the number of anaerobic bacteria present in an animal with diarrhea. Because metronidazole is effective against anaerobes, it may help return the animal's stool to its normal consistency. Additional information on metronidazole and its spectrum of activity appears in Chapter 14.

Stool Loosening

Constipation is a condition in which passage of feces is slowed or nonexistent. A **laxative** is a medicine that loosens the bowel contents and encourages evacuation of stool. Veterinarians use laxatives to help animals evacuate without excessive straining, to treat chronic constipation from nondietary causes and movable intestinal blockages (such as hairballs), and to evacuate the GI tract before surgery, radiography, or proctoscopy. **Cathartics** are harsher laxatives that result in a soft to watery stool and abdominal cramping. A *purgative* is a harsh cathartic, causing watery stool and abdominal cramping. Categories of laxatives include osmotic, stimulant, bulk-forming, and emollients (stool softeners).

Osmotic (or hyperosmolar) *laxatives* include salts or saline products, lactulose, and glycerin. The saline products are composed of sodium or magnesium, and small amounts may be systemically absorbed, causing electrolyte imbalances. Hyperosmolar salts pull water into the colon and increase water content in the feces, increasing bulk and stimulating peristalsis. Use of saline products should be limited in animals with heart failure and renal dysfunction; some (such as Fleet Enema®), are not recommended for cats. Prolonged use of osmotic laxatives can also cause dehydration. Examples of osmotics include:

- lactulose (Cephulac®), also used in liver disease because it eliminates ammonia.
- sodium phosphate with sodium biphosphate (Fleet Enema®, Gent-L-Tip Enema®).
- magnesium sulfate (Epsom Salts).
- magnesium hydroxide (Milk of Magnesia®, Carmilax-Powder®, Magnalax®, Poly Ox II Bolus®).

Stimulant (irritant or contact) *laxatives* increase peristalsis by chemically irritating sensory nerve endings in the intestinal mucosa. Their site of action may vary from affecting only the large or small intestine to affecting the entire gastrointestinal tract. Although most stimulant laxatives come from natural sources, two agents (bisacodyl and phenolphthalein) are synthetic. Many stimulant laxatives are absorbed systemically and can cause a variety of side effects. Examples in this group include:

- bisacodyl (Dulcolax®), a cathartic that comes in enteric-coated and suppository forms. Side effects include cramping and diarrhea. Tablets should not be crushed or chewed, as this will result in intense abdominal cramping. Milk or antacids should not be given within one hour of administration of this drug, as they may dissolve the enteric coating.
- phenolphthalein (Ex-Lax®, Correctol®), which is given orally to loosen stool by altering water and electrolyte absorption. Side effects include cramping, diarrhea, and discoloration of stool. The laxative effect of phenolphthalein may last for a few days.
- castor oil, which has an active ingredient (ricinoleic acid) that is liberated in the small intestine. Ricinoleic acid inhibits water and electrolyte absorption, leading to fluid accumulation in the gastrointestinal tract and increased peristalsis. Side effects include diarrhea, abdominal pain and electrolyte imbalance.

The *bulk-forming laxatives* group consists of natural fibrous substances or semisynthetic compounds that absorb water into the intestine, increase fecal bulk, and stimulate peristalsis, resulting in large, soft stool production. Unlike stimulant laxatives, bulk-forming laxatives tend to produce normally formed stools. Most of these products contain psyllium preparations or plantago seed (the seed coating absorbs water and swells). Most bulk-forming laxatives have minimal effects on nutrient absorption and are not systemically absorbed. Examples include:

- psyllium hydrophilic mucilloid (Metamucil®, Equine Psyllium®, Perdiem®).
- polycarbophil (FiberCon®, Fiberall®).
- bran.

Emollients are stool softeners (which reduce stool surface tension and reduce water absorption through the colon), lubricants (which facilitate passage of fecal material, increasing water retention in stool), and fecal wetting agents (detergent-like drugs that permit easier penetration and mixing of fats and fluid with the fecal mass, resulting in a softer stool that is more easily passed). Emollients are not absorbed systemically, and thus have fewer side effects, as well as decreasing straining during defecation. Side effects are rare but include some abdominal cramping and diarrhea. Examples include:

- docusate sodium (Colace®). This is also known as docusate sodium succinate (DSS).
- docusate calcium (Surfak®).
- docusate potassium (Dialose®).
- petroleum products (solid or liquid). Mineral oil is liquid petroleum, commonly used in horses as a laxative, but also used to decrease absorption of fat-soluble toxins. White petroleum is solid petroleum, commonly used to prevent and treat hairballs in cats (CatLax®, Laxatone®).

TIP

Animals taking bulk-forming laxatives should have free access to water. Water consumption is necessary to assure that the bulk-forming process goes smoothly.

TIP

Enemas are liquids administered directly into the colon. They can elicit a laxative response as well as cleansing the bowel prior to a surgical or diagnostic procedure.

Vomit Stopping

Vomiting or *emesis* is the expulsion of stomach (and sometimes duodenal) contents through the mouth. Vomiting is caused by a variety of things, including viral and bacterial infection, dietary indiscretion, food intolerance, surgery, pain, and other drugs. The act of vomiting is controlled by the vomiting center in the medulla of the brain. Acetylcholine is the neurotransmitter for the vomiting center. The vomiting center

TIP

Mechanoreceptors in the digestive tract detect tension in response to distention and contraction in the gastrointestinal tract. Chemoreceptors in the digestive tract detect chemicals in the intestinal lumen.

gets input from many pathways that signal it to activate. These inputs include equilibrium changes in the inner ear, responses of the higher centers of the cerebral cortex due to pain or fear, intracranial pressure changes, vagus nerve stimulation in the gastrointestinal tract, and activity in the chemoreceptor trigger zone (CRTZ). Figure 11-3 shows the areas that affect the vomiting center.

TIP

Keep in mind that not all animals vomit; horses, rabbits, and rats do not. In ruminants, abomasal content can go into the forestomach, but ejection of content from the mouth does not occur.

TIP

Many times vomiting can be controlled by withholding food and water (NPO) and correcting dehydration.

Vomiting occurs secondary to stimulation of receptors in the central nervous system and/or gastrointestinal tract. In the central nervous system, the CRTZ (which lies near the medulla), senses chemical changes. The CRTZ is outside the blood-brain barrier and therefore responds to stimuli (toxins, drugs, etc.) from either the cerebrospinal fluid (CSF) or the blood. This detection of chemical changes stimulates activation of the vomiting center. Dopamine is the neurotransmitter for the CRTZ and stimulation of the CRTZ results in dopamine release and stimulation of the vomiting center. Some sensory impulses, such as odor and taste, are transmitted directly to the vomiting center from higher cerebral cortical centers. Impulses from the inner ear related to equilibrium are also sent directly to the vomiting center. Peripheral irritation to the gastrointestinal tract by infectious agents, foreign material, chemicals, and overdistention sends impulses to the vomiting center via the vagus nerve. This mechanism may be protective by inhibiting absorption of toxic materials.

The concerns with vomiting are dehydration, electrolyte loss (the animal's electrolyte balance will be shifted due to loss of sodium, potassium, and chloride ions), and acid-base changes. Ideally, the cause of any vomiting will be identified and treated; however, **anti-emetics** (drugs that control vomiting) may help alleviate discomfort and help control electrolyte balance. Most anti-emetics are given parenterally, as the patient may vomit the medication before it can be absorbed through the gastrointestinal tract. Types of anti-emetics include the following.

Phenothiazine derivatives. This group of drugs works by inhibiting dopamine in the CRTZ, thus decreasing the stimulation to vomit. Side effects include hypotension and sedation. Phenothiazines lower the seizure threshold and therefore should not be used in epileptic animals. Examples in this group include:

- acepromazine (PromAce®).
- chlorpromazine (Thorazine®, Laragactil®, Thor-Prom®).

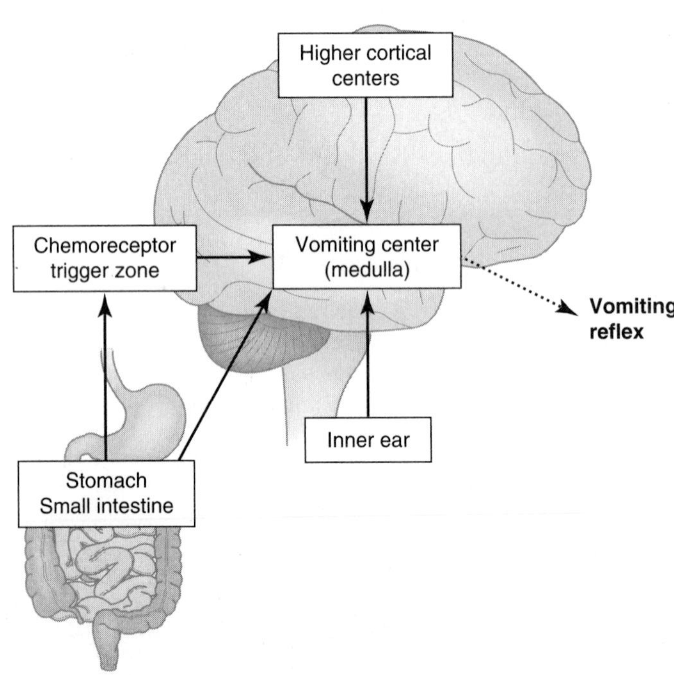

FIGURE 11-3 The vomiting center and factors that control vomiting

- prochlorperzine (Compazine®; Darbazine® has isopropamide added to the prochlorperzine).
- perphenzaine (Trilafon®).

Antihistamines. Antihistamines are used to control vomiting in small animals when the vomiting is due to motion sickness, vaccine reactions, or inner ear problems. They work by blocking input from the vestibular system to the CRTZ. Antihistamines can cause sedation. Examples include:

- trimethobenzamide (Tigan®).
- dimenhydrinate (Dramamine®).
- diphenhydramine (Benadryl®).

Anticholinergics. Anticholinergics block acetylcholine peripherally, which decreases intestinal motility and secretions. These drugs also decrease gastric emptying, which may increase the tendency to vomit. Side effects include dry mouth, constipation, urinary retention, and tachycardia. These drugs are contraindicated in animals with glaucoma or pyloric obstruction. Examples include:

- aminopentamide (Centrine®) (may also block CRTZ).
- atropine (may also block CRTZ).
- propantheline (Pro-Banthine®).

Procainamide derivatives. This group of anti-emetics works both centrally, by blocking the CRTZ (as a dopamine antagonist), and peripherally, by speeding gastric emptying, strengthening cardiac sphincter tone (decreasing gastroesophageal reflux), and increasing the force of gastric contractions. These anti-emetics are not recommended in animals with GI obstructions, GI perforation, or hemorrhage, because of the peripheral effects (stimulation of gastric motility, which is harmful in these cases). However, this effect also makes them acceptable for the treatment of gastric motility disorders. One example in this group is metoclopramide (Reglan®), used widely in human medicine to control the vomiting associated with cancer chemotherapy treatment. It stimulates motility of the upper gastrointestinal tract without stimulating the production of gastric, biliary, or pancreatic secretions.

Serotonin receptor antagonists. Serotonin receptor antagonists work selectively on 5-HT_3 receptors, which are located peripherally (on nerve terminals of the vagus nerve) and centrally (in the CRTZ). It is believed that some chemicals (especially chemotherapeutic agents and certain anesthetic agents such as propofol) cause vomiting because they increase serotonin release from cells in the small intestine. Blocking the release of serotonin from these sites controls vomiting. These drugs are expensive, can cause gastrointestinal upset, and are reserved for vomiting associated with specific chemicals. Drugs in this category include ondansetron (Zofran®) and dolasetron (Anzemet®).

Vomit Producing

Emetics are drugs that induce vomiting. Emetics are used in the treatment of poisonings and drug overdoses. Keep in mind that vomiting should not be induced if caustic substances have been ingested, such as ammonia, lye, bleach, or other damaging products. Always check with poison control before inducing vomiting. Emetics may be centrally acting (working on the CRTZ) or peripherally acting (working on receptors locally).

TIP

Emetics should not be used in unconscious, seizuring, or compromised animals.

Centrally acting emetics include apomorphine, which stimulates dopamine receptors in the CRTZ and thus induces vomiting. It is a morphine-derived emetic that is given SQ, IM, or topically in the conjunctival sac. Vomiting occurs rapidly, usually within 5 to 10 minutes when given peripherally or 10 to 20 minutes when given subconjunctivally. It is the emetic of choice for dogs and is available generically and

TIP

Ipecac syrup should not be given with activated charcoal, as the charcoal may absorb the emetic portion of the ipecac. However, if vomiting does not occur, it is recommended to give activated charcoal to absorb ipecac and the ingested toxin.

through compounding pharmacies. Side effects include protracted vomiting, CNS depression, and restlessness. Xylazine (Rompun®, Gemini®, AnaSed®) induces vomiting in cats as a side effect of its use as a sedative. The mechanism of emetic action is not fully understood for xylazine. It is considered the emetic of choice for cats. Xylazine does not induce vomiting in horses, cattle, sheep, or goats, due to their inability to vomit. Side effects include bradycardia and decreased respiratory rate.

The best-known peripherally acting emetic is ipecac syrup, which is made from roots and rhizomes of plants. It is an over-the-counter medication that is given orally. Two alkaloids, emetine and cephaeline, cause irritation to the gastric mucosa and centrally stimulate the CRTZ. Contents of both the stomach and small intestine are evacuated within 10 to 30 minutes. It is usually used in dogs and cats. Ipecac can cause cardiovascular problems with higher doses.

Some home remedy emetics that have been used to induce vomiting include hydrogen peroxide, salt and water, mustard and water, and salt followed by food. The results of these home remedies vary from case to case, and they are considered less reliable methods of inducing emesis.

Activated charcoal is given in poisoning cases when emesis is contraindicated. Activated charcoal absorbs many chemicals and drugs in the upper gastrointestinal tract, reducing their absorption. It comes in liquid form or powder form for reconstitution with water. Side effects include constipation or diarrhea and blackening of the feces. Trade names of activated charcoal products include SuperChar® Vet Powder, SuperChar® Vet Liquid, and Toxiban® Granules (Toxiban® products have activated charcoal and kaolin in them).

Ulcer Stopping

Ulcers are erosions of mucosa and are named according to the site of involvement: gastric ulcer, duodenal ulcer, and esophageal ulcer, for example. Ulcers form for many reasons, including metabolic disease, drug therapy, and stress. Most cases are due to increased release of hydrochloric acid from the parietal cells of the stomach. Histamine, gastrin, and acetylcholine influence the parietal cells. A thick layer of mucus separates and protects the mucosal lining from gastric secretions. Two sphincter muscles, the cardiac (located at the upper portion of the stomach) and the pyloric (located at the lower portion of the stomach), act as barriers to prevent reflux of acid into the esophagus and the duodenum. Signs of ulcers may include anorexia, melena, abdominal pain, and hematemesis.

Antiulcer drugs help prevent the formation of ulcers. Figure 11-4 summarizes antiulcer drug actions. Categories of antiulcer drugs include the following.

Antacids promote ulcer healing by neutralizing HCl and reducing pepsin activity. They do not coat the ulcer. Antacids can interact with other drugs by adsorption or binding of the other drugs (decreasing oral absorption of bound drug), by increasing stomach pH (causing a decrease in absorption of certain drugs), and by increasing urinary pH (may inhibit elimination of drugs that are weak bases). There are two types of antacids: systemic (those absorbed into the blood) and nonsystemic (those that remain primarily in the GI tract).

Systemic antacids are readily soluble in stomach fluid and, once dissolved, are readily absorbed. Most systemic antacids have a rapid onset and short duration of action. Examples of systemic antacids are sodium bicarbonate and calcium carbonate. Sodium bicarbonate can cause sodium excess and water retention and is not frequently used as an antacid in veterinary medicine. Calcium carbonate is more effective than sodium bicarbonate in neutralizing acid, but can result in excessive calcium levels, and is not frequently used as an antacid in veterinary medicine.

TIP

Because only a small portion of a nonsystemic antacid is absorbed, most of the drug stays in the GI tract and will not affect electrolyte levels.

Nonsystemic antacids remain primarily in the GI tract. They are composed of alkaline salts of aluminum (aluminum hydroxide) and magnesium (magnesium hydroxide). Magnesium hydroxide has greater neutralizing power than aluminum hydroxide. Magnesium may cause diarrhea and aluminum may cause constipation. A combina-

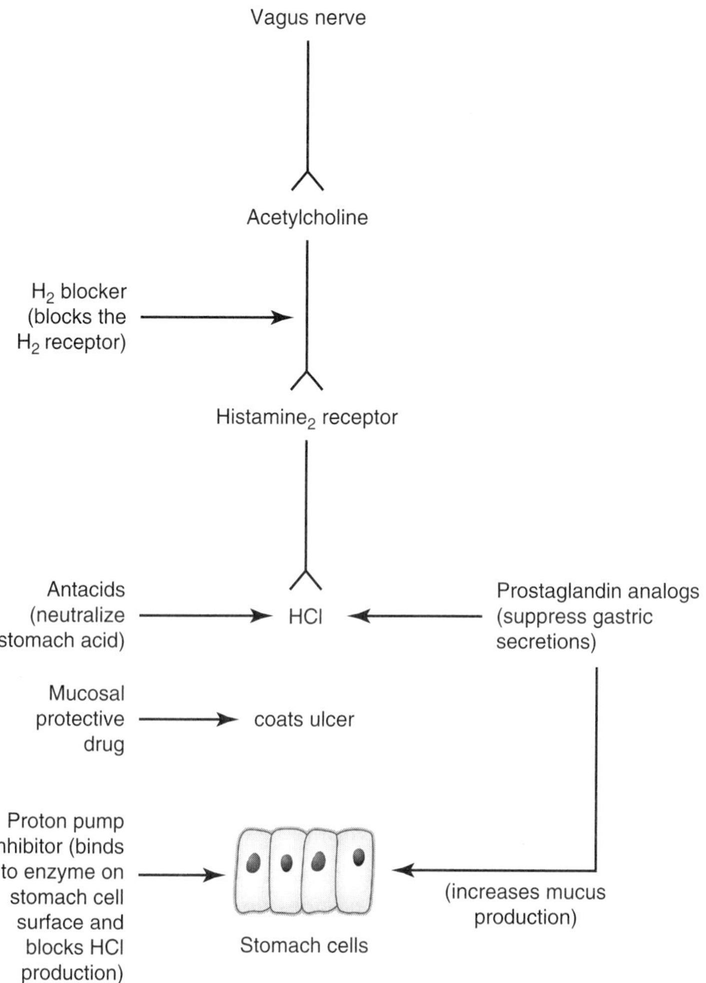

FIGURE 11-4 Mechanism of action of antiulcer drugs

tion of these two compounds provides acid neutralization with minimal GI side effects. Sometimes simethicone, an antigas agent, is added to these antacids.

Antacid use in animals has decreased due to difficulty of administration and the introduction of histamine-2 blockers (see next section). Examples include:

- magnesium hydroxide (Magnalax®, Rulax II®, Milk of Magnesia®). In ruminants, magnesium hydroxide is used to increase rumen pH and as a laxative to treat rumen acidosis (grain overload). Magnesium-containing antacids are contraindicated in animals with kidney disease because they may develop electrolyte imbalances.
- aluminum/magnesium hydroxide (Maalox®, Mylanta®).
- aluminum hydroxide (Amphojel®). Aluminum hydroxide has also been used to lower high serum phosphate levels because it binds and thereby depletes phosphorus.

> **TIP**
> To minimize interactions between antacids and other drugs, animals should not be given other oral medication for one to two hours after antacid administration.

Histamine-2 (H_2) or *receptor antagonists blockers* prevent acid reflux by competitively blocking the H_2 receptors of the parietal cells in the stomach, thus reducing gastric acid secretion. Side effects are rare, but may include diarrhea and inhibition of liver enzymes that affects metabolism of other drugs and waste products. Examples from this group include:

- cimetidine (Tagametor®), the first H_2 blocker developed. It comes in tablet, oral solution, and injectable forms.

- ranitidine (Zantac®), which is more potent and has a longer duration of action than cimetidine. It is also available in tablet, oral solution, and injectable forms.
- famotidine (Pepcid®), which is more potent and has fewer side effects than ranitidine, but is less bioavailable. It is available in coated tablets, oral powder, and injectable forms.

TIP

Sucralfate does little to neutralize stomach acid.

Mucosal protective drugs, also known as *pepsin inhibitors*, are typified by the drug sucralfate (Carafate®). Sucralfate is a chemical derivative of sucrose that is nonabsorbable and combines with protein to form an adherent substance that covers the ulcer and protects it from stomach acid and pepsin. Sucralfate comes in 1-gram tablets and its only side effect is constipation. Because sucralfate binds to ulcers in an acid environment, it should not be given at the same times as H_2 receptor antagonists.

Prostaglandin analogs appear to suppress gastric secretions and increase mucus production in the GI tract. An example of a prostaglandin analog is misoprostol (Cytotec®), an oral tablet that is usually given to animals taking nonsteroidal anti-inflammatory drugs (NSAIDs). Pregnant women should be cautioned about handling this medication, because of its prostaglandin effects. Side effects include GI signs such as diarrhea, vomiting, and abdominal pain.

Proton pump inhibitor drugs bind irreversibly to the H^+-K^+-ATPase enzyme on the surface of parietal cells of the stomach. This inhibits hydrogen ion transport into the stomach so that the cell cannot secrete HCl. When this enzyme is blocked, acid production is decreased, and this allows the stomach and esophagus to heal. Drugs in this group include:

- omeprazole (Prilosec®, Gastrogard®). In horses and foals more than four weeks of age, omeprazole is used to heal gastric ulcers and to prevent recurrence. Gastric ulcers in horses form secondary to feeding problems (too little hay intake), training (intense exercise increases gastric acid production), and changes with growth (a developing stomach may be injured by acid and enzymes). Gastric ulcers in horses interfere with performance.
- lansoprazole (Prevacid®), used to treat gastroesophageal reflux disease and to help heal gastric and duodenal ulcers. It blocks the last step of gastric acid production and is used extra-label in animals. Lansoprazole is less likely to react with other drugs than omeprazole.

Foam Stopping

Antifoaming agents are drugs that reduce or prevent the formation of foam. As a gastrointestinal drug, antifoaming agents are used in ruminants whose rumens are subject to acute frothy bloat. *Frothy bloat* (also known as pasture bloat or legume bloat) is a condition in which the rumen is distended with a gas that mixes with fluid to form a froth; the froth can asphyxiate the animal by blocking its ability to eructate (belch). Antifoaming agents make this foam less stable, breaking it up to promote gas release through belching. Poloxalene and polymerized methyl silicone accomplish this when administered as solutions by stomach tube directly into the forestomach. Trade names of antifoaming agents include Therabloat®, Bloat Guard®, Bloat-Pac®, and Bloat Treatment®.

Motility Enhancing

Prokinetic agents increase the motility of parts of the GI tract to enhance movement of material through it. Parasympathomimetic agents, dopaminergic antogonists, and serotonergic agents may act as prokinetics.

Parasympathomimetic agents include acetylcholinesterase inhibitors; neostigmine (Prostigmin®) is the best drug example in this category. It works by competing with acetylcholine for acetylcholinesterase, resulting in increased intestinal tone and salivation. It is used in ruminants for treatment of rumen atony. It is also used for diagnosis of myasthenia gravis in dogs. Side effects are cholinergic in nature (vomiting, diarrhea, and increased salivation).

Cholinergics mimic the parasympathetic nervous system because they make a precursor to acetylcholine. An example of a drug in this category is dexpanthenol (d-panthenol injectable, Ilopan®). Dexpanthenol is used to treat intestinal distention or atony, paralytic ileus, and colic. It is also used after surgery to help animals increase gastric motility.

Dopaminergic antagonists stimulate gastroesophageal sphincter, stomach, and intestinal motility by sensitizing tissues to the action of the neurotransmitter acetylcholine. Side effects are behavioral in nature. Examples of dopaminergic antagonists include metoclopramide (Reglan®) and domperidone (Motilium®).

Serotonergic agents stimulate motility of the gastroesophageal sphincter, stomach, small intestine, and colon. An example is cisapride (Propulsid®), used for the treatment of constipation, gastroesophageal reflux, and ileus. Side effects may include diarrhea, megacolon, and abdominal pain.

TIP

Because metoclopramide interferes with acetylcholine action, anticholinergic drugs completely inhibit the action of metoclopramide.

Enzyme Supplementing

Pancreatic exocrine insufficiency (PEI) is a disease in which the pancreas does not produce **digestive enzymes**. These enzymes can be supplemented in the diet through the use of pancrealipase. Pancrealipase (Viokase®-V Powder, Pancrezyme®) contains primarily lipase (enzyme that digests fats), but also has amylase (enzyme that digests starch) and protease (enzyme that digests proteins) to help with digestion of fats, starch, and protein. Side effects include gastrointestinal problems such as diarrhea and abdominal pain. Care should be taken when handling this drug, as it can be irritating to the skin on contact and to nasal passages upon inhalation.

Dental Prophylaxis and Treatment

Advances in veterinary dental care have greatly increased both attention to dental care and dental product use. The types of products available for dental care include enzymes, antiseptics, and fluoride products. Human dental products should not be used on animals, as human products may have higher concentrations of certain chemicals that can lead to problems in animals. Some of the products and their uses are listed in Table 11-1. Table 11-2 summarizes the drugs covered in this chapter.

SUMMARY

The gastrointestinal tract is a long, muscular tube that plays a central role in bringing life-sustaining nutrients into the body and taking waste products out of it. Gastrointestinal drugs play a role in maintaining the unobstructed and regulated flow of elements in and out of the body.

Gastrointestinal drugs include antisialogues (drugs that decrease salivary flow), antidiarrheals (drugs that decrease peristalsis to reverse diarrhea), laxatives (drugs that loosen stool), anti-emetics (drugs that control vomiting), emetics (drugs that cause vomiting), antiulcer drugs (drugs that heal or prevent ulcers), antifoaming agents (drugs used to prevent frothy bloat in ruminants), prokinetic agents (drugs that increase motility of parts of the GI tract), digestive enzyme supplements (drugs used to replace digestive enzymes), and dental products (products to clean teeth and prevent dental disease).

TABLE 11-1 Veterinary Dental Products and Their Uses

Product Type	Trade Name	Considerations
Cleansing products	• C.E.T.® Dentifrice • Oxydent® • Nolvadent® Oral Cleansing Solution • Oral Dent® • C.E.T.® Oral Hygiene Spray • C.E.T.® Chews • C.E.T.® Toothbrush • CHX® Oral Cleansing Solution • MaxiGuard® Oral Cleansing Gel • Hill's t/d Diet® • C.E.T.® Prophypaste (polishing paste)	• These products are used to clean oral surfaces with or without antiseptic components. • These products may help with plaque removal and breath freshening. • Most of these products need not be rinsed. • These products contain either enzymes such as lactoperoxidase or antiseptics such as chlorhexidine. • Dentifrices are substances that clean the teeth.
Fluoride-containing products	• C.E.T.® Oral Hygiene Spray with Fluoride • FluroFom®	• Fluoride-containing products designed for animals have lower concentrations of fluoride than human products. • Fluoride desensitizes exposed dentin, strengthens tooth enamel, and stimulates remineralization of the enamel.

TABLE 11-2 Drugs Covered in This Chapter

Category	Examples
Antisialogues	• Anticholinergics: glycopyrrolate, atropine
Antidiarrheals	• Anticholinergics: atropine, aminopentamide, isopropamide, propantheline, methscopolamine • Protectants/adsorbents: bismuth subsalicylate, kaolin/pectin, activated charcoal • Opiate-related agents: diphenoxylate, loperamide, paregoric
Laxatives	• Osmotics: lactulose, sodium phosphate with sodium biphosphate, magnesium sulfate, magnesium hydroxide • Stimulant: bisacodyl, phenolphthalein, castor oil

TABLE 11-2 *Continued*

Category	Examples
Laxatives *(cont.)*	• Bulk-forming: psyllium hydrophilic mucilloid, polycarbophil, bran • Emollients: docusate sodium, docusate calcium, docusate potassium, petroleum products
Anti-emetics	• Phenothiazine derivatives: acepromazine, chlorpromazine, prochlorperaine, perphenzaine • Antihistamines: trimethobenzamide, dimenhydrinate, diphenhydramine • Anticholinergics: aminopentamide, atropine, propantheline • Procainamide derivatives: metoclopramide • Serotonin receptor antagonists: ondansetron, dolasetron
Emetics	• Centrally acting: apomorphine, xylazine • Peripherally acting: ipecac syrup, various home remedies
Antiulcer drugs	• Systemic antacids: sodium bicarbonate, calcium bicarbonate • Nonsystemic antacids: magnesium hydroxide, aluminum/magnesium hydroxide, aluminum hydroxide, magnesium hydroxide • H_2 receptor antagonists: cimetidine, ranitidine, famotidine • Mucosal protective drugs: sucralfate • Prostaglandin analogs: misoprostol • Proton pump inhibitors: omeprazole, lansoprazole
Antifoaming agents	• Defoaming agents: poloxalene, polymerized methyl silicone
Prokinetic agents	• Parasympathomimetic agents: neostigmine, dexpanthenol • Dopaminergic antagonists: metoclopramide, domperidone • Serotonergic agents: cisapride
Digestive enzyme supplements	• Digestive enzyme replacement: pancrealipase

CHAPTER REFERENCES

Delmar's A-Z NDR-97 Nurses drug reference. Clifton Park, NY: Thomson Delmar Learning.

Reiss, B., & Evans, M. (2002), *Pharmacological aspects of nursing care* (6th ed., revised by Bonita E. Broyles, pp. 306–355). Clifton Park, NY: Thomson Delmar Learning.

Veterinary pharmaceuticals and biologicals (12th ed.). (2001). Lenexa, KS: Veterinary Healthcare Communications.

Veterinary values (5th ed.). (1998). Lenexa, KS: Veterinary Healthcare Communications.

<www.thomsonhc.com/pdrel/librarian>; PDR Electronic Library, Thomson Health Communications.

Wynn, R. (1999). *Veterinary pharmacology*. Cambridge, MA: Harcourt Learning Direct.

CHAPTER REVIEW

Matching

Match the drug name with its action.

1. ______ ipecac syrup
2. ______ apomorphine
3. ______ aminopentamide
4. ______ kaolin/pectin
5. ______ bran
6. ______ poloxalene
7. ______ sodium bicarbonate
8. ______ glycopyrrolate
9. ______ pancrealipase powder
10. ______ docusate calcium

a. centrally acting emetic
b. drug used to reduce salivation
c. digestive enzyme supplement
d. antifoaming agent used to treat frothy bloat in cattle
e. systemic antacid
f. peripherally acting emetic
g. anticholinergic anti-emetic
h. bulk-forming laxative
i. emollient laxative
j. protectant/adsorbent antidiarrheal

Multiple Choice

Choose the one best answer

11. What part of the brain controls vomiting?
    a. cerebrum
    b. cerebellum
    c. brainstem
    d. ventricles
12. What category of drug simulates the parasympathetic nervous system or homeostatic function?
    a. dopaminergic
    b. adrenergic
    c. cholinergic
    d. sympathetic
13. Which group of antidiarrheals controls diarrhea by blocking acetylcholine release from parasympathetic nerve endings?
    a. cholinergic
    b. anticholinergic
    c. sympathetic
    d. sympathetic blocking agents
14. What type of laxative loosens stool by pulling water into the colon and increasing water in the feces?
    a. osmotics
    b. contact
    c. bulk-forming
    d. emollients

15. Which of the following control vomiting by inhibiting dopamine in the CRTZ?
    a. acepromazine
    b. trimethobenzamide
    c. aminopentamide
    d. metoclopramide
16. Which emetic can be given in the subconjunctival sac?
    a. xylazine
    b. ipecac
    c. apomorphine
    d. hydrogen peroxide
17. Which group of antiulcer medications block receptors on parietal cells in the stomach, thereby reducing gastric acid secretion?
    a. mucosal protectives
    b. H_2 receptor antagonists
    c. systemic antacids
    d. nonsystemic antacids
18. What drug is used for rumen atony?
    a. poloxalene
    b. neostigmine
    c. cisapride
    d. metoclopramide

True/False

Circle a. for true or b. for false.

19. Vomiting should be induced whenever an animal ingests a poison.
    a. true
    b. false
20. Fluoride-containing products designed for animals should have a higher concentration of fluoride than human products.
    a. true
    b. false

Case Studies

21. A 2-yr.-old M Labrador retriever (Figure 11-5) is brought to the clinic after ingesting drain cleaner. The veterinarian wants to sedate the dog to perform a physical exam.
    a. Xylazine is typically used at your clinic to sedate animals. Is this a good choice in this case? Why or why not?
    b. To prevent this animal from vomiting, what are good choices for sedation?
    c. What drug can be given to this animal via a stomach tube to absorb toxins if this is indicated by poison control?

FIGURE 11-5 Labrador retriever.

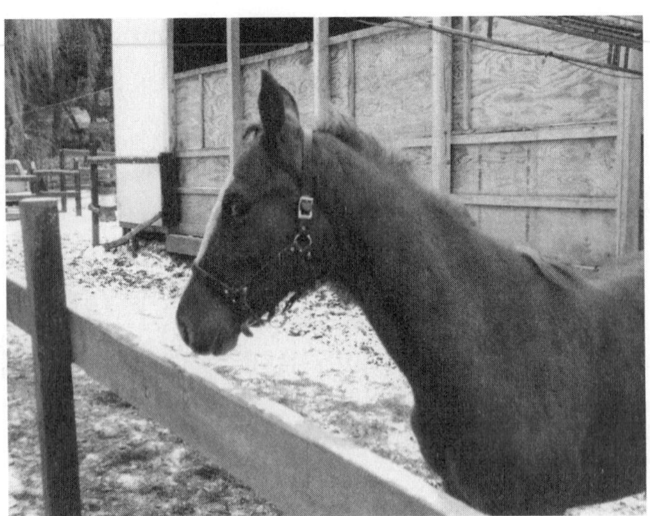

FIGURE 11-6 Saddlebred colt.
Source: Monica Hellenbrand, CVT

22. A 6-month-old saddlebred colt (Figure 11-6) has had diarrhea for the past few days. The owner wants to stop the diarrhea in this horse.

a. What antidiarrheal is commonly used in large animals?

b. What other concerns would you have for this horse that could be easily tested for?

CHAPTER 12

Urinary System Drugs

OBJECTIVES

Upon completion of this chapter, the learner should be able to

- describe the anatomy and physiology of the urinary system.
- outline the biological process of urine formation.
- describe how diuretics affect urine production.
- describe the modes of action of various diuretics.
- list urinary system drugs and describe their modes of action.
- describe the ways various hypertensive drugs affect urine production.

KEY TERMS

diuretics
diuresis
antihypertensives
urolith
urinary incontinence
cholinergic agonist
anticholinergics
adrenergic antagonists

INTRODUCTION

A client calls the clinic and says that her cat is spending a lot of time in his litterbox. When he is in the litterbox, he squats and occasionally starts crying. The client tells you that this has been going on since last night. What questions should you ask this client? What concerns do you have regarding this cat? Does this cat need immediate veterinary care, or can this problem wait until the cat's next physical exam appointment?

BASIC URINARY SYSTEM ANATOMY AND PHYSIOLOGY

The structures of the urinary system include paired kidneys, paired ureters, a single urinary bladder, and a single urethra. Within each kidney are millions of individual structures that do the actual work of the kidneys. These structures, referred to as the functional units of the kidney, are called *nephrons* (Figure 12-1). The nephron consists of the glomerulus, Bowman's capsule, the proximal convoluted tubule, the loop of Henle, a distal convoluted tubule, and a collecting duct. Urine is formed in the nephron via the processes of glomerular filtration, tubular reabsorption, and tubular secretion.

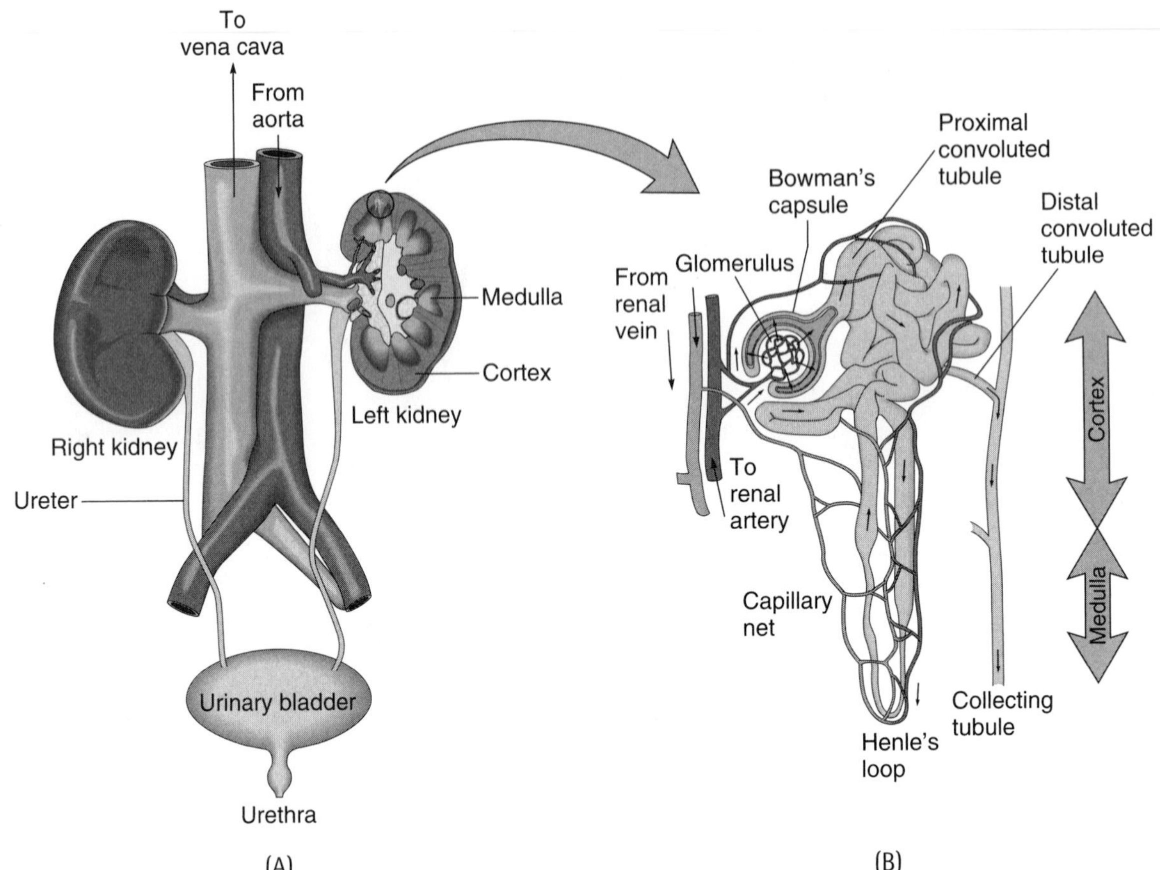

FIGURE 12-1 (A) Structures of the urinary tract (B) Structures of a nephron

A glomerulus is a cluster of capillaries that is responsible for glomerular filtration. Basically, the glomerulus filters substances from blood into the glomerular filtrate (water and dissolved substances). Glomerular filtration is affected by blood pressure (and hence blood volume and flow rate), plasma osmotic pressure (related to particles in the blood), and capsule pressure (which can alter resistance to blood flow).

The tubules are responsible for reabsorption or secretion of substances. The substances needed by the body are reabsorbed from the filtrate (fluid within the tubules) through the tubular cell wall, and reenter the plasma. These substances include water, glucose, amino acids, and ions (sodium (Na^+), potassium (K^+), calcium (Ca^{+2}), chloride (Cl^-) and bicarbonate (HCO_3^-)). If these substances are in excess or are not useful, they remain in the filtrate to be excreted in the urine. Tubular secretion is the process in which substances are carried into the tubular lumen. These substances are secreted by active transport and include K^+, H^+, NH_3^+, creatinine, and drugs. Tubular secretion is helpful in maintaining blood pH and in excreting drugs. Collecting ducts collect urine from the nephron and carry it to the renal pelvis. Figure 4-4 shows elimination of substances from the kidney.

Because the nephron plays an important role in drug excretion, animals with kidney problems will experience diminished function in excretion of drugs. Drug doses for these animals may have to be modified to meet their specific needs.

The main responsibilities of the urinary system are to remove from the body waste products produced during metabolism and to maintain homeostasis, a steady state in the internal environment of the body. The urinary system removes waste products from the body by the constant process of blood filtration. The major waste product of protein metabolism is urea, which is filtered by the kidney and is used in some diagnostic tests to determine the health status of the kidney. Uremia is a condition in which we see waste products in the blood (that should normally be excreted in the urine). Uremia may be seen with many kinds of kidney or renal disease.

URINARY SYSTEM DRUGS

Many different types of drugs are used in the management of renal disease and urinary system disorders. Some urinary drugs directly influence urine production and electrolyte balance. Others maintain blood pressure and reduce urinary system disease.

Urine Producing

Diuretics increase the volume of urine excreted by the kidneys and thus promote the release of water from the tissues. This process, called **diuresis**, lowers the fluid volume in tissues. Many disease conditions make fluid reduction desirable. The two main purposes of diuretic use are to decrease edema and to lower blood pressure. Edema (extracellular fluid accumulation) occurs with conditions such as congestive heart failure and chronic liver and kidney disease. Diuretics effectively reduce the edema associated with these conditions, as well as edemas of nonspecific nature, pulmonary edema, pulmonary congestion, and any pathological accumulation of noninflammatory liquid. Diuretics are summarized in Table 8-4.

The old adage used in the discussion of urine formation is "Where sodium goes, water will follow." The kidneys secrete and reabsorb sodium and chloride ions as they make urine. Diuretics block the reabsorption of these ions, so the sodium has nowhere to go but out of the kidneys and into the urinary bladder. Water then follows the sodium out of the kidneys and into the urinary bladder, and diuresis occurs. Though diuretics come from different chemical families, most work by affecting the reabsorption of sodium and chloride. The classes of diuretics include the thiazides, the loop diuretics, potassium-sparing diuretics, carbonic anhydrase inhibitors, and osmotics.

Thiazides are diuretics that act directly on the renal tubules to block sodium reabsorption and promote chloride ion excretion. Oral administration of thiazides produces diuresis in all animal species, with few documented side effects. This effect remains even with prolonged use, but long-term thiazide use does cause excessive potassium secretion, leading to hypokalemia (potassium deficiency) and cardiac dysfunction. To prevent hypokalemia, veterinarians suggest that potassium-rich diets or potassium supplements accompany thiazide diuretics. Thiazides most often manage edema associated with congestive heart failure. Hydrochlorothiazide (HydroDIURIL®), the standard thiazide drug, is given intravenously, intramuscularly, and orally to small animals, horses, and cattle. Other veterinary thiazides include chlorothiazide (Diuril®), hydroflumethiazide (Saluron®), and bendroflumethiazide (Corzide®).

Loop diuretics get their name from their point of action. The loop of Henle, a U-shaped renal tubule, is the sodium-reabsorbing site that lends its name to this type of diuretic. Loop diuretics influence the reabsorption action at the loop of Henle. Furosemide (Lasix®, Disal®, Diuride®) and ethacrynic acid (Edecrin®), two loop diuretics, are potent and effective drugs that block absorption of the following ions: chloride, potassium, calcium, hydrogen, magnesium, and bicarbonate. The result of blocking reabsorption of all of these electrolytes is tremendous diuresis. Dogs and cats given oral loop diuretics respond within 30 minutes. Parenteral administration produces diuresis immediately on a fully functional kidney. In large animal practice, furosemide is used in dairy cattle with udder edema and in racing horses to control respiratory hemorrhaging, which appears as a nosebleed. The main side effects of loop diuretics are electrolyte imbalances, especially hypokalemia.

Potassium-sparing diuretics are weaker diuretics than thiazides or loop diuretics. Drugs in this group are used as mild diuretics or in combination with other drugs.

Potassium-sparing diuretics act on the distal convoluted tubules to promote sodium and water excretion and potassium retention. These drugs interfere with the sodium-potassium pump that is controlled by aldosterone (a mineralocorticoid produced by the adrenal cortex that affects sodium and potassium levels). Potassium is reabsorbed and sodium is excreted. Drugs in this category include spironolactone (Aldactone®), triamterene (Dyazide®), and amiloride (Midamor®). The main side effect of these drugs is hyperkalemia.

Carbonic anhydrase inhibitors, such as acetazolamide (Diamox®) and dichlorphenamide (Daranide®), block the action of the enzyme carbonic anhydrase. This enzyme is used by the body to maintain acid-base balance (mainly between hydrogen and bicarbonate ions). Inhibition of this enzyme causes increased sodium, potassium, and bicarbonate excretion. Side effects with prolonged use include metabolic acidosis. This group of drugs is mainly used to decrease intraocular pressure with open-angle glaucoma; they are covered in Chapter 18 on ophthalmic and otic drugs.

Osmotic diuretics increase the osmolality (concentration) of the filtrate in the renal tubules. This results in excretion of sodium, chloride, potassium, and water. This group of drugs is used to prevent kidney failure, to decrease intracranial pressure, and to decrease intraocular pressure (i.e., glaucoma). Mannitol (Osmitrol® and generic) and glycerin (Osmoglyn®) are examples of osmotic diuretics. Side effects are uncommon with osmotic diuretics, but can include fluid and electrolyte imbalance and vomiting.

Blood Pressure Lowering

Drugs used to decrease hypertension, called **antihypertensives**, work in a variety of ways. Hypertension is a condition in which an animal's arterial blood pressure is higher than normal. The primary factor in hypertension is increased resistance to blood flow, resulting from the narrowing of peripheral blood vessels. If left untreated, animals with elevated blood pressure are at risk for developing cardiac and renal dysfunction. Some drugs that affect blood pressure include the following.

Diuretics. Diuretics have an antihypertensive effect by promoting sodium and water loss, which causes a decrease in fluid volume and blood pressure. These drugs were covered earlier in this chapter and in Chapter 8.

Angiotensin-converting enzyme inhibitors (ACE inhibitors). The kidneys regulate blood pressure via the renin-angiotensin system. Renin, an enzyme released by the kidneys, stimulates the conversion of angiotensin I to angiotensin II (a potent vasoconstrictor). Angiotensin II causes the release of aldosterone (a mineralocorticoid from the adrenal cortex that promotes retention of sodium and water). Retention of sodium and water increases fluid volume, which elevates blood pressure. ACE inhibitors block the conversion of angiotensin I to angiotensin II, which decreases aldosterone secretion. Clinically, ACE inhibitors are used to treat hypertension. Examples in this category are enalapril (Enacard®, Vasitec®), captopril (Capoten®), lisinopril (Zestril®), and benazepril (Lotensin®).

Calcium-channel blockers block the influx of calcium ions into the myocardial cells, resulting in an inhibition of cardiac and vascular smooth muscle contractility. This decreased resistance to blood flow reduces blood pressure. Side effects include hypotension and edema. Examples include diltiazem (Cardizem®), verapamil (Isoptin®), and nifedipine (Procardia®).

Direct-acting arteriole vasodilators relax smooth muscles of the blood vessels, mainly arteries, causing vasodilation. Side effects include edema due to sodium and water retention. Examples include hydralzaine (Apresoline®) and minoxidil (Loniten®).

Beta-adrenergic antagonists, also known as *beta blockers*, can affect the heart and bronchi. Beta-1 blockers work on the heart and beta-2 blockers work on the bronchial receptors. Nonselective beta blockers will inhibit the activity of beta-1 and beta-2 receptors, resulting in a slowed heart rate and bronchoconstriction. Side effects include decreased blood pressure, decreased cardiac output, and bronchospasm. An example of a nonselective beta blocker is propranolol (Inderal®).

Alpha-adrenergic antagonists block the alpha-1 adrenergic receptors, resulting in vasodilation. This group is covered later in the chapter in Table 12-1.

Urolith Treatment

Uroliths (also known as *urinary calculi*) are abnormal mineral masses in the urinary system. Uroliths are composed of a large amount of crystalline material (organic and inorganic crystalloids) and a small amount of organic matrix (typically mucoid material). The different types of uroliths include struvite, also known as triple phosphate crystals because they contain the three minerals magnesium, ammonium phosphate, and hexahydrate; calcium oxalate; calcium phosphate; urate, also known as ammonium biurate; cystine; and mixed. In some animal species, urolith formation may result from bacterial infections of the urinary tract. The development of urolith formation is not fully understood, but dietary factors are known to be important in some cases. Uroliths in the urinary bladder cause hematuria (blood in the urine) and dysuria (painful urination). Uroliths that lodge in the urethra may cause obstruction, which is a major concern in male cats and male sheep because of the narrow diameter of the urethra.

Diagnosis of uroliths is done via clinical signs (hematuria, dysuria, and possible urethral obstruction when the animal cannot urinate), urinalysis (Figure 12-2 and Figure 12-3), and radiography and/or ultrasound. Treatment includes antibiotic therapy (if warranted), medical dissolution of the uroliths, and possibly surgical removal of the uroliths.

Medical dissolution of uroliths depends on the type of urolith found and the area of the urinary tract in which it occurs. Uroliths found in the urinary bladder can be successfully treated with medical dissolution because the urolith is bathed in urine; changing the composition of the urine can allow the urolith to dissolve. This is not true of kidney uroliths. The pH of the urine in which the urolith is found also plays a role in the type of urolith formed. Animals with uroliths that form in alkaline (basic) urine (struvite uroliths) are fed a urine-acidifying diet that dissolves uroliths (Hill's Prescription Diet s/d®), or are prescribed urinary acidifiers such as methionine (Methio-tabs®, Methigel®) and ammonium chloride (Uroeze®). Once the uroliths are dissolved or removed, animals are then maintained on diets that produce acid urine (Hill's Prescription Diet c/d®, Iams Low pH/s™ (struvite uroliths). Animals with uroliths that form in acid urine (calcium oxalate, cystine, and ammonium urate uroliths) are fed urine-alkalinizing diets (Hills Prescription diet k/d®, Iams Moderate pH/O™ diet (oxalate urolith) or Purina's NF-Formula® diet), or are prescribed urinary alkalinizers such as potassium citrate (Urocit-K®). Animals that have renal disease should not be fed acidifying diets or prescribed urinary-acidifying drugs. Drugs used in the treatment of uroliths are summarized in Table 12-1.

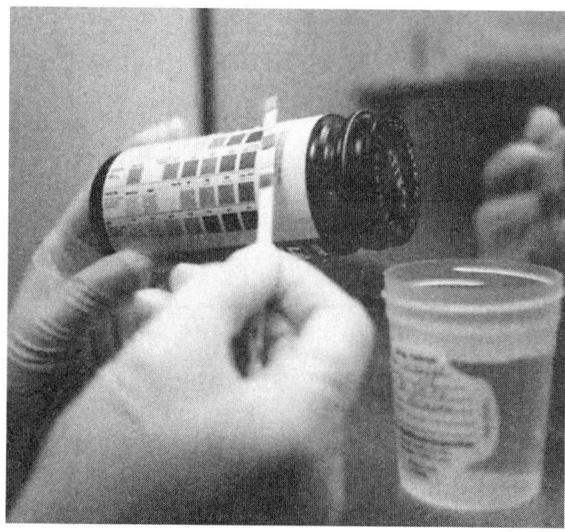

FIGURE 12-2 In the urinalysis, the chemical properties of urine, such as pH, glucose, ketones, and bilirubin, are tested with a dipstick.

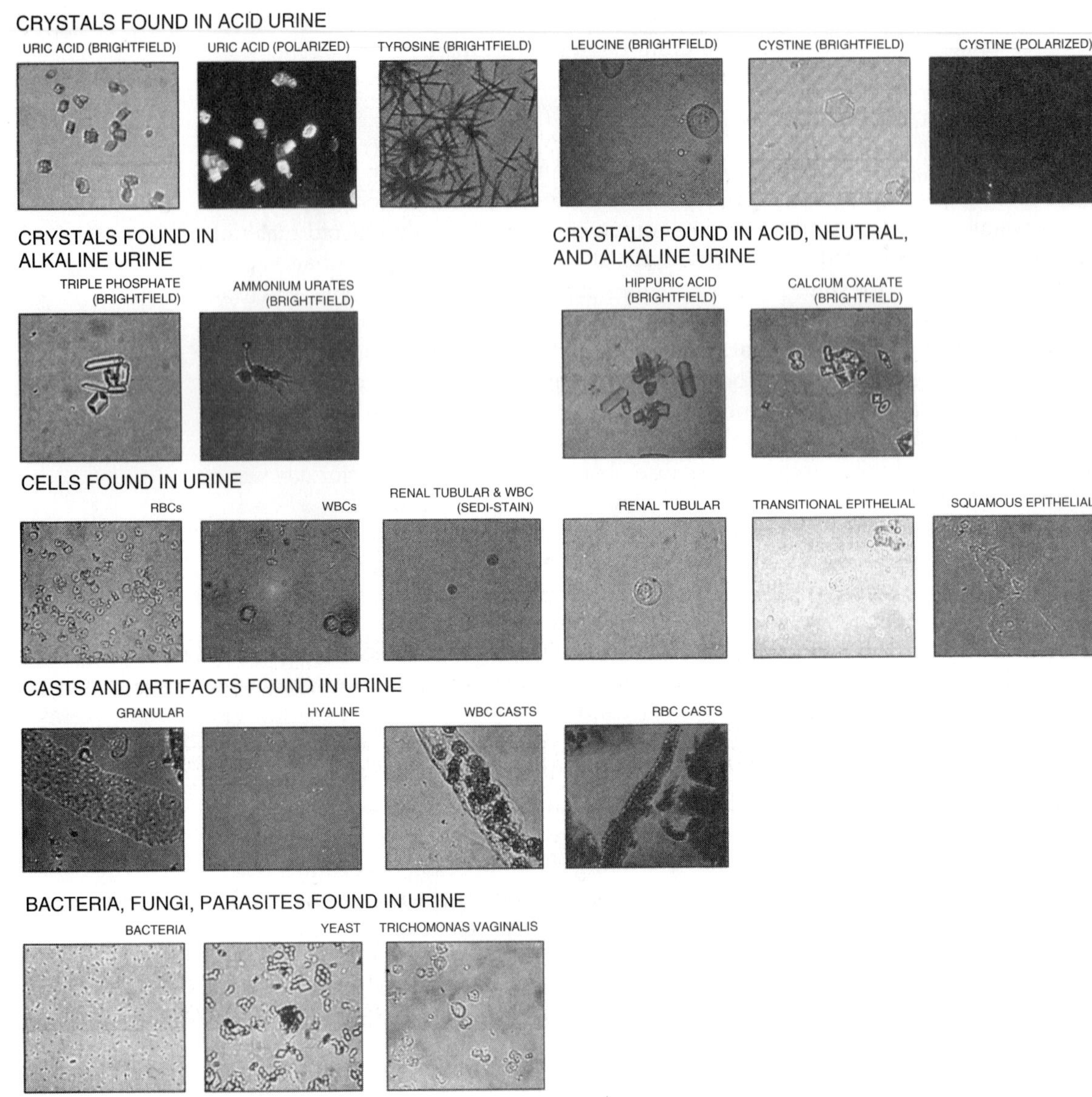

FIGURE 12-3 Crystals, cells, and casts found in urine

> **TIP**
> **Animals that are being fed urinary-acidifying diets should not also be given urinary-acidifying drugs. This combination can cause acid-base disturbances in the animal.**

Some uroliths, such as urate uroliths, also indicate the need for a low-protein, low-purine, low-oxalate diet to prevent recurrence (Hill's Prescription diet u/d®, Purina's NF Kidney Function® diet). Animals with urate uroliths are also often prescribed *xanthine oxidase inhibitors*. This group of drugs decreases the production of uric acid. Decreasing uric acid production helps prevent the formation of ammonium urate uroliths. Allopurinol (Zyloprim®, Lopurin®) is an example of a xanthine oxidase inhibitor. Side effects of allopurinol use are rare.

Urinary Incontinence

Urinary incontinence is the loss of voluntary control of micturition (a two-stage process involving the passive storage of urine and the active voiding of urine). Urinary incontinence can be divided into two main categories: (1) disorders due to neurologic disorders and (2) nonneurologic disorders. The body of the urinary bladder is composed of smooth muscle, the neck of the urinary bladder (also known as the *internal sphincter*) is composed of smooth muscle, and the external urethral sphincter is composed of skeletal muscle (Figure 12-4). Nervous control to the urinary bladder and urethra consists of both autonomic and somatic nervous system components. Autonomic nervous system control is both parasympathetic (the pelvic nerve supplies the smooth muscle of the urinary bladder, called the *detrusor muscle*) and sympathetic (the hypogastric nerve supplies beta-adrenergic fibers to the detrusor muscle and alpha-adrenergic fibers to the smooth muscle of the internal sphincter and part of the urinary bladder). Somatic innervation (pudendal nerve) stimulates the skeletal muscle of the external urethral sphincter. The coordination of these nervous system components results in voluntary micturition.

TIP

Detrusor muscle contraction and external urethral sphincter relaxation are both necessary for micturition to occur.

Neurologic problems may result in urinary incontinence (dribbling) or urine retention (inability to void urine that is in the urinary bladder). Neurologic causes of urinary problems can result from trauma to the spinal cord, tumors affecting the nervous system tracts, or degeneration of the nervous system tracts. Nonneurologic causes of urinary incontinence include hormone-responsive incontinence, stress incontinence, urge incontinence, ectopic ureters (abnormal entry of the ureters into the distal urethra instead of the urinary bladder), or urinary bladder overdistention. The cause of urinary incontinence determines the treatment used in each case. Drugs used to treat urinary incontinence are summarized in Table 12-1.

Neurologically Caused Incontinence Drugs used to treat animals with neurologic causes of urinary incontinence or urine retention include the following.

Cholinergic agonists (parasympathomimetic agents) are used to treat animals with "spinal cord bladders"—that is, damage to the nerves that control relaxation of the urinary bladder outflow sphincters. This nerve damage results in the retention of urine. Cholinergic agonists promote voiding of urine from the urinary bladder. These drugs simulate the action of acetylcholine by direct stimulation of cholinergic receptors. The cholinergic agonist binds to the receptors on smooth muscles, allowing sodium and calcium to enter the cells. This influx of sodium and calcium in turn allows muscle contraction. Tone of the detrusor muscle of the urinary bladder is increased, which may increase detrusor muscle contractions. An example of a cholinergic agonist is bethanechol (Urecholine®, Duvoid®, Urabeth®). Side effects include GI signs such as vomiting and diarrhea.

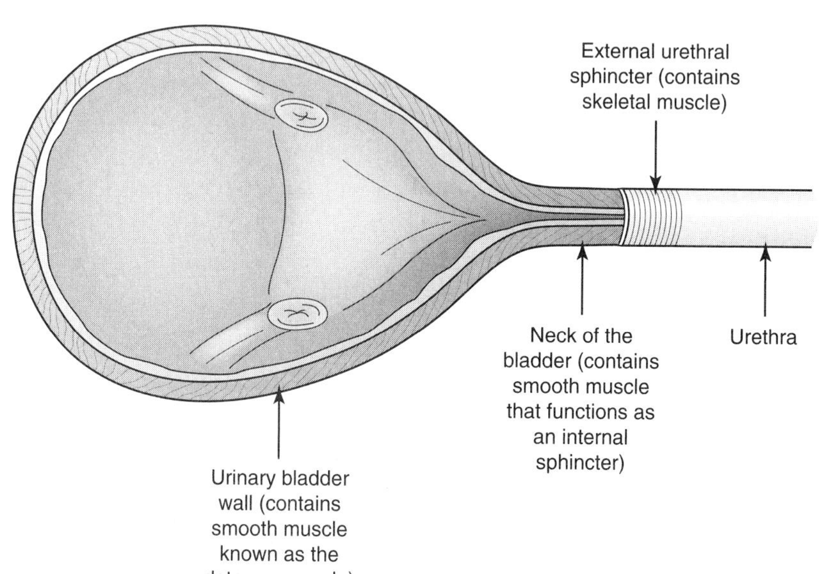

FIGURE 12-4 Urine pathway: Urine travels from the renal pelvis of the kidney through the ureters via peristalsis and is collected in the urinary bladder. During micturition, urine is directed through the neck of the bladder (known as the internal sphincter) and through the external urethral sphincter toward the urethra. Urine does not escape from the urinary bladder while the bladder is filling because the external urethral sphincter is contracted and urine cannot pass into the urethra. When urine is expelled from the urinary bladder, the external urethral sphincter muscle relaxes and the urinary bladder muscles contract.

TABLE 12-1 Urinary Bladder Drugs and their Mechanisms of Action

Category of Urinary Bladder Drug	Mechanism of Action	Examples
Cholinergic agonists or parasympathomimetic agents	Clinically used to promote voiding of urine from the urinary bladder	bethanechol (Urecholine®, Duvoid®, Urabeth®)
Anticholinergics or parasympatholytic drugs	Used to treat urinary incontinence by promoting urine retention in the urinary bladder	propantheline (Pro-Banthine®), dicyclomine (Bentyl®), butyl hyoscine (Buscpan®)
Adrenergic agonist	Increase the tone of the urethral sphincters via both alpha and beta receptors	phenylpropanolamine (Propagest®), ephedrine (available as a generic)
Adrenergic antagonists 1. Alpha-adrenergic antagonists 2. Beta-adrenergic antagonists	1. Clinically used to decrease the tone of urethral sphincters, (needed in the treatment of decreased urinary tone due to overextension of the urinary bladder.) 2. Clinically used for hypertension (see under blood-pressure lowering drugs in this chapter.)	1. phenoxybenzamine (Dibenzyline®), Prazosin (Minipress®), Nicergoline (Sermion®) 2. propranolol (Inderol®)
Estrogen	Sex hormone; helps maintain urethral muscle tone in some animals	diethylstilbestrol (DES)
Testosterone	Sex hormone; helps maintain urethral muscle tone in some animals	testosterone cypionate (Depo-Testosterone®), testosterone propionate (available as a generic)
Alpha and beta adrenergic agonists	Stimulation of these receptors increases urethral tone	phenylpropanolamine (Propagest®, Dexatrim®), ephedrine (available in generic form)
Skeletal muscle relaxants	Clinically used to treat animals that have urge incontinence or urethral obstructions due to increased external urethral sphincter tone; may also be used after unobstructing male cats to limit spasticity of the external urethral sphincter	dantrolene (Dantrium®), aminopropazine (Jenotone®), diazepam (Valium®)
Urinary acidifiers	Clinically used to produce acid urine, which dissolves and helps prevent formation of struvite uroliths; use has decreased with increasing use of urinary acidifying diets	methionine (Methio-tabs®, Methigel®), ammonium chloride (Uroeze®)
Urinary alkalinizers	Clinically used in treatment of calcium oxalate, cystine, and ammonium urate uroliths	potassium citrate (Urocit-K®)
Xanthine oxidase inhibitors	Decreases the production of uric acid, which helps decrease the formation of ammonium urate uroliths	allopurinol (Zyloprim®, Lopurin®)

Anticholinergics (parasympatholytic drugs) are used to treat urinary incontinence by promoting urine retention in the urinary bladder. These drugs work by blocking the binding of acetylcholine to its receptor sites and thereby causing muscle relaxation. Examples of anticholinergics that promote urine retention are propantheline (Pro-Banthine®), dicyclomine (Bentyl®), and butyl hyoscine (Buscpan®). Side effects of propantheline include decreased gastric motility (may affect absorption of other drugs) and other gastrointestinal problems.

Adrenergic antagonists are divided into alpha and beta categories. Beta-adrenergic antagonists, used in the treatment of hypertension, were covered in Chapter 8. Alpha-adrenergic antagonists are used to decrease the tone of internal urethral sphincters; they are useful in the treatment of decreased urinary tone due to overdistention of the urinary bladder. Alpha-adrenergic antagonists work by blocking circulating epinephrine or norepinephrine from binding to their receptors. These drugs are also used to decrease blood pressure. Examples of alpha-adrenergic antagonists include phenoxybenzamine (Dibenzyline®), prazosin (Minipress®), and nicergoline (Sermion®). The main side effect of these drugs is weakness due to decreased blood pressure.

Nonneurologically Caused Incontinence Drugs used to treat nonneurologic causes of urinary incontinence include the following.

Estrogen (diethylstilbestrol or DES) is used to treat hormone-responsive incontinence. Hormone-responsive incontinence is seen primarily in spayed female dogs that are more than eight years old. It is believed that sex hormones contribute to the maintenance of urethral muscle tone; therefore, lack of sex hormones decreases urethral muscle tone and allows urine to dribble from the urethra. This urine dribbling usually occurs when the animal is relaxed or asleep. Side effects of estrogen include bone marrow suppression, endometrial hyperplasia, and pyometra.

Testosterone (testosterone cypionate, testosterone propionate) is used to treat hormone-responsive incontinence in castrated male dogs. Males suffer hormone-responsive incontinence less frequently than spayed female dogs (described previously). Side effects of testosterone include the development of prostatic disorders and behavior changes (including aggression).

Alpha- and *beta-adrenergic agonists* include

- phenylpropanolamine (Propagest®, Dexatrim®), an oral medication used to treat stress incontinence. It may be used before resorting to hormones like estrogen and testosterone in the treatment of hormone-responsive incontinence. It is an alpha- and beta-adrenergic agonist that increases urethral tone. Side effects of phenylpropanolamine include anorexia, restlessness, and hypertension. In 2000 the FDA began taking steps to remove phenylpropanolamine (PPA) from all drug products, based on evidence that PPA increases the risk of hemorrhagic stroke in humans. This action has limited the availability of PPA in some areas.
- ephedrine, an alpha and beta agonist that increases urethral tone. It is used to treat stress incontinence and is available in tablet or injectable form. Side effects are the same as for phenylpropanolamine.

Skeletal muscle relaxants, such as dantrolene (Dantrium®), aminopropazine (Jenotone®), and diazepam (Valium®), are sometimes used to treat animals that have urge incontinence or urethral obstructions due to increased external urethral sphincter tone. These drugs may also be used after unobstructing male cats to limit spasticity of the external urethral sphincter. Side effects of skeletal muscle relaxants include mild tranquilization and sedation. Dantrolene has the addition side effect of hepatotoxicity.

Miscellaneous Drugs

Other groups of drugs, such as antibiotics, are used in management of urinary disease and infection. These groups of drugs are covered in Chapter 14.

SUMMARY

Urinary system disease and dysfunction can be caused by a variety of factors. Drugs used in the management of urinary system disease include diuretics, urinary bladder drugs, and antihypertensive drugs. Diuretics increase the volume of urine excreted by the kidneys and thus promote the release of water from the tissues. The classes of diuretics include thiazides, loop diuretics, potassium-sparing diuretics, carbonic anhydrase inhibitors, and osmotic diuretics. A variety of conditions affect the urinary bladder and either the voiding or retention of urine. Urinary bladder drugs include cholinergic agonists, anticholinergic drugs, alpha-adrenergic agonists, urinary acidifiers, urinary alkalizers, and xanthine oxidase inhibitors. Antihypertensives decrease blood pressure; this class includes diuretics, ACE inhibitors, calcium-channel blockers, direct-acting vasodilators, and beta- and alpha-adrenergic antagonists.

Uroliths or urinary calculi are abnormal mineral masses in the urinary system. Treatment of uroliths includes antibiotic therapy (if warranted), medical dissolution of the uroliths, and possibly surgical removal of the uroliths. Medical dissolution of struvite uroliths in alkaline (basic) urine involves feeding the animal a urine-acidifying diet that dissolves uroliths or using prescribed urinary acidifiers like methionine and ammonium chloride. Animals with uroliths that form in acid urine (calcium oxalate, cystine, and ammonium urate uroliths) are fed urine-alkalinizing diets or are prescribed urinary alkalinizers such as potassium citrate. A low-protein, low-purine, low-oxalate diet helps to prevent urate uroliths from recurring. Animals with urate uroliths are also prescribed xanthine oxidase inhibitors, which decrease the production of uric acid and thus help decrease the formation of ammonium urate uroliths. Allopurinol is an example of a xanthine oxidase inhibitor.

Urinary incontinence is the loss of voluntary control of micturition (a two-stage process involving the passive storage of urine and the active voiding of urine). Urinary incontinence may be caused by neurologic or nonneurologic disorders. Neurologic causes of urinary incontinence include trauma to the spinal cord, tumors affecting the nervous system tracts, or degeneration of the nervous system tracts. Types of nonneurologic urinary incontinence include hormone-responsive incontinence, stress incontinence, urge incontinence, ectopic ureters, and urinary bladder overdistention. The cause of urinary incontinence determines the treatment used in each case.

Drugs used to treat animals with neurologic causes of urinary incontinence include cholinergic agonists (parasympathomimetic agents) that promote voiding of urine from the urinary bladder; anticholinergics (parasympatholytic drugs) that promote urine retention in the urinary bladder; adrenergic agonists, used to treat incontinence due to stress and other nonneurologic disorders; and alpha-adrenergic antagonists, used to decrease the tone of urethral sphincters and balance decreased urinary bladder tone due to overextension.

Drugs used to treat nonneurologic causes of urinary incontinence include estrogen and testosterone, which are used to treat hormone-responsive incontinence; alpha- and beta-adrenergic agonists, such as phenylpropanolamine and ephedrine, which increase urethral tone; and skeletal muscle relaxants, such as dantrolene, aminopropazine, and diazepam, which are sometimes used to treat animals that have urge incontinence or urethral obstructions due to increased external urethral sphincter tone.

CHAPTER REFERENCES

Amundson Romich, J. (2000). *An illustrated guide to veterinary medical terminology.* Clifton Park, NY: Thomson Delmar Learning.

Delmar's A-Z NDR-97 Nurses drug reference. Clifton Park, NY: Thomson Delmar Learning.

Reiss, B., & Evans, M. (2002), *Pharmacological aspects of nursing care* (6th ed., revised by Bonita E. Broyles, pp. 598–627). Clifton Park, NY: Thomson Delmar Learning.

Veterinary pharmaceuticals and biologicals (12th ed.). (2001). Lenexa, KS: Veterinary Healthcare Communications.

Veterinary values (5th ed.). (1998). Lenexa, KS: Veterinary Healthcare Communications.

Wynn, R. (1999). *Veterinary pharmacology.* Cambridge, MA: Harcourt Learning Direct.

Iams Web site <www.iams.com>

Hill's Pet Nutrition Web site <www.hillspet.com>

CHAPTER REVIEW

Matching

Match the drug name with its action.

1. ______ furosemide
2. ______ mannitol
3. ______ propantheline
4. ______ methionine
5. ______ potassium citrate
6. ______ diltiazem
7. ______ enalapril
8. ______ propranolol
9. ______ spironolactone
10. ______ acetazolamide

a. potassium-sparing diuretic
b. urinary acidifier
c. calcium-channel blocker
d. carbonic anhydrase inhibitor
e. urinary alkalinizer
f. beta blocker
g. loop diuretic
h. ACE inhibitor
i. osmotic diuretic
j. anticholinergic that promotes urine retention

Multiple Choice

Choose the one best answer.

11. Which group of urinary drugs is also used for treatment of glaucoma?
    a. loop diuretics
    b. ACE inhibitors
    c. carbonic anhydrase inhibitors
    d. cholinergic agonists
12. What part of the nephron is responsible for filtration?
    a. glomerulus
    b. proximal convoluted tubule
    c. loop of Henle
    d. distal convoluted tubule
13. ACE inhibitors work by
    a. blocking electrolyte channels.
    b. causing vasoconstriction.
    c. pulling fluid into the renal tubules.
    d. blocking the conversion of angiotensin I to angiotensin II.
14. What type of drug promotes water release from tissues?
    a. diuretic
    b. urinary acidifier
    c. urinary alkalinizer
    d. cholinergic

15. What type of urine helps prevent formation of and encourages dissolution of struvite crystals?
    a. acidic
    b. basic
16. What type of diuretic acts directly on the renal tubules to block sodium reabsorption and promote chloride ion excretion?
    a. thiazide
    b. loop
    c. potassium-sparing
    d. osmotic
17. Which category of urinary drug promotes voiding of urine from the urinary bladder?
    a. cholinergics
    b. anticholinergics
    c. sympathomimetic
    d. sympathetic blocking agents
18. Which category of urinary drug decreases urethral sphincter tone and is used to treat animals with an overextended urinary bladder?
    a. cholinergics
    b. anticholinergics
    c. alpha-adrenergic agonists
    d. alpha-adrenergic antagonists
19. Which group of drugs is used in animals with ammonium urate uroliths?
    a. xanthine oxidase inhibitors
    b. loop diuretics
    c. osmotic diuretics
    d. urinary acidifiers
20. Glomerular filtration is affected by
    a. blood pressure.
    b. plasma osmotic pressure.
    c. capsule pressure.
    d. all of the above.

Case Studies

21. A 3-yr.-old M Dalmatian (Figure 12-5) presents to the clinic with what the owner describes as "urinating all the time, but only producing a small amount of urine." The physical examination reveals T = 102°F, HR = 84 bpm, and the dog is panting. The rest of the PE was unremarkable. The veterinarian asks you to collect a clean-catch urine sample on this dog. The urinalysis reveals a urinary tract infection (UTI) and the presence of ammonium urate crystals. The veterinarian prescribes a broad-spectrum antibiotic for this dog to treat the UTI and a low-protein diet to restrict purines.
    a. What other drug should be dispensed to the owner of this dog to help decrease the formation of ammonium urate uroliths?
    b. This dog would also benefit from a urinary alkalizer. What is an example of a urinary alkalizer?

FIGURE 12-5 Dalmation.
Photo by Isabelle Francais

FIGURE 12-6 Llasa Apso.
Photo by Isabelle Francais

22. A 15-yr.-old M/N Llasa Apso (25#) (Figure 12-6) was put on a loop diuretic for his congestive heart failure.

a. What is an example of a loop diuretic?
b. Where do loop diuretics work?
c. What is the main side effect of loop diuretics?
d. Diuresis is rapid and tremendous with loop diuretics. What would you want to warn this owner about?

Drugs Affecting Muscle Function

OBJECTIVES

Upon completion of this chapter, the learner should be able to

- explain the process of skeletal muscle activation.
- describe the role of the neuromuscular junction.
- describe the veterinary use of the following: anti-inflammatory drugs in treating muscle disease, neuromuscular blockers, skeletal muscle spasmolytics, and anabolic steroids.

KEY TERMS

neuromuscular junction
acetylcholine
acetylcholinesterase
inflammation
anti-inflammatory drugs
neuromuscular blockers
competitive nondepolarizers
spasmolytic
anabolic steroid

INTRODUCTION

A client who loves to jog with his dog comes into the clinic to ask some advice. His dog is getting older, but he wants the dog to continue to jog with him. He has already been giving the dog aspirin for pain from running, but he wants to know if he can also give something to ease any muscle pain the dog may be experiencing. He would like to give this medication daily so that the dog is not in pain. What options does he have for treating this dog? Could there be any complications from treating this dog as the owner would like to? What advice do you have for this jogger whose dog is having trouble keeping up with the owner's daily exercise? Are all your recommendations based on drug use, or are there other options you could give this client?

BASIC MUSCLE ANATOMY AND PHYSIOLOGY

Muscles are tissues that contract to produce movement. Muscles are made up of long, slender cells called *muscle fibers*. Muscle fibers are encased in a fibrous sheath. Muscle cells are categorized into three types, based on their appearance and function: skeletal, smooth, and cardiac. The focus here is on skeletal muscles. Skeletal muscles are those muscles that are attached to the skeleton and are activated by voluntary control.

A motor nerve that originates in the spinal cord and terminates in fibers connected to muscle cells activates skeletal muscle. The point at which a motor nerve fiber connects to muscle cells is known as a **neuromuscular junction** (Figure 13-1). When an electrical impulse of sufficient strength travels from the spinal cord to the neuromuscular junction, it causes release of the cholinergic neurotransmitter acetylcholine. **Acetylcholine** moves across the neuromuscular junction and binds to specialized receptor sites on the muscle opposite the nerve ending. As the electrical charge travels from the receptor sites across the length of the muscle fiber, depolarization of the muscle occurs, calcium is released, and the muscle contracts. The acetylcholine that causes this action is inactivated by **acetylcholinesterase**, an enzyme that breaks down acetylcholine and readies the muscle fibers for the next nerve impulse.

DRUGS AFFECTING MUSCLE FUNCTION

Drugs that affect skeletal muscle include anti-inflammatories (drugs that counteract inflammation), neuromuscular blockers (drugs that produce paralysis), skeletal muscle spasmolytics (drugs that reduce muscle spasms), and anabolic steroids (whose tissue-building effects can reverse muscular atrophy or wasting).

Inflammation Reducing

Inflammation is a normal response to injury, infection, or irritation of living tissue. Redness, pain, swelling, heat, and decreased range of motion are all signs of inflammation. **Anti-inflammatory drugs** are used to relieve these signs. Anti-inflammatory drugs include steroidal and nonsteroidal varieties. Steroidal anti-inflammatory drugs (corticosteroids) are chemically related to the naturally occurring hormone cortisone,

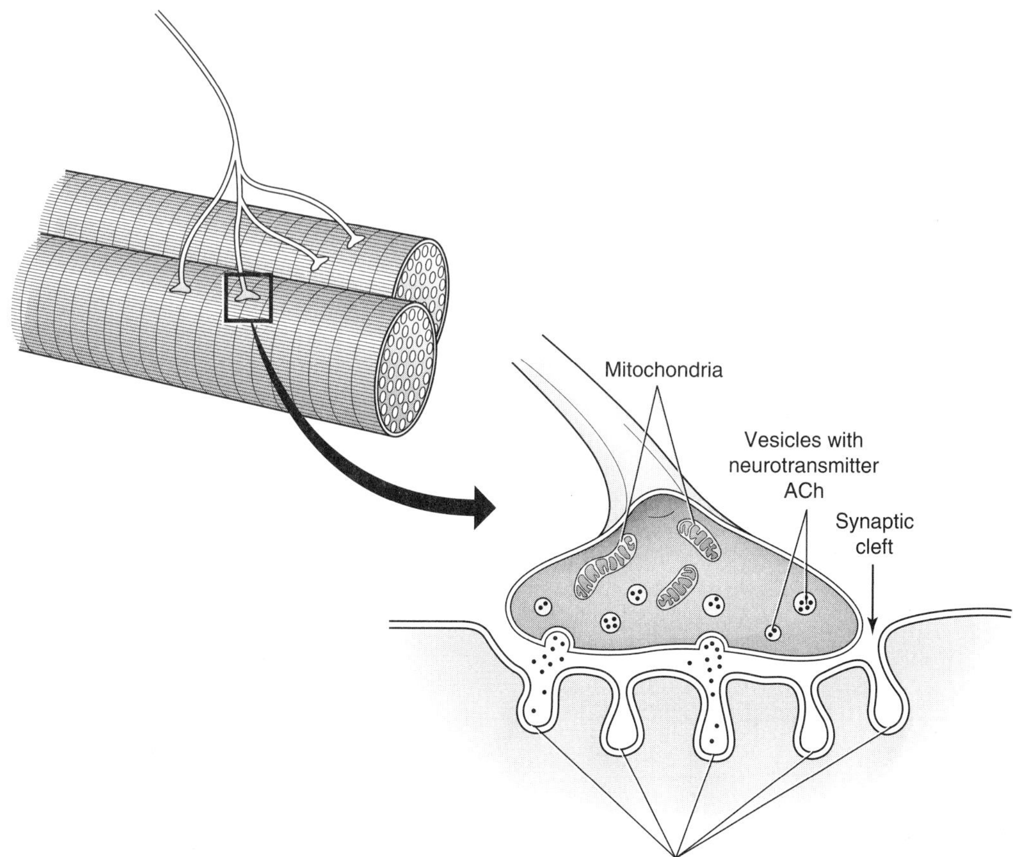

FIGURE 13-1 The neuromuscular junction is where the axon terminal meets a muscle fiber. The axon terminal does not touch the sarcolemma of the muscle, but rather fits into a depression in the cell membrane.

which is secreted by the adrenal cortex. Corticosteroids have potent anti-inflammatory effects and cause a variety of metabolic effects, including modification of the body's immune system. Nonsteroidal anti-inflammatory drugs (NSAIDs) are synthetic products that are unrelated to the substances produced by the body. These agents are widely used in the treatment of inflammation and pain reduction. The anti-inflammatory mechanism of action of NSAIDs is the result of inhibition of prostaglandin synthesis. These drugs are covered in more detail as to their mechanisms of action and side effects in Chapter 16.

TIP

Neuromuscular blockers do not provide analgesia or sedation. Animals receiving neuromuscular blocking agents must be monitored closely, because they may develop respiratory paralysis and/or cardiac collapse.

TIP

Atracurium and pancuronium should be refrigerated. Do not store these drugs in plastic containers/syringes, because the plastic may absorb them.

Muscle Paralyzing

Veterinarians need to relax the muscles of animals that are undergoing surgery and to prevent or treat muscle spasms. To do this, they may use **neuromuscular blockers**. These agents, administered intravenously, paralyze muscles by disrupting the transmission of nerve impulses from motor nerves to skeletal muscle fibers.

All but one of the commonly used neuromuscular blockers are **competitive nondepolarizers**, which are neuromuscular blocking agents that compete with acetylcholine for the same receptor sites, thus inhibiting the effects of acetylcholine. These drugs are also called *curarizing agents* because they resemble curare alkaloids, toxic botanical extracts originally used as arrow poisons in South America. Examples of competitive nondepolarizers are pancuronium (Pavulon®) and atracurium (Tracrium®). Pancuronium is used as an adjunct to general anesthesia to produce muscle relaxation and to facilitate endotracheal intubation. When given IV, it causes muscular relaxation in 2 to 3 minutes, and the effect lasts for about 45 minutes. Side effects include increased heart rate and blood pressure. Atracurium is used for the same purposes as pancuronium. It is given IV, and takes 2 to 3 minutes to cause muscle relaxation that lasts about 25 minutes. It causes minimal cardiovascular side effects and can be used in animals with kidney disease.

The competitive nondepolarizers require antidotes once their paralytic effect is no longer needed. Veterinarians administer such drugs as neostigmine (Prostigmin®, Stiglyn®), pyridostigmine (Mestinon®), and edrophonium (Enlon®, Tensilon®, Reversol®), which allow more acetylcholine into the muscle site and thus reverse the paralyzing effect of competitive nondepolarizers. Neostigmine competes with acetylcholine for acetylcholinesterase, allowing acetylcholine to accumulate in the neuromuscular junction and prolong the cholinergic effect. Neostigmine is also used to treat rumen atony (see Chapter 11) and as a diagnostic agent and treatment for myasthenia gravis. Side effects include vomiting, diarrhea, and excess salivation. Pyridostigmine competes with acetylcholine for attachment to acetylcholinesterase. This competition allows acetylcholine to accumulate. It is used to treat myasthenia gravis and its side effects include vomiting and diarrhea. Edrophonium is a very short-acting agent that attaches to acetylcholinesterase, thereby hindering its breakdown of acetylcholine. Side effects related to acetylcholine buildup include miosis (pupillary constriction), bronchoconstriction, and excessive salivation. Edrophonium is used for the diagnosis of myasthenia gravis.

Other examples of neuromuscular blocking agents include gallamine (Flaxedil®), vecuronium (Norcuron®) and metocurine (Metubine®). Because all these agents are cholinergic in nature, atropine (an anticholinergic) is sometimes given with these drugs to decrease the profound cholinergic effects.

One muscle paralyzer that is not a competitive nondepolarizer is succinylcholine (Anectine®, Quelicin®, Sucostrin®). It is known as a depolarizing drug, but rather than competing with acetylcholine, succinylcholine works independently of the transmitter. Depolarizing drugs mimic the action of acetylcholine in muscle fibers, and because they are not destroyed by acetylcholinesterase, their action is prolonged. Succinylcholine binds to the cholinergic receptors to produce depolarizations that continue as long as sufficient amounts of succinylcholine remain. This noncompetitive depolarizer does not need an antidote; it wears off by itself in a short period of time. It is used to reduce the muscle contraction associated with toxicities or pharmacologically induced convulsions. Side effects include excessive salivation, cardiovascular effects, and rash.

Spasm Stopping

Spasmolytics are drugs used to treat acute episodes of muscle spasticity associated with neurological and musculoskeletal disorders, such as malignant hyperthermia, equine postanesthetic myositis, and traumatic injury that results in muscle spasms. Spasmolytics "break down" the muscle spasticity. They work by unknown mechanisms; some are associated with calcium release from the sarcoplasmic reticulum or are responsible for depressing nerve impulse transmission. They include methocarbamol (Robaxin-V®), guaifenesin (Guailaxin®, Gecolate®), diazepam (Valium®), and dantrolene (Dantrium®).

Methocarbamol (Robaxin-V®) is used as an adjunctive therapy for inflammatory and traumatic conditions of skeletal muscle (especially intervertebral disc disease and traumatic injuries). It is also used to reduce muscle spasms. It is a potent, centrally acting muscle relaxant with selective action on the CNS that helps reduce muscle spasms but does not decrease muscle tone. Methocarbamol is available in injectable and oral forms. Side effects are rare, but include sedation, lethargy, and weakness.

TIP

Animals on methocarbamol may develop darker urine—clients should be advised that this is not a concern.

Guaifenesin (Guailaxin®, Gecolate®) is a muscle relaxant that relaxes both laryngeal and pharyngeal muscles, thereby easing intubation. It is used as an adjunct to anesthesia in large and small animals. A benefit of guaifenesin is that it does not affect the diaphragm and thus does not affect respiratory function. Guaifenesin is also covered in Chapter 7 on CNS drugs.

Diazepam (Valium®) is a centrally acting muscle relaxant that decreases the turnover of acetylcholine in the brain. It is a C-IV controlled substance used for muscle relaxation and anxiety control in animals. It is also covered in Chapter 7.

Dantrolene (Dantrium®) is a peripherally acting muscle relaxant that inhibits calcium release from the muscle, making it less responsive to nerve impulses. Dantrolene is used to prevent and treat malignant hyperthermia in various species, to treat urethral obstruction due to increased external urethral sphincter tone in dogs and cats (see Chapter 12), and to treat postanesthetic myositis in horses. The most significant side effect of dantrolene is liver toxicity, followed by sedation and vomiting.

Tissue Building

Anabolic steroids are steroids with a tissue-building effect—simply put, these drugs increase muscle mass. Veterinarians use them to promote growth, counteract postsurgical debility, and treat diseases such as muscular atrophy and orthopedic conditions. These drugs include nandrolone (Deca-Durabolin®), stanozolol (Winstrol-V®), and boldenone (Equipose®). Nandrolone is an injectable anabolic steroid used to stimulate erythropoiesis (red blood cell formation) in animals. It is also used as an appetite stimulant. It is a C-III controlled substance and should not be used in animals that are pregnant or have kidney disease.

Stanozolol is a synthetic anabolic steroid that increases retention of nitrogen and minerals, reverses tissue-depleting processes, and promotes better use of dietary protein. It is used to improve appetite, promote weight gain, and increase strength and vitality. It is a C-III controlled substance available in injectable and tablet forms. It should not be used in animals that are pregnant, have kidney disease, or are intended for use in food.

Boldenone is a long-acting injectable anabolic steroid used to treat debilitated horses. Boldenone increases appetite and improves musculature and haircoat appearance. It is a C-III controlled substance. Most horses respond with one to two treatments. This drug should not be used in animals that are pregnant or intended for use as food.

TIP

To obtain optimal results when giving anabolic steroids, adequate and well-balanced dietary intake is essential. Caution is required, however, because anabolic agents can cause electrolyte imbalance, liver toxicity, behavioral changes, and reproductive abnormalities.

TABLE 13-1 Drugs Covered in This Chapter

Drug Category	Examples
Neuromuscular blockers	gallamine, pancuronium, alcuronium, atracurium, succinylcholine
Spasmolytics	methocarbamol, guaifenesin, diazepam, dantrolene
Anabolic steroids	nandrolone, stanozolol, boldenone

SUMMARY

Motor nerves that affect the neuromuscular junction activate skeletal muscle. Acetylcholine moves across the neuromuscular junction, binds to muscle receptors, and causes muscular contraction for movement. Acetylcholinesterase stops this action. Drugs used to treat muscle disorders include anti-inflammatories, neuromuscular blockers, skeletal muscle spasmolytics, and anabolic steroids. Anti-inflammatory drugs reduce inflammation. Neuromuscular blockers disrupt transmission of nerve impulses from motor nerves to skeletal muscle fibers. Skeletal muscle spasmolytics work via unknown mechanisms to decrease muscle spasticity. Anabolic steroids build up muscle tissue.

Table 13-1 summarizes the drugs covered in this chapter.

CHAPTER REFERENCES

Amundson Romich, J. (2000). *An illustrated guide to veterinary medical terminology*. Clifton Park, NY: Thomson Delmar Learning.

Delmar's A-Z NDR-97 Nurses drug reference. Clifton Park, NY: Thomson Delmar Learning.

Reiss, B., & Evans, M. (2002), *Pharmacological aspects of nursing care* (6th ed., revised by Bonita E. Broyles, pp. 410–425). Clifton Park, NY: Thomson Delmar Learning.

Veterinary pharmaceuticals and biologicals (12th ed.). (2001). Lenexa, KS: Veterinary Healthcare Communications.

Veterinary values (5th ed.). (1998). Lenexa, KS: Veterinary Healthcare Communications.

Wynn, R. (1999). *Veterinary pharmacology*. Cambridge, MA: Harcourt Learning Direct.

CHAPTER REVIEW

Matching

Match the drug name with its action.

1. ______ pancuronium
2. ______ nandrolone
3. ______ guaifenesin
4. ______ dantrolene
5. ______ atracurium
6. ______ edrophonium
7. ______ stanozolol
8. ______ boldenone
9. ______ methocarbamol
10. ______ neostigmine

a. skeletal muscle spasmolytic
b. neuromuscular blocking agent
c. reverses the paralytic effects of a muscle paralyzer
d. anabolic steroid
e. neuromuscular blocking agent antidote

Multiple Choice

Choose the one best answer.

11. What type of drug overrides the stimulative effect of acetylcholine?
    a. anti-inflammatory
    b. competitive nondepolarizer
    c. spasmolytic
    d. anabolic steroid
12. How is acetylcholine normally deactivated at the neuromuscular junction?
    a. It is competitively overridden.
    b. It is inactivated by acetylcholinesterase.
    c. It is degraded by the warmer temperature at the neuromuscular junction.
    d. It has a half-life of two to four minutes.
13. Anabolic means
    a. breakdown.
    b. building.
    c. homeostasis.
    d. deactivation.
14. What mineral is released from the sarcoplasmic reticulum of muscles?
    a. potassium
    b. sodium
    c. chloride
    d. calcium
15. What type of neuromuscular drug would be used to counteract postsurgical debility?
    a. anti-inflammatory
    b. neuromuscular blocker
    c. spasmolytic
    d. anabolic steroid

Case Study

16. A-10-yr.-old F/S Dachshund (18#) (Figure 13-2) presents to the clinic with reluctance to walk and lethargy. The owner states that she thought the dog had hurt herself, because she no longer wants to jump on the bed or walk up stairs. Physical examination reveals T = 103°F, RR = panting, HR = 120 bpm. The dog has pain on palpation of its back and proprioceptive deficits of both hind limbs (seen with spinal cord disease). The veterinarian orders an x-ray of the back, which shows the dog has intervertebral disc disease. The veterinarian recommends surgery; however, the owner is reluctant to have surgery done on her dog. She opts for medical management, which includes the use of glucocorticoids to reduce inflammation of the spinal cord, and cage rest.

FIGURE 13-2 A Dachshund.
Photo by Isabelle Francais

a. What muscular drug may also benefit this dog?
b. The veterinarian prescribes methocarbamol at a dosage of 15 mg/kg po tid. How much would this dog get per dose?
c. Methocarbamol comes in 500 mg tablets that are scored into quarters. How many would this dog get per dose?
d. The veterinarian wants this dog on methocarbamol for seven days. How many pills do you dispense?

CHAPTER 14

Antimicrobials

OBJECTIVES

Upon completion of this chapter, the learner should be able to

- describe the role of antimicrobials in treating animal infections.
- describe the function of antibiotics.
- differentiate between narrow- and broad-spectrum antibiotics.
- differentiate between bactericidal and bacteriostatic antibiotics.
- describe the use of MIC.
- describe various mechanisms by which antibiotics work.
- list key uses, side effects, and differences among the different categories of antibiotics.
- list key uses, side effects, and differences among the different categories of antifungals.
- describe the uses, side effects, and limitations of antiviral drugs.
- differentiate between disinfectants and antiseptics.
- list key uses and differences among the different categories of disinfectants and antiseptics.

KEY TERMS

antimicrobial
antibiotic
spectrum of action
narrow-spectrum antibiotics
broad-spectrum antibiotics
bactericidal
bacteriostatic
resistant
intermediate
sensitive
minimum inhibitory concentration (MIC)
antibiotic resistance
antibiotic residues
beta-lactamase
antifungal
antiviral
disinfectant
antiseptic
germicide
virucidal
fungicidal
sporicidal

INTRODUCTION

A cat owner comes into your clinic requesting a refill of an antibiotic because her cat is sneezing again. The client explains to you that when the cat starts sneezing, she gives him antibiotics. The owner says she only needs about 14 pills, because she only has to give her cat 2 to 3 days worth of pills before he stops sneezing. If you give her 14 pills, that will give her enough antibiotic to treat the cat for a few "bouts" of sneezing. She would also like the "strongest" antibiotic you have. Can you help this client understand that her rationale of antibiotic use is flawed? What are some consequences of this kind of antibiotic use? Can you explain to this client how antibiotics work, what side effects can be expected, and why "stronger" is not a good term to use in describing antibiotics?

ANTIMICROBIAL TERMINOLOGY

Antimicrobials are drugs that counteract infection. An antimicrobial is a chemical substance that has the capacity, in diluted solutions, to kill (biocidal activity) or inhibit the growth (biostatic activity) of microorganisms. Biocidal agents attack something that the microorganisms have that the patient does not have and thereby kill the microorganisms. Biostatic agents attack something that both the microorganism and the patient have, but that the microorganism needs more of. These agents only injure (cripple or starve) the microorganisms, but rely on the patient's immune system to kill the bacteria. Depending on the drug dose and serum level of the drug, certain drugs can be both biocidal and biostatic.

Antimicrobials can be further classified as antibiotics, antifungals, antiviral drugs, and antiparasitic drugs. There are thousands of antimicrobials available to treat infections. This chapter describes some of the important agents used widely by veterinarians to treat bacterial, fungal, and viral infections. Chapter 15 covers antiparasitic drugs.

ANTIMICROBIALS FOR BACTERIA

Antibiotics are chemicals that work only on bacteria. They are described by their **spectrum of action**, which refers to the range of bacteria on which the agent is effective. Most bacteria are classified by Gram stain, a staining and decolorizing procedure that divides them into gram-positive bacteria and gram-negative bacteria. Gram-positive bacterial strains resist decolorization by the Gram stain process and have cell walls less complex in chemical composition than gram-negative bacteria, which are decolorized by the Gram stain process (Figure 14-1). Some of the gram-positive infectious bacteria include *Staphylococcus sp.* and *Streptococcus sp.*; some of the gram-negative types include *Salmonella sp.* and *Proteus sp.* Drugs that act specifically on the gram-positive family or specifically on the gram-negative family of bacteria are referred to as **narrow-spectrum antibiotics**. Those that act on both gram-positive and gram-negative bacteria are called **broad-spectrum antibiotics**.

> **TIP**
>
> Narrow-spectrum antibiotics act on *either* gram-positive or gram-negative bacteria.
>
> Broad-spectrum antibiotics act on *both* gram-positive and gram-negative bacteria (this does not mean that they work on all gram-positive and all gram-negative bacteria).

Antibiotics can also be classified as bactericidal or bacteriostatic. **Bactericidal** means that an agent is capable of killing the bacteria, whereas **bacteriostatic** means that an agent is capable of inhibiting the growth or replication of bacteria. Some bactericidal drugs kill bacteria by damaging key bacterial structures during development; therefore, they are effective against actively dividing bacteria. Other bactericidal drugs disrupt cell membranes or protein synthesis of the bacteria and cause bacterial death in existing and multiplying bacteria. Bacteriostatic drugs work by preventing the division of bacteria.

The goal of antibiotic treatment is to render the bacteria helpless (either by killing them or inhibiting their replication) and not to hurt the animal being treated. This is accomplished by making sure that the infecting bacteria are susceptible to the antibiotic, that the antibiotic reaches the infection site, and that the animal can tolerate the antibiotic. One method to determine if a particular antibiotic is effective against a particular bacterium is the *agar diffusion test* (also known as Kirby-Bauer *antibiotic sensitivity testing*). The agar diffusion test uses a variety of antibiotic-impregnated disks that are placed onto agar plates containing the bacteria being tested. The test uses standard media, standard bacterial counts, and impregnated disks with known amounts of antibiotic. After incubation at the proper temperature for the proper time, zones of inhibition or clear zones where the bacteria have failed to grow are measured (Figure 14-2). These measurements are in millimeters and are compared with a standardized chart to determine R (R = **resistant**, meaning the antibiotic does not work against these bacteria), I (I = **intermediate**, meaning the antibiotic may work against these bacteria), or S (S = **sensitive**, meaning the antibiotic does work against these bacteria). Because

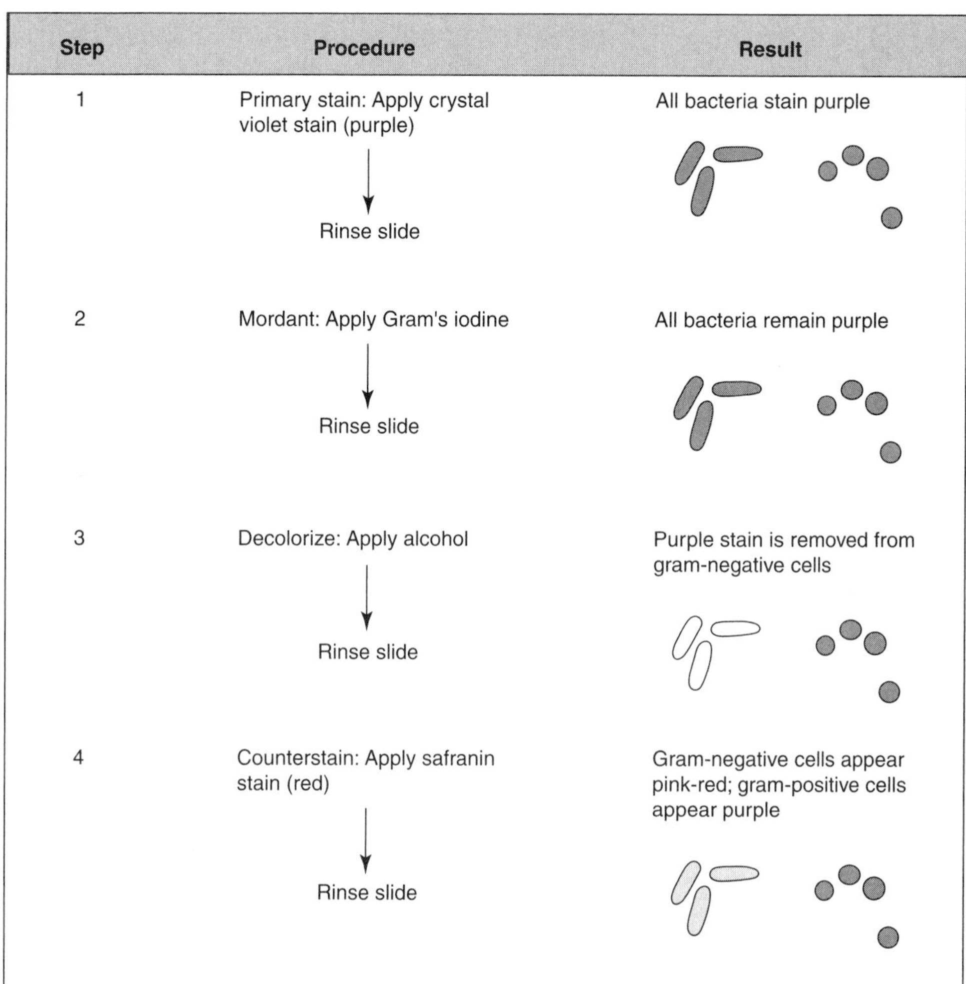

Step	Procedure	Result
1	Primary stain: Apply crystal violet stain (purple) ↓ Rinse slide	All bacteria stain purple
2	Mordant: Apply Gram's iodine ↓ Rinse slide	All bacteria remain purple
3	Decolorize: Apply alcohol ↓ Rinse slide	Purple stain is removed from gram-negative cells
4	Counterstain: Apply safranin stain (red) ↓ Rinse slide	Gram-negative cells appear pink-red; gram-positive cells appear purple

FIGURE 14-1 Gram stain procedure: 1. Smear is covered with crystal violet; the dye is removed. 2. Smear is covered with iodine; the iodine is removed. 3. Smear is washed with a decolorizer; the decolorizer is washed off. 4. A counterstain is used on the smear; the stain is removed.

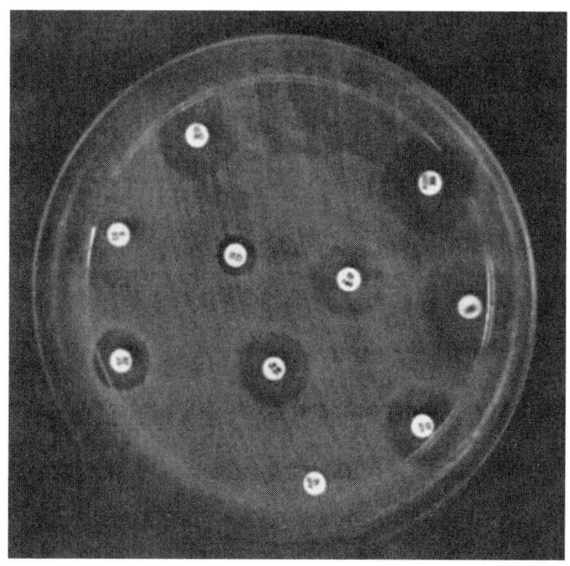

FIGURE 14-2 Zones of inhibition on an agar diffusion sensitivity test plate

the antibiotics diffuse out into the agar at a rate unique to each antibiotic, each antibiotic has its own zone-of-inhibition reference values to determine resistance and sensitivity.

Another method to determine bacterial susceptibility is the *broth dilution method.* In this test, a standard dilution of bacteria is placed into a series of tubes or wells containing different concentrations of a particular antibiotic. The lowest concentration of that particular antibiotic that visually inhibits the growth of bacteria is the **minimum inhibitory concentration** or **MIC** (Figure 14-3). An antibiotic's MIC varies with the bacterial species.

In addition to making sure that bacteria are sensitive to the antibiotic being used and using the proper dose of antibiotic, the veterinarian must make sure that the antibiotic gets to the infection site in high enough concentrations. An antibiotic that is effective against *E. coli* is of no help to an animal with a *E. coli* urinary tract infection if the antibiotic does not get to adequate levels in the urinary tract.

How Antibiotics Work

Antibiotics can work by a variety of mechanisms (Table 14-1). Some of these mechanisms include:

- inhibition of cell wall synthesis. An intact bacterial cell wall keeps the bacterium from filling with water and bursting. Disruption of cell wall synthesis can occur only when bacteria are growing and dividing. Already developed bacteria would not be affected by this mechanism.
- damage to the cell membrane. Damaging the bacterial cell membrane alters the membrane permeability of the bacterium. This allows substances to enter or leave the cell when they are not supposed to.
- inhibition of protein synthesis. Protein synthesis in bacteria occurs at the ribosome. Bacteria make protein by sending an RNA copy of the DNA for a specific protein to these ribosomes. Transfer RNA takes different amino acids to the ribosome where they are sequenced. If the amino acids are not linked properly, normal protein production in the bacterium is disrupted.
- interference with metabolism. Some antibiotics block the action of enzymes or bind to compounds needed by bacteria.
- impairment of nucleic acids production. Some antibiotics damage the production of nucleic acids, so that the bacterium cannot divide or function properly.

Additional Considerations in Antibiotic Use

The use of antibiotics has also brought two concerns to the veterinary community: **antibiotic resistance** and **antibiotic residues**. When bacteria are sensitive to an antibi-

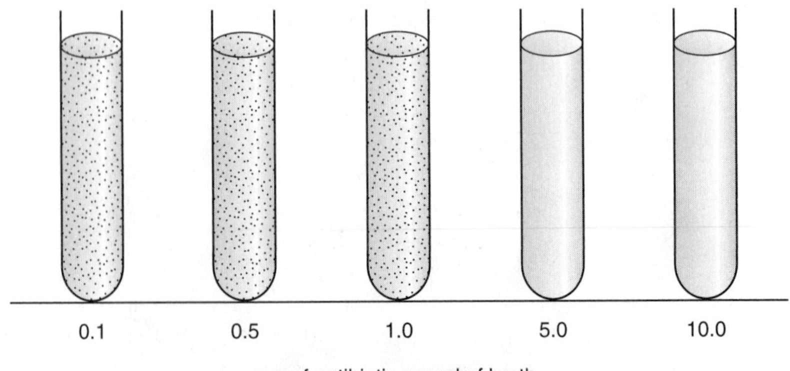

FIGURE 14-3 MIC is the lowest level of antibiotic that will at least inhibit the growth of bacteria. In this example the MIC is 5 micrograms of antibiotic/ml.

TABLE 14-1 Mechanisms of Action of Antibiotic Drugs

Action	Effect	Drug Examples
Inhibition of cell wall synthesis	Bactericidal effect by inhibition of an enzyme in the synthesis of the cell wall	penicillin, cephalosporin, bacitracin, vancomycin
Alteration in cell membrane permeability	Bacteriostatic or bactericidal effect; as the membrane permeability is increased, the loss of cellular substances causes cell lysis	polymyxin B
Inhibition of protein synthesis	Bacteriostatic or bactericidal effects due to interference with bacterial protein synthesis, but not animal cell protein synthesis or inhibition of the steps of protein synthesis	aminoglycosides, tetracyclines, chloramphenicol, florfenicol, macrolides, lincomycins
Interference with metabolism	Bacteriostatic effects due to the deprivation of essential material for bacterial metabolism	sulfonamides
Nucleic acid impairment	Bactericidal effects due to inhibition of nucleic acid enzymes	quinolones, fluoroquinolones

otic, the bacteria are inhibited or destroyed. If bacteria are resistant to an antibiotic, the bacteria survive and continue to multiply, despite administration of that antibiotic. In this instance, antibiotic use promotes the development of antibiotic-resistant bacteria. *Antibiotic resistance* occurs when bacteria change in some way that reduces or eliminates the effectiveness of the agent used to cure or prevent the infection. Bacteria can become resistant to a particular antibiotic when the drug is not used properly (for example, when antibiotics are used in an animal with a viral infection) or when the drug is not administered for the proper length of time or at the proper dosage.

One way bacteria can become resistant is through mutation. Even a single random gene mutation can have a large impact on the bacterium's pathogenicity, and since most bacteria replicate very quickly, they can evolve rapidly. Therefore, a mutation that helps a bacterium survive in the presence of an antibiotic will quickly become predominant throughout the bacterial population.

Another way that bacteria become resistant is by acquiring genes that code for resistance. Such genes can be transferred from members of the same bacterial species or from unrelated bacteria through plasmid transfer (transfer of DNA material from one resistant bacterium to a nonresistant bacterium). Again, the antibiotic-resistant bacterium survives in the presence of the antibiotic and quickly becomes predominant throughout the bacterial population.

Antibiotic resistance forces veterinarians to increase the dose and/or the duration of a course of a particular antibiotic or to abandon use of a particular antibiotic altogether. Cross-resistance can also occur between antibiotics that have similar actions; for example, between penicillins and cephalosporins. Clinics also serve as a fertile environment for antibiotic-resistant bacteria. Close contact among sick patients can lead to the survival of antibiotic-resistant bacteria; therefore, close contact among hospitalized patients should be minimized. To determine if an antibiotic has an effect on a specific bacterium, it is best to perform a culture and sensitivity test (see the preceding description of antibiotic sensitivity testing). Other ways to help prevent the development of antibiotic-resistant bacteria are to avoid administering antibiotics for viral infections, to avoid saving antibiotics used to treat one infection to use again on another infection, to administer antibiotics at the prescribed dose for the prescribed time, and to avoid using one animal's prescription for another animal.

A *residue* is the presence of a chemical or its metabolites in animal tissues or food products. Antibiotic residues in food-producing animals are of great concern because even low levels of these residues can cause allergic reactions in people or can produce resistant bacteria that can be transferred to the people who consume these products. Cooking or pasteurization does not usually degrade antibiotic residues. Withdrawal times for antibiotics are aimed at eliminating antibiotic residues in food-producing animals.

CLASSES OF ANTIBIOTICS

Antibiotics are often categorized by their mode of action such as cell wall agents, cell membrane agents, protein synthesis agents, antimetabolites, nucleic acid agents, and other.

Cell Wall Agents

Many bacteria have cells walls, and animal cells do not. Finding drugs that are active against these structures is thus advantageous, because they will attack the bacteria and not the host. Examples of antibiotics that affect the cell wall are penicillins, cephalosporins, bacitracin, and vancomycin.

Penicillins Penicillin was developed during the 1940s and now comes in a variety of forms. Today the penicillins consist of a group of natural and semisynthetic agents that are active against gram-positive and gram-negative cocci (spheres) and bacilli (rods). Penicillin may be either the natural product of mold (such as *Penicillium notatum*) or a semisynthetic derivative of the *Penicillin* molds. Penicillins can be identified by the *-cillin* suffix in their generic names.

The penicillins have a beta-lactam structure that interferes with bacterial cell wall synthesis among newly formed bacterial cells. Because the newly formed bacterial cells are unable to develop rigid cell walls when affected by penicillin, the rapidly developing cells die as a result of an increase of fluid in the cells.

Penicillin G is the penicillin most commonly used in veterinary practice. It is a narrow-spectrum antibiotic used against *Staphylococci sp., Streptococci sp.,* and some gram-positive bacilli (rod-shaped bacteria). Penicillin G is inactivated by stomach acid and thus is only used parenterally. Penicillin G is available as a sodium or potassium salt, both of which are rapidly absorbed after IM injection and provide peak levels usually within 20 minutes following injection. The sodium or potassium salt of penicillin G should be the only form administered IV. To produce prolonged blood levels of the antibiotic, veterinarians often inject procaine penicillin G or benzathine penicillin G, two penicillin suspensions. Procaine and benzathine extend the duration of penicillin activity by slowing its absorption from IM sites. Procaine penicillin G usually has a 24-hour duration; benzathine penicillin G usually has a 5-day duration. A combination product, penicillin G benzathine with penicillin G procaine, is also available. Side effects of penicillins include anorexia, vomiting and diarrhea when given orally, and hypersensitivity reactions (especially when given parenterally).

TIP

If refrigerated, reconstituted penicillin sodium and penicillin potassium products typically last 14 days. Any drug left after 14 days should be discarded. Other penicillin products may have a shorter shelf life depending upon their chemical properties and storage conditions; always follow the manufacturer's guidelines.

Penicillin V is another narrow-spectrum penicillin that comes in oral tablet and powder forms. Penicillin V is usually commercially available as a potassium salt. Penicillin V potassium is better absorbed from the gastrointestinal tract and is relatively stable in stomach acid; therefore, it is the preferred penicillin for oral administration.

Broad-Spectrum Penicillins. Some other members of the penicillin family are broader-spectrum than the natural penicillins and have activity against some gram-negative bacteria. These broader-spectrum penicillins are semisynthetic and include drugs such as amoxicillin, ampicillin, carbenicillin, ticarcillin, and methicillin. All the penicillins kill susceptible bacteria by interfering with cell wall synthesis.

Beta-Lactamase Resistant Penicillins. One concern that arises with the use of penicillin and its derivatives is the ability of bacteria to produce beta-lactamase. **Beta-lactamase,** or penicillinase, is an enzyme that destroys the beta-lactam structure found in penicillin (Figure 14-4). If the beta-lactam ring is destroyed, the penicillin is useless.

Some forms of penicillin are more resistant to beta-lactamase action than other penicillins. These penicillins are referred to as β-lactamase resistant penicillins; examples are methicillin, oxacillin, dicloxacillin, cloxacillin, and floxacillin. Most penicillins in this group are narrow spectrum.

Potentiated Penicillins. A *potentiated* drug is chemically combined with another drug to enhance the effects of both. The susceptibility of amoxicillin to destruction by beta-lactamase led to the development of a potentiated drug containing amoxicillin and clavulanic acid. Clavulanic acid binds to beta-lactamase, thus protecting the beta-lactam ring of amoxicillin against beta-lactamase destruction. Amoxicillin/clavulanic acid combinations are Clavamox® for animals and Augmentin® for humans.

Cephalosporins Cephalosporins are semisynthetic, broad-spectrum antibiotics produced from chemical manipulation of cephalosporin C (a fungus derivative). Cephalosporins are structurally and pharmacologically related to the penicillins. Like the penicillins, they also interfere with cell wall synthesis and kill susceptible bacteria quickly.

Cephalosporins are classified into four generations: first, second, third, and fourth. First-generation cephalosporins tend to have the greatest bacteriostatic or bactericidal activity against gram-positive and some gram-negative bacteria. Generally, they can be inactivated by the beta-lactamase produced by some bacteria. Second-generation cephalosporins have a broader spectrum of activity against gram-negative bacteria, and slightly less activity against gram-positive bacteria, than first-generation cephalosporins. Third-generation cephalosporins have an even broader spectrum of activity against gram-negative organisms and are resistant to the beta-lactamase produced by some bacteria. Fourth-generation cephalosporins have an expanded spectrum activity like that of the third-generation cephalosporins; however, fourth-generation cephalosporins are active against *Pseudomonas aeruginosa* and certain Enterobacteriaceae (gram-negative rods of the intestinal tract) that are resistant to third-generation cephalosporins. Fourth-generation cephalosporins may be more active against some gram-positive bacteria than some third-generation cephalosporins.

TIP

Oral penicillin (except amoxicillin) should be given on an empty stomach, as food affects its bioavailability.

TIP

Remember that antibiotics given to food-producing animals have withdrawal times that vary with the species, the antibiotic, and the food use (milk or meat). Check the package insert for the most current recommendations.

TIP

Penicillin use can cause fatal diarrhea in guinea pigs, rabbits, and hamsters because these species have normal flora that consists largely of gram-positive bacteria. The penicillin kills off the normal flora of these species, allowing harmful bacteria to populate the gastrointestinal tract.

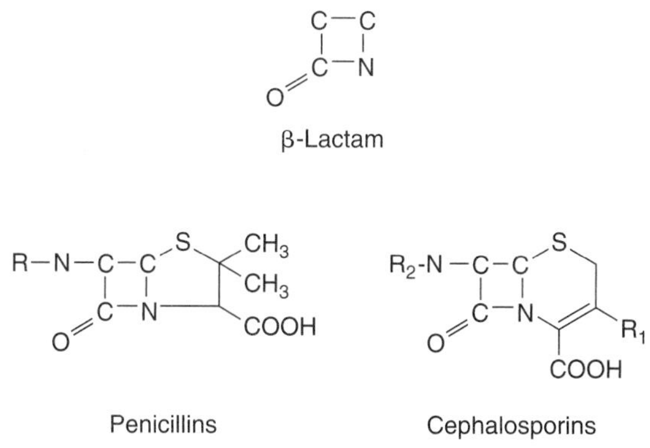

FIGURE 14-4 Beta-lactam ring and beta-lactam ring antibiotics

TIP

As the generation of cephalosporin increases, the spectrum of activity and potential side effects also increase.

TIP

Cephalosporins can be identified by the ceph- or cef- prefix in their generic and/or brand names.

Veterinarians use cephalosporins when first-line antibiotics, such as penicillins, are not effective. Animals allergic to penicillins will probably also be allergic to cephalosporins.

Bacitracin Bacitracin disrupts the bacterial cell wall like penicillin and cephalosporin, but not via the same mechanism. Bacitracin is a polypeptide antibiotic. Polypeptide antibiotics are chemically composed of long chains of amino acids called *polypeptides*. Bacitracin does not have a beta-lactam ring; therefore, it is not susceptible to destruction by beta-lactamase.

Bacitracin works primarily against gram-positive bacteria. Bacitracin is toxic to the kidneys and is most popular as a topical medication for skin, mucous membranes, and ocular surfaces. It is also used as a feed additive to control intestinal pathogens, because it is poorly absorbed from the gastrointestinal tract.

Vancomycin Vancomycin is a glycopeptide bacteriocidal antibiotic that is effective against many gram-positive bacteria. It is used primarily for treatment of infections that are resistant to the toxic antibiotics such as the penicillins or cephalosporins. It is especially useful in the treatment of drug-resistant *Staphylococcus aureus* in humans. It is rarely used in nonhumans to avoid promoting resistance development. Side effects of vancomycin include ototoxicity, nephrotoxicity, and pain on IV injection.

Cell Membrane Agents

Attacking a bacterial cell membrane (also known as the *plasma membrane*) is a more challenging problem than attacking a cell wall, because both bacterial and animal cells have cell membranes. This group of agents can thus cause more toxicity problems in animals. This group of antibiotics includes the polymyxins, which act on the phospholipid bilayer of the cell membrane to alter permeability. Polymyxin is also classified as a polypeptide antibiotic, like bacitracin.

Polymyxin B is effective against gram-negative bacteria only. Polymyxin B is not absorbed when taken orally or applied topically. It is most often applied topically as an ointment or a wet dressing for local treatment. Polymyxin B is often combined with neomycin and bacitracin to create a wide-spectrum topical medication.

Protein Synthesis Agents

Within a bacterial cell, information must be transmitted from DNA to the operational parts of the cell. This process starts with the transfer of information from DNA to messenger RNA. This process is followed by translation, which stimulates the formation of 30S and 50S ribosomal units (these ribosomal units are different in animal cells). Agents that interfere with this process, especially the formation of the 30S and 50S ribosomal units, can affect the production of protein in bacterial cells.

Aminoglycosides The aminoglycosides include any of a group of bactericidal antibiotics derived from various species of *Streptomyces* microbes. Aminoglycosides are specialized antibiotics that work by inhibiting protein synthesis, thereby promoting death of the affected bacteria. Veterinarians use them to treat infections caused by gram-negative bacteria. Because they stay within the GI tract when administered orally, veterinarians inject aminoglycosides to treat systemic infections. Aminoglycosides, identified by the *-micin* or *-mycin* endings in their generic names, include gentamicin, kanamycin, streptomycin, tobramycin, and neomycin. Veterinarians generally inject these antibiotics for short-term treatment of gram-negative bacterial septicemias (bacteria and their toxins released into the blood) and various skin, soft-tissue, bone, joint, respiratory, and postoperative infections. Aminoglycosides can be extremely nephrotoxic, so monitoring of kidney function via serum creatinine levels is recommended. Ototoxicity is another side effect of aminoglycosides. Veterinarians avoid injecting the most toxic aminoglycoside, neomycin, reserving it for topical treatment of localized skin infections caused by gram-negative bacteria. Aminoglycosides are not approved for use in food-producing animals.

Tetracyclines Discovered in 1947 in bacterial soil samples, tetracyclines are broad-spectrum, bacteriostatic antibiotics effective against many gram-positive and gram-negative bacteria (including *Mycoplasma* and *Chlamydia*) and rickettsial agents (including *Erlichia* and *Hemobartinella*). Veterinarians also use tetracyclines to treat Lyme disease (caused by the spirochete bacterium *Borrelia burgdorferi*). Tetracyclines work by inhibiting protein synthesis and are available in oral and parenteral forms.

Tetracyclines can be identified by the *-cycline* ending in their generic names. Tetracyclines can bind to calcium and can be deposited in growing bones (slowing bone development) and teeth (causing yellow discoloration of the teeth). Therefore, tetracycline should not be given to growing or pregnant animals or with dairy products. Tetracyclines can also bind to other minerals, such as magnesium, iron, and copper; therefore, they should not be given with iron supplements, antacids or products containing kaolin, pectin, or bismuth.

TIP

Tetracycline should not be given to pregnant or growing animals. Tetracycline injectable should be used with caution to prevent cardiac problems related to its calcium-binding properties, which may affect muscle contraction. Injectable tetracycline products are painful when given IM.

Chloramphenicol Chloramphenicol is a broad-spectrum antibiotic that works by inhibiting bacterial protein synthesis. Chloramphenicol penetrates tissues and fluids well, including the eye and CNS. This drug is not considered first-line because of its toxicity. A *first-line drug* is the veterinarian's first drug of choice. Drugs not considered first-line are those that have toxic side effects or are considered for use only when the first-line choice cannot be used or is not effective. These drugs are called *second-line drugs*.

Veterinarians suspect that chloramphenicol suppresses blood cell formation by depressing bone marrow function. Because of its potential for bone marrow suppression, chloramphenicol is banned from use in food-producing animals. When using or dispensing chloramphenicol, the veterinary staff and owners should wear gloves for protection and avoid inhalation of chloramphenicol powder. Side effects of chloramphenicol include bone marrow suppression with long-term use, anorexia, vomiting, and diarrhea.

Florfenicol Florfenicol (Nuflor®) is a synthetic, broad-spectrum antibiotic that works by inhibiting bacterial protein synthesis. Florfenicol is available as an injectable solution and is commonly used to treat bovine respiratory disease. Intramuscular injections may result in local tissue reaction and may result in loss of edible tissue at slaughter; therefore, it is recommended to give the injection near the neck area. Other side effects include inappetence, decreased water consumption, and diarrhea. Florfenicol is not approved for use in breeding-age cattle.

Macrolides Macrolides are antibiotics with large molecular structure and many-membered rings. Macrolides work by inhibiting protein synthesis and can cause stomach upset in animals. One example of a macrolide is erythromycin, a broad-spectrum antibiotic produced by a strain of the bacterium *Streptomyces erythreus*. Erythromycin may be bactericidal or bacteriostatic, depending on the concentration of drug given. Erythromycin is used to treat penicillin-resistant infections or in animals that have allergic reactions to penicillin. Erythromycin comes in oral and ointment forms.

Tylosin is another macrolide, structurally similar to erythromycin. It is used mainly in livestock for a variety of diseases. It comes in oral and injectable forms. Tylosin may also be found in feed or water additives (these products usually have other ingredients in the formulations as well). It is not recommended for use in horses, as reports of fatal diarrheas have been documented. If an animal has an adverse reaction to tylosin, it will also react to the other macrolide antibiotics.

Tilmicosin, another macrolide, is used to treat bovine respiratory disease. Tilmicosin is only approved for use in cattle. It is administered subcutaneously and at lower volumes than other antibiotics, so as to cause less muscle damage in meat-producing animals. An advantage to tilmicosin use is that it is administered in a single injection, which remains in the body for approximately three days. Single-injection therapy means less stress on the animal and less labor for the owner. Side effects include swelling at the injection site and tachycardia (increased heart rate).

Lincosamides Lincomycin, and its derivatives clindamycin and pirlimycin, are lincosamide antibiotics produced from a different strain of *Streptomyces* than the macrolides or aminoglycosides. They work primarily by inhibiting protein synthesis and work against gram-positive aerobic bacteria. Veterinarians typically use erythromycin instead of these two antibiotics because they may cause serious gastrointestinal problems. Clindamycin has good efficacy against many pathogenic anaerobes.

Antimetabolites

Another approach to eradicating bacteria is to deprive them of essential material needed for metabolism. These agents are referred to as *antimetabolites*.

Sulfonamides The sulfonamides are a group of drugs that inhibit the synthesis of folic acid, an action that hinders the growth of a wide variety of bacteria. They were the first antibacterial drugs to be used clinically in large patient populations, about seven years before the advent of penicillin. Because they are synthesized in the chemical laboratory, sulfonamides are not considered true antibiotics, but rather synthetic antimicrobials. Some sulfonamides may be designed to stay in the gastrointestinal tract (enteric forms); this form is used to treat coccidia infections. Other sulfonamides are in a form that is absorbed by the gastrointestinal tract and can penetrate tissues (systemic forms). Sulfonamides have a fairly broad spectrum of activity and can treat coccidia and toxoplasma organisms. Sulfonamides can produce such side effects as crystalluria (formation of tiny, sharp-edged crystals in the tubules of the nephron, which can damage the kidneys), keratoconjunctivitis sicca (dry eye), skin rashes, and blood imbalances, so veterinarians need to monitor these possible side effects.

TIP

Sulfonamides precipitate in the kidneys of animals that are dehydrated or have acidic urine. Adequate water intake is important in animals being given sulfonamides.

Sulfonamides may be combined with trimethoprim and ormetoprim to increase their antibacterial effects (these are referred to as *potentiated sulfonamides*). Trimethoprim and ormetoprim inhibit a different step of folic acid synthesis than the sulfonamides. Sulfonamides are bacteriostatic; trimethoprim and ormetoprim are bactericidal; however, if they penetrate tissue sites together they can have a bactericidal effect.

Nucleic Acid Agents

Nucleic acids of bacteria and animal cells are similar; however, some of the enzymes associated with nucleic acids differ between bacterial and animal cells. This makes interference with the bacterial nucleic acids a viable option for treating bacterial infections. A group of drugs that affect bacterial nucleic acids are the quinolones.

TIP

New fluoroquinolones continue to be developed, such as gatifloxacin and moxifloxacin.

Quinolones are synthetic antimicrobials that work by inhibiting DNA function in the bacterium; they do not harm mammalian DNA because of enzyme differences between mammals and bacteria. Quinolones are bactericidal and are effective against both gram-positive and gram-negative bacteria. Nalidixic acid, the first member of this family, has been used to treat urinary tract infections for many years, but is not used in veterinary medicine to any significant extent. Flumequine has been used in many countries to control intestinal infections in livestock. The quinolones are indicated in animals for the treatment of local and systemic infections, particularly against deep-seated infections and intracellular pathogens.

TIP

Do not use quinolones in growing animals, because of potential damage to cartilage.

Newer generations of quinolones, referred to as the fluoroquinolones, include enrofloxacin (Baytril®), ciprofloxacin (Cipro®), orbifloxacin (Orbax®), difloxacin (Dicural®), marbofloxacin (Zeniquin®), and sarafloxacin (SaraFlox®). Fluoroquinolones have fluorine bound to the quinolone base structure, which increases the drug's potency and spectrum of activity as well as improves its absorption. The fluoroquinolones are effective against gram-negative and gram-positive bacteria and can be recognized by the *-floxacin* ending in their generic names. Generally, fluoroquinolones are very safe drugs, but they can cause bubble-like cartilage lesions in growing dogs, and crystalluria has been seen with ciprofloxacin use in humans. Fluoroquinolones given at high doses can cause quinolone-induced blindness in cats. Indiscriminate use of fluroquinolones, for infections that could be treated with other antibiotics, may result in bacterial resistance. These antibiotics tend to be reserved until other antibiotics have been tried, so as to maintain the effectiveness of fluroquinolones on superinfections.

Miscellaneous Antibiotics

Certain antimicrobials do not fit the preceding categories, but do treat specific veterinary diseases effectively.

Nitrofurans These agents have a broad antimicrobial spectrum, but are less potent than traditional antibiotics. Nitrofurans include furazolidone, nitrofurazone, and nitrofurantoin. Nitrofurantoin is rapidly eliminated from the body and the animal may not maintain therapeutic levels for long. Nitrofurantoin is used to treat urinary tract infections because it is filtered unchanged through the kidney, and is applied topically to treat wound infections. Furazolidone and nitrofurazone are found in topical products for wounds and ocular infections. Nitrofuran use in food-producing animals is prohibited because there is evidence that this drug may induce carcinogenic residues in animal tissues. Topical use is allowed because this application does not reach edible tissues. Side effects of nitrofurans include gastrointestinal and liver disturbances.

Nitroimidazoles The nitroimidazole group of drugs have both antibacterial and antiprotozoal activity. Metronidazole is the representative drug of this group. Although there is no approved veterinary form of metronidazole (Flagyl®), it has been used both orally and intravenously for many years to treat *Giardia* and *Trichomonas* (parasitic protozoans), amoebiasis (infection with amoebae), and anaerobic bacteria. Metronidazole is considered by some the drug of choice for canine diarrhea. Metronidazole is believed to work by disrupting DNA and nucleic acid synthesis, and is considered bactericidal. It is recommended that use of this drug be avoided in pregnant animals. It can also cause anorexia, vomiting, diarrhea, and neurologic signs.

Rifampin Rifampin is a rifamycin antibiotic that disrupts RNA synthesis. It is a broad-spectrum antibiotic and is used primarily with erythromycin for the treatment of *Corynebacterium equi* infections in foals. It is bactericidal or bacteriostatic depending on the dose. Owners should be warned that rifampin causes a reddish color to urine, tears, sweat, and saliva.

Table 14-2 summarizes the various groups of antibiotics and gives examples of their trade and generic names.

ANTIMICROBIALS FOR FUNGI

Fungi are divided into two groups: molds and yeast. Fungal infections in animals are either superficial in nature (for example, skin infections such as ringworm) or systemic in nature (for example, internal infections such as blastomycosis). Fungal infections are referred to as *mycoses* and are diagnosed by fungal media or serologic tests. Superficial mycoses tend to be diagnosed with dermatophyte test media (in which the test media change color if the proper fungi are present) and microscopic identification of fungal structures (fungi are examined under a microscope to observe the hyphae and/or

TABLE 14-2 Classes of Antibiotics and Their Effectiveness.

Class of Antibiotic	Action of Antibiotic	Considerations	Examples
penicillins	• Inhibit cell wall synthesis • Bactericidal • Mainly work on gram + bacteria; some gram – with amoxicillin, ampicillin, ticarcillin, carbenicillin, and methicillin	• Oral and injectable forms • Given orally, most absorption occurs in stomach and small intestine • Rapidly distributed • Give 1–2 hours before eating	• penicillin V (V-Cillin K®), penicillin G Procaine (Crystacillin®), penicillin G Benzathine with penicillin G Procaine (Dual Pen®) • amoxicillin (Amoxi-tabs®, Amoxi-drops®, Biomox®, Robamox-V®) • ampicillin (Polyflex®, Omnipen®) • amoxicillin with clavulinic acid (Clavamox®) • ticarcillin (Ticar®) • carbenicillin (Pyopen®, Geocillin®) • cloxacillin (Dari-Clox®, Orbenin-DC®) • dicloxicillin (Dynapen®, Pathocil®) • hetacillin (Hetacin-K®) • nafcillin (Nafcil®)
cephalosporins	• Inhibit cell wall synthesis • Bactericidal • First generation mainly work on gram + bacteria, second through fourth generation work on gram + and gram – bacteria	• Oral and injectable forms • GI absorption not good; usually administered parenterally • Well distributed to tissues, except CNS • Vomiting may occur when given on empty stomach • If animal is allergic to penicillin, it will be allergic to cephalosporin	• First generation: cephapirin (Cefa-Dri®, Cefa-lak®), cephadroxil (Cefa-drops®, Cefa-tabs®) • Second generation: cefoxitin (Mefoxin®), cefaclor (Ceclor®) • Third generation: ceftiofur (Naxcel®) • Fourth generation: cefepime (Maxipime®)
polypeptides	• Inhibit either cell wall or cell membrane synthesis • Bactericidal	• Absorption is poor; used for topical infections or wound lavage	• polymyxcin B (found in Optiprime® ophthalmic ointment) • bacitracin (found in Mycitracin® and Trioptic® ophthalmic ointment)
aminoglycosides	• Inhibit protein synthesis • Bactericidal • Work mainly on gram – bacteria	• Injectable form only • Not absorbed readily from GI tract; usually given parenterally • Nephrotoxicity and ototoxicity concerns • Do not mix with penicillin in the same syringe (makes penicillin inactive)	• gentamicin (Gentocin®, Garacin®) • neomycin (Biosol®, Mycifradin®) • amikacin (Amiglyde-V®, Amikin®) • tobramycin (Nebcin®) • dihydrostreptomycin (Ethamycin®)

TABLE 14-2 *Continued*

Class of Antibiotic	Action of Antibiotic	Considerations	Examples
tetracyclines	• Inhibit protein synthesis • Bacteriostatic • Work on gram + and gram – bacteria, as well as rickettsial bacteria	• Oral and injectable forms • Once given, quickly distributed, sometimes to CNS • Very little metabolism • Bind to calcium, causing side effects (do not give with dairy products or antacids/ antidiarrheal drugs) • Can cause yellow discoloration of teeth due to calcium binding	• tetracycline (Panmycin Aquadrops®, Oxy-Tet 100® injectable, Tetracycline HCl® soluble powder) • oxytetracycline (Terramycin®, Liquamycin®) • chlortetracycline (Aureomycin®) • doxycycline (Vibramycin®, Doxirobe® Gel) • minocycline (Minocin®)
chloramphenicol	• Inhibits protein synthesis • Bacteriostatic • Works on gram + and gram – bacteria as well as rickettsial bacteria	• Oral, injectable, and ointment forms • Readily absorbed into tissues • Side effect of bone marrow suppression makes use not recommended	• chloramphenicol (Chloromycetin®, Viceton®, Amphicol®)
florfenicol	• Inhibits protein synthesis • Bacteriostatic	• Injectable form • Well distributed in body; can achieve therapeutic levels in the CNS	• florfenicol (Nuflor®)
macrolides	• Inhibit protein synthesis • Bactericidal or bacteriostatic	• Well distributed to most body tissues, but not the CNS	• tilmicosin (Micotil®) • tylosin (Tylan®) • erythromycin (Erythro-100®, Erythro-Dry®)
lincosamides	• Inhibit protein synthesis • Bactericidal or bacteriostatic	• Recommended for abscesses and dental infections	• clindamycin (Antirobe®) • pirlimycin (Pirsue®) • lincomycin (Lincocin®)
sulfonamides	• Inhibit folic acid synthesis • Sulfonamides are bacteriostatic • Trimethoprim and ormetroprim are bactericidal • Potentiated sulfas are bactericidal	• Can have anti-inflammatory effects • Well distributed through the body, including eye and CNS and synovial fluid • Can cause increased salivation in cats	• sulfadiazine/trimethoprim (Tribrissen®) • sulfadimethoxine (Albon®) • sulfadimethoxine/ ormetroprim (Primor®)
fluroquinolones	• Inhibit DNA function • Bactericidal	• Readily absorbed into tissues and body fluids after oral and parenteral administration	• enrofloxacin (Baytril®) • orbifloxacin (Orbax®) • difloxacin (Dicural®) • marbofloxacin (Zeniquin®) • sarafloxacin (SaraFlox®) • ciprofloxacin (Cipro®)
nitrofurans	• Inhibit bacterial enzyme systems • Bactericidal	• Eliminated from body quickly; usually used in urinary tract infections	• nitrofurazone (Furazone®, NFZ Puffer®) • nitrofurantoin (Macrodantin®) • furazolidone (Topazon®, Furox®)

(continues)

TABLE 14-2 *Continued*

Class of Antibiotic	Action of Antibiotic	Considerations	Examples
nitroimidazoles	• Disrupt DNA and nucleic acid synthesis • Bactericidal	• Well absorbed after oral administration • Use with caution in pregnant animals	• methronidazole (Flagyl®)
rifampin	• Disrupts RNA synthesis • Bactericidal or bacterioistatic depending on dose	• Relatively well absorbed from GI tract • Can cause red urine, tears, sweat, and saliva • Usually used in combination with other antibiotics	• rifampin (Rifadin®, Rimactane®)

spores). Systemic mycoses are usually diagnosed via serology. **Antifungals** are used to treat diseases caused by fungi.

TIP

Fungal infections are difficult to treat, and it takes a long course of drug treatment to resolve these infections.

Polyene Antifungal Agents

Nystatin and amphotericin B are both polyene antifungal agents. Both drugs are poorly absorbed by the gastrointestinal tract and work by binding to the fungal cell membrane. Nystatin is given orally for proliferation of *Candida albicans* fungi in the gastrointestinal tract, a frequent result of antibiotic drug therapy. It can also be used in a topical ointment, an oral suspension applied to the mouth area for treatment of local infections, and an intravenous injection.

Amphotericin B is given intravenously and is found in creams, lotions, and ointments for topical use. It is used parenterally for the treatment of systemic mycotic infections. Amphotericin B cannot penetrate tissues well, but in the tissues it does penetrate it stays a long time; it is dosed every other day or three times per week. Amphotericin B comes in a vial of powder, and is light and moisture sensitive. It is usually given through a filter system because it can precipitate out of solution. Amphotericin B is used for the systemic infections *Blastomyces*, *Aspergillus*, *Coccidioides*, *Histoplasma*, *Cryptococcus*, *Mucor*, and *Sporothrix*. Amphotericin B is extremely nephrotoxic and must be used with care.

Imidazole Antifungal Agents

Ketoconazole and miconazole are examples of imidazole antifungal agents that work by causing leakage of the fungal cell membrane. These drugs have fewer side effects than amphotericin B, but also have delayed onset of action. Ketoconazole is available in oral and topical forms, whereas miconazole comes in parenteral and topical forms. Both treat systemic and some superficial fungal infections, such as *Blastomyces*, *Coccidioides*, *Cryptococcus*, *Histoplasma*, *Microsporum*, and *Trichophyton*. Although side effects are less serious with these antifungals, hepatotoxicity and cardiotoxicity have been noted.

Itraconazole is an oral imidazole antifungal. Itraconazole is used to treat systemic mycotic infections such as *Candida*, *Aspergillus*, *Cryptococcus*, *Histoplasma*, and *Blastomyces*. It has also been used to treat dermatophyte infections. Itraconazole has gained favor among small-animal veterinarians because of its fewer side effects, though it can cause gastrointestinal signs.

Fluconazole is another imidazole antifungal agent used to treat systemic mycotic infections such as *Cryptococcus*, *Blastomyces*, and *Histoplasma*. It can also be used to treat dermatophyte and superficial *Candida* infections. Fluconazole is fungistatic when given orally or intravenously. Fluconazole is especially useful in treating CNS infections. Side effects of fluconazole include vomiting and diarrhea.

Antimetabolic Antifungal Agents

Flucytosine is an example of an oral antimetabolic antifungal agent that works by interfering with the metabolism of RNA and proteins. Flucytosine is used mainly in combination with other antifungals to treat *Cryptococcus* infections, because it is well absorbed from the gastrointestinal tract. Its main side effects are bone marrow abnormalities.

Superficial Antifungal Agents

Griseofulvin, available only as an oral preparation, treats dermatophyte fungal infections of the skin, hair, and nails. Dermatophytes, which include *Microsporum*, *Trichophyton*, and *Epidermophyton*, are commonly known as *ringworm*. Griseofulvin is fungistatic and works by disrupting fungal cell division. Griseofulvin comes in microsize and ultramicrosize formulas; both forms should be administered with a fatty meal to aid in absorption. The ultramicrosize formulation is better absorbed than the microsize formulation. Griseofulvin can cause gastrointestinal and teratogenic side effects; therefore, it should not be administered to pregnant or breeding animals. Dosing regimens of griseofulvin vary and are usually based on veterinarian preference, ease of owner compliance, and clinical experience.

Another drug that has been used to treat ringworm infections in cats is lufenuron (Program®). Lyme sulfur is a topical product also used in the treatment of ringworm. These drugs are covered in Chapter 15 on antiparasitics.

Examples of antifungal drugs are listed in Table 14-3.

ANTIMICROBIALS FOR VIRUSES

Viruses, unlike bacteria, are intracellular invaders that alter the host cell's metabolic pathways. A typical virus particle contains a strand of RNA or DNA surrounded by a lipid coat. The particle penetrates the host cell and causes new RNA or DNA to be produced therein. The newly formed RNA or DNA carries the genetic traits of the virus particle, and effectively paralyzes the host cell's metabolic machinery. **Antiviral** drugs

TABLE 14-3 Classes of Antifungals and Their Effectiveness

Class	Mechanism of Action	Considerations	Examples
polyenes	Bind to fungal cell membrane	Not well absorbed; fairly toxic	• amphotericin B (Fungizone®) • nystatin (Panalog®)
imidazoles	Cause leakage of fungal cell membrane	Less toxic; used for systemic mycotic infections and some dermatophyte infections	• ketoconazole (Nizoral®) • miconazole (Monistat,® Conofite®) • itraconazole (Sporanox®) • fluconazole (Diflucan®)
antimetabolics	Interfere with RNA and protein synthesis	Used mainly in combination with other antifungals to treat *Cryptococcus* infections	• flucytosine (Ancobon®)
superficials	Disrupt fungal cell division	Used for dermatophyte infections	• griseofulvin (Fulvicin-U/F®, Grifulvin V®)

act by preventing viral penetration of the host cell, or by inhibiting the virus's production of RNA or DNA. Human antiviral drugs include acyclovir, amantadine, idoxuridine, cytosine arabinoside (ara C), adenine arabinoside (ara A), methisazone, interferon, and interferon-inducers; however, only acyclovir and interferon are discussed here, because of the limited use of the other antivirals in veterinary practice.

Acyclovir (Zovirax® tablets, suspension, and injectable) interferes with the virus' synthesis of DNA. It is used to treat ocular feline herpes virus. Side effects include blood disorders such as anemia and leukopenia.

Interferon is a protein substance with multiple roles in the body's natural defenses—chief among them is stimulating noninfected cells to produce antiviral proteins. Researchers have demonstrated interferon's ability to protect host cells from a number of different viruses, and they are developing antiviral drugs called *interferon inducers* that stimulate the production and release of interferon. Interferon (Roferon-A®) has been used orally in veterinary practice for the treatment of feline leukemia virus (FeLV) and ocular herpes infections in cats. Side effects are rarely seen in cats.

INANIMATE VERSUS ANIMATE USE: DISINFECTANTS AND ANTISEPTICS

TIP

Some bacteria can form spores when environmental conditions worsen. Spores are technically called endospores and should not be confused with fungal spores which are reproductive structures.

Disinfection is an important concept in a veterinary hospital, on a farm, or in a research setting. A **disinfectant** kills or inhibits the growth of microorganisms on inanimate objects. **Antiseptics** kill or inhibit the growth of microorganisms on living tissue. Other terms that further describe disinfectants or antiseptics include:

- **germicide**: a chemical that kills microorganisms
- **bactericidal**: a chemical that kills bacteria
- **virucidal**: a chemical that kills viruses
- **fungicidal**: a chemical that kills fungi
- **sporicidal**: a chemical that kills spores, which are especially resistant to chemicals

There are many different kinds of disinfecting agents. Ideally, these agents should: remain stable during storage and use, be easy to apply, not damage or stain the objects they are applied to, be nonirritating to animal tissue, and have the broadest possible spectrum of activity. Low cost is also an advantage, as so much disinfecting product is used in a veterinary facility.

When choosing a disinfecting agent, keep in mind the surface it will be applied to and the range of organisms that you want to eliminate. A disinfecting agent that works well on animal cages would probably not work well on a surgical wound. Likewise, a disinfecting agent that works well on viruses may not work well on bacterial spores.

TIP

When diluting a chemical, always start with the quantity of water and then add the chemical concentrate. This avoids splashing of chemical into your eyes. Latex gloves and protective goggles are also recommended during these procedures.

Some disinfecting agents are less effective in the presence of organic waste (blood, feces, etc.), soap, and hard water. Thorough cleaning of the surface to be treated is an important step in disinfection.

When purchasing disinfecting agents, make sure you read the package insert in regard to dilution recommendations and special use instructions. Some chemicals work on different microorganisms depending upon how and if they are diluted (Figure 14-5). *Contact time*, the amount of time that the agent spends on the surface, is also critical to the efficacy of the chemical.

Always keep or request Material Safety Data Sheets (MSDSs) on all products used in disinfection. The filing of MSDS and container labeling are important components of each clinic's or facility's hazard communication plan. The Occupational Safety and Health Administration (OSHA), a part of the U.S. Department of Labor, enacted the Hazard Communication Standard in 1988 to educate and protect employees who work with potentially hazardous materials. The hazard communication plan includes:

- a written plan that serves as a primary resource for the entire staff. It lists all hazardous materials used in the facility and the name of the person responsible for keeping this list current. The written plan also includes where MSDSs are kept, how the MSDSs are obtained, procedures for labeling materials, a detailed description of employee training, and how independent contractors are informed of hazardous materials in the facility.
- an inventory of hazardous materials on the premises.
- current MSDSs for hazardous materials.
- proper labeling of all materials in the facility (including secondary containers, which are diluted or transferred chemicals that are not in their original containers).
- employee training for every employee working with or exposed to hazardous materials.

The following must be on all MSDSs:

- product name and chemical identification
- name, address, and telephone number of the manufacturer
- list of all hazardous ingredients
- physical data for the product (liquid, solid, gas)
- fire and explosion information
- information on potential chemical reactions when the product is mixed with other materials
- outline of emergency and clean-up procedures
- personal protective equipment required when handling the material
- description of any special precautions necessary when using the material

TIP

Many hazardous materials are labeled with key words to watch for, including Caution, Warning, Danger, Toxic, Flammable, Reactive, and Corrosive.

Types of Disinfecting Agents

Phenols Phenols were the first antiseptics and are found in a variety of products. Phenols are effective against gram-positive bacteria, with some effectiveness against gram-negative bacteria, fungi, and some types of enveloped viruses. Phenols are ineffective against spores and nonenveloped viruses such as parvovirus. Phenols work by destroying the selective permeability of cell membranes, resulting in leakage of cellular material. Phenols should not be used as antiseptics because they are irritating to skin, can be absorbed systemically, and have been linked to neurotoxicity.

TIP

Quaternary ammonium compounds can be recognized by its chemical name having ammonium in it or having an *-nium* in its name.

Quaternary Ammonium Compounds Quaternary ammonium compounds are effective against gram-positive and gram-negative bacteria (better on gram-positive than gram-negative), but are not effective against spores and have limited efficacy on fungi. Quaternary ammonium products work on enveloped viruses, but not nonenveloped viruses such as parvovirus. These compounds work by concentrating at the cell membrane and dissolving lipids in the cell walls and membranes. They act rapidly and are usually not irritating to skin or corrosive to metal. Organic debris, hard water, and soaps inactivate quaternary ammonium compounds; therefore, the site should be free of these materials and dry prior to application.

Aldehydes Aldehydes are organic compounds that contain a functional group -CHO (carbon-hydrogen-oxygen), known as an *aldehyde*. Examples of aldehydes are glutaraldehyde and formaldehyde. Glutaraldehyde works by affecting protein structure. It is a rapid, broad-spectrum antimicrobial that kills fungi and bacteria within a few

ROCCAL®-D PLUS Veterinary Disinfectant

ROCCAL®-D PLUS

NDC 0009-7308-01, **NDC** 0009-7308-05

VETERINARY AND ANIMAL CARE DISINFECTANT

EFFECTIVE IN 400 PPM HARD WATER AS ($CaCO_1$)

DISINFECTS IN 5% ORGANIC SOIL LOAD

ACTIVE INGREDIENTS:

Didecyl dimethyl ammonium chloride 9.2%

Alkyl (C_{12}, 61%; C_{14}, 23%; C_{16}, 11%; C_{18}, 2.5%; C_8 & C_{10}, 2.5%) dimethyl benzyl ammonium chloride 9.2%

Alkyl (C_{12}, 40%; C_{14}, 50%; C_{16}, 10%) dimethyl benzyl ammonium chloride 4.6%

bis-n-tributyltin oxide 1.0%

INERT INGREDIENTS: 76.0%

BACTERICIDE, FUNGICIDE, VIRUCIDE

For Veterinary, Laboratory Animal, Kennel and Animal Breeder Facilities.

DANGER

KEEP OUT OF REACH OF CHILDREN

EPA REG. NO. 65020-12-1023 EPA EST. NO. 65020 GA-1

Roccal-D Plus

- is a complete, chemically balanced disinfectant providing clear use solutions even in hard water.
- is a residual bacteriostat and inhibits bacterial growth on moist surfaces.
- contains rust corrosion inhibitors.
- deodorizes by killing most microorganisms that cause offensive odors.

DIRECTIONS FOR USE IN VETERINARY CLINICS, ANIMAL CARE FACILITIES, ANIMAL RESEARCH CENTERS, ANIMAL BREEDING FACILITIES, KENNELS AND ANIMAL QUARANTINE AREAS.

It is a violation of Federal Law to use this product in a manner inconsistent with its labeling. Roccal-D Plus is a one-step germicide, fungicide, soapless cleaner and deodorant effective in the presence of organic soil (5% serum). It is non-selective and when used as directed, will not harm tile, terrazo, resilient flooring, concrete, painted or varnished wood, glass or metals.

- To clean and disinfect hard surfaces, use ½ fluid ounce of Roccal-D Plus per gallon of water. Apply by immersion, flushing solution over treated surfaces with a mop, sponge, cloth or bowl mop to thoroughly wet surfaces. Prepare fresh solutions daily or when solution becomes visibly dirty.
- To clean badly soiled areas, use up to 1½ fluid ounce per gallon of water.
- To disinfect, allow treated surfaces to remain moist for at least 10 minutes before wiping or rinsing.
- To control mold and mildew growth on previously cleaned, hard nonporous surfaces, use ½ fluid ounce per gallon. Allow to dry without wiping. Reapply as new growth appears.

BOOT BATH: Use 1 fluid ounce per gallon in boot baths. Change solution daily and anytime it becomes visibly soiled. Use a nylon bristled brush to clean soils from boots.

DISINFECTING VANS, TRUCKS AND FARM VEHICLES: Clean and rinse vehicles and disinfect with ½ fluid ounce per gallon Roccal-D Plus. If desired, rinse after 10 minutes contact or leave unrinsed.

Do not use Roccal-D Plus on vaccination equipment, needles or diluent bottles as the residual germicide may render the vaccines ineffective.

Roccal-D Plus should not be mixed with other cleaning or disinfecting compounds or products. BROAD SPECTRUM GERMICIDAL ACTION IN HARD WATER AND UNDER SOIL LOAD CONDITIONS: At ½ fluid ounce per gallon (1:256) in official AOAC Use Dilution and Fungicidal Tests, Roccal-D Plus is effective in water up to 400 ppm hardness (as $CaCO_3$) and an organic soil load of 5% serum against the following organisms.

BACTERIA

Pseudomonas aeruginosa ATCC 15442
Salmonella choleraesuis ATCC 10708
Enterobacter aerogenes ATCC 63809
Pasteurella multocida ATCC 7707
Shigella dysenteriae ATCC 13313
Klebsiella pneumoniae ATCC 4352
Enterococcus faecium ATCC 6569
Salmonella gallinarum ATCC 9184
Serratia marcescens ATCC 264
Bordetella avium ATCC 35086
Streptococcus agalactiae ATCC 27916
Mycoplasma gallisepticum ATCC 15302
Mycoplasma gallinarum ATCC 19708
Actinomyces pyogenes ATCC 19411
Actinobacillus pleuropneumoniae ATCC 27088
Corynebacterium pseudotuberculosis ATCC 19410
Rhodococcus equi ATCC 6939
Streptococcus equi var. zooepidemicus ATCC 43079
Staphylococcus aureus ATCC 6538
Salmonella enteriditis ATCC 4931
Streptococcus pyogenes ATCC 9547
Salmonella pullorum ATCC 9120
Escherichia coli ATCC 11229
Alcaligenes faecalis ATCC 8748
Shigella sonnei ATCC 29930
Salmonella typhosa ATCC 6539
Proteus morganii ATCC 25830
Proteus mirabilis ATCC 25933
Mycoplasma iners ATCC 19705
Mycoplasma hypopneumoniae ATCC 25934
Bordetella bronchiseptica ATCC 19395
Streptococcus equi var. equi ATCC 33398

FUNGI

Aspergillus fumigatus ATCC 10894
Trichophyton mentagrophytes var. interdigitale ATCC 9533
Candida albicans ATCC 18804

VIRUSES

Using accepted virus propagation and hard surface test methods, Roccal D-Plus is effective at 1:256 in 400 ppm hard water and 5% serum against the following viruses.

Avian laryngotracheitis ATCC VR-783
Canine parvovirus ATCC VR-953
Infectious bronchitis ATCC VR-22
Transmissible gastroenteritis ATCC VR-763
Mycoplasma gallisepticum ATCC 15302
Infectious bursal disease (Gumboro) ATCC VR-478
Canine distemper onderstepoort strain
Equine herpesvirus ATCC VR-700
Feline infectious peritonititis strain, DF-2
Swine PRRS virus, Reference strain, NVSL
Newcastle's disease ATCC VR-109
Parainfluenza ATCC VR281
Pseudorabies ATCC VR-135
Avian influenza ATCC VR-798
Canine herpesvirus ATCC VR-552
Equine influenza A ATCC VR-297
Porcine parvovirus ATCC VR-742
Vesicular stomatitis virus, Indiana

FIGURE 14-5 Sample disinfectant package insert. Reprinted with permission from Pharmacia Animal Health, Kalamazoo, MI.

STORAGE AND DISPOSAL: Store only in tightly closed original container in a secure area inaccessible to children. Do not contaminate water, food or feed by storage and disposal. Disposal: Pesticide wastes are acutely hazardous. Improper disposal of excess pesticide spray mixture, or rinsate is a violation of Federal Law. If these wastes cannot be disposed of by use according to label instructions, contact your State Pesticide or Environmental Control Agency, or the hazardous waste representative at the nearest EPA Regional Office for guidance.

Do not reuse empty container. Rinse container thoroughly before discarding in trash.

PRECAUTIONARY STATEMENTS

HAZARDS TO HUMANS AND DOMESTIC ANIMALS
DANGER – KEEP OUT OF REACH OF CHILDREN

CORROSIVE: Causes severe eye and skin damage. Do not get into eyes, on skin or clothing. Wear goggles or face shield and rubber gloves when handling the concentrate. Harmful or fatal if swallowed. Avoid contamination of food. Wash thoroughly with soap and water after handling.

ENVIRONMENTAL HAZARDS: This product is toxic to fish. Do not discharge effluent containing this product into lakes, streams, ponds, estuaries, oceans or other waters unless in accordance with the requirements of a National Pollutant Discharge Elimination System (NPDES) permit and the permitting authority has been notified in writing prior to discharge. Do not discharge effluent containing this product to sewer systems without previously notifying the local sewage treatment plant authority. For guidance contact your State Water Board or Regional Office of the EPA.

PHYSICAL AND CHEMICAL HAZARDS: Do not use or store near heat or open flame.

STATEMENT OF PRACTICAL TREATMENT: In case of contact, immediately flush eyes or skin with plenty of water for at least 15 minutes. For eyes, call a physician. Remove and wash contaminated clothing before reuse. If ingested, drink promptly a large quantity of raw egg whites, gelatin solution, or if these are not available, drink a large quantity of water. Avoid alcohol. Call a physician immediately.

NOTE TO PHYSICIAN: Probable mucosal damage may contraindicate the use of gastric lavage.

1/IGL

Manufactured Exclusively for

Pharmacia & Upjohn Company
Kalamazoo, MI 49001, USA **816 538 002**

minutes and spores in about three hours. Viruses are inactivated in a short period of time as well. Organic debris and hard water do not inactivate glutaraldehydes. However, they are not used much in veterinary medicine due to high cost, instability, and toxicity issues (fumes are toxic and proper ventilation is essential).

Formaldehyde can be used in gas or solution form. As a gas, it can disinfect a large area like an incubator; as a liquid, it can disinfect instruments. In the aqueous form it is referred to as *formalin*. It works by affecting nucleic acids of microbes. Formaldehyde's extreme toxicity and classification as a carcinogen limit its clinical use. Read the MSDS before using formalin. Use formalin only in areas with good ventilation and avoid skin and eye contact and inhalation of vapors. Formalin is most commonly used in fixation of tissue biopsies for pathological examination.

Ethylene Oxide Ethylene oxide, a gas at room temperature, works by destroying DNA and proteins. Ethylene oxide is one of the only gases used for chemical sterilization (often referred to as gas sterilization). This gas is rather penetrating but slow acting. It may require 90 minutes to 3 hours for complete sterilization. Products must then be exposed to air for several hours to get rid of as much residual gas as possible. Ethylene oxide is explosive and rated as a potent carcinogen, so it must be handled and contained carefully. Its benefit is that it can sterilize objects that cannot withstand heat, such as rubber products.

Alcohols Alcohols, either 70% ethyl alcohol or 50% or 70% isopropyl alcohol, are used alone in aqueous solutions or as solvents in other chemicals. Alcohols work by coagulating proteins and dissolving membrane lipids. Alcohols are nonirritating, nontoxic, and inexpensive. Alcohols work well on both gram-positive and gram-negative bacteria and enveloped viruses, but are ineffective on spores and nonenveloped viruses. To be antifungal, alcohols must be in contact with fungi for several minutes. Alcohols must be applied in sufficient quantity, at the proper concentration, and for adequate time (several seconds to minutes) to be effective. Alcohol is not recommended as an antiseptic, because of the pain it causes and the denaturing effect it has on proteins. Dirt and organic debris must be removed prior to alcohol application for the alcohol to be effective. It is commonly used to clean skin surfaces prior to injections and when taking blood samples.

Halogens Chlorine and iodine agents represent the halogens. They work by interfering with proteins and enzymes of the microbe. Chlorine kills bacteria, fungi, viruses, and spores (effectiveness depends upon concentration, exposure time, and formulation). Chlorine, which is found in bleach, is inexpensive and easy to purchase. The disadvantages of chlorine use are bleaching of fabric, its corrosiveness to metal surfaces, and vapor that can be irritating to eyes and mucous membranes. Bleach is routinely used in a 1:10 dilution. It is easily inactivated by organic material and unstable if exposed to light.

> **TIP**
>
> **OSHA recommends bleach more than any other disinfecting solution in medical-based practices. Bleach solutions have a short potency span and must be replaced every two to three days, or mixed up fresh when needed.**

Iodine is a black chemical that forms a brown solution when mixed with water or alcohol. Most classes of microbes are killed by iodine if proper concentration and exposure times are used (activity against bacterial spores is not consistent). Iodine compounds are commonly used as topical antiseptics. Iodophors are complexes of iodine and a neutral polymer such as polyvinyl alcohol (PVA). This formulation causes slow release of iodine and increases its ability to penetrate skin. Iodophor compounds are marketed as scrubs, solutions, or tinctures. *Scrubs* have soap products added to them, *solutions* are dilutions of iodine and water, and *tinctures* are iodine and dilute alcohol. Iodine products can be corrosive to metals if left in contact for long periods of time, can be irritating to skin in high concentrations, and can stain fabrics and other materials.

Biguanides Chlorhexidine, a biguanide, is one of the most commonly used disinfects and antiseptics in veterinary practice. Chlorhexidine works by denaturing proteins. It is relatively mild, nontoxic, and fast acting. It is effective against bacteria, some fungi, and enveloped viruses such as feline infectious peritonitis (FIP) and FeLV, but does not work on nonenveloped viruses and spores. Chlorhexidine is commonly used as a surgical scrub, for cleaning wounds in lower concentrations, and as a teat dip. It can have residual activity of 24 hours because it binds to the outer surface of the skin.

Other Agents Hydrogen peroxide has been used to kill anaerobic bacteria in deep wounds such as puncture wounds. This chemical causes oxygen to be released when it reacts with cellular products and compromised tissue. Because hydrogen peroxide damages proteins and thus can damage animal tissue, its use should be limited. It is commonly used for oral infections and surface wound management.

Soaps or detergents have limited bactericidal activity. Their main functions are mechanical removal of microbes. They may contain ingredients effective against some bacteria, but do not work on spores and have limited use on viruses.

Table 14-4 summarizes the various disinfectants.

TABLE 14-4 Types of Disinfectants and Antiseptics

Disinfectant Group	Product Examples	Use	Action	Comments
phenols	ortho-phenylpenol (Lysol®, Amphyl®), hexachlorophene (Phisohex®)	Laundry, floors, walls, equipment	Moderately bactericidal, virucidal, and fungicidal	• Action not affected by organic material; • Used as a 2%–5% solution on contaminated objects
quaternary ammonium compounds	didecyl dimethyl ammonium chloride (Roccal®-D), benzalkonium chloride (Zephiran®)	Instruments, rubber, inanimate objects	Moderately bactericidal, virucidal, and fungicidal	• Action not affected by hard water
aldehydes	gluteraldehydes (Cidex®, Glutarol®)	Instruments	Highly bactericidal, virucidal, and fungicidal	• Action not affected by organic material or hard water
ethylene oxide		Rubber goods, blankets, lensed instruments	Highly bactericidal, virucidal, and fungicidal	• "Gas sterilization" for objects that cannot withstand heat • Carefully read MSDS prior to handling • Keep away from flames and sparks
alcohols	70% isopropyl, 50% ethyl alcohol	Instruments, thermometers	Highly bactericidal, some virucidal action, and poor fungicidal action	• 70% solution usually used • Affected by organic material and dirt
halogens	• chlorines: (Chlorox®) • iodophors: (Betadine®, Povidine®)	Chlorines: floors, cages Iodophors: presurgical scrub, thermometers	Moderately to highly bactericidal, highly virucidal, moderately to highly fungicidal and some sporicidal activity	• Corrosive to surfaces; • Vapors can be irritating • Iodine tincture is about 2%
biguanide	chlorhexidine (Nolvasan®, Hibiclens®, Virosan®)	Skin wounds, presurgical scrub, oral cleaning solutions, cages	Highly bactericidal, moderately virucidal, and poorly fungicidal	• Residual action of about 24 hours due to binding to skin

SUMMARY

Antimicrobials, sometimes referred to as anti-infectives, include the drug categories antibiotics, antifungals, and antivirals. Disinfectants and antiseptics also work on these microorganisms.

Antibiotics work on bacteria. Antibiotics may have a narrow spectrum of activity (working only on gram-negative or gram-positive bacteria) or a broad spectrum of activity (working on both gram-negative and gram-positive bacteria). Antibiotics can also be described as bactericidal (actually kill the bacteria) or bacteriostatic (inhibit the growth of bacteria). Bacteria are either resistant, sensitive, or intermediate as to a particular antibiotic. The sensitivity of bacteria to a particular antibiotic is determined by agar diffusion testing (antibiotic sensitivity testing). Concerns regarding antibiotic use include bacterial resistance and antibiotic residues. Antibiotics work by a variety of mechanisms, including inhibition of bacterial cell wall synthesis, damage to the bacterial cell membrane, interference with bacterial metabolism, and impairment of bacterial nucleic acids.

Antifungals work against fungal infections. Fungal infections may be superficial or systemic. Antifungal drugs work in a variety of ways, including binding to the fungal cell membrane, causing leakage of the fungal cell wall, interfering with fungal metabolism, and disrupting fungal cell division.

Antiviral drugs work by preventing viral penetration of the host cell or by inhibiting the virus's production of RNA or DNA. Antiviral drugs have been used in veterinary practice for feline herpes virus and feline leukemia virus.

Disinfectants are chemicals that kill or inhibit the growth of microorganisms on inanimate objects. Antiseptics are chemicals that kill or inhibit the growth of microorganisms on animate or living tissue. Each disinfectant and antiseptic has its own spectrum of activity, its own dilution values, and its own application instructions; therefore, labels should be read prior to use.

CHAPTER REFERENCES

Adams, H.R. (2001). *Veterinary pharmacology and therapeutics* (8th ed.). Ames, IA: Iowa State University Press.

Delmar's A-Z NDR-97 Nurses drug reference. Clifton Park, NY: Thomson Delmar Learning.

Gilman, A.G., et al. (1985) *Goodman and Gilman's The pharmacological basis of therapeutics* (7th ed.). New York: Macmillan.

Rice, J. (1999). *Principles of pharmacology for medical assisting* (3rd ed.). Clifton Park, NY: Thomson Delmar Learning.

Shimeld, L.A. (1999). *Essentials of diagnostic microbiology.* Clifton Park, NY: Thomson Delmar Learning.

Veterinary pharmaceuticals and biologicals (12th ed.). (2001). Lenexa, KS: Veterinary Healthcare Communications.

Veterinary values (5th ed.). (1998). Lenexa, KS: Veterinary Healthcare Communications.

Wynn, R. (1999). *Veterinary pharmacology.* Cambridge, MA: Harcourt Learning Direct.

CHAPTER REVIEW

Matching

Match the drug name with its action.

1. ______ interferon
2. ______ enrofloxacin
3. ______ sulfonamide
4. ______ nystatin
5. ______ erythromycin
6. ______ doxycycline
7. ______ cephalexin
8. ______ gentamicin
9. ______ procaine penicillin G
10. ______ ketoconazole

a. antifungal used to treat *Candida* infections
b. used to treat viral infections
c. fluoroquinolone antimicrobial
d. antimicrobial used widely before the advent of penicillin
e. aminoglycoside antibiotic that may cause nephro- and ototoxicity
f. long-acting penicillin form given only by injection
g. antibiotic known as a cephalosporin
h. tetracycline antibiotic
i. macrolide antibiotic
j. drug used orally and topically to treat fungus infections by causing leakage of the fungal cell membrane

Multiple Choice

Choose the one best answer.

11. Which antibiotics contain the beta-lactam ring?
   a. penicillins and aminoglycosides
   b. quinolones and tetracyclines
   c. macrolides and quinolones
   d. penicillins and cephalosporins

12. Which antibiotics are not recommended for young animals?
   a. penicillins and aminoglycosides
   b. quinolones and tetracyclines
   c. macrolides and quinolones
   d. penicillins and cephalosporins

13. Which antibiotic many cause bone marrow suppression if taken systemically or handled improperly?
   a. penicillin
   b. chloramphenicol
   c. enrofloxacin
   d. tetracycline

14. Which antibiotics are used only topically?
   a. penicillins and cephalosporins
   b. polymyxin B and bacitracin
   c. erythromycin and doxycycline
   d. gentamicin and sulfonamides

15. Which antifungal is only given IV?
   a. nystatin
   b. amphotericin B
   c. ketoconazole
   d. flucytosine

16. Griseofulvin is used to treat
   a. gram-positive bacterial infections.
   b. gram-negative bacterial infections.
   c. systemic mycotic infections.
   d. dermatophyte infections.

17. Chlorine and iodine agents represent which group of disinfectants/antiseptics?
   a. phenols
   b. aldehydes
   c. halogens
   d. alcohols

18. Which antiseptic is used commonly as a surgical scrub, for cleansing wounds, and as a teat dip?
   a. hydrogen peroxide
   b. alcohols
   c. chlorhexidine
   d. formaldehyde

19. What chemical is used to sterilize rubber goods?
   a. phenol
   b. ethylene oxide
   c. hydrogen peroxide
   d. alcohol

True/False

Circle a. for true or b. for false.

20. All disinfectants are sporicidal.

a. true
b. false

Case Studies

21. A 1200-lb. Holstein cow (Figure 14-6) freshened (gave birth) 4 days ago, which was 2 weeks earlier than her due date. She refuses to eat her feed today. On physical examination, the cow shows an elevated temperature of 104.2°F and is lethargic. Vaginal examination reveals that she has retained fetal membranes. The veterinarian decides to treat her with procaine penicillin at a dosage of 15,000 IU/kg/day.

a. What is one reason why procaine penicillin is used?
b. What would the dose be for this cow in IU?
c. The concentration of procaine penicillin is 300,000 IU/ml. How many ml will this cow get per dose?

22. A 6-month-old M Beagle puppy (25#) (Figure 14-7) had a history of tick infestation from 2 months ago. The dog is now lame and is running a fever.

a. What disease should be considered in this dog?
b. What is the usual treatment for this dog?
c. Can this drug be used in this dog?

FIGURE 14-6 A Holstein cow. *Source:* Deborah Bohn, CVT

FIGURE 14-7 Beagle. Photo by Isabelle Francais

23. An owner of a cattery comes into the clinic and says that her cats have had a positive culture for ringworm (Figure 14-8). She would like to use a drug on all her cats.

a. One drug that works on ringworm is griseofulvin. What do you want to discuss with this owner before the veterinarian prescribes griseofulvin?

b. What other drug choice do you have in this situation?

FIGURE 14-8 Cat.

CHAPTER 15

Antiparasitics

OBJECTIVES

Upon completion of this chapter, the learner should be able to

- differentiate between the two main groups of parasites and describe how treatment for each varies.
- differentiate between nematodes, cestodes, and trematodes.
- describe various treatment and prophylactic strategies for nematodes, cestodes, and trematodes.
- describe treatment and prophylactic strategies for coccidia.
- explain why anticoccidial drugs are called coccidiostats and how their nature affects treatment.
- describe the various types of heartworm preventatives and treatments.
- describe the various forms in which ectoparasite products are available.
- describe the various chemicals used in ectoparasite control.

KEY TERMS

endoparasite
ectoparasite
anthelmintic
antinematodal drug
microfilaricide
anticestodal drug
antitrematodal drug
coccidiostat
anticoccidial drug
antiprotozoal drug
adulticide

INTRODUCTION

A farmer has 200 calves that he would like to start on a deworming program. He is interested in giving his calves something that will cover a lot of different parasites and is easy to administer. He would like to know what dewormers are available and whether they come in easy-to-administer forms. Are dewormers safe for cattle of all ages? What side effects does he need to watch for? Understanding the types of parasites that have to be controlled or treated, the routes of administration used in parasite control and treatment, and the side effects of these drugs is critical to your clients. Are you able to explain antiparasitic drugs to clients?

TYPES OF PARASITES

Parasite control in animals is an important function for the veterinary staff. Clients want to know about the safest and most effective medicines to use in the treatment and prevention of parasitic disease in their pets and livestock. The two major groups of parasites are endoparasites and ectoparasites. **Endoparasites** live within the body of the host and cause internal parasite *infections*; **ectoparasites** live on the body surface of the host and cause external parasite *infestations*.

Parasites can be contracted in a number of different ways, including animal-to-animal contact, ingestion of contaminated food or water, insect transmission, or direct contact with the parasite (walking, lying, or rolling on infected soil). Some parasites that are present on or within the host may not cause clinical signs in the animal, whereas other parasites present on or within the host actively harm the animal. Most intestinal parasites are diagnosed by microscopic fecal examination. Keep in mind that clinical signs may develop with some parasitic infections before eggs are detected in clinical samples.

INSIDE: ENDOPARASITES

Helminths are worms found primarily in the gastrointestinal tract, liver, lungs, and circulatory system. The two major helminth groups are nematodes and platyhelminths (Figure 15-1). *Nematodes* are cylindrical (like a slender tube) and nonsegmented, and may be referred to as *roundworms*. Nematodes inhabit the stomach and intestines of domestic animals, wild animals, and birds. Filarial nematodes are transmitted by arthropods and are covered later in this chapter. *Platyhelminths* are flattened and may be referred to as *flatworms*. Flatworms are further divided into *cestodes* (sometimes referred to as *tapeworms*) and *trematodes* (sometimes referred to as *flukes*). There are many types of cestodes and trematodes, and they inhabit various locations in animals. Some cestodes can live in body tissues in the immature form, while other cestodes can live in the intestinal tract in the adult form. Some trematodes live in the bile ducts of ruminants and cause considerable losses to the livestock industry. Table 15-1 summarizes the types of helminths usually seen in veterinary medicine.

Antiparasitic drugs are categorized by the type of parasite they work against. **Anthelmintics** kill worm parasites. Anthelmintics are further categorized as antinematodal, anticestodal, or antitrematodal (Table 15-2). These drugs are given in a variety of ways (Table 15-3). The drug's solubility largely dictates the route of administration. Water-insoluble anthelmintics are usually given orally (suspension, paste, granules), whereas more soluble compounds can be given orally as a solution, topically as a pour-on, or via injection. The drug particle size also plays a role in the route of administration. Generally, smaller particles are more easily absorbed from the gastrointestinal tract, larger particles (especially those that are insoluble) have minimal absorption from the gastrointestinal tract and therefore have lower toxicities.

Antinematodal Drugs

Antinematodal drugs work against nematodes. There are six groups of antinematodal drugs. Some drugs are in several of these groups, as they may be effective against more than one type of parasite.

The *benzimidazole* category of antinematodal drugs are excellent against nematode infestations. They are believed to work by interfering with energy metabolism of the worm. The main benzimidazole is thiabendazole, whose effectiveness and low toxicity made its introduction a landmark in helminth therapy. The other benzimidazoles work at lower dose levels and with broader-spectrum activity. Animals take all the benzimidazoles orally, either as a paste, a granulated powder, or a solution. Repeated doses of benzimidazoles are advantageous because of their slow killing process. Side effects are rare with benzimidazoles, but may include vomiting, diarrhea, and lethargy. Other benzimidazoles include oxibendazole, mebendazole, fenbendazole, and oxfendazole (covered in Table 15-2).

TIP

Ectoparasiticides treat arthropod infestations (insects and arachnids), anthelmintics treat worm infections, antiprotozoals treat protozoan parasitic infections, and endectocides treat internal parasitic infections and external parasitic infestations. (Insects include flies, mosquitoes, bots, cuterebra, lice, and fleas. Arachnids include spiders, scorpions, ticks, and mites.)

TIP

Antiparasitic drugs should be used with caution in old, young, pregnant, or debilitated animals. Consult package inserts or the *VPB* before applying antiparasitics to these groups of animals.

TIP

Most anthelmintics have withdrawal times; consult reference material for these times.

TIP

Benzimidazole drugs have the *-azole* ending in their generic names; however, not all *-azole* drugs function as anthelmintics (for example ketoconazole, works on fungi).

TIP

The most common target for antiparasitic drugs to attack is the parasite's nervous system.

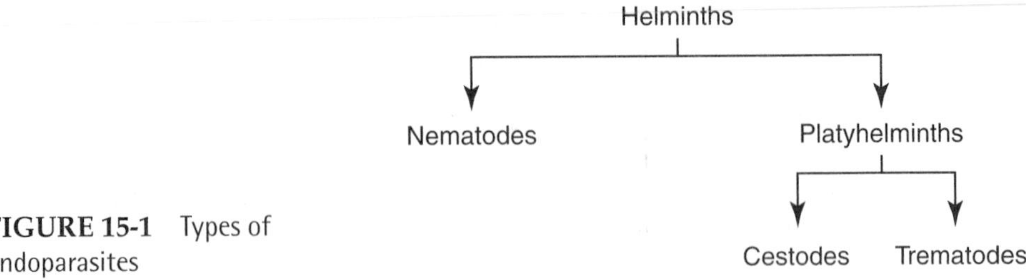

FIGURE 15-1 Types of endoparasites

TIP

Resistance to benzimidazoles is a problem in treating small strongyles in horses and abomasal trichostrongyles in sheep. In time, resistance issues may arise in other species.

- Thiabendazole works against ascarids and strongyles. It also has antifungal and anti-inflammatory effects that make it suitable for otic preparations (e.g., Tresaderm Otic®).
- Oxibendazole is used as a horse dewormer and in combination products for dogs. In dogs, liver toxicity has been noted with this product. An example is Anthelcide EQ Equine Wormer Paste®.
- Mebendazole is a granular powder used in dogs and horses to treat ascarid, hookworm, and cestode infections. An example is Telmintic®. Liver toxicity has been noted with this product.
- Fenbendazole is used in small animals, food animals, and horses. It has a wide spectrum of activity, including nematodes, hookworms, whipworms, and the cestode *Taenia pisiformis* in small animals. Fenbendazole may also be used to treat metronidazole-resistant giardiasis. Vomiting and diarrhea have been caused by this drug. It is not approved for use in lactating dairy animals, because of withdrawal issues. An example is Panacur®, which is found in a variety of forms including granules, suspensions, and pastes.

A drug in a related group, the probenzimidazoles, is febantel. Probenzimidazoles are examples of probiotics, which are metabolized in the animal's body to true benzimidazoles and then work the same as the benzimidazoles. Febantel is usually added to another antiparasitic. Drontal Plus® tablets is a trade name for a febantel, pyrantel pamoate and praziquantel combination.

The *imidazothiazoles* are a second group of antinematodal drugs, of which levamisole is an example. Levamisole stimulates the nematode's cholinergic nervous system, leading to paralysis of the parasite. Because levamisole works on the nematode's nervous system, it is not ovicidal. Levamisole expels most nematodes within 24 hours, though the worms may be passed alive. Levamisole comes in oral forms, such as pellets, powder to mix into drinking water, suspensions, and pastes. Levamisole is effective against ascarids, strongyles, whipworms, and hookworms and has been used as a **microfilaricide** in heartworm treatment regimens in dogs (this use is no longer common). Levamisole also has anti-inflammatory and immunostimulant properites. Levamisole toxicity in the host animal is due mainly to its cholinergic effects, which may cause salivation, ataxia, and muscle tremors. Trade name examples of levamisole are Levasole® and Tramisol®.

The *tetrahydropyrimidines* include pyrantel pamoate, pyrantel tartrate, and morantel tartrate. These drugs are cholinergic agonists that mimic the action of acetylcholine. This causes initial stimulation and then paralysis of the worm, allowing it to be excreted by the intestine. These drugs have a somewhat pleasant taste and are very safe, with few side effects. This group is effective at treating ascarids, pinworms, strongyles, and hookworms. Trade name examples of this group include Nemex® and Strongid-T®.

TIP

Organophosphates are becoming less popular with the development of safer products.

Organophosphates (OPs) are used against both endoparasites and ectoparasites. Organophosphates have a narrow range of safety and are generally given as a feed supplement to beef and dairy cattle. Organophosphates are one of the few drugs effective against bots in horses. OPs inhibit cholinesterase, causing acetylcholine to remain active in the neuromuscular junction of the parasite. They are neurotoxic to

TABLE 15-1 Helminths Encountered in Veterinary Medicine

Main Category	Location of Parasite	Scientific and Common Names
Nematodes	Abomasal worms in ruminants or stomach worms in monogastric animals	• Barberpole worm *Haemonchus contortus, H. placei* (R) • Brown stomach worm *Ostertagia ostertagi* (R) • Small stomach worm or hairworm *Trichostrongylus axei* (R, H) • *Hyostrongylus rubidus* (Sw) • Large-mouth stomach worm *Habronema muscae* (H)
	Intestinal worms	• Small intestinal worms *Cooperia punctata, C. oncophora, C. mcmasteri* (R) • Hookworms *Bunostomum phlebotomum* (R), *Ancylostoma sp* (D, F) • Nodular worms *Oesophagostomum spp.* (R, Sw) • Thread-necked intestinal worm *Nematodirus helvetianus* (R) • Bankrupt worm *Trichostrongylus colubriformis* (R) • Large strongyles *Strongylus vulgaris, S. edentatus, S. equinus, Triodontophorus spp.* (R, E) • Small strongyles *Cyathostomum spp., Cylicocyclus spp., Cylicostephanus spp., Cylicodontophorus spp.* (R) • Whipworms *Trichuris suis* (Sw), *Trichuris vulpis* (D) • Threadworms *Strongyloides ransomi* (Sw), *Strongyloides westeri* (E), *Strongyloides stercoralis* (D) • Ascarids *Parascaris equorum* (H), *Toxocara canis* (D), *Toxocara cati* (F), *Toxascaris leonina* (D, F) • Pinworms *Oxyuris equi* (E)
	Circulatory systems worms	• Heartworms *Dirofilaria immitis* (D, F)
	Lungworms	• *Dictyocaulus spp.* (R, H) • *Prostostrongylus rufescens* (S, G) • *Muellerius capillaris* (S, G) • *Metastrongylus spp.* (Sw)
	Kidney worms	• *Stephanurus dentatus* (Sw)
	Gastrointestinal and skin worms	• *Habronema spp.*, worms *Draschia spp., Onchocerca spp.* (H)
Cestodes (tapeworms)		• *Moniezia benedeni* (R) • *Taenia spp.* (R, D, F)

(continues)

TABLE 15-1 *Continued*

Main Category	Location of Parsite	Scientific and Common Names
Cestodes (tapeworms) (*cont.*)		• *Echinococcus granulosus* (R is intermediate host, D) • *Dipylidium caninum* (D, F) • *Anophocephala perfoliata*, *A. magna* (H) • *Paranoplocephala mamillana* (H)
Trematodes (flukes)	Liver fluke Deer liver fluke Lung fluke	• *Fasciola hepatica* (R) • *Fascioloides magna* (R) • *Paragonimus kellicotti* (D, F)
Protozoa		• Coccidia *Eimeria spp.* (R, Sw, H) • Coccidia *Isospora spp.* (Sw, D, F) • *Toxoplasma gondii* (F) • *Giardia spp.* (R, H, D, F) • *Cryptosporidium spp.* (R, H) • *Sarcocystis spp.* (R in muscle; D, F shed in stool) • *Tritrichomonas foetus* (C in reproductive tract)

R = ruminants, C = cattle, S = sheep, G = goats, Sw = swine, D = dogs, F = cats, H = horses

parasites and can cause some neurologic side effects in the host. OPs should not be given to heartworm-positive dogs, because they may cause dyspnea (difficulty breathing) and death due to sudden worm kill-off. OPs should be used with caution in all species; if the OP is not labeled for a particular species, it is better to use an antiparasitic drug that is labeled for that species, rather than using an OP extra-label. OPs are effective against bots, nematodes, strongyles, and pinworms in horses; strongyles in ruminants; hookworms, nematodes, and whipworms in dogs and cats; and ascarids, whipworms, strongyles, and nodule worms in swine. Some generic and trade names of OPs are dichlorvos (Task®), and coumaphos (Baymix®).

TIP

Ivermectin and related compounds are not effective against cestodes or trematodes.

TIP

Most parasite infections are diagnosed via microscopic identification of eggs or oocysts. Egg identification is important so that the most appropriate antiparasitic agent can be used to treat the affected animal. Repeating the fecal examination after treatment (fecal recheck times vary with the parasite) is equally important to ensure that the treatment was effective.

Piperazine compounds consistently eradicate only ascarids in dogs, cats, horses, ruminants, and poultry. They are anticholinergic drugs that block neuromuscular transmission in the parasite. These drugs go in food or drinking water; dogs and cats usually get tablets. Clients should be made aware that treating animals that have ascarid infections with piperazine will often result in intact worms being vomited or passed in the stool. Piperazines are practically nontoxic and are sold over the counter under such trade names as Pipa-Tabs® and Hartz Health Measures Once-a-Month Wormer® for Puppies.

The representative drug of the *avermectins* (also known as *macrocyclic lactones*) is called ivermectin. Avermectins bind to certain chloride channels in invertebrate nerve and muscle cells, causing paralysis and death of the parasite. Low doses of ivermectin, given orally and parenterally, work well against a wide range of internal and external parasites. The low dose also allows it to be injected subcutaneously. Occasionally ivermectin can cause rapid death of parasites, resulting in allergic reactions, inflammation, and swelling in the area of the body where the parasite dies. Ivermectin has also produced adverse reactions in collies and collie crosses. Trade names for ivermectin products are Heartgard®, Heartgard Plus®, and Ivomec®. The "Plus" products contain other antiparasitic agents to broaden their spectrum of activity. Another avermectin is moxidectin. This is the chemical used in ProHeart®-6, the six-month injectable heartworm

TABLE 15-2 Types of Antiparasitic Drugs

Benzimidazoles (nematodes) • thiabendazole • oxibendazole • mebendazole (also works on tapeworms) • fenbendazole (also works on tapeworms and flukes) • albendazole (also works on tapeworms, flukes, and giardia) • oxenfendazole (also works on tapeworms and flukes) • febantel (a probenzimidazole used in combination with other products to broaden its spectrum of activity)		
Imidazothiazoles (nematodes) • levamisole		
Tetrahydropyrimidines (nematodes) • pyrantel pamoate	• pyrantel tartrate	• morantel tartrate
Organophosphates (nematodes; ectoparasites, including bots) • trichlorfon	• dichlorvos	• coumaphos
Piperazines (nematodes)	• piperazine dihydrochloride	• piperazine sulfate
Macrocyclic lactones (nematodes; heartworm prevention; ectoparasites, such as bots and grubs) • ivermectin • moxidectin	• eprinomectin • milbemycin oxine	• selamectin • doramectin
Pyrazine derivatives • praziquantel (tapeworms, flukes) • epsiprantel (tapeworms)		
Benzenesulfonamide (flukes) • clorsulon		
Coccidiostats (coccidia) • sulfadimethoxine • nicarbazine	• amprolium • monensin	• decoquinate • robenidine
Nitroimidazoles (*Giardia*) • metronidazole		

TABLE 15-3 Administration Routes for Anthelmintics

Main Route	Example Route	Description	Concerns
Oral	Tablets	May be hard tablet, chewable tablet, or bolus (large pill) that animal ingests or is given	• Allows greater control over amount of drug given • Palatability is important (especially with cats and horses)
	Liquid	May be in the form of a solution, suspension paste, paste syringe, or drench	• Drug must be shaken well to ensure adequate mixing of chemical throughout the liquid • Paste syringes contain a precalibrated amount of paste that need not be shaken

(continues)

TABLE 15-3 *Continued*

Main Route	Example Route	Description	Concerns
Oral (*cont.*)			• Drenches are liquid forms given by mouth that force the animal to drink • Allows greater control over amount of drug given
	Feed additives	May be in feed, added to mineral mixes, added to drinking water or added to salt blocks	• Treats parasites in large numbers of animals • Allows little control over amount of drug ingested by an individual animal • Is stress-free for the animals • Saves expense of rounding up livestock
	Sustained-release	Device is implanted in rumen to allow slow release of the drug over time (especially helpful in treating later stages of parasite larvae)	• Saves time in retreating of animals • Can treat animals over a period (like the complete grazing season)
Injectable	Solution	Given SQ usually (if given IM, may affect the carcass)	• Easy way to administer • Local reaction is possible, but is rare • Allows greater control over amount of drug given • Achieves higher blood levels rapidly • Requires livestock to run through chutes or be crowded into restricted spaces for injection with multidose syringes for treatment • Can control amount of drug given
Topical (pour-ons)	Solution	Absorbed through the skin via the sebaceous glands and hair follicles	• Used in large animals • Achieves high blood levels rapidly • Easy to administer • Reduces stress for treated animals • Used for internal and external parasites • Pour-ons are highly concentrated dose forms that may be hazardous

preventative, and in Quest 2% Equine Oral Gel®. Side effects of this group of antiparasitics include stupor and convulsions. (This group of drugs is also covered under heartworm treatments later in this chapter.)

Anticestodal Drugs

Anticestodal drugs are used for the treatment of cestode infections in animals. Cestodes consist of a scolex (head), neck, and proglottids (segments). Cestodes increase in length by producing new proglottids from the neck area; therefore, the oldest proglottids are at the distal end of the worm. A cestode will attach itself to the intestinal wall by its scolex, so dislodging the scolex is an important part of cestode treatment. Some of these products are used in combination with the antiparasitic drugs already discussed.

Praziquantel works by increasing the cestode's cell membrane permeability, resulting in disintegration of the worm's outer tissue covering. This disintegration includes the scolex, so the cestode cannot re-lodge itself into the bowel wall and begin producing new proglottids. Because the cestode disintegrates, proglottid segments are not observed in the stool. Praziquantel works on all of the cestode species, but flea control is needed to eliminate the cestode *Dipylidium caninum*. Praziquantel is effective against adult cestodes. Side effects include anorexia, vomiting, diarrhea and lethargy. Droncit® is a trade name of praziquantel that is available as an oral tablet or IM injection (injection stings, so the oral route is preferred).

Epsiprantel, known by its trade name Cestex®, is effective against *Taenia* and *Dipylidium* cestodes, but not *Echinococcus* species. Epsiprantel causes disintegration of the cestode, so proglottid segments are observed in the stool.

Fenbendazole, a benzimidazole covered earlier, is effective against *Taenia spp.* of cestodes, but not *Dipylidium caninum*. Unless other parasites that can be treated with fenbendazole are present in the animal, praziquantel or epsiprantel are recommended.

TIP

Ruminant cestodes rarely cause problems and can be treated with fenbendazole. Cestodes in horses (*Anoplocephala spp.*) can be pathogenic and should be treated with praziquantel or a double dose of pyrantel pamoate.

Antitrematodal Drugs

Antitrematodal drugs are used in the treatment of *trematodes*, the flat, leaf-shaped helminths whose bodies lack segmentation. Trematodes are commonly referred to by the area they infect (for example, the liver trematode). Trematodes have intermediate hosts, the first of which is almost always a snail. Snails are found in wet, rainy environments; therefore, managing trematode infections requires more than just drug treatment. Clorsulon, albendazole, and praziquantel are all antitrematodal drugs.

Clorsulon, a benzene sulfonamide, inhibits the trematode's enzyme systems used in energy production and robs the trematode of its energy. Clorsulon is a drench and is effective against the adult and immature form of the liver trematode, *Fasciola hepatica*, in cattle. Curatrem® is an over-the-counter product of clorsulon. Withdrawal times in milk have not been established, therefore, use of clorsulon is not recommended in dairy animals.

Albendazole is effective against a variety of parasites, including *Fasciola hepatica* adults in cattle. It is a benzimidazole and it works by interfering with the energy metabolism of the worm. Albendazole is also effective against some nematodes. Valbazen® is a trade name albendazole product that is a broad-spectrum anthelmintic (it also works on cestodes and nematodes). Both albendazole and clorsulon have withdrawal times associated with their use. Milk withdrawal times for albendazole have not been established, therefore, it is not approved for use in lactating animals.

Praziquantel was covered earlier in the anticestode section. It may also be used to treat lung trematodes in dogs and cats.

Anticoccidials and Other Antiprotozoals

Protozoa are single-celled parasites found in many species of animals. *Coccidiosis* is a protozoan (*Coccidia*) infection that causes various intestinal disorders—some serious and even fatal—in various species. Coccidiosis (always prevalent in the poultry industry) can be managed by adding **coccidiostats** (**anticoccidial drugs**) to the daily feed ration. Coccidia have a complex life cycle that makes them difficult to treat. Coccidia infections tend to involve management problems with sanitation procedures. Coccidiostats do not kill the coccidia parasite, so cleaning of contaminated waste from housing facilities and proper disinfection are key in removing coccidia from an environment. Providing clean food and water and avoiding overcrowded conditions also help to keep coccidia organisms in check. Sulfa drugs produced in the 1940s were the first effective coccidiostats; however the protozoans developed sulfa-resistant strains. One widely used sulfa drug, sulfadimethoxine (Albon®), is still used in poultry, dogs, and cats for treatment of coccidia. Sulfadimethoxine does not eliminate the infection

TIP

Coccidia infections in carnivores are caused by *Isospora spp.* and coccidia infections in herbivores are caused by *Eimeria spp.* If dogs have *Eimeria* infections, it is due to their ingestion of herbivore feces. "

TIP

"Controlled infection" is the goal for treating coccidial infections. Drugs and sanitation play prominent roles in this control.

that the animal already has, but it reduces the spread of disease by reducing the number of oocysts shed. This reduced shedding reduces the spread of coccidia to other animals. More effective drugs introduced through the 1970s included nicarbazine (Maxiban 72® for chickens), amprolium (Corid® for calves), monensin (Coban 60® for poultry, Rumensin® for cattle), decoquinate (Deccox® for cattle and goats), and robenidine (Robenz Type A Medicated Article® for chickens). These drugs mainly work by affecting some aspect of the protozoan's metabolism (amprolium competes with the protozoa for thiamine, monensin binds with ions to affect mitochondrial function). These drugs work at various stages in the coccidian life cycle. Most of these drugs, when given to animals that are actively shedding oocyts, will not cure those animals, but may reduce the number of oocysts shed into the environment. Most of these drugs are given orally in feed, water, or by liquid or tablets.

In addition to coccidia, other parasitic protozoa cause disease in animals. *Giardia* is another protozoan parasite seen in animals for which **antiprotozoal drugs** are used. For the organism *Giardia lamblia*, there is a preventative vaccine called GiardiaVax®, as well as drugs for treating the organism. GiardiaVax® is an annual vaccine labeled to prevent infection by giardia; it also reduces the shedding of cysts. Initially, two injections are given SQ about two to four weeks apart. Treatment of giardia includes metronidazole (Flagyl®), fenbendazole (Panacur®), and albendazole (Valbazen®). Metronidazole is believed to work by entering the protozoal cell and interfering with its ability to function and replicate. Fenbendazole and albendazole were covered earlier in this chapter. References should be consulted prior to using these drugs in pregnant or lactating animals, to assess the drug's safety in these animals. Metronidazole has been experimentally identifed as causing mutations.

A blood protozoan, *Babesia sp.*, is transmitted by ticks. It can be treated subcutaneously or intramuscularly with imidocarb (Imizol®). Imidocarb appears to have a cholinergic effect on the protozoan. It is slowly metabolized and therefore is not used in food-producing animals, because of increased withdrawal times. Side effects of imidocarb include salivation, dyspnea, and restlessness. Using tick prevention can reduce infections with *Babesia sp.*

TIP

Microfilariae are prelarval stages of *Dirofilaria immitis*. The term microfilaria is not synonymous with larva.

IN THE HEART: HEARTWORM MEDICATIONS

Heartworm disease is a parasitic disease of dogs and other canines, but has also been identified in cats, ferrets, and occasionally humans. This disease is caused by the filarial nematode *Dirofilaria immitis*, transmitted by mosquitoes. Adult heartworms usually live in the pulmonary artery, but can also be found in the right ventricle and right atrium and, rarely, in the vena cava. Female adult heartworms produce microfilariae. Microfilariae circulate in the bloodstream and are taken up by mosquitoes when they feed on an infected animal's blood. In the mosquito, the microfilariae undergo changes until they become infective third-stage larvae. When these mosquitoes bite other animals, the third-stage larvae are transmitted to the new host. These larvae develop into adults and ultimately reach the aforementioned locations.

Medications to prevent and treat this disease, as well as guidelines issued by the American Heartworm Society, have been improved and monitored to find the best methods of curtailing this disease. Managing this disease involves preventing third-stage larvae from reaching maturity, eliminating adult heartworms, and eliminating circulating microfilariae.

Third-stage larvae can be kept from reaching maturity by the use of preventative medication. When given to a heartworm-negative dog, such a drug will prevent any larvae transmitted by a mosquito vector from reaching the adult stage. When given at higher doses, these medications also kill circulating heartworm microfilariae in heartworm-positive dogs. Examples of preventative drugs include:

- ivermectin (Heartgard®, Heartgard Plus®, Heartgard® for Cats, Iverhart™ Plus, Tri-Heart™ Plus). This monthly preventative is given orally to eliminate the tissue stage of heartworm microfilaria. Ivermectin revolutionized heartworm prevention in dogs, allowing dosing to go from once daily to once monthly (there are now other drugs that work monthly or semiannually). Heartgard Plus®, Tri-Heart™ Plus, and Iverhart™ Plus also contain pyrantel pamoate, which is effective against nematodes and hookworms. At higher dosages, ivermectin is used after **adulticide** treatment in heartworm-positive dogs. Ivermectin is not recommended for puppies less than six weeks of age, and it should be used with caution in collies or collie mixes. Collies and collie mixes should be observed for at least eight hours after administration of ivermectin for development of side effects. Side effects for all dogs include neurologic signs such as salivation, ataxia, and depression.

  Ivermectin effectiveness varies with dosage. At low dosages (6 mcg/kg), it is used as a heartworm preventative in dogs and cats that acts by blocking larval development. At moderate dosages (50 mcg/kg), it is used to kill microfilariae in dogs. At high dosages (200 mcg/kg), it will kill most internal parasites. Ivermectin doses must be calculated accurately. It is given in doses of *mcg/kg*, which is a thousandfold lower than most drugs. Most calculation errors occur because the mcg unit is overlooked.
- milbemycin (Interceptor®, Sentinel®). This is a monthly preventative, given orally that eliminates the tissue stage of the heartworm microfilariae by interfering with invertebrate neurotransmission. Both products are also effective against hookworms, nematodes, whipworms, and *Demodex* mites in dogs. Milbemycin has been used as a heartworm preventative in cats. Sentinel® also contains lufenuron for flea control. Side effects from either product are uncommon.
- selamectin (Revolution®). This monthly preventative is applied topically and absorbed systemically for the prevention of heartworm disease, and the prevention and control of fleas, ear mites, and sarcoptic mange in dogs. It has some activity against the American dog tick (*Dermacentor variabilis*), but not the deer tick (*Ixodes scapularis*), which is the carrier of Lyme disease. It is effective for treatment of nematodes and hookworms in cats. Side effects include alopecia at the application site, vomiting, and diarrhea.
- moxidectin (ProHeart®-6). This six-month injection interrupts early larval development and thus prevents heartworm disease. Its side effects include neurologic and gastrointestinal signs. If used to replace another heartworm preventative, the first dosage of moxidectin must be given within one month of stopping the original preventative. ProHeart®-6 was withdrawn from the market in September 2004 due to increased reports of adverse side effects such as liver and bleeding abnormalities.
- diethylcarbamazine (Nemacide®, Carbam®, Filaribits®, Filaribits Plus®). Diethylcarbamazine or DEC is a daily preventative given orally to eliminate the tissue stage of heartworm disease. It may be in tablet, chewable tablet, or liquid form. Nemacide® also controls ascarid infections. Filaribits Plus® has oxibendazole for the control of hookworms, whipworms, and ascarids. Side effects include vomiting, which can be controlled by giving the medication after eating or with a meal. Filaribits Plus® has been associated with liver disease. If you are replacing DEC with a monthly product, give the first dose of the monthly product within one month after stopping DEC.

Adult heartworms may develop in the right ventricle because of improper or no use of preventatives. Adulticides are used to kill the adult heartworm in heartworm-positive dogs. Melarsomine (Immiticide®) is an arsenical drug given by deep IM injection once daily for two days. The drug is given in the epaxial muscles on either side of the vertebral column, in the area between the L3 and L5 vertebrae.

TIP

All manufacturers of heartworm preventative recommend that dogs be microfilaria-negative before they are started on a preventative. This is especially important with DEC.

TIP

Glucocorticoids and aspirin are sometimes given after adulticide treatment to prevent emboli; however, the use of these drugs is currently discouraged because of their paradoxical protective effect on adult female worms.

Melarsomine appears to be less toxic than thiacetarsamide (Caparsolate®, the former adulticide treatment) because it contains a different form of the arsenic compound. It is not recommended for animals with large numbers of heartworms in the right ventricle, right atrium, or vena cava (referred to as *caval syndrome* when the vena cavae are involved). Coughing, gagging, and lethargy are side effects of this drug. Nephrotoxicity and hepatotoxicity may occur with melarsomine, but less so than with thiacetarsamide.

Any circulating microfilariae that appear after infection with adult heartworms must be eliminated. Drugs given for this purpose are given six weeks after administration of the adulticide and include:

- ivermectin (discussed previously). As a microfilaricide, it is given at a higher dose orally and monitored for neurologic side effects. The dog then returns in three weeks for a blood test; if positive, the dog is reevaluated for adult heartworms. If negative, the dog is started on a monthly preventative.
- milbemycin (discussed previously).
- levamisole (Levasole®, Tramisol®). This drug is given orally for a week or longer depending on the dosage. It is now used infrequently, due to the increasing use of ivermectin and milbemycin.

OUTSIDE: ECTOPARASITES

Ectoparasites live on the outside of the animal's body, but can still cause disease and loss of revenue in farm animals. Accurate identification of the ectoparasite is necessary to select the most appropriate drug to use for control. Ectoparasites include flies (bots and maggots), grubs, lice, fleas, mites, ticks, and mosquitoes.

Ectoparasites are controlled in many different ways. Ectoparasiticides are externally applied through the use of sprays (prediluted and concentrated), dips, pour-ons, ear tags, collars, spot-ons, shampoos, dusts, and foggers. Ectoparasiticides can also be delivered as oral medications. Table 15-4 summarizes various forms of ectoparasiticides.

TABLE 15-4 Application Methods for Ectoparasiticides

Type of Product	Advantages	Disadvantages
Prediluted sprays (include sprays for animals and premise sprays)	• Convenient and easy to use (apply from head to tail, avoiding eyes, mouth, and nose) • Usually has quick kill • May have residual effects • Available for animal and environment	• Water-based sprays do not penetrate oily coats or fabrics well • Alcohol-based sprays may be drying and irritating to skin
Concentrated Sprays	• Concentrated form may offer cost savings • Can be diluted at different concentrations for different ectoparasites	• Error in dilution may occur • Diluted product may not have long shelf life
Yard spray/Kennel spray	• Offer residual effects	• Can only be used on environment • Efficacy varies
Dips	• Offer residual effects	• Must be diluted properly • Animal should be shampooed first • Animal must dry with dip product on—cannot rinse

TABLE 15-4 *Continued*

Type of Product	Advantages	Disadvantages
Pour-ons	• Can ensure that an individual animal is treated • May treat many animals at a time with proper application devices	• Activity of drug may be limited if applied to unclean animal (e.g., animal with caked mud or manure on its hide) • May be applied incorrectly, resulting in limited value of the treatment or development of toxicity (application varies; may be along the backline from shoulders to the hip bones or in single spot)
Shampoos	• Rinse well • May contain medication effective against parasites	• May only contain products for cleaning the coat • No residual effect even if medication present • May have to be diluted before use • Must leave on animal for a specified time prior to rinsing
Dusts or powders	• Can be used in animals that do not tolerate sprays	• Do not provide quick kill • May irritate and dry skin
Foggers	• Work well in large, open rooms • Quick method for environmental control	• Product does not get everywhere needed can be improved if (corners, furniture); this premise spray is used with the fogger • Can be toxic to fish; must cover food products when applying
Oral products	• No mess • Works for a period of time	• May not kill all stages of the ectoparasite • May have systemic effects • Ectoparasite may have to take a blood meal for the medication to be effective
Topical long-acting (Spot-ons)	• Long-lasting • May work for multiple parasites • May work for different stages of parasite development • Work by providing area of repellent near application site	• May cause skin problem at site of application • Causes oiliness at site of application • Animal should avoid bathing or swimming with some products • May not be usable on young, old, or sick animals
Injectables	• Long-acting in some cases • Easy for owners who do not want to do the treatment themselves	• May cause adverse reaction at injection site • Must be given by veterinary staff

TIP

Bot flies are host-specific and site-specific parasites in the larval (commonly called bot) stage. Bots in horses are caused by *Gastrophilus sp.*, and the larval form is found in the horse's stomach.

TIP

Spot-on describes medication applied topically in small amounts to a localized area. Some spot-ons, such as older organophosphate flea prevention products, may be absorbed systemically. Others work topically and are absorbed systemically (flea, tick, and heartworm prevention products), and still others only work topically (flea and tick prevention products). Spot-ons that are absorbed systemically have the potential to cause more side effects in animals.

Always read product labels to determine if you need to follow any safety procedures when using these products. You may need to wear aprons and waterproof gloves when using some products, and proper disposal of excess or unused product may have to be done in accordance with state pesticide and federal (EPA) guidelines, to avoid groundwater or wildlife contamination. Animals treated with residual products may have to have their activity limited to avoid environmental contamination (for example, cattle may not be permitted to enter lakes, ponds, or streams for a designated time, to avoid wildlife contamination). Proper ventilation and/or disposal of bedding material in the area in which the product is applied may also be recommended. Always consult MSDS information, package inserts, or the *VPB* prior to use.

Table 15-5 lists chemical agents for ectoparasite treatment in animals. Table 15-6 lists drugs used for monthly or longer prevention of fleas, ticks, and heartworm disease.

Chemicals used to treat ectoparasites may also have adverse effects on the animal being treated. Animals should always be observed for the development of local reactions to the chemicals being applied. Some chemicals, such as carbamates and organophosphates, may cause additional problems such as excessive salivation, vomiting, diarrhea, and muscle tremors in treated animals. Animals may need to be a certain age before a particular chemical can safely be used on them. Animals (and humans) may be able to ingest certain products, such as collars, that may result in toxicity. Concurrent use of other medications may also cause toxicity in animals treated with ectoparasiticides. All of these factors must be considered before any product is dispensed. Always consult the animal's medical record and MSDS information, package inserts, or the *VPB* prior to use and/or dispensing of these products.

TABLE 15-5 Chemical Products for Ectoparasite Control

Product	Trade Name Examples	Efficacy
Pyrethrins and pyrethroids • Names end in *-rin* or *-thrin* • Pyrethrins are natural plant products • Pyrethroids are synthetic pyrethrins	• pyrethrin (Mycodex Shampoo®) • d-trans allethrin (Duocide Spray®) • permethrin (ProTICall®)	• Very safe • Quick kill • Often manufactured with other products • May have limited residual effects • Form labeled for dogs may be too high a concentration for cats • Commonly used in sprays, dips, fogger, insecticidal ear tags, and premise sprays
Insect growth regulators (IGR): Include insect development inhibitors and juvenile hormone mimics	• methoprene (Ovitrol® and Siphotrol®) • pyriproxyfen (Nylar®) • fenoxycarb; pulled from the market because it breaks down to formaldehyde, a carcinogen	• Products with IGR provide the flea with high levels of IGR, which mimics the insect's juvenile hormone (JH). Fleas need low levels of JH to molt to the next stage; high levels interrupt normal molting, so the insect stays in the larval stage and eventually dies

TABLE 15-5 *Continued*

Product	Trade Name Examples	Efficacy
Insect growth regulators (IGR) (cont.)		• Do not have adulticide activity • Found in sprays and flea collars
Chitin synthesis inhibitor	• lufenuron (Program®)	• Chitin is an insect protein that gives strength and stiffness to its body; chitin synthesis inhibitors prevent proper formation of this protein • Lufenuron is given orally to dogs and orally or SQ to cats • Fleas that feed on blood containing lufenuron continue to lay eggs, but the eggs fail to develop normally
Neonicotinoid	• nitenpyram (Capstar®)	• Neonicotinoid compound that binds and inhibits nicotinic acetylcholine receptors • Tablet that kills adult fleas within 30 minutes • Can safely give a dose as often as one per day • Can use on puppies and kittens older than 4 weeks and weighing more than 2 pounds
Carbamates	• carbaryl (Mycodex shampoo with carbaryl®, Sevin Dust®, Adams Flea and Tick Dust®)	• Act as a cholinesterase inhibitor, causing continual stimulation of the postsynaptic neuron • May cause SLUDDE in animals (salivation, lacrimation, urination, defecation, dyspnea, and emesis) • Found in dusts, sprays, shampoos, and collars
Organophosphates	• phosmet (Paramite dip®) • chlorpyrifos (Adams Flea and Tick Dip® and Spray®) • cythioate (Proban® tablets and liquid) • diazinon (Escort®)	• Act as cholinesterase inhibitors (see carbamates) • Found in sprays, dips, dusts, and systemic medications • With oral products, fleas must bite animal to get medication
Formamidines	• amitraz (Mitaban® Dip, PrevenTIC® Collar, Taktic®)	• Used for treatment of demodectic mange in dogs • Animals may show sedation for 24–72 hours following treatment • Toxic to cats and rabbits • Use gloves and protective clothing when applying to animals • Use in well-ventilated area
Synergists	• piperonyl butoxide • N-octyl bicycloheptene dicarboximide	• Have no activity against arthropods; however, they increase the efficacy of pyrethrins and pyrethroids

(continues)

TABLE 15-5 *Continued*

Product	Trade Name Examples	Efficacy
Imidacloprid	• imidacloprid (Advantage®)	• Acts as an insect neurotoxin • Marketed for dogs and cats • Applied topically at the back of the neck, but is not absorbed into the blood • Kills adult fleas on contact • Has four-week residual effect
Imidacloprid and permethrin	• imidacloprid and permethrin (K9 Advantix®)	• Works synergistically to rapidly paralyze and kill parasites • Kills fleas, mosquitoes, ticks (deer, American dog, Brown dog, lone star)
Lime sulfur	• lime sulfur (Lym Dyp®)	• Provides antimicrobial and antiparasitic activity • Used in the treatment of sarcoptic mange • Also effective for the treatment of ringworm • May stain light-colored animals • Used as a rinse or dip
Fipronil	• fipronil (Frontline®, Frontline Plus®)	• Overstimulates insect's nervous system, causing death • Is applied topically, but is not absorbed into the blood • Kills newly emerged adult fleas before they can lay eggs • Residual activity even if dog is bathed • Labeled for fleas and ticks
Repellents	• DEET (Blockade®) • butoxypolypropylene glycol (VIP® Fly Repellent Ointment)	• Used to repel mosquitoes, flies, and gnats • May be used in combination with pyrethrins and pyrethroids • Include sprays, ear tags, and topicals for ear tips
Rotenone	rotenone (Ear Miticide®, Mitaplex-R®)	• Used in dips and pour-on liquids • Toxic to fish; consider runoff possibilities when using
Ivermectin	ivermectin (Ivomec®)	• Injectable or oral solution used for some ectoparasites and endoparasites • Blood-feeding ectoparasites (such as fleas and lice) are killed much better with ivermectin than superficial, nonblood feeders (like *Cheyletiella spp.*)
D-limonene	D-limonene (VIP Flea Dip and Shampoo®)	• Extract of citrus peel that has some insecticidal activity • Provides quick kill • No residual • Pleasant smell • Used with other products

TABLE 15-5 *Continued*

Product	Trade Name Examples	Efficacy
Selamectin	Revolution®	• In cats works on adult fleas and eggs, heartworms, ear mites, hookworm, roundworms • In dogs works on adult fleas and eggs, heartworms, ear mites, sarcoptic mange, American dog tick

TABLE 15-6 Drugs for Monthly (or Longer) Prevention of Fleas, Ticks, and Heartworm Disease

Effective for Fleas	Effective for Ticks	Effective for Prevention of Heartworm Disease
permethrin (ProTICall®)	permethrin (ProTICall®)	
imidacloprid (Advantage®)		
imidacloprid and permethrin (K9 Advantix®)		
selamectin (Revolution®)	selamectin (Revolution®) (labeled for American dog tick only)	selamectin (Revolution®) (also works on ear mites)
nitenpyram (Capstar®)		
lufenuron (Program®)		
fipronil (Frontline®)	fipronil (Frontline®)	
	amitraz(PrevenTIC® collar)	
		ivermectin (Heartgard®, Heartgard Plus®, Tri-Heart Plus® and Iverhart™ Plus) (Plus products also work on some intestinal parasites)
		milbemycin (Interceptor®) (also works on some intestinal parasites)
		milbemycin (Sentinel®) (also works on some intestinal parasites)
		Moxidectin (ProHeart®-6 works for six months)

SUMMARY

Antiparasitic drugs work on endoparasites and/or ectoparasites. Endoparasites are found in the body and include parasitic worms such as nematodes, cestodes, and trematodes. Coccidia, a type of protozoan, are also endoparasites treated in veterinary medicine.

Antinematodal drugs include benzimidazoles, imidazothiazoles, tetrahydropyrimidines, and organophosphates. Anticestodal drugs include praziquantel and epsiprantel. Antitrematodal drugs include clorsulon, albendazole, and praziquantel. Anticoccidial drugs, also known as coccidiostats because they inhibit but do not kill coccidia, include sulfadimethoxine, nicarbazine, amprolium, monensin, decoquinate, and robenidine. Antiprotozoal drugs include metronidazole, febendazole, albendazole, and imidocarb.

Heartworm disease is caused by the parasite *Dirofilaria immitis*. Multiple strategies exist for controlling heartworm disease, including preventing third-stage larvae from reaching maturity (prophylaxis), eliminating adult heartworms (adulticide therapy), and eliminating circulating microfilariae (microfilaricide therapy).

Ectoparasites live on the outside of animals and include flies, grubs, lice, fleas, mites, and ticks. There are a variety of ways to apply ectoparasiticidic chemicals and a wide variety of products used to control ectoparasites.

CHAPTER REFERENCES

American Heartworm Society: <www.heartwormsociety.org>.

Delmar's A-Z NDR-97 Nurses drug reference. Clifton Park, NY: Thomson Delmar Learning.

Nolan, T. (1999). *Clinical parasitology.* Cambridge, MA: Harcourt Learning Direct.

Reiss, B., & Evans, M. (2002), *Pharmacological aspects of nursing care* (6th ed., revised by Bonita E. Broyles). Clifton Park, NY: Thomson Delmar Learning.

Veterinary pharmaceuticals and biologicals (12th ed.). (2001). Lenexa, KS: Veterinary Healthcare Communications.

Veterinary values (5th ed.). (1998). Lenexa, KS: Veterinary Healthcare Communications.

Wynn, R. (1999). *Veterinary pharmacology.* Cambridge, MA: Harcourt Learning Direct.

CHAPTER REVIEW

Matching

Match the drug name with its action.

1. ______ pyrethroids
2. ______ fenbendazole
3. ______ avermectin
4. ______ organophosphate
5. ______ clorsulon
6. ______ sulfadimethoxine
7. ______ melarsomine
8. ______ IGR
9. ______ carbamate
10. ______ rotenone

a. found in ear mite solution
b. chemical that mimics juvenile hormone of insects
c. antitrematodal
d. quick-kill chemical found in flea products
e. coccidiostat
f. commonly used antinematodal that includes the brand name product Panacur®
g. cholinesterase inhibitor
h. cholinesterase inhibitor that is used against endo- and ectoparasites and has a narrow safety margin
i. category of antiparasitic that works against endo- and ectoparasites and includes the drug ivermectin
j. adulticide used to treat heartworm disease

Multiple Choice

Choose the one best answer.

11. What chemical is used to treat demodectic mange in dogs?
    a. amitraz
    b. fipronil
    c. imidacloprid
    d. methoprene
12. What is a disadvantage of using oral forms of flea prevention?
    a. These products work for a specified period of time.
    b. The flea needs to take a blood meal to get the medication.
    c. They provide quick knockdown effects.
    d. These products must be diluted prior to use.
13. Cleaning of contaminated waste is important when treating an animal with
    a. antinematodals.
    b. anticestodals.
    c. antitrematodals.
    d. coccidiostats.
14. Ivermectin can be used in treating all stages of heartworm disease except
    a. the third-stage larvae.
    b. the adults.
    c. the microfilariae.
    d. the juveniles.
15. Antitrematodal drugs include all of the following *except*
    a. clorsulon.
    b. albendazole.
    c. epsiprantel.
    d. praziquantel.
16. Ectoparasiticidic chemicals used to quickly treat the indoor environment are known as
    a. yard sprays.
    b. dusts.
    c. foggers.
    d. topical long acting.
17. Treatment of *Giardia lamblia* infections should include
    a. metronidazole.
    b. praziquantel.
    c. epsiprantel.
    d. ivermectin.

18. Heartworm preventative can be given at the following intervals:
    a. daily.
    b. monthly.
    c. six months.
    d. all of the above are correct depending on the drug used.

True/False

Circle a. for true or b. for false.

19. All heartworm preventatives also control intestinal parasites.
    a. true
    b. false

20. All flea products also control ticks.
    a. true
    b. false

Case Studies

21. A 4-month-old F Springer Spaniel (20#) (Figure 15-2) is brought to the clinic for a puppy check. The owner is concerned about the dog having worms, because she also has young children. She would like to get a dewormer for this dog.
    a. What dewormer do you recommend?
    b. The owner brings in a stool sample and cestode eggs are found. What medication should be given to this client?

22. An owner calls the clinic and wants to know what is available for prevention of heartworm disease in his dogs. What do you tell him?

FIGURE 15-2 Springer Spaniel.
Photo by Isabelle Francais

CHAPTER 16

Anti-Inflammatory and Pain-Reducing Drugs

OBJECTIVES

Upon completion of this chapter, the learner should be able to

- explain the role of inflammation in protecting the body from damage.
- describe the inflammatory pathway involving arachidonic acid.
- describe how steroidal and nonsteroidal anti-inflammatory drugs affect the inflammatory pathway.
- describe the advantages and disadvantaes of using corticosteroid drugs.
- describe how the negative feedback pathway affects glucocorticoid production and how this affects use of corticosteroid drugs.
- describe the use of glucocorticoid drugs in animals.
- describe the use of NSAIDs in animals.
- differentiate between the types of NSAIDs and list the advantages and disadvantages of each.
- explain the role of antihistamines in controlling inflammation.

KEY TERMS

inflammation
anti-inflammatory drugs
steroidal anti-inflammatory drug
nonsteroidal anti-inflammatory drug (NSAID)
corticosteroids
glucocorticoids
mineralocorticoids
iatrogenic
antipyretic
antihistamine
analgesic

INTRODUCTION

A horse owner calls the clinic, requesting an exam on his five-year-old quarterhorse gelding, which has been showing signs of lameness for two days. On physical exam, the veterinarian noted that the horse is pointing his left front foot (heel is off the ground). A nerve block is done to determine the location of the injury. Radiographs are also taken and it is determined that the horse has degenerative changes in the navicular bone. The horse is put on phenylbutazone and referred to a farrier for proper shoeing. The owner asks you if the phenylbutazone will fix his horse's lameness. Will it? Do analgesics and anti-inflammatory drugs cure disease? Can they cause additional problems for the animal?

INFLAMMATION

Inflammation is a natural response of living tissue to injury and infection. It is a useful and normal process that occurs in the body as a protective mechanism. A series of events is triggered when inflammation occurs, including vascular changes and releases of chemicals that help the body destroy harmful agents at the injury site and repair damaged tissue. Inflammation may be caused by chemical, physical, or biological agents; however, the response is fairly uniform regardless of the cause. Beneficial effects, such as vasodilation and increased permeability of blood vessels, allow blood substances and fluid to leave the plasma and go to the injured site in the *early* or *vascular phase* of inflammation. Following the brief vascular phase is the *delayed* or *cellular phase*, which can last hours to days after the injury occurs. Accumulation of leukocytes (white blood cells) in inflamed tissue, reduced blood flow, and widespread tissue damage occur during the cellular phase. Chemicals such as histamine, prostaglandin, and bradykinin are released, which can result in bronchodilation, shock, and pain in the cellular phase. The activity of neutrophils, monocytes, and lymphocytes also increases during this phase.

Clinically, inflammation presents as redness, heat, swelling, pain, and decreased range of motion. These signs of inflammation are summarized in Table 16-1 and Figure 16-1.

The inflammatory pathway is a response by tissues to initiate the healing process. When tissue is damaged, phospholipids in the cell membrane are broken down by the enzyme phospholipase. This breakdown of phospholipids by phospholipase produces arachidonic acid. The enzyme cyclooxygenase breaks down arachidonic acid to prostaglandins and thromboxanes. Another enzyme, lipoxygenase, can also break down arachidonic acid, producing leukotrienes. Prostaglandins and leukotrienes are chemicals that induce inflammation. Thromboxanes cause aggregation of platelets. Figure 16-2A shows the inflammatory pathway.

TABLE 16-1 Signs of Inflammation

Sign	Explanation
Pain	Pain is due to tissue swelling and release of chemicals such as prostaglandin
Heat	Heat is due to increased blood accumulation and pyrogens (fever-producing substances) that interfere with temperature regulation
Redness	Redness occurs in the early phase of inflammation due to blood accumulation in the area of tissue injury from chemical release (such as prostaglandins and histamine)
Swelling	Swelling occurs in the delayed phase of inflammation because kinins dilate arterioles and increase capillary permeability. This increased capillary permeability allows plasma to leak into the interstitial tissue at the injury site
Decreased range of motion	Function is lost due to fluid accumulation at the injury site Pain also decreases mobility to an area

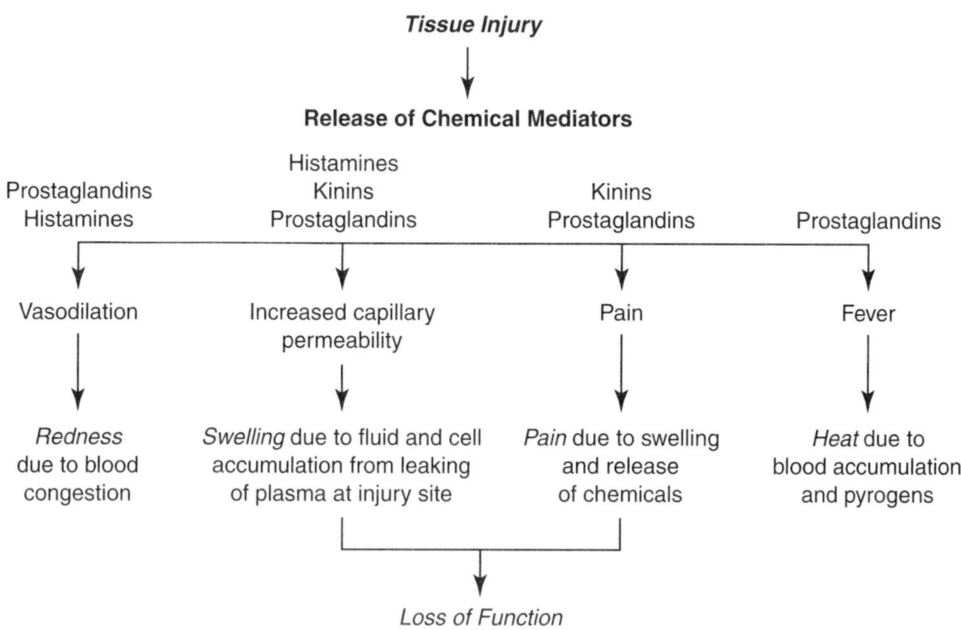

FIGURE 16-1 Tissue response to injury

INFLAMMATION REDUCERS

Although inflammation is a useful protective response to injury, there are times when inflammation should be reduced. Drugs that reduce inflammation are known as **anti-inflammatory drugs**. In an attempt to control inflammation when necessary, two main groups of drugs are used: steroidal anti-inflammatory drugs and nonsteroidal anti-inflammatory drugs. **Steroidal anti-inflammatory drugs** work by blocking the action of the enzyme phospholipase. **Nonsteroidal anti-inflammatory drugs (NSAIDs)**, work by blocking the action of the enzyme cyclooxygenase (Figure 16-2B). A third drug category used in controlling inflammation is the antihistamines, which counteract the action of histamine, a chemical produced in inflammation.

TIP

Steroidal and nonsteroidal anti-inflammatory drugs both have anti-inflammatory effects, but they are not chemically related.

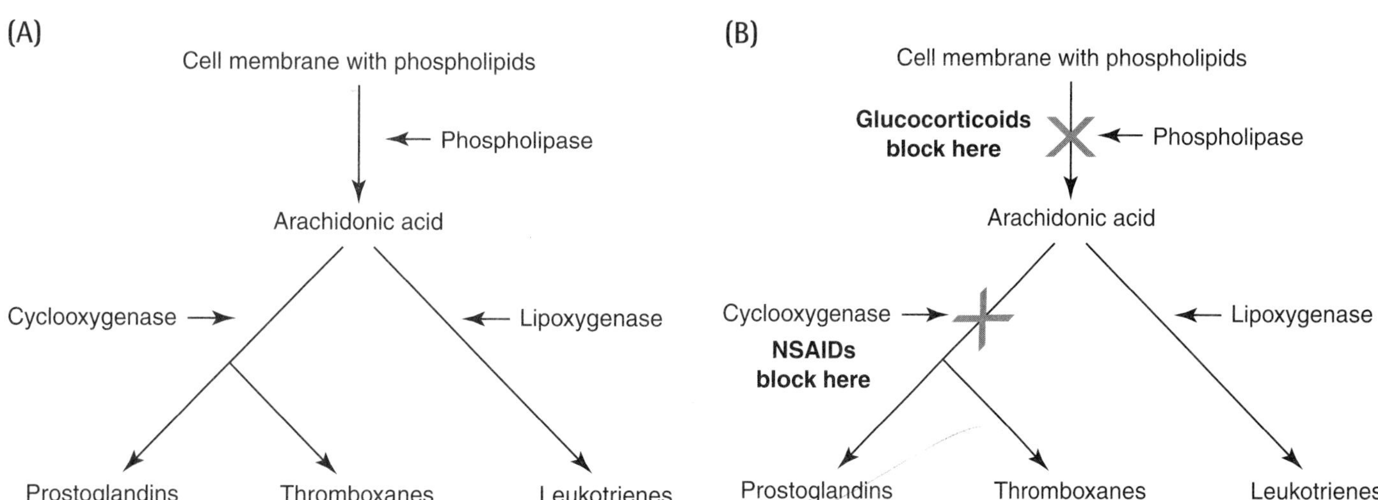

FIGURE 16-2 Tissue response to injury (A) The inflammatory pathway (B) Drug therapy and the inflammatory pathway

Steroidal Inflammation Reducers

Corticosteroids are hormones produced by the adrenal cortex, the outer part of the adrenal gland. A *hormone* is a chemical substance produced in one part of the body that is transported to another part of the body where it influences and regulates cellular and organ activity. Two groups of corticosteroids synthesized by the adrenal cortex have important veterinary applications: **glucocorticoids** (cortisol is the main form in animals) and **mineralocorticoids** (aldosterone is the main form in animals). The mineralocorticoids primarily help the body to retain sodium and water. Mineralocorticoids also help maintain the fluid and electrolyte balance crucial for body functions. Glucocorticoids have anti-inflammatory effects because they inhibit phospholipase, an enzyme that damages cell membranes (thus, glucocorticoids help stabilize cell membranes). Glucocorticoids raise the concentration of liver glycogen and increase blood glucose levels. Glucocorticoids also affect carbohydrate, protein, and fat metabolism, as well as muscle and blood cell activity. This section examines the anti-inflammatory applications of glucocorticoids.

Glucocorticoids are produced by the adrenal cortex in response to adrenocorticotropic hormone (ACTH) secretion by the anterior pituitary gland. The anterior pituitary gland secretion of ACTH is controlled by a releasing factor from the hypothalamus. These substances work in conjunction with each other through a process called *negative feedback* (see Chapter 10). When blood levels of glucocorticoid are low, a signal is sent to the hypothalamus to secrete releasing factor. This releasing factor then signals the anterior pituitary gland to secrete ACTH. ACTH in turn signals the adrenal cortex to produce glucocorticoid. Likewise, if blood levels of glucocorticoid are high, a signal is sent to the hypothalamus to stop secreting releasing factor and start secretion of inhibiting factor. This decreased secretion of releasing factor and increased secretion of inhibiting factor causes the anterior pituitary to produce less ACTH, resulting in less glucocorticoid production by the adrenal cortex. This check-and-balance system (Figure 16-3) keeps some hormones, including glucocorticoid levels, at the proper blood concentration.

Glucocorticoid drugs are used to treat many conditions, including inflammatory conditions, allergic responses, and systemic diseases. Inflammatory conditions that may require glucocorticoid drugs are autoimmune disorders, shock, ocular inflammation, musculoskeletal inflammation, spinal cord inflammation due to intervertebral disc disease, and lameness in horses (Figure 16-4). Allergic responses such as drug reactions, contact dermatitis, pruritus, and anaphylaxis may require glucocorticoid drugs. Systemic diseases such as Addison's disease (adrenocortical insufficiency) and treatment of some forms of cancer may benefit from glucocorticoid drug use. Glucocorticoid drugs can also be used to terminate late-stage pregnancies.

The benefits of glucocorticoid drugs include:

- They reduce inflammation, therefore they reduce pain (they are the "feel-good" drugs).
- They relieve pruritus (itching).
- They reduce scarring by delaying healing.
- They reduce tissue damage.

The drawbacks to glucocorticoid drugs include:

- They delay wound healing.
- They increase the risk of infection by decreasing attraction of white blood cells to the damaged site.
- They may cause gastrointestinal ulceration and bleeding.
- They increase the risk of corneal ulceration if corneal damage already exists.
- They can induce abortion in some species.

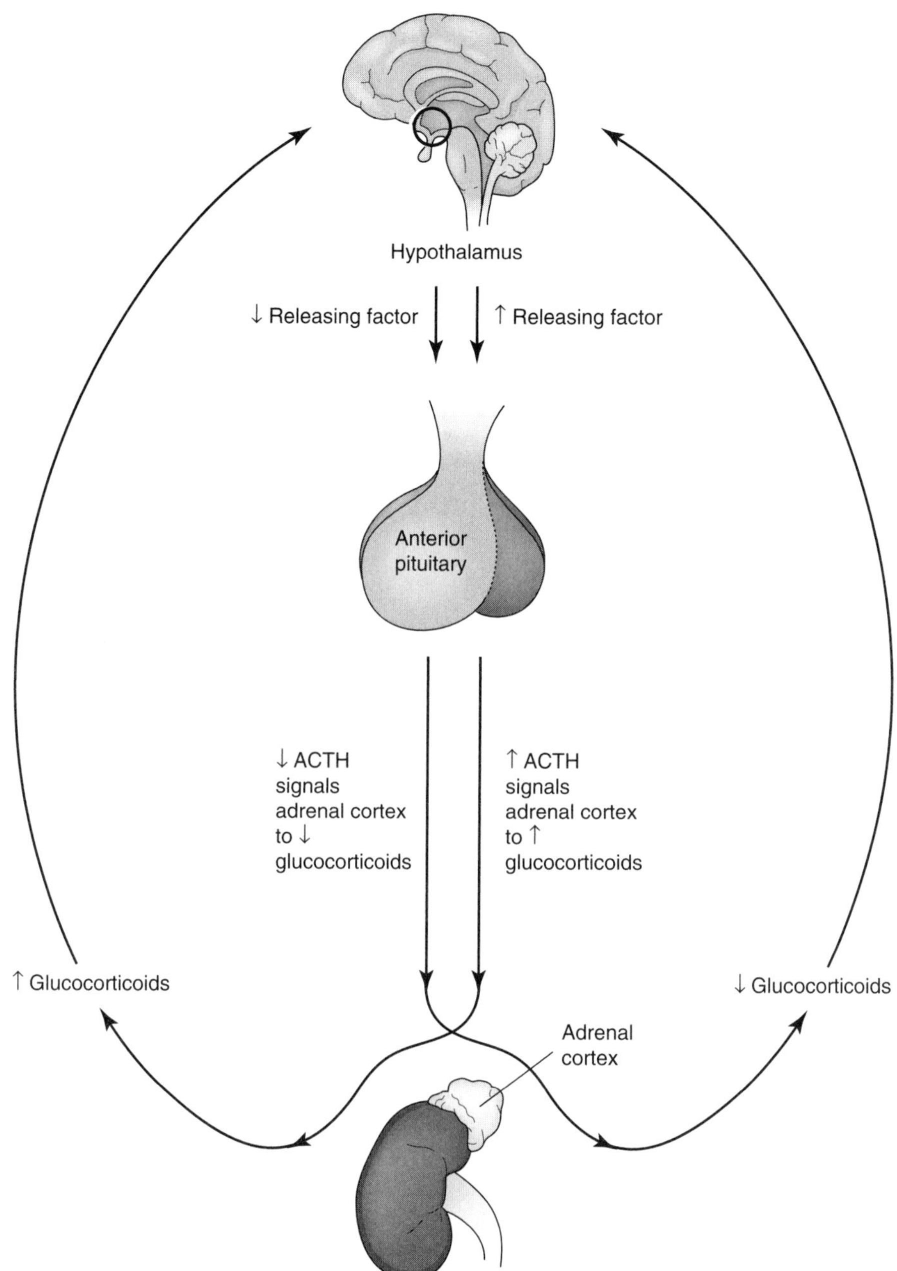

FIGURE 16-3 Negative feedback mechanism for glucocorticoid production

A wide variety of glucocorticoid drugs, often referred to simply as *steroids* or *cortisone drugs*, are synthetically produced (Figure 16-5). The glucocorticoid drugs manufactured in the laboratory have significantly greater anti-inflammatory effects than adrenally produced glucocorticoids. Glucocorticoid drugs can be short-acting (duration of action less than 12 hours), intermediate-acting (duration of action between 12 and 36 hours), or long-acting (duration of action more than 48 hours). Glucocorticoid drugs come in many formulations, including aqueous solutions, alcohol solutions, suspensions, and tablets. These drugs have several routes of administration: oral, parenteral, and topical. The water-soluble formulations can be injected IV, IM, or SQ. Repository or long-acting depot products are not given IV.

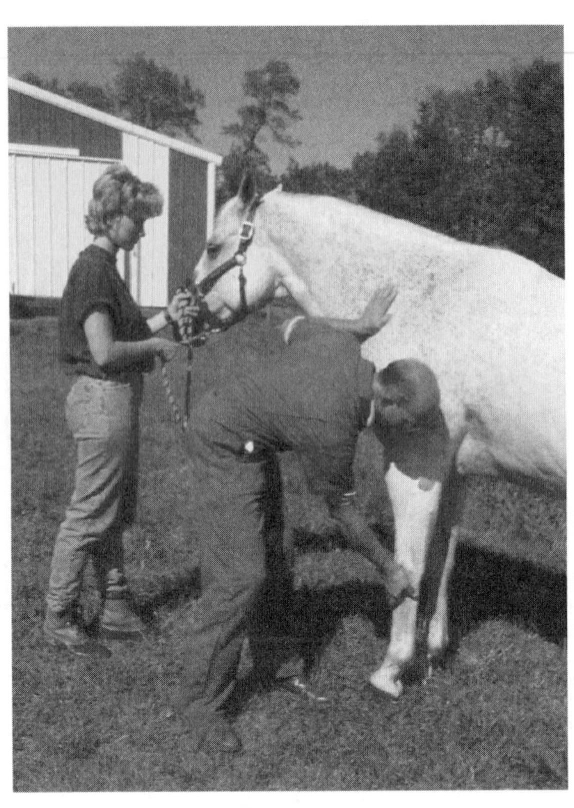

FIGURE 16-4 Veterinarians use corticosteroids to treat acute lameness in horses. Photograph courtesy of Cappy Jackson

TIP

A glucocorticoid can usually be recognized by the *-sone* ending in the generic name.

TIP

Prednisone or prednisolone? Prednisone must be converted to prednisolone by the liver; therefore, animals with liver dysfunction should be prescribed prednisolone.

Glucocorticoid drugs may also be found in preparations containing antibiotics and antifungals. Some of the glucocorticoids used in veterinary medicine are listed in Table 16-2.

Even though glucocorticoid drugs are widely used in veterinary medicine, their use should be monitored for side effects. Some side effects of steroid use include polyuria, polydipsia, polyphagia, suppressed healing, gastric ulcers, thinning of skin, and muscle wasting. Another important consequence of steroid use is the development of iatrogenic Cushing's disease (hyperadrenocorticism). Cushing's disease is a condition caused by overproduction of glucocorticoids (due to an adrenal tumor or pituitary disease). Animals with Cushing's disease are polyuric, polydipsic, and lethargic; have bilaterally symmetrical alopecia (hair loss); and have a pendulous abdomen caused by the catabolic effects of glucocorticoids. **Iatrogenic** means caused by the treatment ordered by physicians/veterinarians. Iatrogenic hyperadrenocorticism occurs when an animal has been on steroids for a period of time or on excessive doses of steroids. Animals with iatrogenic Cushing's disease appear clinically similar to animals with Cushing's disease.

In healthy animals, when glucocorticoid levels rise, the hypothalamus releases an inhibitory factor that signals the anterior pituitary gland to decrease ACTH production. This in turn slows glucocorticoid production. When glucocorticoid supplements are given, the animal's body responds the same way: by signaling the hypothalamus to release inhibitory factor that tells the anterior pituitary gland to decrease ACTH production. As long as the treatments are being given, the adrenal cortex is no longer

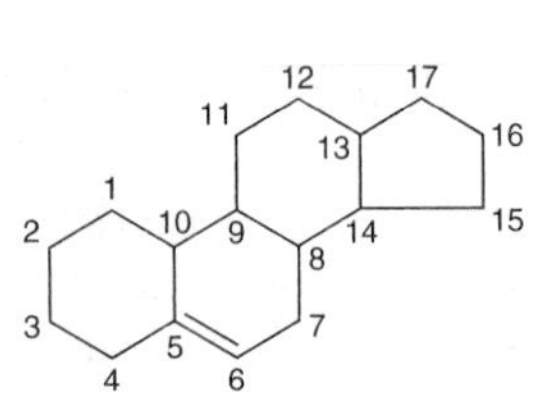

FIGURE 16-5 Steroid molecule

TABLE 16-2 Corticosteroids Used in Veterinary Medicine

Generic Name	Example(s) of Trade Names	Preparation	Duration of Action
cortisone	Cortone®	Tablets, injection, ophthalmic ointment/suspension	Short
hydrocortisone	Cortef® Solu-Cortef®	Tablets, oral suspension, topical cream, lotion, ointment, injection	Short
prednisone/prednisolone	Deltasone® Meticorten®	Tablets, injection, ophthalmic ointment	Intermediate
prednisolone sodium succinate	Solu-Delta-Cortef®	Injection	Intermediate
methylprednisolone	Medrol®	Tablets	Intermediate
methylprednisolone acetate	Depo-Medrol®	Suspension for injection	Intermediate
triamcinolone	Vetalog®	Tablets, suspension for injection, topical cream	Intermediate
dexamethasone	Azium®	Tablets, injection, ophthalmic suspension	Long
betamethasone	Betasone®	Injection, topical	Long
fluocinolone	Synalar®	Topical cream and solution	Long

stimulated to produce glucocorticoid, and in time the adrenal cortex can shrink or atrophy. The atrophied adrenal cortex cannot immediately start making its own glucocorticoids if glucocorticoid treatment is suddenly stopped. The animal's decreased blood levels of glucocorticoid will decrease further when the treatments are stopped and the body fails to make its own glucocorticoids. Decreased blood levels of glucocorticoid result in weakness, lethargy, vomiting, and diarrhea. This shortage of glucocorticoid results in iatrogenic Addison's disease, a condition caused by insufficient glucocorticoid in the animal's system.

Some key points about glucocorticoid or steroid therapy are as follows:

- The anti-inflammatory effects of glucocorticoid drugs do not cure any specific disease.
- The anti-inflammatory effects of glucocorticoid drugs will help disseminate any infectious disease that may be present in the body. Adequate antibiotic therapy may prevent this problem.
- The anti-inflammatory effects of glucocorticoid drugs help chronic diseases that flare up periodically and in the absence of a known cause: arthritis, tendonitis, bursitis, conjunctivitis, and various forms of dermatitis.
- Caution should be used when giving high dosages of glucocorticoid drugs to pregnant animals (one of their uses is to terminate late-stage pregnancies).
- Whenever possible, glucocorticoid drugs should be applied locally to avoid disturbing the hormonal balance of adrenally produced steroids. Even though some absorption may occur, the low systemic steroid level maintains normal adrenal cortex function while also maintaining a high local level of the drug.
- Use alternate-day dosing at the lowest possible dose whenever possible, to prevent iatrogenic Cushing's disease (hyperadrenocorticism).

- Taper animals off glucocorticoid drugs to prevent iatrogenic Addison's disease (hypoadrenocorticism).
- Glucocorticoid drugs should not be used in animals with corneal ulcers.

NSAIDs

In addition to steroidal anti-inflammatory agents, veterinary pharmacology also employs anti-inflammatory drugs known as *NSAIDs* (nonsteroidal anti-inflammatory drugs). Veterinarians use many kinds of NSAIDs: the phenylbutazone family; the relatively new agents in the ibuprofen family; and the classical salicylates such as aspirin. Table 16-3 lists some nonsteroidal anti-inflammatory drugs used in animals.

TIP

NSAIDs are also referred to as prostaglandin inhibitors.

NSAIDs work by inhibiting cyclooxygenase, which has two forms: cyclooxygenase-1 (commonly called cox-1) and cyclooxygenase-2 (commonly called cox-2). Cox-2 is believed to be more involved in inflammation, whereas cox-1 is believed to be more involved with the stomach (hence the gastrointestinal side effects often associated with NSAIDs). NSAIDs tend to have fewer side effects than glucocorticoid drugs and produce analgesia, inhibition of platelet aggregation, and fever reduction (which is not true of glucocorticoid drugs). Some side effects of NSAID use include gastric ulceration and bleeding, bone marrow suppression, and tendency to bleed. The following are the most common NSAID types.

TIP

Buffered or coated aspirin still carries a risk of causing gastrointestinal ulceration and bleeding.

Salicylates. The salicylates include aspirin, also known as *acetylsalicylic acid* or *asa*. Aspirin is the oldest anti-inflammatory agent and is a potent inhibitor of prostaglandin synthesis (by inhibition of cyclooxygenase). Prostaglandins accumulate at tissue injury sites, causing pain and inflammation. Aspirin is an analgesic, **antipyretic** (fever-reducing), and anti-inflammatory, and a reducer of platelet aggregation. Aspirin is used for the treatment of degenerative joint disease (commonly known as *osteoarthritis*), as an adjunct to postadulticide treatment of heartworm parasites, to give pain relief, to reduce fever, and in the treatment of some forms of shock. Cats cannot metabolize aspirin as rapidly as other species, so it must be used with caution in cats. If aspirin is used to treat a cat's condition, it is given at lower dosages and less frequently (e.g., every 48 to 72 hours). Aspirin comes in many forms, including adult and baby strengths,

TABLE 16-3 NSAIDs Used in Animals

Salicylates	Pyrazole Derivatives	Propionic Acid	Miscellaneous
aspirin	phenylbutazone (Buazolidin®, Equipalazone®, Butatron®)	• ibuprofen (Motrin®, Advil)® • ketoprofen (Ketofen®, Orudis®) • carprofen (Rimadyl®) • naproxen (Equiproxen®)	• flunixin meglumine (Banamine®, Finadyne®) • DMSO (Domoso®) • etodolac (EtoGesic®) • meclofenamic acid (Arquel®) • deracoxib (Deramaxx®) • meloxicam (Metacam®) • tepoxalin (Zubrin®)

buffered forms, enteric-coated forms, and combination forms. Care must be taken so as not to dispense a product that contains other drugs (such as a steroid, calcium, or other pain medication) if the other drug is not also prescribed. Some gastric side effects can be decreased if aspirin is given with food.

Pyrazolone derivatives. Phenylbutazone, an inhibitor of prostaglandin synthesis by the inhibition of cyclooxygenase, is the representative drug in this category. Phenylbutazone (or "bute" as it is commonly referred to by horse owners) is frequently used in equine medicine for pain associated with the musculoskeletal system. The liver metabolizes phenylbutazone and it is highly protein bound, so care must be taken when bute is given to an animal with low albumin (blood protein) levels. Phenylbutazone is a mild to moderate analgesic, antipyretic, and anti-inflammatory. It comes in paste, tablet, powder, and bolus forms for oral use, as well as gel and injection forms.

Dipyrone (Novin®) is another pyrazolone derivative that is used mainly as an antipyretic, but also has analgesic and slight anti-inflammatory properties. This product is no longer available because of side effects such as blood disorders.

Propionic acid derivatives. This is the *-fen* group of NSAIDs, which includes ibuprofen (Motrin®, Advil®), ketoprofen (Ketofen® and Orudis®), carprofen (Rimadyl®), and naproxen (Equiproxen®). These products block both cyclooxygenase and lipoxygenase and are used for their analgesic properties. Some of these products are also antipyretic and anti-inflammatory. Ketprofen and naproxen are approved for use in horses, and carprofen is approved for use in dogs, so the use of human over-the-counter products is no longer recommended. Carprofen is believed to target cox-2 (cyclooxygenase-2), which acts in only inflammation and not on gastrointestinal receptors, so it has limited gastrointestinal side effects. However, liver toxicities to Rimadyl® have been noted, especially in Labrador retrievers. Before beginning Rimadyl® therapy, it is recommended that blood work be done to assess liver enzymes.

Flunixin meglumine. This NSAID is a potent inhibitor of cyclooxygenase and is labeled for horses and cattle as Banamine® and Finadyne®. It is used extra-label in other species with extreme caution and typically only for a single dose. Flunixin is an injectable (IV or IM) analgesic, antipyretic, and anti-inflammatory drug commonly used for musculoskeletal and colic pain. It is also used for the treatment of shock, intervertebral disc disease, and pain secondary to surgery and parvovirus infection. Flunixin is a potent analgesic: the onset of analgesia can begin in about 15 minutes.

Dimethyl sulfoxide. Dimethyl sulfoxide (DMSO) is an anti-inflammatory drug, but is also well known for its ability to penetrate skin and serve as a carrier of other drugs. DMSO also causes vasodilation. DMSO works by inactivating the superoxide radicals produced by inflammation. This mechanism traps oxygen radicals, reducing cellular damage. DMSO is labeled for reduction of swelling via topical application, but has been used extra-label for a variety of things, such as IV for treatment of CNS swelling and topically following perivascular injection of irritating solutions. This drug is available as a gel and solution and is also found in a combination ear product containing steroid (Synotic®). Caution should be used when applying DMSO, as it can cause skin irritation, can leave a garlic taste in the mouths of both patient and medical personnel, and can cause birth defects in some species. Always wear rubber gloves when applying DMSO. A bandage may be applied over the area of application to prevent owner and animal contact with DMSO; however, this may cause burning of the treated area.

Miscellaneous Anti-Inflammatory Drugs

Indol acetic acid derivative. The representative drug in this group is etodolac (EtoGesic®). It is labeled as an anti-inflammatory agent and analgesic for osteoarthritis in dogs. It works by inhibiting cyclooxygenase and is believed to be a more selective inhibitor of cyclooxygenase-2 (cox-2) than cyclooxygenase-1 (cox-1). It also has the benefit of once-daily oral dosing.

Fenamates. Meclofenamic acid (Arquel®) is the representative drug of the fenamate group and is an inhibitor of cyclooxygenase. It is used as an analgesic and anti-inflammatory in equine medicine for osteoarthritis. It comes in granule form that is mixed in the feed.

Cox-2 inhibitors. Newer NSAIDs now on the market were developed to be more specific in their inhibition of prostaglandins by targeting cox-2 without interfering with the protective cox-1. One drug in this group is deracoxib (Deramaxx®), which is a member of the coxib class of NSAIDs that works by inhibiting cox-2. Cox-2 is responsible for the synthesis of inflammatory mediators. Deracoxib does not inhibit cox-1, which is an enzyme responsible for physiological processes such as platelet aggregation and protection of the gastric mucosa. Deracoxib is indicated for the control of postoperative pain and inflammation associated with orthopedic surgery in dogs weighing more than four pounds. It is available in a chewable tablet that should be given after eating to increase its bioavailability. Side effects include vomiting, anorexia, diarrhea, and blood abnormalities such as elevated kidney and liver values. Dogs should have predrug blood testing prior to the use of deracoxib.

Meloxicam (Metacam®) is an NSAID of the oxicam group that acts by inhibition of cox-2, which in turn inhibits prostaglandin synthesis. It also inhibits leukocyte infiltration into the inflamed area. Meloxicam has anti-inflammatory, analgesic, and antipyretic effects. In cattle, it is used as an injection to treat acute respiratory infection, in combination with appropriate antibiotic therapy to reduce clinical signs in affected animals. In dairy cattle, it is also used to treat mastitis. In dogs, it is available as an oral suspension that is applied to the animal's food and is used for the alleviation of inflammation and pain in both acute and chronic musculoskeletal disorders. In cats, it is approved as a one-time SQ injection prior to surgery for control of postoperative pain and inflammation associated with surgery.

Meloxicam should not be used in pregnant or lactating animals, or in animals with liver, cardiac, kidney, or gastrointestinal problems. Meloxicam should not be used in combination with other NSAIDs or steroidal drugs. Side effects include anorexia, vomiting, diarrhea, and lethargy.

Dual-pathway NSAID. Tepoxalin (Zubrin®) is a rapid-disintegrating tablet used for the control of pain and inflammation associated with osteoarthritis in dogs. Tepoxalin tablets dissolve in the dog's mouth within seconds, which enhances owner compliance and helps ensure complete dosing. Tepoxalin blocks both parts of the arachidonic acid cycle (the cyclooxygenase and lipoxygenase pathways). Tepoxalin decreases prostaglandin production at the site of inflammation (cox-1 inhibition), decreases edema and pain (cox-2 inhibition), and decreases gastric irritation and ulceration (lipoxygenase inhibition). Side effects of tepoxalin include gastrointestinal and renal problems.

Other Osteoarthritis Treatments

Anti-inflammatory drugs are used frequently in the treatment of inflammation secondary to osteoarthritis (degenerative joint disease). Other drugs used in the treatment of osteoarthritis include the following:

Glycosaminoglycans. Glycoproteins, known as *proteoglycans*, form part of the extracellular matrix of connective tissue like cartilage. Proteoglycans are high-molecular-weight substances and contain many different polysaccharide side chains. These polysaccharides make up most of the proteoglycan structure; hence, proteoglycans resemble polysaccharides more than they do proteins. The polysaccharide groups in proteoglycans are called *glycosaminoglycans* or *GAGs*. GAGs include hyaluronic acid, chondroitin sulfate, dermatan sulfate, keratan sulfate, and heparin sulfate. All of the GAGs contain derivatives of glucosamine or galactosamine. Glucosamine derivatives are found in hyaluronic acid and heparin sulfate. Galactosamine derivatives are found in chondroitin sulfate.

- Hyaluronic acid is normally part of the joint fluid in animals. When hyaluronic acid is given intra-articularly (it is FDA approved via this route), it is believed to help cushion degenerating joints. This cushioning helps relieve joint pain and improve mobility. Some animals may develop local reactions to hyaluronic acid, but such reactions usually subside within 24 to 48 hours. A trade name drug is Hyalovet®.
- Polysulfated glycosaminoglycans (PGAGs) appear in a drug that is a semisynthetic mix of glycosaminoglycans from bovine cartilage. It is known as Adequan® and can be given intra-articularly or IM in horses and dogs. It is reported to promote the production of joint fluid from the synovial membrane and has some anti-inflammatory action.
- Glucosamine and chondroitin sulfate are believed to play a role in the maintenance of cartilage structure and function. Glucosamine may be found in products alone (Maxi GS®) or in combination with chondroitin sulfate (Cosequin®). Glucosamine/chondrotin sulfate is a neutraceutical that is approved by the USDA for its analgesic and anti-inflammatory activity. The combination product comes in tablet and food supplement forms. Chondroitin sulfate in combination with hyaluronic acid is a FDA-approved drug.

Orgotein. This superoxide dismutase drug inactivates superoxide radicals (similar in action to DMSO). Palosein® is an example of orgotein and is used IM in horses for joint disease. It can also be given intra-articularly to horses and SQ to dogs.

ANTIHISTAMINES

Antihistamines are drugs that counteract the effect of histamine. Histamine causes bronchoconstriction and inflammatory changes when it is released from mast cells. When histamine combines with tissue receptors, it causes dilation of blood vessels, increased capillary permeability, smooth muscle spasming, and increased glandular secretion. Antihistamines, or H_1 blockers, compete with histamine for receptor sites, thus preventing a histamine response. Antihistamines do not have an effect after histamine attaches to its receptor site. H_1 blockers are also called *histamine antagonists*. There are two types of histamine receptors: H_1, which constrict extravascular smooth muscles when stimulated, and H_2, which increase gastric secretions when stimulated (see Chapter 11). Side effects of antihistamines include drowsiness, dry mucous membranes, and stimulation of the central nervous system in high doses.

H_1 blockers are used to treat pruritus, laminitis in large animals, motion sickness, anaphylactic shock, and some upper respiratory conditions. Examples of H_1 blockers are listed in Table 16-4.

TIP

The generic names of some antihistamines end in *-amine*.

TIP

Antihistamines do not affect the release of histamine, but act instead by competing with histamine for its receptor sites. Antihistamines are similar in structure to anticholinergic drugs; therefore, some of their side effects are anticholinergic in nature.

IMMUNOMODULATORS

Immunomodulation is the adjustment of the immune response to a desired level. This adjustment could include immunopotentiation (enhancement of the immune response), immunosuppression (reducing the immune response), and induction of the immunologic tolerance (nonreactivity to an antigen).

TABLE 16-4 Antihistamines Used in Veterinary Medicine

Generic Name	Trade Name Examples
diphenhydramine	• Benadryl® • Histacalm Shampoo®
dimenhydrinate	• Dramamine®
chlorpheniramine	• Chlor-Trimeton®
pyrilamine maleate	• Histavet-P®
triplelennamine hydrochloride	• PBZ® • PBZ-SR®
terfenadine	• Seldane®
hydroxyzine HCl	• Atarax®
meclizine	• Bonine® • Antivert® • D-Vert®

Cyclosporine is a selective immunomodulator that suppresses T-lymphocyte activity. Cyclosporine is considered one of the most effective immunosuppressant agents available. It has anti-inflammatory and anti-pruritic properties. Atopica® is a cyclosporine developed for the treatment of atopic dermatitis in dogs. Initially, it is given orally daily until clinical improvement is seen (usually 4–8 weeks) and then reduced to treatment every second day. Once clinical signs of atopic dermatitis are controlled, it is given every 3–4 days. Side effects include gastrointestinal disturbances such as vomiting and diarrhea. It is recommended that bacterial and fungal infections be treated before using cyclosporine because the inhibition of T-lymphocytes may decrease the immune response. Cyclosporine is also covered in Chapter 20 (antineoplastic and immunosuppressive drugs).

PAIN RELIEVERS

Analgesics are drugs that relieve pain without causing loss of consciousness. The proper analgesic is valuable in treating pain. It should also be kept in mind that complete masking of pain might be undesirable, because it eliminates a way of monitoring disease progression and returns an animal to its normal functions before an injury or illness is healed.

Selection of an analgesic is based on many things, including:

- effectiveness of the agent. Mild to moderate pain can be controlled by NSAIDs, but severe pain may have to be controlled by potent analgesics such as opioid products.
- duration of action. An analgesic with a short duration of action may be sufficient for brief postsurgical pain; moderate to severe chronic pain may require use of an analgesic with a longer duration of action.
- duration of therapy. Some analgesics are highly effective but harmful with prolonged use. Long-term safety must be considered when long-term therapy is needed.
- available routes of drug administration. For long-term pain relief, oral therapy is the most convenient form for owners. Injectable routes that can be used in the clinic may be time-consuming and expensive for the client.

Analgesics fall into two categories: narcotics (opioid) and non-narcotics. Narcotic analgesics were covered in Chapter 7. Non-narcotic analgesics include aspirin, pyrazolone derivatives (phenylbutazone), propionic acid derivatives (ibuprofen drugs), flunixin meglumine, indol acetic acid derivatives (etodolac), and meclofenamic acid. These drugs and their actions were covered in the anti-inflammatory section of this chapter.

Another non-narcotic analgesic is acetaminophen. This drug class includes the drug Tylenol®, which has analgesic effects but limited, if any, antipyretic and anti-inflammatory effects. Most consider acetaminophen to have no anti-inflammatory effects, and it is rarely used in veterinary medicine. Acetaminophen works by reducing the perception of pain through its effect on the hypothalamus of the brain. Gastrointestinal side effects of acetaminophen are rare due to its mechanism of action, though it can cause liver and kidney dysfunction in all animals. Acetaminophen toxicity is treated with acetylcysteine. Acetaminophen is contraindicated in cats at any dosage. Cats lack the enzymes to process acetaminophen, which leads to a build-up of toxic metabolites that causes blood disorders in cats.

Table 16-5 summarizes the drugs covered in this chapter.

16-5 Drugs Covered in This Chapter

Drug Category	Example
Glucocorticoids	• cortisone • prednisone • hydrocortisone • prednisolone • methylprednisolone • dexamethasone • betamethasone • triamcinolone • flucinolone
NSAIDs	• aspirin • phenylbutazone • ibuprofen, ketoprofen, carprofen, naproxen • flunixin meglumine • DMSO
Miscellaneous anti-inflammatory/ antipyretic agents	• etodolac • meclofenamic acid • hyaluronic acid • PGAGs • glucosamine • chondroitin sulfate • orgotein
Antihistamines	• diphenhydramine • dimenhydrinate • pyrilamine maleate • triplelennamine hydrochloride • terfenadine • hydroxyzine • meclizine
Analgesics	• narcotics (see Chapter 7) • non-narcotic (see anti-inflammatories section)

SUMMARY

There are two main categories of anti-inflammatory drugs: glucocorticoids (glucocorticoid is a type of corticosteroid) and nonsteroidal anti-inflammatories. Glucocorticoid drugs work by inhibiting phospholipase and most NSAIDs work by inhibiting cyclooxygenase. Side effects are more severe with glucocorticoid drugs and include PU/PD, polyphagia, suppressed healing, gastric ulcers, thinning of skin, and muscle wasting. Iatrogenic Cushing's disease may be seen in animals given glucocorticoid drugs for long periods of time. Care must be taken when stopping glucocorticoid drug use so that the negative feedback mechanism of the hypothalamus-anterior pituitary-adrenal gland is given time to reestablish itself thus preventing the development of iatrogenic Addison's disease.

Glucocorticoid drugs can be short-, intermediate-, or long-acting. They can usually be recognized by the *-sone* ending of their generic names.

Most NSAIDs are good analgesic, antipyretic, and anti-inflammatory drugs. The main side effect of their use is gastrointestinal ulceration and bleeding, which may be diminished with the use of specific cox-2 inhibitors. Examples of NSAIDs include aspirin, ibuprofen-like products, phenylbutazone, and flunixin meglumine.

Antihistamines work by blocking histamine attachment to H_1 receptors. Antihistamines are used as anti-inflammatory drugs for many diseases, including pruritus and laminitis. Some generic names of antihistamines end in *-amine*.

Analgesics relieve pain without causing loss of consciousness. Analgesics fall into two categories: narcotic (opioid) and non-narcotic. Acetaminophen reduces the perception of pain through its effect on the hypothalamus of the brain. Acetaminophen has limited, if any, antipyretic and anti-inflammatory effects.

CHAPTER REFERENCES

Delmar's A-Z NDR-97 Nurses drug reference. Clifton Park, NY: Thomson Delmar Learning.

Medical economics. *PDR for nutritional supplements*. (2001). Montvale, NJ: Thomson Healthcare.

Reiss, B., & Evans, M. (2002), *Pharmacological aspects of nursing care* (6th ed., revised by Bonita E. Broyles, pp. 271–294). Clifton Park, NY: Thomson Delmar Learning.

Veterinary pharmaceuticals and biologicals (12th ed.). (2001). Lenexa, KS: Veterinary Healthcare Communications.

Veterinary values (5th ed.). (1998). Lenexa, KS: Veterinary Healthcare Communications.

Wynn, R. (1999). *Veterinary pharmacology*. Cambridge, MA: Harcourt Learning Direct.

CHAPTER REVIEW

Matching

Match the drug name with its action.

1. ______ dexamethasone
2. ______ prednisone
3. ______ dimethyl sulfoxide (DMSO)
4. ______ ibuprofen
5. ______ phenylbutazone
6. ______ aspirin
7. ______ flunixin meglumine
8. ______ triamcinolone
9. ______ methylprednisolone
10. ______ diphenhydramine

a. nonsteroidal anti-inflammatory drug
b. glucocorticoid drug or "steroid"
c. antihistamine

Multiple Choice

Choose the one best answer.

11. Which type of anti-inflammatory drug has the greatest amount of anti-inflammatory effects?
    a. Glucocorticoid drugs, because they block phospholipase, which occurs first in the inflammatory pathway.
    b. NSAIDs, because they block cyclooxygenase, which occurs first in the inflammatory pathway.
    c. Antihistamines, because they block histamine, which is the only inflammatory mediator.
    d. All of the categories work equally.
12. Iatrogenic disease is caused by
    a. endocrine disease.
    b. the treatment.
    c. steroids.
    d. glucocorticoid drugs.
13. Propionic acid derivatives are NSAIDs that
    a. can be identified by the *-fen* ending of their generic names.
    b. block both cyclooxygenase and lipoxygenase.
    c. are analgesic, antipyretic, and anti-inflammatory.
    d. all of the above are correct.
14. Which NSAID works by inhibiting cyclooxygenase-2 and not cyclooxygenase-1?
    a. flunixin meglumine
    b. DMSO
    c. deracoxib
    d. aspirin
15. Indol acetic acid derivatives like etodolac (EtoGesic®) have the benefit of
    a. being labeled for use in all species.
    b. being available in chewable and injectable forms.
    c. being available in gel and solution forms.
    d. once-daily oral dosing.
16. What ending on the generic name of some antihistamines may indicate that they are antihistamines?
    a. *-hist*
    b. *-ane*
    c. *-amine*
    d. *-ate*
17. What type of anti-inflammatory drug is sometimes used for motion sickness in animals?
    a. corticosteroid drugs
    b. NSAIDs
    c. PGAGs
    d. antihistamines
18. What NSAID is commonly used for pain relief of lameness and colic in horses?
    a. aspirin
    b. phenylbutazone
    c. ibuprofen
    d. etodolac

19. What anti-inflammatory drug readily penetrates skin and must be handled cautiously to avoid absorption of the drug by the person giving the treatment?
    a. flunixin meglumine
    b. cortisone
    c. DMSO
    d. PGAG

20. What NSAID is commonly used and labeled for use in large animals, and should be used with caution (if at all) in small animals?
    a. flunixin meglumine
    b. etodolac
    c. DMSO
    d. carprofen

Case Studies

21. A 10-yr.-old M/N German Shepherd (95#) (Figure 16-6) is brought to the clinic for ongoing lameness. His physical exam is normal except for pain on palpation of the hips. The veterinarian requests an x-ray of the dog's hips. The x-rays reveal that this dog has hip dysplasia. The owner would like some pain medication for this dog.
    a. What are some choices for pain and inflammation relief in this dog?
    b. The owner would like to try aspirin in this dog. What side effects do you warn her about?
    c. The dog develops diarrhea on aspirin. The owner would now like to try carprofen. What test would you recommend that she have done on her dog prior to starting carprofen?
    d. The owner would like to give the dog medication only once daily. What is her best choice?

22. The owner of a 10-yr.-old M/N Dachshund calls the clinic to ask for advice about her dog. Another veterinarian had put this dachshund on a 10-day course of prednisone for intervertebral disc disease. The owner wants to stop giving the dog prednisone immediately, because the dog is urinating in the house.
    a. What could happen as a result of suddenly withdrawing high doses of corticosteroids from this dog?
    b. How can we prevent this from occurring?

FIGURE 16-6 German Shepherd.
Photo by Isabelle Francais

CHAPTER 17

Drugs for Skin Conditions

OBJECTIVES

Upon completion of this chapter, the learner should be able to

- describe the anatomy and physiology of the skin.
- explain basic conditions of skin diseases.
- describe topical antibacterial agents.
- describe topical antifungal agents.
- describe topical nonsteroidal and steroidal antipruritics.
- describe topical antiseborrheics.
- list commonly used astringents, antiseptics, soaks, and dressings used in veterinary medicine.

KEY TERMS

topical
pruritus
antipruritic
seborrhea
seborrhea oleosa
seborrhea sicca
antiseborrheics
keratolytics
astringent
antiseptic
soak
dressing
caustics
counterirritants

INTRODUCTION

A farmer tells you that one of his children's fair calves has a skin infection and that he has been treating it with the ointment you once gave him for his other cattle. He says it looked exactly like the bacterial infection he saw in his cattle last year, so he thought this ointment would be an easy way to get rid of the infection. The fair is rapidly approaching, and he wants to clear this infection before the calf is judged. He does not want to treat the calf with oral antibiotics because there is a meat withdrawal time for these drugs. What can he give this calf? Are there additional tests you want to do or questions you want to ask this farmer?

BASIC SKIN ANATOMY AND PHYSIOLOGY

The integumentary system consists of the skin and its appendages, such as glands, hair, fur, wool, feathers, scales, claws, nails, and hooves. Considered one of the largest organs in the body, the integumentary system is involved in many processes.

Skin plays a role in the immune system (acting as a physical barrier to infection and containing special antigen-processing cells), waterproofs the body, prevents fluid loss, and provides a site for vitamin D synthesis. Sebaceous glands lubricate the skin and discourage bacterial growth on the surface. Sweat glands regulate body temperature and excrete wastes through sweat. Hair helps control body heat loss and is a sense receptor. Nails, hooves, and claws protect the surface of the distal phalanx.

The skin is made up of three layers: the epidermis, dermis, and subcutaneous layers (Figure 17-1). The outermost or most superficial layer is the *epidermis*. The epidermis is several cell layers thick and does not contain blood vessels. The epidermis is dependent on the deeper layers for nourishment. The thickness of the epidermis varies greatly from region to region in any animal and varies from species to species. The epidermis is made up of stratified squamous epithelium. The base of the epidermal layer is known as the *basal layer*. Cells multiply and push upward in an orderly fashion from the basal layer. As the epidermal cells are pushed further to the surface from the underlying blood supply, they die and their cytoplasm is converted to keratin. The basal layer also contains melanocytes, which produce and contain the black pigment, *melanin*, that gives color to the skin and hairs.

The *dermis* (or *corium*) is the layer directly deep to the epidermis. The dermis is composed of blood vessels, lymph vessels, nerve fibers, and the accessory organs of the skin (glands and hair follicles), all situated in a tight mesh of collagen strands.

The subcutaneous layer, or *hypodermis*, is located deep to or under the dermis and is composed of connective tissue. The subcutaneous (SQ) layer contains a large amount of fat.

DRUGS USED IN THE TREATMENT OF SKIN DISORDERS

Many drugs used in the treatment of skin disorders have been discussed in previous chapters. This section addresses the use of **topical** treatments (agents applied to a surface that affect the area to which they are applied) and their role in skin disease management. *Alopecia*, or hair loss resulting in bald patches, has been addressed in preceding sections on the endocrine system (Chapter 10), antimicrobial drugs (Chapter 14), and antiparasitic drugs (Chapter 15). The role of anti-inflammatory agents was discussed in Chapter 16.

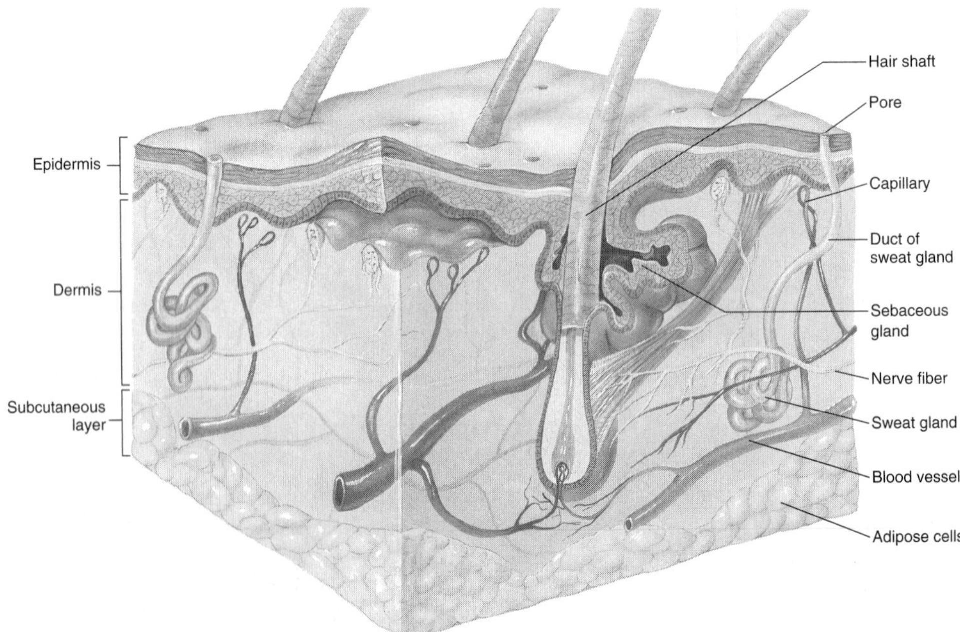

FIGURE 17-1 Skin layers and structures

On Top of Bacteria and Fungi

Topical antibacterial agents are used to prevent infection associated with minor skin abrasions and to treat superficial skin infections caused by susceptible bacteria. Topical use of antibiotics can lead to the development of sensitivity to the drug being used, so any reactions to topical medication must be documented in the animal's medical record. Several antibiotic agents are often combined in a single product to take advantage of the different antibacterial spectrum of each drug.

Antifungal drugs are most commonly employed in the treatment of two types of fungal infections of the skin: dermatophytes and yeast. *Dermatophyte* infections are commonly referred to as *ringworm* infections and include the genuses *Microsporum, Trichophyton* and *Epidermophyton*. Dermatophyte organisms can live only on dead keratin tissue and can be successfully eliminated only if the infected area is free of fungus. Therapy for dermatophytes typically involves oral and/or topical treatments for several weeks.

Topical antibacterials and antifungals are listed in Table 17-1. This is only a partial list of products available and their spectrum of activity should be reviewed in Chapter 14 on antimicrobials. Other topical treatments are covered in Chapter 18 on ophthalmic and otic medications.

TIP

Use caution when applying antibiotics to extensively damaged skin, as appreciable amounts of drug may be absorbed systemically.

TIP

Ringworm is a contagious and zoonotic disease. Ideally, affected animals should not have contact with unaffected animals.

TIP

Topical antibiotics are usually not recommended for treatment of puncture wounds.

On Top of Itching

Pruritus, or itching, may be associated with many skin and systemic diseases. Topical **antipruritics** provide slight to moderate relief of itching, and are usually used in conjunction with oral medications such as antihistamines and corticosteroids. Some products listed in other categories may have antipruritic effects, but control of itching is not their main function. Topical antipruritics include the following.

TABLE 17-1 Topical Antibacterials and Antifungals

Category	Examples
Topical antibacterials contain	• nystatin: Panalog®, Animax® (both nystatin/neomycin/thiostrepton and triamcinolone ointment) • thiostrepton (gram + spectrum of activity): see list above • bacitracin and polymyxin: Mycitracin®, Vetro-Biotic® • nitrofurazone: Fura-zone®, Fura-ointment®, Furazolidone® aerosol powder • mupirocin (broad-spectrum antibiotic that inhibits bacterial protein synthesis): Bactoderm® • gentamicin: found in topical sprays like gentamicin sulfate with betamethasone and Gentacin Spray®
Topical antifungals	• miconazole: Conofite®, Micaved Spray 1%®, Resizole Leave-in Conditioner® • clotrimazole: Clotrimazole Solution 1%® • nystatin: Panalog® • copper naphthenate: Kopertox® (used to treat thrush in horses)

Nonsteroidal Topical Antipruritics. These products give temporary relief of itching and are usually used in combination with oral products.

Topical anesthetics. Local anesthetics inhibit the conduction of nerve impulses from sensory nerves, thereby reducing pain and pruritus. They are generally used topically to minimize discomfort associated with allergies, insect bites, and burns. Local anesthetics are poorly absorbed from intact skin, but can be absorbed through damaged skin. Most topical anesthetic drugs can be recognized by the *-caine* ending in the name, and include such agents as lidocaine, tetracaine, benzocaine, and pramoxine. Trade name examples include DermaCool® with Lidocaine, Xylocaine® (lidocaine), Dermoplast® (benzocaine), ResiPROX® (pramoxine with oatmeal), and Pontocaine® (tetracaine).

Soothing Agents/Colloidal Oatmeal Shampoos. Oatmeal is believed to have soothing and anti-inflammatory effects when applied topically. Trade name examples include Epi-Soothe® and Oatderm Soothing Shampoo.®

Antihistamine. Antihistamines are antipruritics that provide temporary relief of pain and itching associated with allergic reactions and sensitive skin. Products contain diphenhydramine (an antihistamine to control itching) and oatmeal (to cleanse and soothe irritated skin). Examples of this category are Histacalm Shampoo® and Histacalm Spray.®

Topical Corticosteroids. Topically applied corticosteroids are very effective in alleviating inflammatory signs. Corticosteroids have anti-inflammatory, antipruritic, and vasoconstrictive action. When applied to the skin, they interfere with normal immune responses and reduce redness, itching, and edema. They also slow the rate of skin cell production.

The effectiveness of a topical corticosteroid depends on the potency of the drug used, the vehicle used to carry the corticosteroid to the skin, the thickness and integrity of the skin at the application site, and the amount of moisture present in the skin. Damaged skin at the application site may increase the amount of drug absorbed into the bloodstream and result in systemic side effects.

The least potent topical corticosteroid is *hydrocortisone*. It is suitable for long-term topical use. Topical corticosteroids containing a fluorine atom in their structure are among the most potent products (for example, fluocinolone). These products should be used sparingly.

Another difference between corticosteroid drugs is the duration of action. Short-acting corticosteroids, such as hydrocortisone and cortisone, have a duration of action of less than 12 hours. Intermediate-acting corticosteroids, including prednisone, prednisolone, triamcinolone, and methylprednisolone, have a duration of action between 12 and 36 hours. Long-acting corticosteroids, such as betamethasone, flumethasone, and dexamethasone, have a duration of action greater than 48 hours.

TIP

Corticosteroids slow the rate of skin cell production; hence, they slow healing time of wounds.

Topical corticosteroid products may be combined with other ingredients such as antibiotics to broaden their action. Examples of products containing topical corticosteroids include:

TIP

Animals that lick topically applied medication increase their risk of systemic side effects.

- Gentocin Topical Spray® (betamethazone and gentamicin)
- Vetalog Cream® (triamcinolone)
- Panolog Cream® (triamcinolone and nystatin, neomycin and thiostrepton)
- DermaCool-HC® (hydrocortisone)
- Synalar Cream® (fluocinolone cream) and Neo-Synalar® (fluocinolone/ neomycin cream)

On Top of Seborrhea

Seborrhea dermatitis is a skin condition characterized by abnormal flaking or scaling of the outermost layer of the epidermis. Seborrhea accompanied by increased production of sebum is referred to as **seborrhea oleosa**. Seborrhea without increased production of sebum (oil) is referred to as **seborrhea sicca**. A variety of products are used to treat seborrhea (**antiseborrheics**). One important group is the **keratolytics**, which remove excess keratin and promote loosening of the outer layers of the epidermis. Keratolytics break down the protein structure of the keratin layer, permitting easier removal of this compacted material. Keratolytics are found in medicated shampoos to help in the treatment of seborrhea. Topical antiseborrheics include:

- sulfur, which is keratolytic, antipruritic, antibacterial, antifungal, and antiparasitic. It tends to be nonirritating and nonstaining. Sulfur is used to treat seborrhea sicca. Products include SebaLyt Shampoo®, Sebolux Shampoo®, Allerseb-T Shampoo®, and NuSal-T Shampoo®.
- salicylic acid, which is keratolytic, antipruritic, and antibacterial. Salicylic acid is used to treat seborrhea sicca and hyperkeratotic skin disorders. Products include SebaLyt Shampoo® and KeraSolv Gel®.
- coal tar, which is keratolytic and degreasing. It is irritating and may stain light-colored haircoats. Coal tar is used to treat seborrhea sicca and may be irritating in cats. Products include Mycodex Tar and Sulfur Shampoo®, and LyTar Shampoo®.
- benzoyl peroxide, which is keratolytic, antipruritic, antibacterial, and degreasing. It is used to treat seborrhea dermatitis in which overproduction of oil occurs. It is also used to treat moist dermatitis (hot spots), pyoderma, stud tail, and a variety of skin lesions that are moist and/or are contaminated with bacteria. Products include Pyoben Gel®, Sulf/OxyDex Shampoo®, and OxyDex Gel®.
- selenium sulfide, which is keratolytic, degreasing, and antifungal. It is used to treat seborrhea and eczema. Use of selenium sulfide may result in subsequent irritation. Products include Seleen Plus Medicated Shampoo® and Selsun Blue Shampoo®.

TIP

Medicated shampoos are used to treat skin conditions and should not be used for routine bathing. For maximum effectiveness they typically should be left on the skin for 5 to 15 minutes before rinsing.

TIP

"Hot spots" is a common term for acute moist dermatitis. Hot spots are treated with multiple drugs, including corticosteroids, antibiotics, and benzoyl peroxide, in combination with shaving of the affected area.

Miscellaneous Topical Drugs

Table 17-2 lists other categories of drugs used topically to treat skin disorders. Table 17-3 reviews the drugs covered in this chapter.

TABLE 17-2 Miscellaneous Topical Drugs

Category	Action	Examples
Astringents: agents that constrict tissues	• Stop discharge by precipitating protein • Have some antibacterial properties • Used to treat moist dermatitis and other moist skin lesions	• Stanisol® (contains salicylic acid, tannic acid and boric acid) • Tanni-Gel® (contains isopropyl alcohol, salicylic acid, tannic acid and benzocaine)

(continues)

TABLE 17-2 *Continued*

Category	Action	Examples
Antiseptics: substances that kill or inhibit the growth of micro-organisms on living tissue	• Alcohols are bactericidal, astringent, and cooling • Chlorhexidine is bactericidal, fungicidal, and partially virucidal • Propylene glycol is antibacterial and antifungal • Acetic acid is effective against *Pseudomonas sp.* • Iodine is bactericidal fungicidal, virucidal, and sporicidal • Benzalkonium Chloride is antibacterial and antifungal	• 70% or 90% isopropyl alcohol, 70% ethyl alcohol • Chlorhexidine products include Nolvasan® Solution, Scrub, and Ointment; Chlorhexi-Derm Shampoo® • Propylene glycol products are typically used as solvents or vehicles for other drugs (goes by generic name) • Acetic acid is usually found in ear preparations; products include Fresh-Ear®, Oti-Clens®, and ClearEar Cleansing Solution® • Iodine is found in Betadine® Solution and Scrub, Tincture of Iodine, Lugol's® Solution, and Xenodine® Spray • Benzalkonium chloride is found in Myosan Cream® and Dermacide®
Soaks and dressings: substances applied to areas to draw out fluid or relieve itching	• Aluminum acetate is drying and mildly antiseptic. It is used as a soak to relieve itching and inflammatory discharge • Magnesium sulfate is used in wound dressings to draw fluid out of tissues	• Aluminum acetate is found in Domeboro® powder and tablets • Magnesium sulfate is found in Epsom salts
Caustics: substances that destroy tissue	• Destroy tissue • Used to treat warts and tissue (proud flesh in horses)	• Silver Nitrate Stick Applicators®, Stypt-Stix® excessive granulation (silver nitrate). Note: silver nitrate products can stain • Equine HoofPro® Copper Suspension®, HoofPro+® (copper sulfate)
Counterirritants: substances that produce irritation and inflammation in areas of chronic inflammation	• Thought to increase blood supply to the area, which in turn brings WBC, antibodies, and so on to the area to stimulate healing	• Lin-O-Gel® (alcohol, camphor, menthol, and iodine) • Equi-Gel® (camphor, menthol, thymol, witch hazel, and isopropyl alcohol)

TABLE 17-3 Drugs Covered in This Chapter

Drug Category	Example
Topical antibiotics	nystatin, thiostrepton, nitrofurazone, mupirocin, gentamicin, bacitracin, polymixin
Topical antifungals	miconazole, clotrimazole, nystatin, copper naphthenate
Nonsteroidal topical antipruritics	Topical anesthetics: lidocaine, benzocaine, tetracaine, pramoxine Soothing agents: oatmeal Antihistamines: diphenhydramine
Topical corticosteroids	hydrocortisone, betamethazone, triamcinolone, fluocinolone
Antiseborrhea agents	sulfur, salicylic acid, coal tar, benzoyl peroxide, selenium sulfide
Astringents	Products containing: salicylic acid, tannic acid, boric acid, isopropyl alcohol
Antiseptics	alcohols, chlorhexidine, propylene glycol, acetic acid, iodine, benzalkonium chloride
Soaks and dressings	aluminum acetate, magnesium sulfate
Caustics	silver nitrate, copper sulfate
Counterirritants	Products contain alcohol, camphor, menthol, iodine

SUMMARY

Topical medications are used to treat a variety of skin conditions, including pruritus, scaling, and infection. Types of topical medications are antibiotics, antifungals, antiparasitics, antipruritics, antiseborrheals, and miscellaneous. Topical antibiotics and antifungals are commonly used in combination products to widen their spectrum of activity. Topical antiparasitic products are covered in Chapter 15. Topical antipruritic drugs tend to be used in conjunction with systemic antipruritic medications; this class includes corticosteroids and nonsteroidals (topical anesthetics, oatmeal-based products, and antihistamines). Ingredients in topical antiseborrheals include sulfur, salicylic acid, coal tar, benzoyl peroxide, and selenium sulfide. Miscellaneous topical medications include astringents, antiseptics, soaks and dressings, *caustics*, and *counterirritants*. The spectrum of activity of antibiotics and antifungals is covered in Chapter 14.

CHAPTER REFERENCES

Amundson Romich, J. (2000). *An illustrated guide to veterinary medical terminology.* Clifton Park, NY: Thomson Delmar Learning.

Delmar's A-Z NDR-97 Nurses drug reference. Clifton Park, NY: Thomson Delmar Learning.

Reiss, B., & Evans, M. (2002), *Pharmacological aspects of nursing care* (6th ed., revised by Bonita E. Broyles, pp. 279–818). Clifton Park, NY: Thomson Delmar Learning.

Veterinary pharmaceuticals and biologicals (12th ed.). (2001). Lenexa, KS: Veterinary Healthcare Communications.

Veterinary values (5th ed.). (1998). Lenexa, KS: Veterinary Healthcare Communications.

Wynn, R. (1999). *Veterinary pharmacology*. Cambridge, MA: Harcourt Learning Direct.

CHAPTER REVIEW

Matching

Match the drug name with its action.

1. ______ triamcinolone
2. ______ sulfur
3. ______ benzoyl peroxide
4. ______ iodine
5. ______ acetic acid
6. ______ magnesium sulfate
7. ______ chlorhexidine
8. ______ coal tar
9. ______ selenium sulfide
10. ______ salicylic acid

a. keratolytic, antipruritic, antibacterial shampoo ingredient and gel used to treat hot spots, pyoderma, stud tail, and other moist lesions
b. keratolytic, degreasing shampoo ingredient that may be irritating and stain light-colored haircoats; used to treat seborrhea sicca
c. corticosteroid used to treat pruritus
d. keratolytic, antipruritic, antimicrobial shampoo ingredient that is nonirritating and nonstaining
e. keratolytic, antipruritic, antibacterial shampoo ingredient that is used to treat hyperkeratotic skin disorders
f. keratolytic, degreasing, antifungal shampoo ingredient used to treat seborrhea and eczema
g. antiseptic that is bactericidal, fungicidal, and works on some viruses
h. antiseptic that is bactericidal, fungicidal, virucidal, and sporicidal
i. chemical that is effective against *Pseudomonas sp.* and is found in ear preparations
j. component of Epsom salts

Multiple Choice

Choose the one best answer.

11. An example of a component found in nonsteroidal topical antipruritics is
    a. miconazole.
    b. polymyxin.
    c. fluocinolone.
    d. oatmeal.
12. Which of the following shampoo ingredients is used as a keratolytic to treat seborrhea?
    a. sulfur
    b. oatmeal
    c. chlorhexidine
    d. propylene glycol
13. What group of chemicals works by destroying tissue at the site of application?
    a. keratolytics
    b. antifungals
    c. antiseborrheals
    d. caustics

14. What chemical is found in products like Betadine® and Lugol's Solution®?
   a. chlorhexidine
   b. iodine
   c. propylene glycol
   d. acetic acid
15. What group of drugs works by drawing fluid out of tissues?
   a. caustics
   b. counterirritants
   c. dressings
   d. antipruritics
16. Which group of skin products works by precipitating protein, thus stopping discharge?
   a. antiseptics
   b. caustics
   c. astringents
   d. soaks
17. Which of the following products has an antibacterial, antifungal, and corticosteroid drug in it?
   a. Nitrofurazone®
   b. Conofite®
   c. Kopertox®
   d. Panalog®

True/False

Circle a. for true or b. for false.

18. The epidermis does not contain blood vessels.
   a. true
   b. false
19. Keratolytics promote loosening of the outer layers of the epidermis.
   a. true
   b. false
20. Counterirritants work by calming the inflammatory response or counteracting inflammation.
   a. true
   b. false

Case Studies

21. A 5-yr.-old M/N Dachshund (15#) (Figure 17-2) presents to the clinic with oily and scaly ear margins. On physical examination the dog appears normal and has normal TPR. The veterinarian recommends a biopsy of the affected area, which is performed. The biopsy report reveals that this dog has a form of seborrhea.
   a. What shampoo ingredients could be used on this dog's ears?
   b. The owner wants to know how long to leave these products on. What do you tell her?

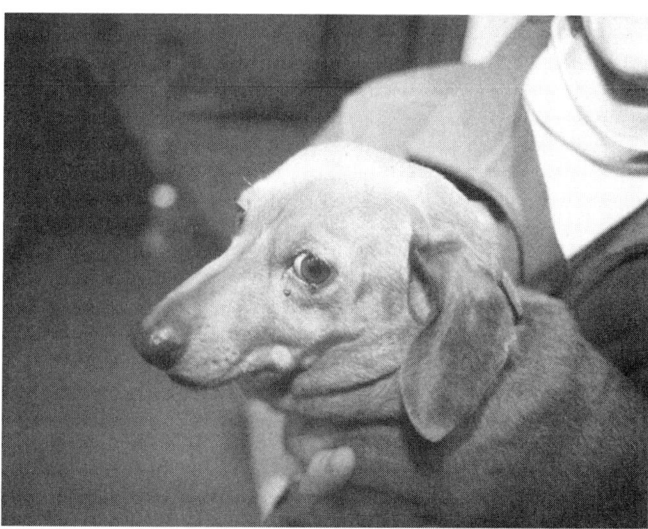

FIGURE 17-2 Dachshund

22. A 5-month-old F Labrador retriever (40#) presents to the clinic with excessive licking of the ventral abdomen. On physical exam, the TPR are normal, the dog is excited, and is in good flesh. Examination of the ventral abdomen reveals multiple bumps, some of which appear to have pus in them, and pink skin. The veterinarian thinks that this dog has puppy pyoderma.
    a. What would you, as a veterinary technician, want to check for in this dog?
    b. The observation you made to answer the preceding question was negative. The veterinarian decides to treat this dog with a topical antibiotic. Is there another ingredient that would be helpful to have in this product?
    c. What is a disadvantage to using a topical product in this case?
    d. What is an advantage to using a topical product in this case?

Ophthalmic and Otic Medications

OBJECTIVES

Upon completion of this chapter, the learner should be able to

- describe the anatomy and physiology of the eye.
- explain the indications for ophthalmic medication, including diagnostics, miotics, mydriatics, intraocular pressure reducers, and other topical forms.
- describe the anatomy and physiology of the ear.
- explain the indications for otic medications.

KEY TERMS

miotics
mydriatic
cycloplegic
glaucoma
carbonic anhydrase inhibitor
beta-adrenergic blocker
osmotic diuretic
keratoconjunctivitis sicca (KCS)
lacrimogenic
immunomodulating agent
otitis externa
otitis media
otitis interna
drying agent
dewaxing agent

INTRODUCTION

A farmer comes into the clinic and says that some of his calves have developed pinkeye (a bacterial eye infection). He remembers that he has ointment at home that he used to treat his cat when the cat got a scratch on the cornea during a fight. He wonders if he can use the same ointment for his calves. Can he? Are there any questions you want to ask him? Can antibiotics in ophthalmic ointments be absorbed systemically, thereby raising concerns about withdrawal times with these calves?

BASIC ANATOMY AND PHYSIOLOGY

The ocular system is responsible for vision and is comprised of the eyes and *adnexa* (surrounding structures). The eyes are the receptor organs for sight. The eye consists primarily of a multilayered sphere called the *globe*. The fibrous outer layer of the globe is the *sclera*. The sclera maintains the shape of the eye. The anterior portion of the sclera is transparent and is called the *cornea*. The cornea provides most of the focusing power of the eye.

The *choroid* is the opaque middle layer of the globe. The choroid contains blood vessels and supplies blood to the entire eye. The choroid consists of the *iris* (the pigmented, muscular diaphragm that helps regulate the amount of light entering the pupil), the *pupil* (a circular opening in the center of the iris), the *lens* (the clear, flexible, curved capsule located behind the iris and pupil), and the *ciliary body* (the thickened extension of the choroid that assists in accommodation of the lens). The term *uvea* refers to the iris, ciliary body, and choroid.

The inner layer of the globe is the *retina*. The retina is the nervous tissue layer of the eye that receives images. The retina contains specialized cells called *rods* and *cones* that convert visual images to nerve impulses that travel from the eye to the brain via the optic nerve. The *optic disk* is the region of the eye where nerve endings of the retina gather to form the optic nerve. Figure 18-1 shows the structures of the globe.

The eye is divided into compartments as well: the *anterior compartment* (also known as the *aqueous chamber*) is the anterior one-third of the globe in front of the lens and is divided into the anterior and posterior chambers by the iris. The *vitreous compartment* (also known as the vitreous chamber) is the posterior two-thirds of the eye behind the lens. The anterior compartment contains watery aqueous humor and the vitreous compartment contains the gelatinous vitreous body. Figure 18-2 shows the compartments of the eye.

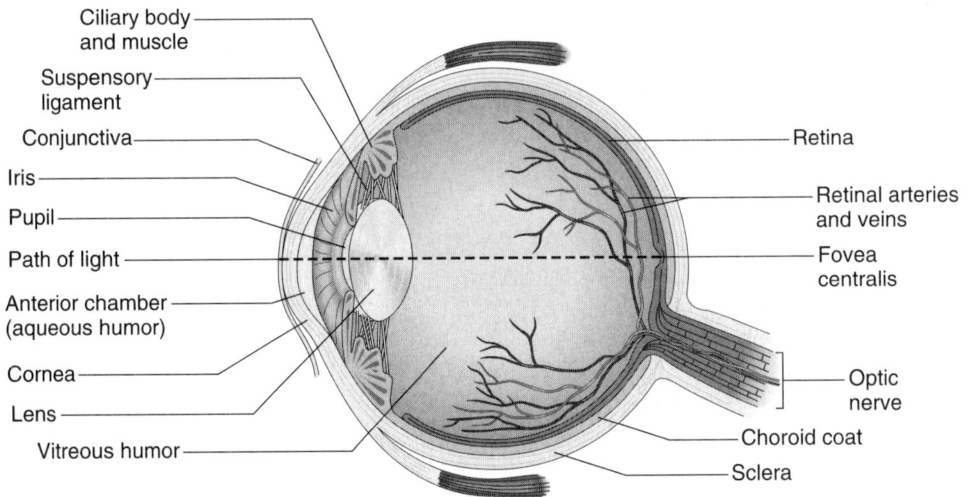

FIGURE 18-1 Cross-section of the globe of the eye

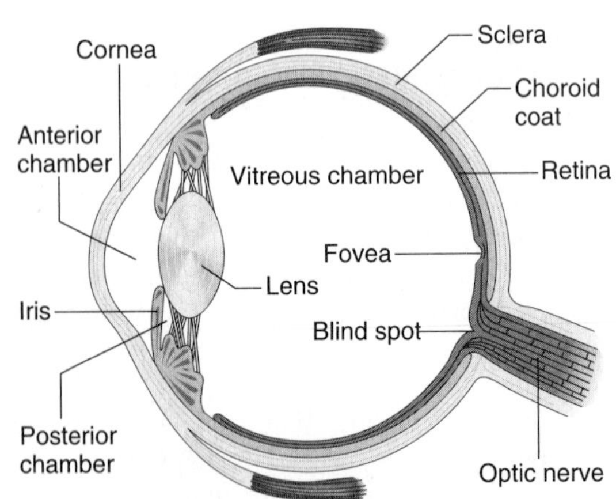

FIGURE 18-2 Chambers of the eye

The adnexa, or accessory organs, of the eye include:

- the orbit (the bony cavity of the skull that contains the globe).
- the eye muscles (seven muscles that control eye movement).
- the eyelids (upper and lower lids that protect the eye from injury, foreign material, and excessive light and the nictitating membrane or third eyelid).
- the eyelashes (hairlike structures that protect the eye).
- the conjunctiva (the mucous membranes that line the underside of each eyelid and cover the front surface of the eyeball).
- the lacrimal apparatus (structures that produce, store, and remove tears).

Figure 18-3 shows the adnexa of the eye.

OPHTHALMIC DRUGS

The ophthalmic drugs covered in this chapter are mainly topical medications. Topical ophthalmic drugs take the form of either eye drops or ointments. When using topical ophthalmic drugs, veterinarians take into account drug penetration, frequency of drug application, and ability of the owner to apply the drug formulation.

Drug penetration. Topical ophthalmic drugs tend to be absorbed into the anterior chamber and have little effect in the posterior or vitreous chambers. Water-soluble drugs penetrate the corneal stroma layer and lipid-soluble drugs penetrate the corneal epithelium. Therefore, a combination of water- and lipid-soluble properties in a topical ophthalmic preparation is desirable.

Frequency of drug application. Ointments are usually administered less frequently than eye drops. Owners may prefer to apply eye medication less frequently, because of schedule issues and apprehension about putting medication in the animal's eye; therefore, owners should be questioned about their preference before ophthalmic medication is dispensed. Ointments tend to blur an animal's vision and should be administered appropriately to avoid excessive vision problems.

Ease of application. Some clients have better success at administering ointments and other clients have better success with drops. The client's preference may be taken into account in these cases. Figure 18-4A and B demonstrate administration of ophthalmic ointment and drops into the eye.

TIP

Water-soluble drugs (such as atropine and pilocarpine) penetrate the cornea well; highly polar drugs (such as water-soluble antibiotics) penetrate the cornea poorly.

TIP

Ophthalmic ointments tend to last longer than ophthalmic drops; therefore, they can be administered less frequently.

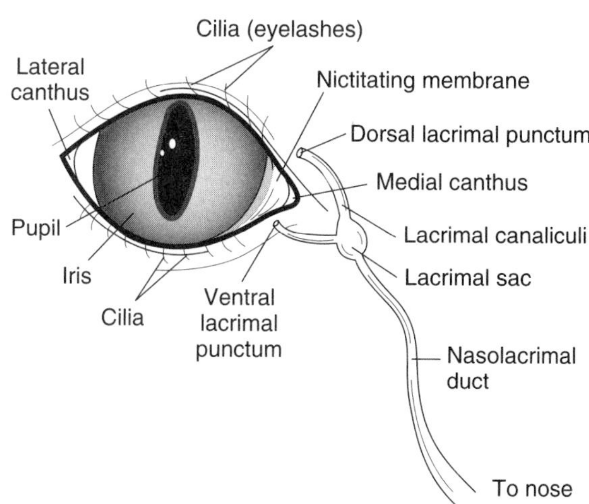

FIGURE 18-3 Adnexa of the eye

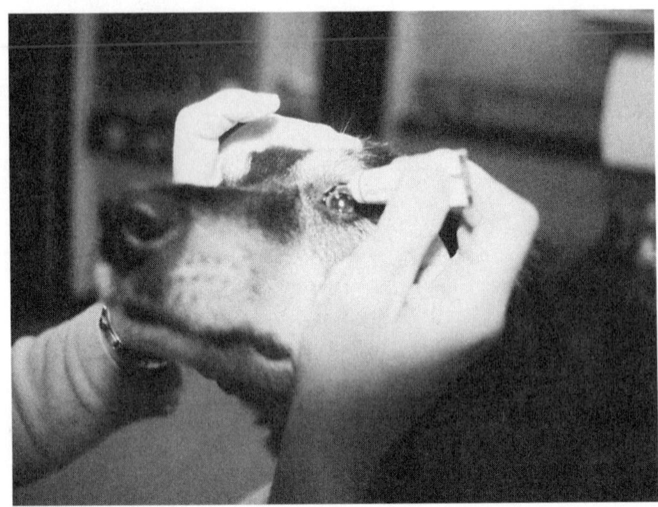

FIGURE 18-4A Administering ophthalmic ointment in the eye

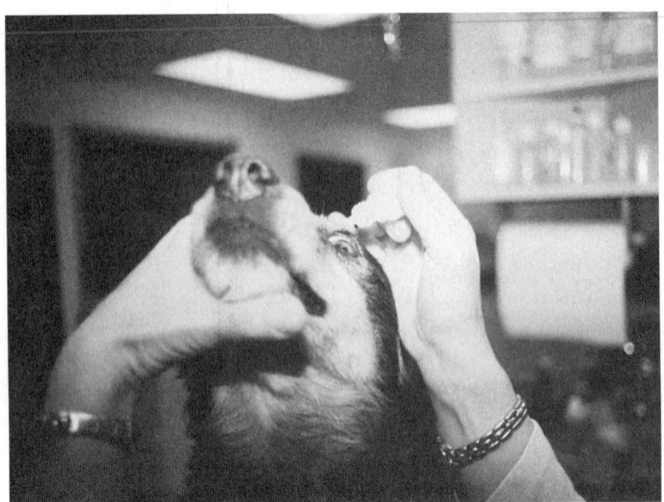

FIGURE 18-4B Administering ophthalmic drops in the eye. When giving ophthalmic medications, it is important not to touch the tip of the applicator to the eye, to avoid contamination of the drug.

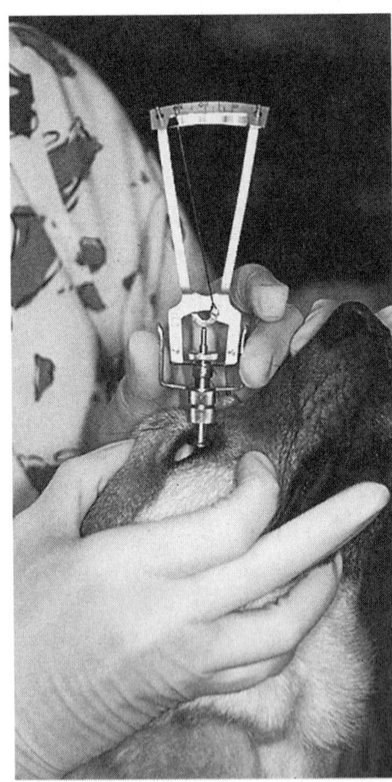

FIGURE 18-5 Detection of intraocular pressure by Shiotz tonometer. Before intraocular pressure readings are taken, topical anesthetic is applied to the cornea, to allow the tonometer to be placed directly on the cornea.

Diagnostic

Diagnostic drugs are used to locate lesions or foreign objects in the eye. Diagnostic drugs include topical anesthetics and fluorescein sodium.

Topical anesthetics are used during removal of foreign bodies from the eye and performance of comprehensive eye examinations in which instruments are applied to the cornea to measure intraocular pressure (Figure 18-5). Corneal anesthesia is accomplished in about one minute and lasts for about ten minutes. Examples include proparacaine hydrochloride (Ophthaine®, Ophthetic®) and tetracaine hydrochloride (Pontocaine®). Opened bottles should be stored in the refrigerator and any discolored solutions should be discarded.

Fluorescein sodium is used to detect corneal scratches (the stain is orange until it adheres to a corneal defect, where it appears green), foreign bodies (which are surrounded by orange), and patency of the nasolacrimal duct (dye appears in the nasal secretions). Fluorescein impregnates an individually wrapped sterile strip to which sterile saline may be added to allow drops to fall onto the cornea for application to the eye (Figure 18-6). After a few seconds, excess fluorescein is flushed from the eye with sterile saline. Stain will be retained in areas of full-thickness corneal epithelial loss.

Fluorescein adheres to areas of full-thickness epithelial loss because it is water-soluble. Because the outer layer of the cornea is fat-soluble, the drug cannot penetrate or adhere to an intact cornea. The stroma layer beneath the outer layer is water-soluble. If the outer layer is damaged due to corneal injury or corneal ulceration, the fluorescein adheres to the stroma layer when the dye is rinsed with sterile saline.

Pupil Closing

Miotics are cholinergic drugs that constrict the pupil. They are used to treat open-angle glaucoma because they lower the intraocular pressure by increasing outflow of aqueous humor. Miotics generally promote the outflow of aqueous humor. Pilocarpine, a topical cholinergic drug, is the most commonly used miotic. Systemic absorption of pilocarpine

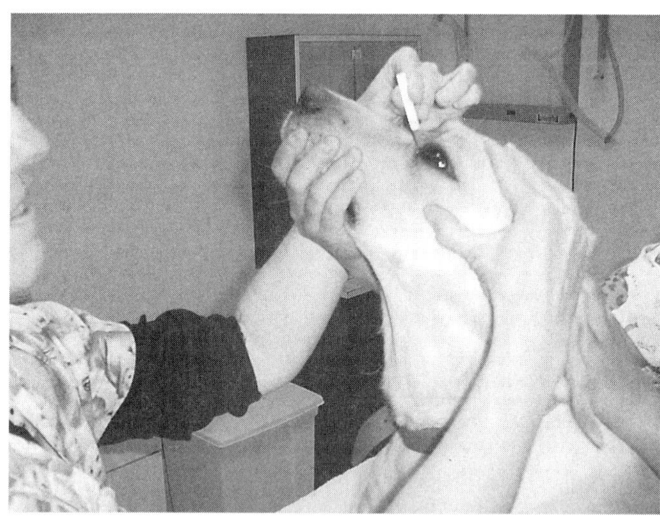

FIGURE 18-6
Fluorescein dye is applied to the cornea to allow visualization of corneal defects. *Source:* Poynette Veterinary Service

is possible, but uncommon. The onset of action is usually 10 to 30 minutes and duration of action is from 4 to 8 hours. Side effects of pilocarpine include local irritation and redness. Trade names of pilocarpine include Piloptic® and Isopto-Carpine®.

Pupil Opening

Mydriatics are drugs that dilate pupils. **Cycloplegics** are drugs that paralyze the ciliary muscles (ciliary muscles function in controlling the shape of the lens) and may minimize pain due to ciliary spasm. These drugs may be used in combination or alone to achieve desired outcomes. Examples include:

- atropine, an anticholinergic drug used for the treatment of acute inflammation of the anterior uvea and as an aid in examination of the retina (by dilating the pupil to allow visualization of the retina). Atropine produces mydriasis and cycloplegia by blocking the effect of acetylcholine on the sphincter muscle of the iris and the muscles of the ciliary body. Peak effect for mydriasis is 30 to 40 minutes and 1 to 3 hours for cycloplegia. Side effects include salivation. Its use is contraindicated in animals with glaucoma, because it increases intraocular pressure, and in animals with keratoconjunctivitis sicca (KCS or dry eye), because it decreases tear production. Atropine is usually obtained generically and is available as an ophthalmic solution and ointment.
- homatropine, used for eye examination and treatment of uveitis. It has a faster onset and shorter duration of action than atropine. It produces mydriasis and cycloplegia, but less so than atropine. The side effects are the same as for atropine. Trade names include Isopto Homatropine® and Homatrocel Ophthalmic®.
- phenylephrine, a sympathomimetic drug used to evaluate eye diseases such as uveitis and Horner's syndrome. It may be used prior to conjunctival surgery to decrease hemorrhage. It produces mydriasis, but not cycloplegia. It also produces vasoconstriction. Side effects include ocular discomfort, tearing, and rebound miosis. Trade name examples are Mydfrin 2.5%® and Neo-Synephrine®.
- tropicamide, used for fundic examination. It is a rapid-acting mydriatic and has slight cycloplegic effect. It has a more rapid onset and shorter duration of action than atropine. Its side effects include local discomfort and salivation. It is contraindicated in animals with glaucoma or KCS. Mydriacyl® and Opticyl® are trade names of tropicamide.
- epinephrine, a sympathomimetic drug that reduces intraocular pressure, produces mydriasis, and aids in the diagnosis of Horner's syndrome. It is used to

TIP

Topical anesthetics may cause slight discomfort to the patient when first applied.

TIP

Topical anesthetics can be recognized by the *-caine* ending in their generic names.

TIP

Fluorescein may temporarily stain fur, skin, and clothing. The fluorescein strip should not be allowed to touch the cornea, as it will cause stain retention at the site of contact. This will result in false positive readings.

TIP

Mydriatics are used to aid in eye exams, to relieve inflammation associated with uveitis (inflammation of the iris, ciliary body, and choroid) and keratitis (inflammation of the cornea), to break up or prevent adhesions between the iris and the lens, and to prepare an animal for ocular surgery.

TIP

Mydriatics are contraindicated in an animal with glaucoma because they relax iris muscles and hinder aqueous humor outflow from the eye.

TIP

Miotics are cholinergic or sympatholytic drugs. Mydriatics are sympathomimetic or anticholinergic drugs.

prevent glaucoma in the unaffected eye, but should not be used in cases of closed-angle glaucoma. Epinephrine may cause ocular discomfort when applied to the eye. Epifrin® is a trade name of epinephrine.

Pressure Reducing

Glaucoma is a group of diseases characterized by increased intraocular pressure. The disorder can be caused by an acquired structural defect within the eye (primary glaucoma); it may be the consequence of another ocular disease or trauma (secondary glaucoma); or it may be the result of a genetic defect (congenital glaucoma). If left untreated, the increase in intraocular pressure can damage the nervous tissue of the eye, mainly the retina and optic nerve, resulting in blindness.

In healthy eyes, aqueous humor is constantly being produced by the ciliary process located behind the iris. Its production is controlled by enzyme systems, mainly carbonic anhydrase. Once the aqueous humor enters the eye, it passes from the posterior chamber through the pupil and into the anterior chamber. From there it is drained from the eye through a sponge-like substance called the *trabecular meshwork*. Figure 18-7A demonstrates aqueous humor flow through a healthy eye.

When intraocular pressure increases, the outflow mechanism for aqueous humor is blocked. If the iris occludes the trabecular meshwork, normal outflow of aqueous humor is prevented and the animal is said to have *narrow-angle glaucoma* (Figure 18-7B). If there is no change in the chamber angle of the eye, but aqueous humor outflow is impeded because of degenerative changes, the animal is said to have *open-angle glaucoma*.

Drugs used to decrease intraocular pressure as a result of glaucoma include:

- miotics, which cause constriction of the pupil (discussed previously in this chapter).
- **carbonic anhydrase inhibitors**, which interfere with the production of carbonic acid and thus lead to decreased aqueous humor formation. The decrease in aqueous humor decreases intraocular pressure. Carbonic anhydrase inhibitors were originally developed as diuretics and are given orally and parenterally. Side effects include vomiting, diarrhea, and weakness. Examples of carbonic anhydrase inhibitors are acetazolamide (Diamox®), dichlorphenamide (Daranide®), and methazolamide (Neptazane®). Side effects of these drugs include gastrointestinal disturbances such as anorexia, dehydration, and urinary crystal development.

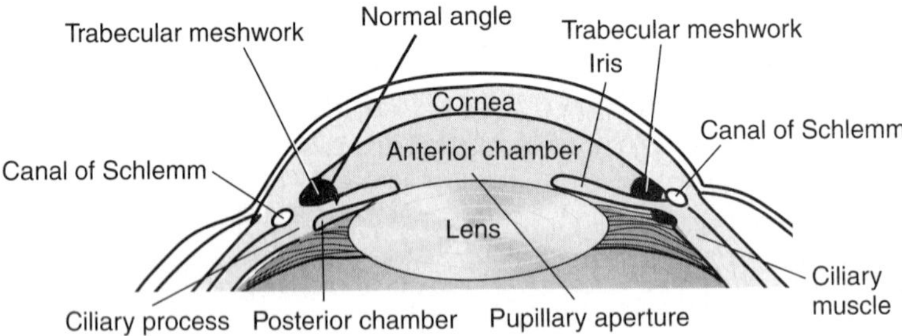

FIGURE 18-7A The normal flow of aqueous humor.

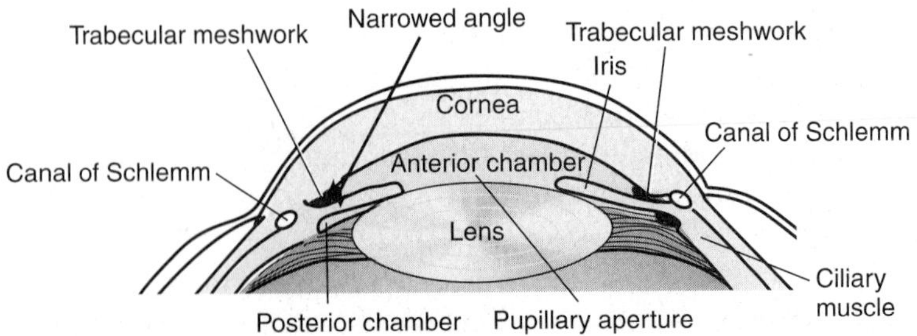

FIGURE 18-7B In narrow-angle glaucoma, the outflow of aqueous humor is impeded, resulting in an increase in intraocular pressure.

- **beta-adrenergic blockers**, which decrease the production of aqueous humor and thus decrease intraocular pressure. Heart and respiratory monitoring is indicated because of the systemic side effects of this drug category (bradycardia, hypotension, and bronchospasms). This category of drug tends to be used with primary glaucoma to prevent development of disease in both eyes. It is used topically and may cause blurred vision, so owners should be made aware of this possibility. Examples in this group include timolol maleate (Timoptic®) and betaxolol hydrochloride (Betoptic®).
- **Osmotic diuretics**, used prior to surgery or as an emergency treatment of glaucoma. They are given IV to decrease vitreous humor volume and rapidly decrease intraocular pressure. Side effects of osmotics include electrolyte imbalances, cardiovascular problems, and gastrointestinal problems such as vomiting. Examples include mannitol (Osmitrol®) and glycerin (Glyrol®, Osmoglyn®).

Dry Eye Repairers

Keratoconjunctivitis sicca (**KCS** or dry eye) is a disease in which tear production is decreased, resulting in persistent mucopurulent conjunctivitis and corneal scarring and ulceration. KCS is common in dogs and is thought to be immune mediated (Figure 18-8). Treatment for KCS includes the following:

- artificial tears, isotonic solutions that are pH-buffered to serve as a lubricant for dry eyes and to alleviate eye irritation from KCS. There are a large variety of over-the-counter artificial eye products (Table 18-1).
- antibiotic-steroid preparations. These combination products may be used if corneal ulcers do not exist. These products are also listed in Table 18-1.
- **lacrimogenics**, which increase tear production. The lacrimal glands are under parasympathetic control, so giving drugs that stimulate the parasympathetic nervous system will increase tear production. Pilocarpine is a lacrimogenic that was formerly used topically and orally in food to increase tear production in animals with KCS. Side effects, such as vomiting and diarrhea, have limited its use for KCS treatment.

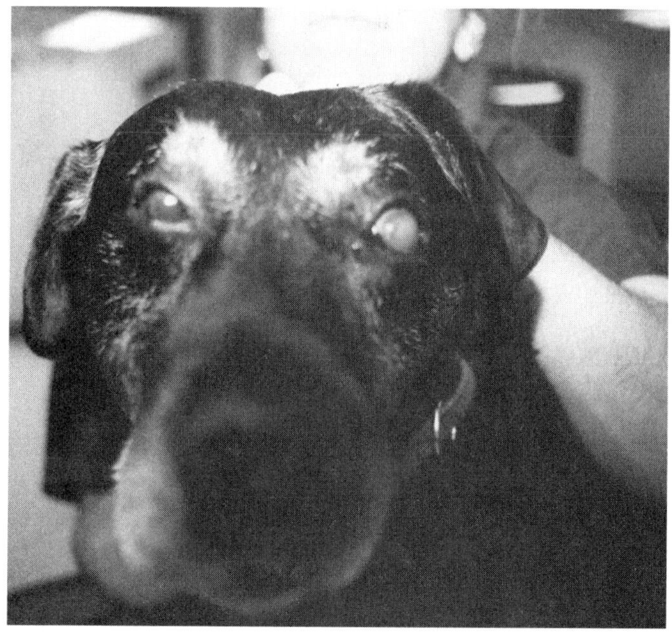

FIGURE 18-8 Animal exhibiting KCS

TABLE 18-1 Ophthalmic Anti-Infectives, Anti-Inflammatories, and Tear Supplements

Category	Example	Action
Topical antibacterial drugs (read labels to determine if corticosteroids are used in these preparations)	bacitracin (Mycitracin®, Trioptic-P®, Vetropolycin®, Neosporin® Ophthalmic)	Works against gram + organisms; usually found in combination with neomycin and polymyxin B
	polymyxin B (Mycitracin®, Trioptic-P®, Vetropolycin®, Neosporin® Ophthalmic)	Works against gram – organisms; usually used in combination with bacitracin and neomycin
	oxytetracycline (Terramycin®)	Broad-spectrum antibiotic that is effective against *Chlamydia sp.*; may be used in combination with other drugs
	tetracycline (Achromycin®)	Broad-spectrum antibiotic that is effective against *Chlamydia sp.*
	aminoglycosides • gentamicin (Gentocin®, Garamycin®, Genoptic®) • tobramycin (Tobrex®) • neomycin (Mycitracin®, Trioptic-P®, Vetropolycin®, Neosporin® Ophthalmic)	Work against *Staphylococcal* and gram – organisms, including *Pseudomonas sp.* ; may be formulated alone or with corticosteroids; neomycin is a broad-spectrum antibiotic usually used in bacitracin and in combination with polymyxin B
	erythromycin (Ilotycine Ophthalmic®)	Broad-spectrum antibiotic; usually used for gram + infections
	fluoroquinolones • ciprofloxacin (Ciloxan®) • norfloxacin (Chibroxin®) • ofloxacin (Ocuflox®)	Broad-spectrum antibiotic
	chloramphenicol (Bemacol®, Chlorbiotic®, Chloricol®, Vetrachloracin®, Chlorasol®)	Broad-spectrum antibiotics; handle with care due to human side effects; cannot use in food-producing animals
Topical antifungal drugs	natamycin (Natacyn-Ophthalmic®)	Works in treating mycotic keratitis (mainly *Fusarium sp.*, *Candida sp.*, *Aspergillus sp.*)
Topical antiviral drugs	• idoxuridine (Stoxil®) • trifluridine (Viroptic®) • vidarabine (Vir-A Ophthalmic®)	Used to treat viral infections of the eye, mainly in cats (ocular herpes). These drugs interrupt viral replication and are virostatic, not virucidal; treatment must be continued past clinical resolution.
Topical corticosteroid drugs	• prednisolone acetate drops (Pred Mild®, Econopred®) • prednisolone sodium phosphate drops (generic brands) • dexamethasone drops and ointment (Decadron Phosphate®, Maxidex®) • triple antibiotic with hydrocortisone (Neobacimyx H®, Trioptic-S®, Vetropolycin HC®) • neomycin and isoflupredone acetate (Neo-Predef®) • gentamicin with betamethasone (Gentocin Durafilm®) • chloramphenicol and prednisolone (Chlorasone®)	Corticosteroids are used to treat inflammation of the conjunctiva, sclera, cornea, and anterior chamber. Penetration to the vitreous chamber and eyelids is poor. Corticosteroids delay healing and should not be used in patients with corneal ulcers, fungal infections, or viral infections.

TABLE 18-1 *Continued*

Category	Example	Action
Topical nonsteroidal anti-inflammatory drugs	• flurbiprofen (Ocufen®) • ketorolac tromethamine (Acular®) • diclofenac (Voltaren®)	Topical NSAIDs are used to treat inflammation, usually after surgery.
Tear supplements	• artificial tears (Bion Tears®, Liquifilm Tears®, Hypotears®, Adsorbotear®) • lubricants (Lacri-Lube S.O.P.®, Akwa Tears®)	Artificial tears are isotonic, pH-buffered solutions that lubricate dry eyes and provide eye irrigation. Lubricants are petrolatum-based products that lubricate and protect eyes (mainly used during anesthesia, in which the eyes may remain open while tear production is reduced).

- immunomodulators, which adjust the immune response to a desired level. Cyclosporine is an **immunomodulating agent** that interferes with interleukin production by T-lymphocytes. This interference stops local inflammation, resulting in improved tear production after several weeks of treatment. Commercial products containing cyclosporine are mixed with oil to enable topical application. Systemic absorption is not seen following topical treatment. A trade name of cyclosporine is Optimmune®.

All the Rest

Ophthalmic drugs are used to treat bacterial and fungal eye infections, inflammatory and allergic conditions, pruritus (itching), and pinkeye, known medically as *infectious keratoconjunctivitis*. Small animals receive most of the ophthalmic drugs veterinarians use, because ophthalmic preparations used on food-producing animals may accumulate in tissues. Because they are most often used to treat multiple infections, most ophthalmic preparations are mixtures of more than one drug. A typical ophthalmic preparation may contain an antibacterial such as neomycin, a second antibacterial such as polymyxin B, and an anti-inflammatory such as prednisolone. Table 18-1 summarizes those drugs.

INTRODUCTION

A client brings in a kitten that she adopted from a farm. The kitten has been scratching at her ears. The cat is examined and determined to be healthy otherwise. Upon examination of her ears, you see brownish ear discharge and redness due to scratching. You can see white mites moving when you look in the ears with otoscopic magnification (the warmth of the otoscope causes them to move). It is determined that the kitten has ear mites (*Otodectes cynotis*). What should be done prior to treatment? Ear cleaning? Determination of whether the tympanic membrane is intact? Questioning the owner about other animals in the household? Do you know?

BASIC EAR ANATOMY AND PHYSIOLOGY

The ear is the sensory organ that allows hearing and helps maintain balance. The ear is divided into three parts: outer, middle, and inner. The outer ear consists of the *pinna*

TIP

All dogs and cats can get otitis, but those dogs with long pendulous ears tend to have more chronic ear problems.

TIP

The only liquid that is safe to use for flushing ears with ruptured tympanic membranes is sterile isotonic saline solution. All disinfectants should be avoided. Otic drugs that cannot be used with ruptured tympanic membranes include aminoglycosides, chlorhexidine, chloramphenicol, and iodine compounds.

(also known as the *auricle*) and the external auditory canal. The pinna catches sound waves and transmits them into the external auditory canal. The external auditory canal transmits the sound from the pinna to the tympanic membrane.

The middle ear begins with the *tympanic membrane* or *eardrum*. The middle ear contains the tympanic membrane, auditory ossicles, eustachian tube, oval window, and round window. The sound waves are transmitted from the tympanic membrane to the auditory ossicles, which are three small bones in the middle ear. The auditory ossicles transmit sound waves past the eustachian tube (involved in air-pressure equilibrium) to the oval window. From the oval window, the sound waves enter the inner ear.

The inner ear is a series of labyrinths or canals. The vestibule, cochlea, and semicircular canals make up the bony labyrinth. The vestibule and semicircular canals are responsible for sensing balance and equilibrium and the cochlea is responsible for hearing. The structures of the inner ear receive the sound waves that are then relayed to the brain. Figure 18-9 shows the structures of the ear.

A common ear problem seen in veterinary medicine is inflammation. **Otitis externa** is inflammation of the pinna and external auditory canal. It is commonly seen in dogs and cats. Many things, including bacteria, parasites, yeast, allergies, systemic disease, and neoplasia, can cause otitis externa. Clinical signs include head shaking, ear scratching, and discharge from the ear. **Otitis media** is inflammation of the middle ear. Otitis media may be difficult to diagnose because it may be clinically silent (have no signs) or have signs of purulent discharge and head shaking. Otitis media may develop

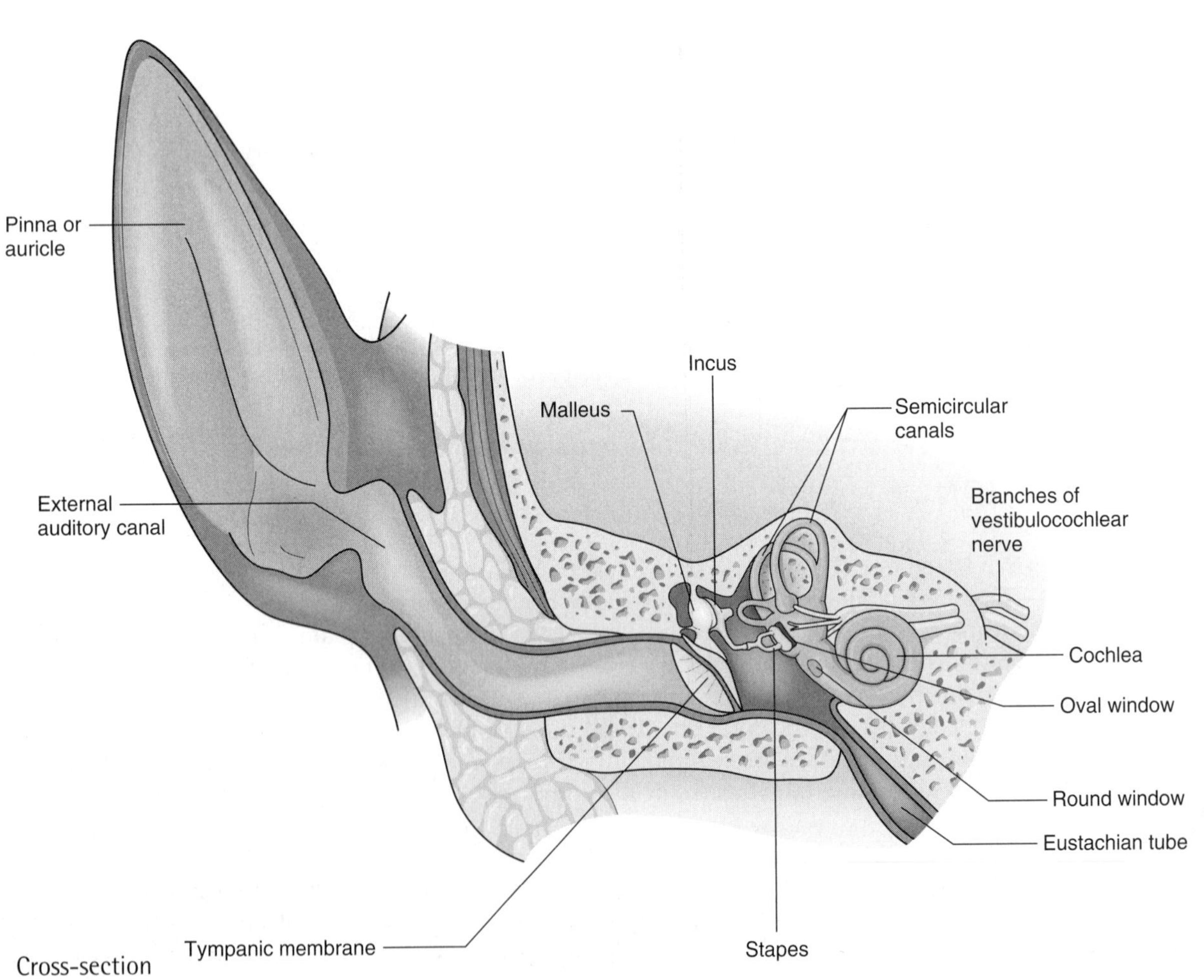

FIGURE 18-9 Cross-section of ear structures

secondary to extension of otitis externa through a ruptured tympanic membrane or extension up the eustachian tube following an upper respiratory infection. Other causes of otitis media include polyps, trauma, and infection. **Otitis interna** is an inner ear infection. Signs of otitis interna include head tilt toward the affected side, ataxia, and possibly nausea and vomiting.

It is important to identify the underlying cause of otitis before treatment, so that therapy can be designed to achieve optimal results. Examination usually includes otoscopic examination, ear cytology, and culture-and-sensitivity testing if indicated by the suspected presence of infection.

OTIC MEDICATIONS

Among the many drug combinations veterinarians use to alleviate ear diseases are antifungal agents to eradicate fungi, corticosteroids to reduce inflammation and pruritus, antibiotics to treat bacterial infections, antiparasitics to treat ear mites, and local anesthetics to reduce pain. Other types of otic preparations include cleansers, drying agents, and *cerumen* (earwax) dissolvers that are used in combination with otic medications to treat or prevent disease. Many otic preparations are a combination of these ingredients. Table 18-2 lists the various otic drugs.

TIP

It is important to wash your hands after applying topical antifungal agents, to avoid the spread of fungal infections.

TABLE 18-2 Otic Drugs

Category	Example	Action
Topical antibiotic otic drugs	• aminoglycosides: gentamicin (Otomax®, Getocin Otic Solution®, Tri-Otic®, GentaVed Otic®); neomycin sulfate (Tresaderm®, Panalog®, Tritop®)	Broad-spectrum antibiotics that are usually combined with a corticosteroid, antifungal agent, and/or antiparasitic agent. Neomycin sulfate products also contain thiostreptin, a gram + antibiotic; tympanic membrane should be intact when using this medication
	• thiostreptin (Tresaderm®, Panalog®, Tritop®)	Gram + antibiotic that is usually combined with a corticosteroid, antifungal agent, and/or antiparasitic agent
	• chloramphenicol (Liquichlor®, Chlora-Otic®)	Broad-spectrum antibiotic usually combined with a corticosteroid and/or topical anesthetic. Do not use in food-producing animals. Handle this drug with caution.
	• fluoroquinolones: enrofloxacin (Baytril Otic®)	Broad-spectrum antibiotic that is combined with silver sulfadiazine (has both antifungal and broad spectrum antibacterial properties)
Topical antiparasitic otic drugs	• thiabendazoles: (Tresaderm®)	Used to treat ear mites in dogs and cats. Thiabendazole has antiparasitic and antifungal properties. Preparation may contain an antibiotic (such as neomycin) and a corticosteroid (such as dexamethasone)
	• pyrethrins (Mita-Clear®, Cerumite®, Aurimite®)	Used to clear mite infestations of the ear; treatment should continue for at least three weeks

(continues)

TABLE 18-2 *Continued*

Category	Example	Action
Topical antiparasitic otic drugs (*cont.*)	• rotenone (Ear Miticide®, Mitaplex-R®)	Used to clear mite infestations of the ear; treatment should continue for at least three weeks
	• milbemycin oxime (MilbeMite®)	Used to treat ear mite infestations in cats and kittens four weeks of age and older. Treatment consists of administering solution from one tube per ear as a single treatment. MilbeMite® is available in a foil pouch that contains two tubes of solution (one tube for each ear).
	• ivermectin: injectable (Ivomec®) topical (Acarexx 0.01% suspension®)	Injectable treatment that is given SQ extra-label to treat ear mites in dogs and cats Otic solution is labeled for use in cats and kittens (more than four weeks of age) and is packaged in ampules
	• selamectin (Revolution®)	Once-monthly treatment for ear mites in cats; applied to dorsal cervical skin; (also protects against roundworms, hookworms, and fleas)
Topical otic antifungal agents	• clotrimazole (Otibiotic®, Otomax®)	Works against *Malassezia*, *Microsporum*, *Trichophyton*, *Epidermophyton*, and *Candida* fungi; antifungal agent is combined with gentamicin and betamethasone
	• nystatin (Dermagen®, Panalog®, Derma-Vet®)	Works against *Candida* fungi; antifungal agent is combined with neomycin sulfate, thiostreptin, and triamcinolone
	• miconazole (Conofite®)	Works against *Microsporum* and *Trichophyton*; local irritation may be seen with miconazole treatment
	• thiabendazole (Tresaderm®)	Works against *Microsporum* and *Trichophyton*; also contains antiparasitic, antibiotic, and corticosteroid
Topical otic drying agents	Various products that contain salicylic acid, acetic acid, boric acid, or tannic acid. Examples include Dermal Dry®, VetMark Ear Powder®, OtiRinse Cleansing/Drying Ear Solution®, Oti-Care-B®	Reduces moisture in the ear to help prevent or treat certain infections of the ear. Ears should be cleaned prior to putting in drying agents.
Topical otic cleansing agents	Various products contain antibiotics, antiseptics, and/or anesthetics/soothing agents. Examples include Solvaprep®, Epi-Otic®, Oti-Clens®, Fresh-Ear®	Used to clean ears and control odor. Also used for gentle flushing of the ear using a bulb syringe or tubing. Ears should be dried thoroughly after use.
Topical otic dewaxing agents	Various products that contain cerumen softeners or drying agents such as benzyl alcohol, cerumene, and similar chemicals. Examples include Cerulytic® and Cerumene®.	Used to remove debris and wax before treatment with topical medications, and to aid in wax removal by flushing of the ear with a bulb syringe or tubing.
Topical anti-inflammatories	• fluocinolone + DMSO (Synoptic®)	DMSO enhances percutaneous absorption of steroids. Avoid contact with human skin to reduce risk of absorbing the drug.

SUMMARY

Ophthalmic drugs are used to treat conditions of the eyes; they include miotics, mydriatics, and drugs to decrease intraocular pressure. Miotics cause pupillary constriction and are used for treatment of glaucoma. Mydriatics cause pupillary dilation and are used in the treatment of inflammatory disorders, as well as an aid in ocular exams. Drugs that decrease intraocular pressure are used to treat glaucoma, and include miotics, carbonic anhydrase inhibitors, beta-adrenergic blockers, and osmotic diuretics. Keratoconjunctivitis sicca or dry eye is thought to be immune-mediated and is treated with artificial tears, antibiotic-steroid preparations, and immunomodulators. Another large category of ophthalmic drugs includes topicals, such as antibiotics, antifungals, antivirals, anti-inflammatories (both steroidal and nonsteroidal); and artificial tears and lubricants.

Otic drugs are used to treat conditions of the ear—mainly bacterial, fungal, or parasitic infections. Drying agents, cleansing solutions, and dewaxing agents can aid in the treatment and prevention of ear infections.

Tables 18-3 and 18-4 summarize the drugs covered in this chapter.

TABLE 18-3 Opthalmic Drugs Covered in This Chapter

Category	Example
Topical ophthalmic anesthetic	• proparacaine hydrochloride • tetracaine hydrochloride
Diagnostic stain	• fluorescein
Miotic	• pilocarpine
Mydriatic	• phenylephrine
Mydriatic/Cycloplegic	• atropine • homatropine • tropicamide • epinephrine
Carbonic anhydrase inhibitor	• acetazolmide • dichlorphenamide • methazolamide
Beta-adrenergic blocker	• timolol maleate • betaxolol
Osmotic diuretic	• mannitol • glycerin
Immunomodulator	• cyclosporine
Topical antibacterial ophthalmic drug	• bacitracin • neomycin • polymyxin B • oxytetracycline • tetracycline • aminoglycosides • gentamicin • tobramycin • erythromycin • fluoroquinolones • ciprofloxacin • norfloxacin • ofloxacin • chloramphenicol

(continues)

TABLE 18-3 *Continued*

Category	Example
Topical antifungal ophthalmic drug	• natamycin
Topical antiviral ophthalmic drug	• idoxuridine • trifluridine • vidarabine
Topical corticosteroid ophthalmic drug	• prednisolone acetate drops • prednisolone sodium phosphate drops • dexamethasone drops and ointment • triple antibiotic with hydrocortisone • neomycin and isoflupredone acetate • gentamicin with betamethasone • chloramphenicol and prednisolone
Topical nonsteroidal anti-inflammatory ophthalmic drug	• flurbiprofen • ketorolac tromethamine • diclofenac
Tear supplement	• artificial tears • lubricants

TABLE 18-4 Otic Drugs Covered in This Chapter

Category	Example
Topical antibiotic otic drugs	• gentamicin • neomycin sulfate • thiostreptin • chloramphenicol • enrofloxacin
Topical antiparasitic otic drugs	• thiabendazole (neomycin/thiabendazole/dexamethasone solution) • pyrethrins • rotenone • milbemycin oxime • ivermectin • selamectin
Topical otic antifungal agents	• clotrimazole • nystatin • miconazole • thiabendazole
Topical otic drying agents	• salicylic acid, • acetic acid • boric acid • tannic acid
Topical otic cleansing agents	antibiotics, antiseptics, and/or anesthetics/soothing containing agents used to clean the ear
Topical otic dewaxing agents	cerumen softeners or drying agents containing chemicals such as benzyl alcohol, cerumene, and similar chemicals
Topical anti-inflammatories	• fluocinolone + DMSO

CHAPTER REFERENCES

Amundson Romich, J. (2000). *An illustrated guide to veterinary medical terminology.* Clifton Park, NY: Thomson Delmar Learning.

Delmar's A-Z NDR-97 Nurses drug reference. Clifton Park, NY: Thomson Delmar Learning.

Reiss, B., & Evans, M. (2002), *Pharmacological aspects of nursing care* (6th ed., revised by Bonita E. Broyles, 504–531). Clifton Park, NY: Thomson Delmar Learning.

Veterinary pharmaceuticals and biologicals (12th ed.). (2001). Lenexa, KS: Veterinary Healthcare Communications.

Veterinary values (5th ed.). (1998). Lenexa, KS: Veterinary Healthcare Communications.

Wynn, R. (1999). *Veterinary pharmacology.* Cambridge, MA: Harcourt Learning Direct.

CHAPTER REVIEW

Matching

Match the drug name with its action.

1. ______ fluorescein
2. ______ mannitol
3. ______ cyclosporine
4. ______ pilocarpine
5. ______ topical otic drying agents
6. ______ pyrethrin
7. ______ atropine
8. ______ chloramphenicol
9. ______ acetazolamide
10. ______ ivermectin

a. mydriatic used to treat acute inflammation of the anterior uvea
b. antiparasitic otic solution used to treat ear mites
c. extra-label ear mite medicine given SQ to dogs and cats
d. dye used to detect corneal defects such as scratches and ulcers
e. broad-spectrum antibiotic that should not be used in food-producing animals
f. carbonic anhydrase inhibitor used to treat glaucoma
g. products containing salicylic acid, acetic acid, boric acid, or tannic acid
h. stimulates tear production and used to treat KCS
i. miotic used to treat open-angle glaucoma
j. osmotic diuretic used as an emergency treatment for glaucoma

Multiple Choice

Choose the one best answer.

11. Otomax® is the trade name of a topical otic solution containing a corticosteroid, antifungal, and

a. chloramphenicol.
b. neomycin sulfate.
c. gentamicin.
d. enrofloxacin.

12. When treating patients with topical antibiotic otic drugs, it is important to
    a. refrigerate all medications.
    b. determine whether the tympanic membrane is intact.
    c. handle the drugs with caution and not use them in food-producing animals.
    d. treat for parasites as well.
13. Proparacaine and tetracaine are examples of
    a. miotic drugs.
    b. topical anesthetics.
    c. mydriatic drugs.
    d. osmotic drugs.
14. Which group of ophthalmic drugs promotes the outflow of aqueous humor by lifting the iris away from the filtration angle area?
    a. miotic drugs
    b. topical anesthetics
    c. mydriatic drugs
    d. osmotic drugs
15. Which group of ophthalmic drugs causes pupillary dilation, allowing examination of the retina?
    a. miotic drugs
    b. topical anesthetics
    c. mydriatic drugs
    d. osmotic drugs
16. Trifluridine works on what category of microorganisms?
    a. bacteria
    b. fungi
    c. parasites
    d. viruses
17. Which drugs should not be used in patients with corneal ulcers or scratches?
    a. cyclosporine
    b. flurbiprofen
    c. corticosteroids
    d. lubricants
18. What is one advantage of using ophthalmic ointment rather than ophthalmic drops?
    a. You can get more doses from a tube of ointment.
    b. Ointments always contain corticosteroids to reduce inflammation.
    c. Ointments last longer and therefore the client has to treat the animal less frequently.
    d. Ointments make the animal's vision blurry, making it rest more while it is sick.

True/False

Circle a. for true or b. for false.

19. All mydriatics are also cycloplegics.
    a. true
    b. false
20. Most otic treatments for ear mites are effective with a single application.
    a. true
    b. false

Case Studies

21. A 3-yr.-old F/S Poodle (8#) (Figure 18-10) comes into the clinic with a history of rubbing her eyes. On physical exam her TPR is normal; she is timid, but otherwise normal according to her owner. The veterinarian wants to perform an eye exam. The veterinarian already has Schirmer tear test strips (to check for tear production). All of the ophthalmic equipment is already in the clinic.
    a. What products would you want to get for the veterinarian?
    b. It is determined that this dog has a corneal scratch from playing with the owner's cat. What product helped diagnose the corneal scratch?
    c. What product is contraindicated in this poodle?
    d. The veterinarian wants you to dispense an antibiotic for the owner to apply to the dog's eye. The owner works all day. What product form do you recommend?

FIGURE 18-10 Poodle. Photo by Isabelle Francais

FIGURE 18-11 DSH kitten. *Source:* Janet Romich, DVM, MS

22. A 3-month-old M DSH kitten (Figure 18-11) presents to the clinic with excessive ear scratching. A physical exam shows that the kitten is thin, but has normal TPR and does not appear to have signs of a respiratory infection. The kitten is not current on his vaccinations.

a. Knowing the age of the kitten, what would be a good thing to check him for?
b. The kitten is positive for the observation you performed to answer the preceding question. What are some treatment options?

CHAPTER 19

Fluid Therapy and Emergency Drugs

OBJECTIVES

Upon completion of this chapter, the learner should be able to

- describe body fluid composition, location, and percentages in animals.
- describe estimation of the level of dehydration based on physical exam findings.
- describe various routes of fluid administration.
- differentiate between crystalloids and colloids.
- give examples of isotonic, hypotonic, and hypertonic crystalloid solutions and describe their uses.
- give examples of colloid solutions and describe their uses.
- list various fluid additives and describe their uses.
- practice calculating fluid therapy volumes based on maintenance values, rehydration values, and ongoing fluid loss values.
- determine the rate of fluid administration based on adult and pediatric administration sets.
- describe various administration sets and fluid delivery systems.
- detail the importance of monitoring fluid therapy and the parameters to be monitored.
- describe basic emergency protocol for an animal in respiratory or cardiac arrest.
- list various emergency drugs and their uses in emergency procedures.

KEY TERMS

intracellular fluid (ICF)
extracellular fluid (ECF)
crystalloids
colloids
osmolality
tonicity
osmotic pressure
solutes
isotonic
hypotonic
hypertonic
colloid solution
rehydration
maintenance
ongoing fluid loss
adult administration set
pediatric administration set

INTRODUCTION

New clients bring in a six-week-old puppy for examination because he has developed diarrhea. The owners say that the dog has seemed lethargic, but assumed that was because he was in new surroundings. The dog has not been vaccinated nor have any fecal parasite exams been performed on his stool. On physical exam, the veterinarian notes that the puppy is moderately dehydrated, quiet, and slightly pained by abdominal palpation. During the physical examination, the puppy defecates a moderate amount of very watery, blood-tinged diarrhea. The veterinarian orders a fecal examination and blood work (including a parvovirus titer) to be done on this dog. Pending the results of the tests, the dog is hospitalized, isolated from other animals, and fluid therapy is initiated. The owners ask you why fluids have to be given to this puppy, because he is so small and does not drink a lot of water anyway. Can you explain to the owners why this puppy needs fluids, where he is losing fluids, and what type of fluids he will get? Is this puppy more at risk of developing dehydration than an older animal? What is the quickest way to get fluids into this puppy? Can you explain the theory of fluid therapy to this owner?

BASICS OF BODY FLUID

The cells that make up body tissues exist in an environment chemically constant yet physiologically dynamic. The chemical consistency that is achieved through fluid, electrolyte, and acid-base balance is essential to maintaining homeostasis. *Homeostasis* is the stable state produced by physical processes that keep the physical and chemical properties of body fluid relatively constant.

Water is the primary body fluid. Body water is distributed among three types of "compartments": cells, blood vessels, and tissue spaces between blood vessels and cells that are separated by membranes. Body water in each type of compartment may be *intracellular* (within the cell), *intravascular* (within the blood vessels) or *interstitial* (in the tissue spaces between blood vessels and cells). Fluid within the cell is classified as **intracellular fluid (ICF)**, whereas intravascular fluid and interstitial fluid are classified as **extracellular fluid (ECF)**. About two-thirds of body water is intracellular and is found mainly in skeletal muscle, blood cells, bone cells, and adipose tissue. The remaining one-third of body water is extracellular and is found in plasma (about 25 percent) and in the interstitial fluid between cells (about 75 percent).

In healthy animals, a state of equilibrium exists between the amount of water taken in and the amount of water lost in normal physiologic processes. Fluid loss results from urinary, gastrointestinal, respiratory, and skin losses; fluid intake comes from ingestion of liquids and food and from metabolism. The kidneys are the primary regulators of the volume of water within the body. When body water is insufficient, urine volume diminishes and the animal becomes thirsty. When animals drink an excessive amount of water, their urinary output increases.

Body water contains *solutes*, substances that dissolve in a *solvent* (which is water in biological systems). *Electrolytes* are substances that split into ions (charged particles) when placed in water. *Cations* are positively charged ions and *anions* are negatively charged ions. The electrolytes found in intracellular fluid and extracellular fluid are essentially the same; however, their concentrations vary between these two compartments. In an effort to establish equilibrium, body water moves from a less concentrated solution (one with fewer solute particles per unit of solvent) to a more concentrated solution (more solute particles per unit of solvent). This is known as *movement along a concentration gradient*.

The primary ions in the body are sodium, potassium, chloride, phosphate, and bicarbonate. Sodium is the primary extracellular cation and potassium is the primary intracellular cation. Chloride is the primary extracellular anion and phosphate is the primary intracellular anion. Bicarbonate ions are also extracellular.

Electrolytes are balanced when the concentration of individual electrolytes in the body fluid compartments is normal and remains constant. Because electrolytes are dissolved in body fluids (mainly water), electrolyte and fluid balance are closely related. When fluid volume changes, the electrolyte concentration changes.

The primary regulation of electrolyte balance is through reabsorption of cations. Anions follow the cations; water follows the ions. Because sodium and potassium are the predominant cations, they are important in regulating electrolyte balance. If an

TIP

Fluid balance depends on electrolyte balance.

TIP

Remember that diffusion is the movement of molecules or solutes from an area of high concentration to an area of low concentration. Osmosis is the movement of water through a selectively permeable membrane from an area of lesser solute concentration to an area of greater solute concentration.

animal has an elevated sodium concentration in the extracellular fluid, osmotic pressure will increase, causing water to move from the intracellular compartment to the extracellular compartment. *Osmotic pressure* is the pressure or force that develops when two solutions of different concentrations are separated by a selectively permeable membrane. Think of osmotic pressure as the force that draws water across a selectively permeable membrane. To establish osmotic equilibrium, water moves from the less concentrated solution into the more concentrated solution.

Extracellular fluid shifts between the intravascular space (blood vessels) and the interstitial space (tissues) to maintain a fluid balance within the ECF compartment. Fluid exchange occurs only across the walls of capillaries, not across other blood vessels. The capillary membrane acts as a selectively permeable membrane by permitting free passage of crystalloids. **Crystalloids** are diffusible substances that dissolve in solution. The amount of osmotic pressure that develops at the selectively permeable membrane is due mainly to the concentration of nondiffusible substances or **colloids**. In addition, fluid flows only when there is a difference in pressure between the intravascular fluid and the interstitial fluid.

TIP

If the serum osmolality is not within the normal range, a fluid imbalance should be suspected.

Plasma osmolality is the ratio of solute to solvent (water) in the body. Cells are affected by the osmolality of the fluid that surrounds them. In the body, **osmolality** measures the number of dissolved particles regardless of their size per kilogram of water. Sodium is the largest contributor of particles to osmolality. Plasma osmolality is expressed in units called *osmols* or milliosmols per kilogram of water. A solution that has a higher concentration of solute will have a greater plasma osmolality.

FLUID THERAPY

In normal, healthy animals, fluid and electrolytes are balanced. In sick animals, this balance of fluid and electrolytes may be disrupted, making fluid therapy necessary.

TIP

One liter of water weighs one kilogram, so you can determine that a 100-lb adult dog (about 45 kg) has 60 percent or 27 kg of water in its body.

In healthy adult animals, approximately 60 percent of the body weight is water; in healthy neonates, approximately 80 percent of the body weight is water. Because neonates have a proportionally larger percent of body weight as water, dehydration is a significant problem in young animals.

The basis for fluid therapy rests on the animal's hydration status. Hydration status can be determined by assessing the patient's history, physical exam status (checking body weight, skin turgor, pulse rate and quality, capillary refill time, moistness of mucous membranes, and sunken eyes), and laboratory findings (such as TP (total protein), BUN (blood urea nitrogen), PCV (packed cell volume), and urine specific gravity). Table 19-1 provides guidelines for estimating level of dehydration.

TABLE 19-1 Estimating Level of Dehydration

Dehydration Percentage	Physical Exam Findings
<5%	History of vomiting or diarrhea, but no abnormalities noted on PE
5%	Dry mucous membranes
6–8%	Mild to moderate decrease in skin turgor; dry mucous membranes; slight tachycardia (increased heart rate)
10–12%	Marked decrease of skin turgor; dry mucous membranes; weak and rapid pulse; tachycardia; slow capillary refill time; sunken eyes; mild CNS depression

Fluid therapy can replace water, sodium, potassium, and chloride electrolytes, and restore hydrogen ion balance disrupted by ill health, disease, or trauma. Dehydration from water loss can result from systemic diseases (especially those causing *polyuria*, the formation and excretion of a large volume of urine), diarrhea, or excessive vomiting. Sodium deficiency (*hyponatremia*) results from reduced sodium intake or excessive sodium loss through urination. Potassium deficiency (*hypokalemia*) results from reduced intake or excessive loss of potassium in gastrointestinal fluids or urine. Chloride loss results from increased secretion and subsequent loss of gastric juice or intestinal fluids. Excessive loss of these electrolytes disturbs the balance of hydrogen ions. Fluid therapy, in addition to maintaining hydration and electrolytes, can deliver nutrients and medications.

HOW DO WE GET IT THERE?

Fluids can be administered by a variety of routes. Each route has its advantages and disadvantages. The main routes of fluid administration are listed here.

Oral (po) fluids are used for short-term illness and in small animals and neonates. The oral route is the safest route of fluid administration. Oral fluids can be administered by stomach tube, dosing syringe, bottle, nasogastric tube, or gastrostomy tube. Disadvantages of oral fluid administration include possible aspiration pneumonia, the inability to use them in vomiting animals, and less rapid absorption when compared to other methods. Oral fluids are indicated for animals that are anorexic or have diarrhea without vomiting, and for neonatal dehydration.

Subcutaneous (SQ) fluids are used to correct mild to moderate dehydration in non-critically ill patients. In dehydrated animals or animals in shock, peripheral vasoconstriction limits the distribution of fluids from the subcutaneous space to where they are needed. Isotonic fluids are used SQ (other solutions may cause skin sloughing) and they are usually administered by gravity flow through an 18- or 20-gauge needle. SQ fluids are given in a variety of locations, including the flank region and dorsally along the back between the scapulae. Many prefer to give SQ fluids in the flank region to allow more efficient drainage of fluid in case of infection. The volume of fluid that can be administered SQ is limited by the animal's skin elasticity and may range from 10 to 150 ml per site depending on the animal species being treated. Typically, 5 to 10 ml of fluid per pound of body weight is given per injection site in small animals. Animals differ in their ability to tolerate the infused load comfortably, and multiple sites may be needed to administer the total amount of fluids required. All SQ fluids tend to be absorbed in six to eight hours. Disadvantages of SQ fluid administration include the possibility of infection (especially when given dorsally along the back), subcutaneous edema, slower absorption rate than other routes, and the inability to use hypertonic, hypotonic, or irritating solutions.

Intravenous (IV) fluids are the preferred route for moderately to severely dehydrated and hypovolemic animals. IV routes are best for correcting hypotension because they provide rapid delivery at a precise dosage. Various tonicities of fluids can be used. An injection site is prepared using aseptic technique and a sterile IV catheter is seated in the vein. Sterile technique is then used in the administration of fluids or drugs through the IV catheter. Disadvantages of IV fluid therapy include the possibility of fluid overload, injury to and inflammation of the affected blood vessels, and potential extravascular placement of fluids. IV fluid administration requires close monitoring, asepsis, and catheter care.

Intraperitoneal (IP) fluids are given when IV access is not available. Isotonic fluids are used with 16- to 20-gauge needles. The injection site must be as aseptic as possible and is usually located just lateral to ventral midline, between the umbilicus and pelvis. Disadvantages of IP fluid administration include the possibility of sepsis, the inability to use IP routes in animals awaiting abdominal surgery, and the inability to use hypertonic solutions. Relatively rapid absorption is obtained with IP fluid administration.

Intraosseous (IO) fluids are particularly useful in small animals, birds, and pocket pets because they provide direct access to the vascular space. It is also useful when access to a vein is compromised. IO fluids are given via the bone marrow, using sterile technique, through a needle proportionally sized to bone size. A bone marrow needle and stylet are used for mature animals and a spinal needle and stylet are used for younger animals. IO fluids are usually given in the femur (through a site prepared over the trochanteric fossa) or in the humerus (through a site prepared over the greater tubercle). Fluids, whole blood, plasma, and/or drugs can be given IO. IO fluids are absorbed rapidly. Many veterinarians do not practice this route of administration often, so they do not possess confidence in their skills and may be reluctant to try it. Other disadvantages include the possibility of bone infection and the need to avoid growth plates.

Rectally (pr) administered fluids may be a good route of fluid administration in young animals, unless diarrhea is present. Electrolyte absorption is good and absorption is rapid, but this route is not commonly utilized.

WHAT CAN WE GIVE?

Crystalloids and colloids are the two categories of fluids used in fluid therapy. *Crystalloids* are sodium-based electrolyte solutions or solutions of glucose in water that are commonly used to replace lost fluid and electrolytes. The composition is similar to plasma fluid. Crystalloids are further described by their tonicity and are categorized as isotonic, hypotonic, or hypertonic.

Tonicity is based on a measurement called osmolality. Recall that osmolality is the osmotic pressure of a solution based on the number of particles per kilogram of solution. **Osmotic pressure** is the ability of **solutes** (particles) to attract water (causing osmosis). Not all particles contribute to osmolality. Sodium and glucose provide most of the particles to determine osmotic pressure. Normal osmolality of blood and extracellular fluid is 290 to 310 mOsm/kg. The osmolality of **isotonic** solutions is the same as blood and extracellular water; thus isotonic solutions produce no significant changes in the blood. **Hypotonic** solutions have osmolality lower than that of blood and can cause the red blood cells to swell. However, in proper concentrations hypotonic solutions can shift fluid out of the intravascular space without swelling the red blood cells. The osmolality of **hypertonic** solutions is greater than that of blood and can cause the red blood cells to shrink. However, in proper concentrations hypertonic solutions can shift fluid into the intravascular space without shrinking the red blood cells. Figure 19-1 summarizes solution tonicity.

Isotonic fluids have the same sodium concentration as the extracellular fluid that distributes and expands the extracellular space. About one-quarter to one-third of the total fluid volume infused will remain in the vascular space with isotonic fluid. Examples of isotonic fluid include 0.9% sodium chloride (also known as *isotonic saline*, *normal saline* (NS), or *physiologic saline solution* (PSS)), lactated Ringer's solution, Normosol®, and Plasmalyte®.

Isotonic saline (0.9% sodium chloride) contains only sodium and chloride ions and is used to expand plasma volume and to correct hyponatremia (decreased sodium levels) or metabolic alkalosis. Because of its sodium content, saline should not be used in patients with heart failure or those that have sodium retention due to liver disease. The fluid of choice in these cases is 0.45% saline. Saline solutions may or may not contain dextrose.

Lactated Ringer's solution (LRS) is the fluid of choice in many disease situations. LRS is a saline and lactate solution with electrolytes added. LRS contains sodium, potassium, chloride, calcium, and lactate ions. It has a reduced sodium con-

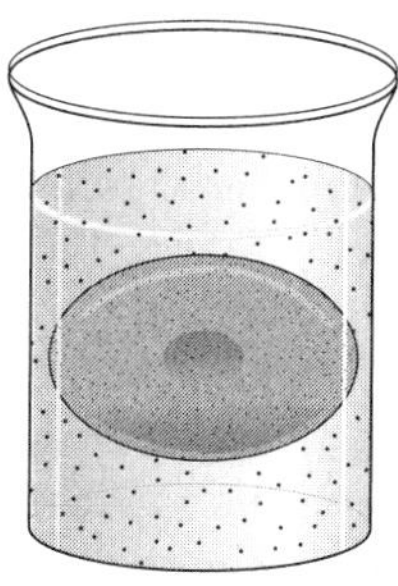

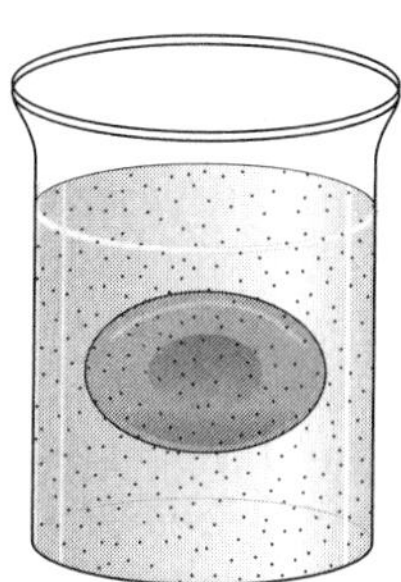

FIGURE 19-1 Solution tonicity. Blood cells placed in hypertonic solution lose fluid in an attempt to equalize the osmolality in the cell to the solution. Blood cells placed in hypertonic solution gain fluid in an attempt to equalize the osmolality in the cell to the solution. Blood cells placed in isotonic solution have neither a net gain or loss of fluid.

tent as compared to 0.9% sodium chloride. The lactate molecule found in LRS is broken down in the liver to bicarbonate and is helpful in the treatment of acidosis. Acidosis is seen with severe dehydration, because as cellular breakdown occurs, acid metabolites (such as lactic acid) are released from cells.

Normosol® is a solution with less sodium, more potassium, more magnesium, less chloride, and no calcium as compared to LRS. It is an all-purpose replacement fluid and has acetate as a buffer, rather than the lactate used in LRS.

Plasmalyte® has less chloride, more magnesium, and no calcium as compared to LRS. Plasmalyte® uses acetate as a base that is broken down in skeletal muscle to bicarbonate. Plasmalyte® cannot be given SQ because it is irritating to tissues.

Hypotonic crystalloid solutions are fluids with less sodium concentration than extracellular fluid; thus, they dilute the extracellular sodium and cause a portion of the fluid to move into the intracellular space. Distribution of hypotonic fluid leaves less fluid in the vascular space. An example of hypotonic solution is 5% dextrose in water (D_5W), which is given IV and divides between the intracellular and extracellular fluid compartments. Two-thirds of the fluid goes to intracellular fluid and one-third goes to the extracellular fluid. D_5W is used to treat hypernatremia, as a carbohydrate source, and as a fluid supplement in patients that cannot tolerate sodium. It should not be given SQ because extracellular fluid already in the body will move toward the D_5W, increasing fluid depletion and possibly causing shock. Other examples of hypotonic solutions include ¼ NS (0.25% normal saline) and ½ NS (0.45% normal saline).

Hypertonic fluids have high sodium concentrations that will increase the sodium concentration of the extracellular fluid and cause water to move out of the cells into the extracellular space. Examples of hypertonic solutions are 0.9% normal saline with 5% dextrose, 10% dextrose in water, and 3% normal saline. Hypertonic solutions can move fluid into the intravascular space, thus replacing fluid volume rapidly within blood vessels. Hypertonic solutions can aid in the treatment of shock and edema reduction.

Colloid solutions are fluids with large molecules that enhance the oncotic force of blood, causing fluid to move from the interstitial and intracellular spaces into the vascular space. Colloids do not diffuse across cell membranes, because of their large size. Colloids are used for vascular space expansion in treating hypovolemic shock (permitting peripheral blood flow) and for treating severe chronic disease in which hypovolemia and hypoproteinemia are seen. Natural colloids include plasma, albumin, and whole blood. Synthetic plasma substitutes include dextrans, hydroxyethyl starch, and oxypolygelatin.

Whole blood is commonly used to treat severe anemia and cases of severe blood loss. Whole blood contains all the cellular (RBC, WBC, and platelets) and plasma components of blood. It provides RBCs for carrying oxygen to tissues and plasma proteins for oncotic volume expansion. If used within 8 to 12 hours of collection, coagulation factors and platelets may still be viable in whole blood.

TIP

Crystalloid solutions can be isotonic (have about the same sodium chloride concentration as blood), hypotonic (have less sodium chloride concentration than blood), or hypertonic (have greater sodium chloride concentration than blood).

The *hematocrit*, also known as the *packed cell volume*, is the volume of red blood cells in proportion to the intravascular fluid volume, expressed as a percentage. It is one indication of fluid loss or gain. A decreased hematocrit reading may be due to blood loss, increased RBC destruction, or failure to produce RBCs. Causes of decreased hematocrit readings include anemia of hemorrhage, nonregenerative anemia, autoimmune hemolytic anemia, hypoproteinemia, coagulopathies (clotting abnormalities), and liver disease. These causes of low hematocrit readings are indications for whole blood use (transfusions). An increased hematocrit reading can indicate extracellular fluid loss because the total number of RBCs is not changing, yet the proportion of RBCs to extracellular fluid is increasing. Therefore, the amount of extracellular fluid must be decreasing. In these cases, use of whole blood is not indicated.

TIP

Colloids are frequently called volume expanders, because they function like plasma proteins in blood and help maintain oncotic pressure.

Whole blood is collected and stored in any of several anticoagulants, including citrate phosphate dextrose adenine (CPSA-1), acid citrate dextrose (ACD), storage medium for blood (SMB), and heparin. Whole blood can be collected in glass vacuum bottles, plastic bags, and plastic syringes. Glass is inert to stored blood, yet is breakable, more expensive, and requires more storage space than plastic bags. It is usually recommended that blood not be stored for more than 20 days. Blood group typing and testing of donors for infectious diseases transmissible via blood is recommended. Blood is usually collected via the jugular vein, although the femoral artery and cardiac collection are also used. Blood is typically given using a 20- or 23-gauge indwelling catheter. As a guide, 2.2 ml/kg of blood raises the packed cell volume by 1 percent when the packed cell volume of the transfused blood is 40 percent.

Plasma is a natural colloid solution that is easy to collect and can be stored frozen for long periods of time (for years if at –40 to –70°C). Plasma contains albumin and globulins (important plasma proteins) that can aid in the treatment of liver disease and diseases causing hypoproteinemia. Plasma can also be used for volume replacement. Fresh frozen plasma is prepared by separating it and freezing it within a few hours after whole blood collection. Fresh plasma and fresh frozen plasma, when used as a source of coagulation factors, can be given at a rate of 6 to 10 ml/kg IV at least 3 times daily for control of bleeding caused by coagulopathies. Frozen plasma should be used within six hours of thawing to maintain the clotting proteins.

Albumin is the main protein in the blood and constitutes about 50 percent of the blood protein. The main function of albumin is to maintain the colloid osmotic pressure of blood. Without albumin (and its role in maintaining colloid osmotic pressure), fluid would accumulate in the tissues and swelling would occur. Albumin concentrate is helpful in restoring body protein and in expanding the plasma volume. Too much albumin or albumin administered too rapidly can cause fluid to be retained in the vessels of the lung.

Dextran, in saline or dextrose, is a synthetic colloid solution with large polysaccharides derived from sugar beets. Dextran comes in two concentrations, 40 and 70. The stronger concentration is a colloid hypertonic solution. Dextran can be used to treat cases of shock, but its use is limited by allergic reactions and clotting problems seen in animals. To restore blood volume, dextran 40 is given at a rate of 0.7 g/lb/day and dextran 70 is given at a rate of 0.9 g/lb/day.

Hydroxyethyl starch or *hetastarch* (Hespan®) is a synthetic colloid that expands plasma volume and lasts longer, with fewer side effects, than dextrans. Hetastarch is a combination of hydroxyethyl starch and normal saline and is used to treat hypovolemic shock and hypoproteinemia. The rate of hetastarch administration is 10 to 20 ml/kg/day. Possible side effects include allergic reactions and coagulopathies. It is hypertonic and must be administered by slow infusion to avoid rapid fluid shifts.

Oxypolygelatin (Rapidvet Plasm.ex®) is a gelatin suspension in sodium chloride. Oxypolygelatin is used IV in companion animals to reestablish and maintain circulatory equilibrium following hemorrhagic shock, hypovolemia without blood loss, burns, and for shock prevention following surgery. Oxypolygelatin is stable at room temperature for at least four years and under extreme conditions for at least one year.

The benefits of this product are that it does not affect coagulation or platelets. Side effects are rare, but may include anaphylaxis during and after infusion, hives, and cardiovascular overload.

Hemoglobin glutamer-200 (Oxyglobin Solution®) is a hemoglobin-based, oxygen-carrying fluid that increases plasma and total hemoglobin (Hb) concentration, resulting in increases in arterial oxygen content. The hemoglobin in this product distributes oxygen via the plasma instead of the red blood cells. Following infusion of hemoglobin glutamer-200, the plasma and total hemoglobin concentrations increase, but the hematocrit may decrease because of the hemodilution resulting from this product's colloidal properties. Therefore, special considerations are taken into account when assessing oxygen-carrying capacity. The standard hemoglobin-to-hematocrit ratio of 1:3 is not accurate following administration of hemoglobin glutamer-200. A theoretical hematocrit is calculated based on the following formula:

theoretical Hct = Hb from RBCs (1/3 Hct) + plasma Hb) × 3.

Oxyglobin Solution® is used as a one-time dose IV for the treatment of anemia in dogs. Oxyglobin® use does not require cross-matching of blood type prior to administration. This product should be administered via a dedicated IV line and should not be shaken prior to use (shaking results in foaming of the product). It is compatible with any other IV fluid, but it should not be combined with other products in its bag. Oxyglobin Solution® is stable for up to 36 months when stored at 2° to 30°C, but should be used within 24 hours after its foil overwrap is opened. Animals should be adequately hydrated prior to administration of Oxyglobin Solution®. Side effects include mucous membrane and urine discoloration, vomiting, circulatory overload (signs include coughing, dyspnea (difficulty breathing), and pulmonary edema), and fever. Oxyglobin Solution® is contraindicated in dogs with congestive heart failure.

WHAT CAN WE ADD?

Special additives in the crystalloid fluids described earlier may help improve patient recovery and reversal of disease signs. When adding additives to the fluid bag, it is important to remember to withdraw and discard an amount of fluid equal to the amount of additive being supplemented. Some additives may precipitate with some types of fluids, so read the product inserts before you add supplements to fluid bags. The following additives are commonly added to fluids.

Fifty percent dextrose. When a patient is prone to hypoglycemia, dextrose may be added to the fluids. Dextrose supplementation may be needed by patients with increased metabolic needs, such as anorexic patients with sepsis or hyperthyroidism. Dextrose is not added to the fluids as a calorie source, but serves as an energy source for the brain. Dextrose is typically added to fluids to make a 2.5 to 5 percent solution. If adding 50% dextrose to a 1000 ml bag of fluids to make a 2.5 percent solution, 50 ml of 50% dextrose must be added (remember to withdraw 50 ml of fluid from the bag first). If adding 50% dextrose to a 1000 ml bag of fluids to make a 5 percent solution, 100 ml of 50% dextrose must be added. A solution containing 5% dextrose should be administered into a central vein via a large-bore catheter.

Potassium. Potassium is usually supplemented to anorexic and/or diuretic patients. Potassium is obtained from food and is easily lost in urine. Hypokalemia may result in lethargy, muscle weakness, and vomiting. Potassium is given IV via a slow drip to avoid cardiac problems. The normal serum potassium level is 3.5 to 5.5 mEq/L. The term milliequivalents (mEq) is used to express the number of ionic charges of each electrolyte on an equal basis. It measures the chemical activity of ions. A milliequivalent is one one-thousandth of an equivalent, which is the weight (g) of an element that will combine with 1 g of H^+. Milliequivalents are based on the molecular weight, valence number, and milligrams of element. When calculating the amount of potassium to give, use mEq/ml as a concentration to figure out the dose. The supplement used to add potassium is KCl and is available in a concentration of 2 mEq/ml. Table 19-2 can be used as a guide to potassium supplementation.

Sodium bicarbonate. Sodium bicarbonate is added to fluids to correct acidosis, as sodium bicarbonate has basic properties. Sodium bicarbonate is available in 8 percent (1 mEq/ml) and 5 percent (0.6 mEq/ml) concentrations. The amount of sodium bicarbonate added is based on the bicarbonate deficit. The bicarbonate deficit is determined by subtracting the normal serum bicarbonate value (24 mEq/L) from the patient's serum bicarbonate value. The bicarbonate deficit is then multiplied by 0.6 and the animal's weight (kg), which yields the number of mEq of sodium bicarbonate to administer. Side effects with sodium bicarbonate administration include development of alkalosis and other electrolyte abnormalities. Replacement should be given slowly over several hours to avoid side effects.

Calcium. Calcium is given to patients with hypocalcemia due to diseases such as milk fever, eclampsia, and endocrine disorders. Calcium may be in the form of calcium gluconate or calcium chloride, or in combination with other electrolytes such as magnesium, potassium, phosphorus, and dextrose. Calcium is available as calcium gluconate 10 percent for injection, calcium gluconate 23 percent (for large animals), calcium chloride 10 percent for injection, and combination products like Cal-Dextro® and Norcalciphos®. Calcium supplementation is calculated using specific dosages and usually given to effect. Cardiac and respiratory rate and rhythm should be monitored when giving calcium.

Vitamins. Water-soluble vitamins may be added to fluids because they are lost rapidly by anorexic or debilitated animals. Vitamin C and B complex are frequently added to fluids. In general, 0.5 ml of B complex is given per 250 ml of fluid. This dose may be doubled in severely debilitated animals. B vitamins can cause pain at the injection site if given SQ, so proper restraint is important when administering fluids containing B vitamins. Vitamin C may also be added to the fluids. It is light sensitive and does not have as long a shelf life as some of the other vitamins.

TABLE 19-2 Guide to Potassium Supplementation in Animals

Serum Potassium Levels	Amount of Potassium to Add to a 250 ml Fluid Bag
3.5–5.5 (normal)	5 mEq
3.0–3.4	7 mEq
2.5–2.9	10 mEq
2.0–2.4	15 mEq
<2.0	20 mEQ

HOW MUCH DO WE GIVE?

When calculating fluid replacement therapy, keep in mind that animals require fluids for the following:

- **rehydration** (to correct body water loss due to dehydration).
- **maintenance** (to replace body water lost via normal body functions).
- replacement of **ongoing fluid losses** (to replace body water lost through vomiting and diarrhea).

Rehydration Volumes

Dehydration in animals can cause many health problems. Things to remember with dehydration include:

- Dehydration affects younger animals much more rapidly than older animals.
- Older patients with chronic disease require more fluids than other animals.
- Animals need more fluids if they are active or if the weather is hot or humid.
- Drugs such as corticosteroids and diuretics will alter fluid and electrolyte requirements.
- Animals that have been anaesthetized may require additional water for a few days.

Clinically, the amount of fluid needed to correct dehydration deficits can be determined from the degree of skin turgor, capillary refill time (CRT), and pulse rate and quality. These parameters were listed in Table 19-1.

The amount of fluid needed to rehydrate an animal is based on the estimated percent of dehydration (the amount by which the animal is dehydrated). To calculate this value, you take the estimated percent dehydration and multiply it times the weight of the animal in kilograms. This will give you deficit in liters. If you want the value in milliliters, take the estimated percent dehydration, multiply it times the animal's weight in kilograms, and then multiply it by 1000. When you do this calculation, the percent dehydration should be in decimal form (for example, 10% = 0.10, 5% = 0.05).

It is difficult to replace all hydration deficits in a 24-hour period. If all of the fluid loss due to dehydration is replaced in 24 hours, urinary output may increase in an attempt to compensate for the large volume of fluid being given. These urinary losses could cause further dehydration in the animal. Therefore, it is recommended that only 75 to 80 percent of the rehydration volume be given during the first 24 hours.

Maintenance Fluid Volumes

Maintenance fluid is the volume of fluid needed by the animal on a daily basis to maintain body function. Maintenance volumes can be divided into sensible and insensible losses. Sensible losses, which are body water lost in urine and feces, can be measured. Insensible body water losses are normal but not easily measured; such losses occur through sweating, ventilation, and mucous membrane evaporation.

Many different values are used for the determination of maintenance fluid volumes; therefore, consulting with the veterinarian about the preferred volume is necessary. One value typically used is 50 ml/kg/day in adult animals and 110 ml/kg/day in young animals. These values include both sensible and insensible body water losses. These values are multiplied by the animal's weight in kg to determine the volume needed in ml.

Ongoing Fluid Loss Volumes

Animals that are losing additional fluid amounts due to vomiting or diarrhea need to have this fluid loss replaced. This number is determined by estimation. If you have a vomiting dog in the clinic, you can estimate the amount of fluid loss by monitoring the quantity and frequency of vomiting episodes. This number is then added to the other numbers described earlier.

EXAMPLES OF FLUID CALCULATIONS

The following are some fluid calculation examples to illustrate the above descriptions of fluid replacement therapy.

Example 1: An adult dog weighing 50 lb needs maintenance fluids. Calculate the amount of fluids this dog needs per 24 hours.

Step 1: Convert 50 lb to kg.

50 lb ÷ 2.2 lb/kg = 22.7 kg

Step 2: Multiply the weight in kg by the maintenance fluid value (in this case, use 50 ml/kg/day).

22.7 kg × 50 ml/kg/day = 1136 ml per day

Example 2: An adult 14 lb cat with 3 percent dehydration (mild to no evidence of clinical dehydration, but a history of fluid loss) comes into the clinic. The cat is to be kept npo (nothing by mouth). Calculate a fluid dose for this cat.

Step 1: Convert 14 lb to kg.

14 lb ÷ 2.2 lb/kg = 6.4 kg

Step 2: Multiply the weight in kg by the maintenance fluid value (in this case, use 50 ml/kg/day).

6.4 kg × 50 ml/kg/day = 318 ml per day

Step 3: Calculate replacement for dehydration.

3% = 0.03
0.03 × 6.4 kg = 0.192 L
0.192 L × 1000 ml/L = 192 ml
192 ml × 0.8 (decimal value of 80%) = 154 ml to replace on day 1
Remember only 75% – 80% of the rehydration volume is given during the first 24 hours.

Step 4: Add all values together to determine total daily fluids for this cat.

318 ml + 154 ml = 472 ml of fluid for day 1

Example 3: The cat in example 2 stays at your clinic overnight and begins vomiting. You estimate that the cat vomited about 100 ml over the evening. Recalculate this cat's fluid needs for the next day, assuming that the cat is still 3 percent dehydrated.

Step 1: Take the amount of fluid calculated in example 2 and add the volume lost through vomiting.

472 ml + 100 ml = 572 ml of fluid for day 2

Example 4: A mature, 450 kg horse with severe obstructive colic is 8 percent clinically dehydrated when presented to your clinic. This horse needs fluids stat (immediately). Calculate a daily fluid replacement volume for this horse.

Step 1: The animal's weight is already in kg, so no conversion is needed.

Step 2: Multiply the animal's weight in kg by the maintenance fluid value (in this case, use 50 ml/kg/day)

$$450 \text{ kg} \times 50 \text{ ml/kg/day} = 22{,}500 \text{ ml}$$

Step 3: Calculate replacement for dehydration.

$$8\% = 0.08$$
$$0.08 \times 450 \text{ kg} = 36 \text{ l}$$
$$36 \text{ l} \times 1000 \text{ ml/l} = 36{,}000 \text{ ml}$$
$$36{,}000 \text{ ml} \times 0.8 = 28{,}800 \text{ ml}$$

Step 4: Add all calculated values to determine total daily fluids for this animal.

$$22{,}500 \text{ ml} + 28{,}800 \text{ ml} = 51{,}300 \text{ ml or } 51.3 \text{ l}$$

HOW FAST DO WE GIVE IT?

The rate of fluid replacement parallels the severity of dehydration. In a clinical setting, the rate fluid replacement tends to be based on clinical judgment, but some general rules apply.

- Fluids should be replaced rapidly at first, especially in cases of shock, and then tapered to a maintenance dose.
- Rate of replacement of the deficit depends on the cardiovascular and renal status of the patient. If these two systems are not functioning properly, the rate of replacement may have to be decreased so as to not overload these systems.
- Fluid input should not necessarily equal output initially; transient imbalance is frequently the optimal course of therapy. Hour-by-hour evaluation is necessary to adjust fluid rates until the patient is stable.

Ideally, fluids should be given over a 24-hour period. Fluids are administered by fluid administration sets that deliver a constant number of drops/ml of fluid. With **adult administration sets**, this value is typically 15 gtt/ml. With **pediatric administration sets**, this value is typically 60 gtt/ml. Always check the administration set you are using, because the number of drops per ml can vary with the manufacturer (some are 10 gtt/ml, others are 20 gtt/ml). Using the preceding examples, calculate the drip rate.

Example 1: We calculated that this dog needs 1136 ml of fluid per day.

Step 1: If we use an adult administration set that delivers 15 gtt/ml, we take this value and multiply it times the fluids needed per day.

$$1136 \text{ ml/24 hrs} \times 15 \text{ gtt/ml} = 17{,}040 \text{ gtt/24 hrs}$$

It is difficult to count drops for 24 hours; therefore, hours should be converted to minutes

Step 2: Convert the value for 24 hours to minutes.

$$17{,}040 \text{ gtt/24 hrs} \times 1 \text{ hr/60 min} = 17{,}040 \text{ gtt/1440 min}$$

Step 3: Take the two values and divide to get drops per minute.

$17{,}040 \text{ gtt} \div 1440 \text{ min} = 11.8 \text{ gtt/min}$ or (rounded up) 12 gtt/min

These values can be reduced to time fractions that you feel comfortable with. For example, 12 gtt/min = 6 gtt/30 sec = 3 gtt/15 sec

Example 2: This cat needs 472 ml of fluid per day. Using a pediatric drip set, calculate the fluid drip rate.

Step 1: If we use a pediatric administration set that delivers 60 gtt/ml, we take this value and multiply it times the fluids needed per day.

$472 \text{ ml}/24 \text{ hrs} \times 60 \text{ gtt/ml} = 28{,}320 \text{ gtt}/24 \text{ hrs}$

Step 2: Convert the value for 24 hours to minutes.

$30{,}600 \text{ gtt}/24 \text{ hrs} \times 1 \text{ hr}/60 \text{ min} = 28{,}320 \text{ gtt}/1440 \text{ min}$

Step 3: Take the two values and divide to get drops per minute.

$28{,}320 \text{ gtt} \div 1440 \text{ min} = 19.7 \text{ gtt/min}$ or (rounded up) 20 gtt/min

The drops in the pediatric administration set are smaller, so we need more of them to deliver 1 ml.

Example 3: This cat needs 572 ml per day. Use a pediatric administration set that delivers 60 gtt/ml to calculate the fluid drip rate.

Step 1: Take this value and multiply it times the fluids needed per day.

$572 \text{ ml}/24 \text{ hrs} \times 60 \text{ gtt/ml} = 34{,}320 \text{ gtt}/24 \text{ hrs}$

Step 2: Convert the value for 24 hours to minutes.

$34{,}320 \text{ gtt}/24 \text{ hrs} \times 1 \text{ hr}/60 \text{ min} = 34{,}320 \text{ gtt}/1440 \text{ min}$

Step 3: Take the two values and divide to get drops per minute.

$34{,}320 \text{ gtt} \div 1440 \text{ min} = 23.8 \text{ gtt/min}$ or (rounded up) 24 gtt/min

Example 4: This horse needs 51,300 ml per day. Use an adult administration set that delivers 15 gtt/ml to calculate the fluid drip rate.

Step 1: Take this value and multiply it times the fluids needed per day.

$51{,}300 \text{ ml}/24 \text{ hrs} \times 15 \text{ gtt/ml} = 769{,}500 \text{ gtt}/24 \text{ hrs}$

Step 2: Convert the value for 24 hours to minutes.

$769{,}500 \text{ gtt}/24 \text{ hrs} \times 1 \text{ hr}/60 \text{ min} = 769{,}500 \text{ gtt}/1440 \text{ min}$

Step 3: Take the two values and divide to get drops per minute.

$769{,}500 \text{ gtt} \div 1440 \text{ min} = 534.4 \text{ gtt/min}$ or (rounded down) 534 gtt/min

Step 4: Take the value per minute and divide by 60 to get drops per second.

$534.4 \text{ gtt/min} \times 1 \text{ min}/60 \text{ sec} = 8.9 \text{ gtt/sec}$ or (rounded up) 9 gtt/sec

This is still pretty difficult to measure, but is more manageable than 534 gtt/min. Estimation will be needed in this case.

Additional problems on fluid volumes and rates of administration can be found in Chapter 10 (IV Calculations) on the Accu-Calc CD®.

WHAT DO WE USE TO GIVE IT?

Fluids are administered by fluid bags and bottles attached to administration sets or by fluid infusion pumps. Fluid bags and bottles deliver fluids by gravity, and the rate can be adjusted by the diameter of the administration line delivering the fluids. Administration sets may have roller clamps that can be adjusted to increase or decrease the amount of fluid delivered. Administration sets may also have screw clamps or slide clamps to control the diameter of the administration line. *Infusion pumps* are machines on which flow rates are set and the total amount to be given is entered. These pumps then give the desired amount of fluid at the desired rate. They can be readjusted when fluid delivery is either too fast or too slow.

KEEPING WATCH

Fluid administration should be monitored to make sure the animal is not getting too much or too little fluid for its needs. Additionally, fluid needs change with the changing health status of the animal. Physical findings such as nasal secretions (increased serous secretions indicate too much fluid), heart rate (tachycardia indicates too much fluid), lung sounds (harsh lung sounds indicate edema from too much fluid), and demeanor of the animal (restlessness may indicate too much fluid) should be monitored regularly. Laboratory values such as hematocrit and total protein can be measured in the clinic to make sure the animal's hydration status is satisfactory.

The amount of fluid given can be monitored in a variety of ways. Fluid bags and bottles have milliliter increments on them that should be monitored to make sure the animal is receiving the correct amount of fluid over time. White tape may be applied to the bags/bottles to keep track of fluid volume delivered per hour, which is extremely helpful if more than one person is monitoring the fluids. Some administration sets have volume control chambers to allow easier measurement of fluid volumes. Some chambers can be filled with the exact amount of fluid to be delivered for that treatment; when the chamber is empty, the animal has received that volume. Examples of fluid bags and administration sets are shown in Figure 19-2A, B, and C.

After the patient has been rehydrated and seems to be improving, the fluid plan is readjusted and only maintenance fluids may be given. Eventually all fluid administration will be stopped, but abruptly stopping fluids may lead to the redevelopment of dehydration due to the increased urine production still being performed by the kidneys. Tapering the fluids allows the animal's body to adjust to the decrease in fluids provided for it.

EMERGENCY DRUGS

A variety of conditions are labeled emergencies, and they cannot all be covered in this chapter. This section does cover dogs and cats brought into a clinic for primary respiratory or cardiac arrest. Cardiac arrest in dogs and cats is usually secondary to a cardiac arrhythmia, so an ECG must be set up. Time is important, and seconds can make a difference, so a plan and emergency work area should be part of all clinic set-ups.

When presented with an animal in respiratory or cardiac arrest, keep in mind the basic life support ABCs:

A = establish *airway*

B = *breathe* for the animal

C = maintain *circulation* with thoracic compressions and IV fluids

The goal of emergency treatment is to maintain adequate oxygenation of vital organs. Restoring ventilation and correcting tissue hypoxia and acidosis are key factors in helping animals survive emergency situations.

TIP

When making first contact with the owner of an animal that needs emergency care, obtain the following information:

- nature of the illness/injury
- condition of the animal
- time injury/event occurred or was noticed
- any preexisting illness
- age, breed, sex, and weight of animal, if available

TIP

Always wear gloves when handling an animal that is bleeding. It is difficult to assess whether the blood is from the animal or from a person who may have been bitten or scratched while helping the animal.

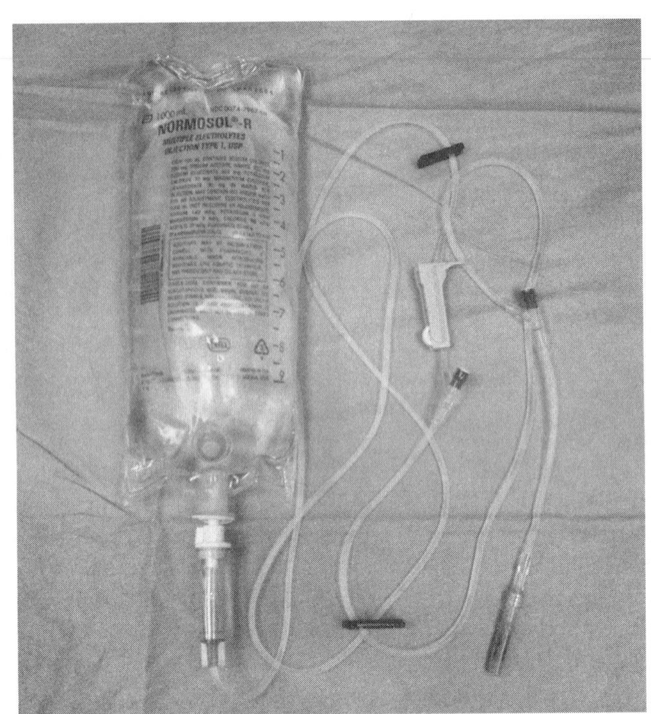

FIGURE 19-2A Fluid bags and administration sets
Photo courtesy of Linda Kratochwill, DVM

ADMINISTRATION THROUGH A SPECIAL ADMINISTRATION CHAMBER

(e.g., Soluset® or Buretrol®)

1. Wash your hands.
2. Follow the manufacturer's directions for priming the setup. After priming the setup, clamp the administration tubing below the drip chamber.
3. Allow 10–15 ml of the fluid being administered intravenously to flow into the drug administration chamber.
4. Close the clamp between the bag and administration chamber.
5. Cleanse the injection site on the administration chamber with alcohol.
6. Inject the medication to be administered into the chamber.
7. Open the clamp between the bag and drug administration chamber and add the appropriate amount of fluid to the administration chamber.
8. Clamp the tubing above the administration chamber.
9. Gently agitate the drug administration chamber to mix the fluids.
10. Open the clamp below the chamber.
11. Establish the flow rate appropriate to permit administration of the required amount of medication within the specified time period.
12. Once the medication has been administered, open the clamp above the administration chamber to resume administration of the fluid as ordered.
13. Chart the procedure including the date, time, medication, dosage, amount of fluid inflused, and client's reaction to the procedure.

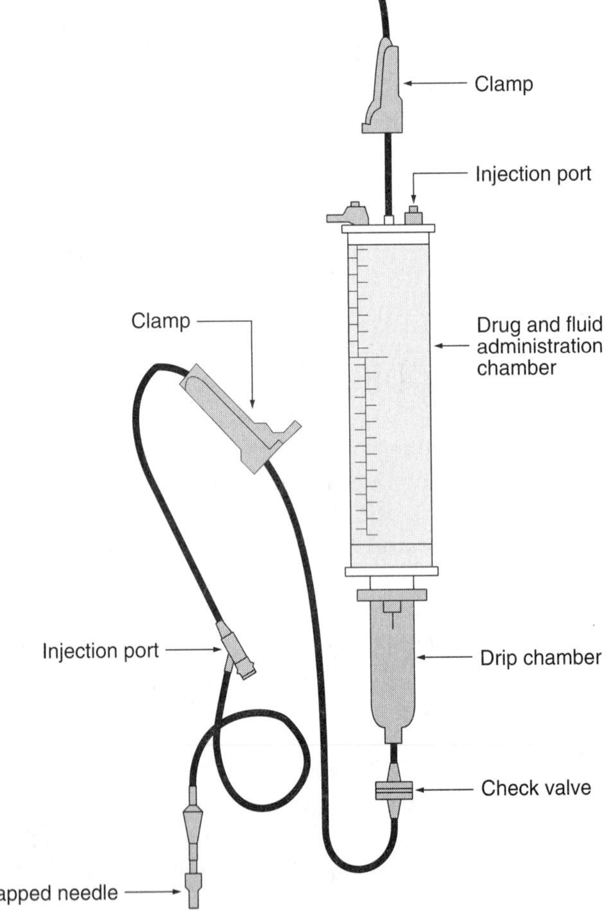

FIGURE 19-2B A volume control set

A = Airway

Tissue hypoxia and acidosis are minimized if ventilatory support is begun immediately. Establishing an airway may include passing an endotracheal tube, suctioning to clear an airway, pulling the tongue out and extending the head and neck, or performing a tracheostomy. The airway must be free of blood, vomit, dirt, and mucus.

B = Breathing

It is best to deliver 100 percent oxygen by positive-pressure ventilation via an endotracheal tube. Manual artificial resuscitators (such as Ambur® bags) deliver room air, which is only about 21 percent oxygen. Mouth-to-tube delivery is only about 17 percent oxygen. Initially, the rate of oxygen delivery should include two breaths of one to one and a half seconds duration to see if spontaneous respiration will begin. If spontaneous respiration does not occur, artificial respiration is initiated. Oxygen is usually delivered at a rate of 25 to 30 breaths/minute, which is a hyperventilatory rate. The ventilation is continued simultaneously with every three to five chest compressions, without a pause in the compression procedure. The volume of oxygen delivered should produce a normal chest expansion.

C = Circulation

If there is no pulse, external cardiac compression is begun. Small animals weighing less than 15 kilograms are placed in lateral recumbency and compressions are done

TIP

The term positive pressure ventilation covers all methods of providing O_2 to a patient. "Bagging an animal," either with a manual artificial resuscitator or anesthetic machine, provides positive pressure during inhalation. Continuous positive airway pressure (CPAP) provides positive pressure during inhalation and exhalation. Positive end expiratory pressure (PEEP) provides positive pressure during exhalation.

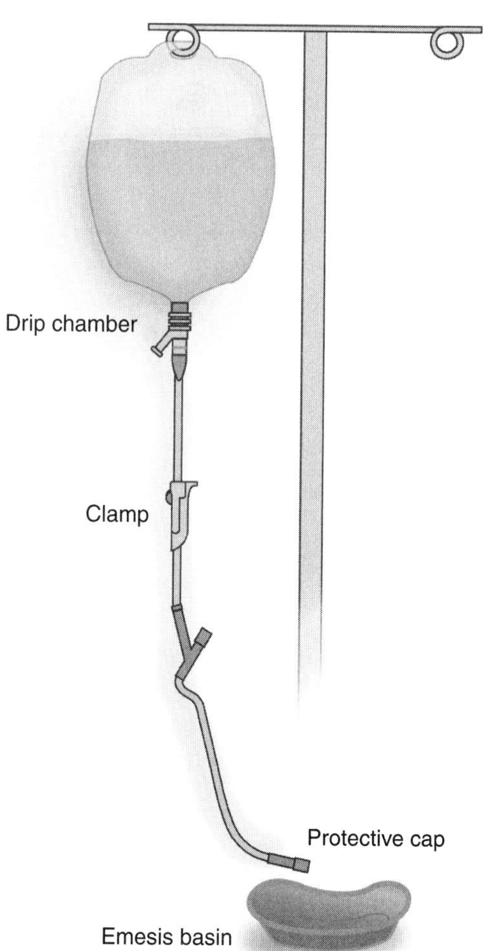

FIGURE 19-2C Priming the intravenous infusion equipment To prepare the administration set for IV infusion, close all clamps, fill half of the drip chamber, remove the protective cap, release the clamps, and allow the solution to clear all air from the tubing. Be certain that all air has been removed from the tubing. Then reclamp the tubing and replace the cap.

over the heart (at about the fourth to fifth intercostal space at the chondrocartilage junction). Large dogs of more than 15 kilograms are placed in dorsal recumbency and compressions are done over the sternum at the widest point. The rate of compressions is usually about 80 to 120 compression per minute, with enough force to displace the chest wall by about 30 percent. If CPR fails to produce a peripheral pulse, blood flow can be augmented by internal cardiac compression, abdominal/pelvic wraps, and IV fluids.

If the preceding methods do not appear to be working successfully, the use of drugs is initiated. These drugs should be given intravenously, intratracheally, or intraosseously. ECG monitoring and blood gas determination, if available, should be done to monitor drug therapy. Emergency drugs include, but are not limited, to those listed in Table 19-3.

Cardiovascular and pulmonary function must be closely monitored for several hours following a successful resuscitation to make sure that cardiac arrest does not recur. Complications of cardiopulmonary resuscitation can cause the patient additional health concerns and must be addressed as they arise.

SUMMARY

Body fluids make up about 60 percent of an adult animal's body weight (in neonates, about 80 percent of body weight is fluid). Body water is divided into intracellular (within cells) and extracellular (in plasma and interstitial fluid).

TABLE 19-3 Emergency Drugs

Drug	When Used	Dosage
epinephrine	If no heartbeat	0.2 mg/kg IV 0.4 mg/kg IV
atropine	If slow heartbeat	0.01–0.02 mg/kg IV
sodium bicarbonate	With prolonged CPR (>15 minutes) to minimalize acidosis	0.5–1 mEq/kg IV every 10 minutes for a maxium of 2 doses
dexamethasone	With prolonged CPR or unconscious >30 minutes, or suspected cerebral edema	4 mg/kg IV
prednisone sodium succinate	With prolonged CPR or unconscious >30 minutes, or suspected cerebral edema	30 mg/kg IV, repeat at 2 hours at 15 mg/kg IV, then every 6 hours at 15 mg/kg for 48 hours
mannitol	With prolonged CPR or unconscious >30 minutes, or suspected cerebral edema	250 mg/kg IV
hemoglobin glutamer-200 (Oxyglobin®)	If animal is anemic, this drug is used to increase systemic oxygen content. This product is a hemoglobin-based, oxygen-carrying fluid that increases plasma and total hemoglobin concentration.	One-time dose of 10–30 ml/kg IV at a rate of up to 10 ml/kg/hr If given too rapidly or to animal with congestive heart failure, may result in circulatory overload

Fluid therapy is needed when an animal cannot maintain fluid and electrolyte balances. Fluid therapy can be administered orally, SQ, IP, IV, and IO.

The types of fluids used in fluid therapy are categorized as either crystalloid (sodium-based electrolyte solutions or solutions of glucose in water) or colloid (fluids with large molecules that enhance the oncotic force of blood, causing fluid to move into the vascular space). The ability of a solution to cause water movement is referred to as its tonicity. Isotonic fluids have solute concentrations similar to that of blood; therefore, fluid has no net movement in either direction. Hypotonic fluids have less solute concentration than blood, causing cells to take in fluid and swell; the net movement of fluid is into the cell. Hypertonic fluids have more solute concentration than blood, causing cells to give up fluid and shrink; the net movement of fluid is out of the cell. Additives can be administered with fluids in certain disease states. Addititives include 50% dextrose, potassium, sodium bicarbonate, calcium, and water-soluble vitamins. When putting additives in the fluid bag, it is important to withdraw and discard an amount of fluid equal to the amount of additive being supplemented.

The amount of fluid to be given to an animal is based on rehydration values, maintenance values (sensible and insensible fluid values), and ongoing fluid loss values. You will not have to calculate all these values for every animal. The rate of fluid administration parallels the severity of dehydration and is usually determined for a 24-hour period. Rate of fluid administration is determined based on the use of either an adult or a pediatric administration set. Fluid administration should be monitored and adjusted based on the health and recovery status of the patient.

In emergency situations, airway patency, assessment of breathing, and maintenance of circulation are key in an animal's treatment. Various protocols for emergency treatments are available.

CHAPTER REFERENCES

Garvey, M. Fluid and electrolyte balance in critical patients. *Veterinary clinics of North America, 19*(6), pp. 1021–1057.

Haskins, S, (1992). Cardiopulmonary resuscitation. In *The compendium collection: Emergency medicine and critical care in practice.* Trenton, NJ: Veterinary Learning Systems.

Kee, J., & Paulanka, B. (1994). *Fluids and electrolytes with clinical application: A programmed approach* (5th ed.). Clifton Park, NY: Thomson Delmar Learning.

Kirby, R. et al. (1989). Critical Care. R. Kirby and C. Stamp (Eds.). *Veterinary Clinics of North America, 19*(6), pp 1009–1019.

Pichler, M., & Turnwald, G. (1992). Blood transfusion in the dog and cat part I: Physiology, collection, storage, and indications for whole blood therapy. In *The compendium collection: Emergency medicine and critical care in practice.* Trenton, NJ: Veterinary Learning Systems.

Pichler, M., & Turnwald, G. (1992). Blood transfusion in the dog and cat part II: Administration, adverse effects, and component therapy. In *The compendium collection: Emergency medicine and critical care in practice.* Trenton, NJ: Veterinary Learning Systems.

Veterinary Values (5th ed.). (1998). Lenexa, KS: Veterinary Healthcare Communications.

CHAPTER REVIEW

Matching

Match the fluid name with its action.

1. ______ 0.9% sodium chloride
2. ______ D_5W
3. ______ whole blood
4. ______ LRS
5. ______ 0.9% normal saline with 5% dextrose
6. ______ dextran
7. ______ plasma protein solutions
8. ______ Normosol®
9. ______ hetastarch
10. ______ albumin

a. natural colloid
b. synthetic colloid
c. isotonic crystalloid
d. hypotonic crystalloid
e. hypertonic crystalloid

Multiple Choice

Choose the one best answer.

11. What happens to red blood cells when placed in a hypotonic solution?
    a. They shrink.
    b. They swell.
    c. They remain the same.
    d. They become nucleated.
12. What happens to red blood cells when placed in a hypertonic solution?
    a. They shrink.
    b. They swell.
    c. They remain the same.
    d. They become nucleated.
13. What happens to red blood cells when placed in an isotonic solution?
    a. They shrink.
    b. They swell.
    c. They remain the same.
    d. They become nucleated.
14. The best way to get oxygen to a patient is via
    a. a manual artificial resuscitator
    b. positive-pressure ventilation from an anesthetic machine.
    c. mouth-to-tube delivery.
    d. all of the above are about equal.
15. Which emergency drug is given if there is no heartbeat?
    a. epinephrine
    b. sodium bicarbonate
    c. mannitol
    d. dexamethasone
16. What emergency drug is given when the patient has a slow heartbeat?
    a. mannitol
    b. atropine
    c. prednisone sodium succinate
    d. sodium bicarbonate

True/False

Circle a. for true or b. for false.

17. When dextrose is added to fluids it is meant to serve as a calorie source.
    a. true
    b. false
18. When calculating rehydration fluid values, the percent dehydration should be in decimal form.
    a. true
    b. false
19. Adult administration sets always deliver fluid at 15 gtt/ml and pediatric administration sets always deliver fluid at 60 gtt/ml.
    a. true
    b. false
20. CPR methods for small animals (<15 kg) and larger animals (>15 kg) are the same.
    a. true
    b. false

Case Studies

21. A mature, 500 kg horse with severe obstructive colic and 8 percent clinical dehydration presents to the clinic. This horse needs fluids stat (immediately).
    a. Calculate fluid therapy for this animal. Assume a maintenance value of 50 ml/kg/day.

    You are to give this horse its daily fluid replacement with the following:
    - 2.5% dextrose. This comes in 50% dextrose in 500 ml bottles.
    - 0.9% NaCl. This comes in granular form.
    - 20 mEq/l KCl. This comes in 30 ml bottles that contain 2 mEq/ml.

    Using the volume concentration method or dimensional analysis method (unit cancellation), calculate these additives. Remember that percents equal g/100 ml (for example, 5% = 5 g/100ml).
    b. How much NaCl will you need in grams?
    c. How much dextrose will you need in grams and milliliters? How many bottles is this?
    d. How much KCl will you need? How many bottles is this?
    e. How much sterile water do you need?
    f. What must you remember to do before adding these additives?
22. The epinephrine available in your clinic is at a concentration of 1:1000. Make up 50 ml of 1:10,000 concentration of epinephrine, which is the form utilized for emergency treatment. How much 1:1000 epinephrine and how much diluent will you use to make the desired volume?

Antineoplastic and Immunosuppressive Drugs

OBJECTIVES

Upon completion of this chapter, the learner should be able to

- explain the theory behind antineoplastic drug use.
- describe the five stages of the cell cycle.
- differentiate between CCNS and CCS antineoplastic drugs.
- give examples of CCNS and CCS antineoplastic drugs.
- explain growth fraction and doubling time as they relate to cancer cells.
- describe the use of biologic response modifiers.
- list various biologic response modifiers.
- describe the role of immunosuppressive drugs on the cell cycle.

KEY TERMS

anticancer drugs
antineoplastic agents
chemotherapeutic agents
cell-cycle nonspecific (CCNS)
cell-cycle specific (CCS)
growth fraction
doubling time
biologic response modifiers (BRMs)
immunosuppressive drugs

INTRODUCTION

A client's boxer has been diagnosed with lymphosarcoma. Owner and dog are sent to a referral clinic for a complete workup and discussion of treatment options. The client calls your clinic after taking his dog to the veterinary oncologist, who offered him a treatment plan consisting of two antineoplastic agents. The owner would like to have the treatments given at your clinic so that he will not have to drive 100 miles to the referral clinic. The oncologist is willing to prescribe the medication and let the client have the treatments done at your clinic. Do you know how to handle antineoplastic agents? Do you know how to administer these agents? What would you tell this client?

CANCER-FIGHTING DRUGS

Anticancer drugs, also called **antineoplastic agents** and **chemotherapeutic agents**, stop the cancerous activity of malignant cells. Clinically useful anticancer drugs act against characteristics unique to malignant cells, including their rapid cell division and growth, their different rate of cellular drug uptake, and their increased cellular response to selected anticancer drugs. Unfortunately, some factors present in malignant cells also occur in other parts of the body. For example, rapid cell division and growth occur in the bone marrow, gastrointestinal tract, reproductive organs, and hair follicles, making these body systems vulnerable to the effects of antineoplastic agents.

Some antineoplastics act at certain phases of the cell cycle (Figure 20-1). The five phases of the cell cycle are:

1. G_1 phase: enzymes that are needed for DNA synthesis are produced.
2. S phase: DNA synthesis and replication.
3. G_2 phase: RNA and protein synthesis.
4. M phase: mitosis phase involving cell division.
5. G_0 phase: resting phase.

TIP

The treatment of cancer is generally most successful when the cancerous cells are localized and have not been disseminated throughout the body. When cancerous cells are localized and accessible (such as some forms of skin cancer), the optimal treatment may be surgical removal of the affected tissue, with or without chemotherapy.

Nonspecific versus Specific

Cancer cells move more quickly through the phases of the cell cycle than do normal cells. Some antineoplastic agents work during any phase of the cell cycle (known as **cell-cycle nonspecific** or **CCNS**); others act during a specific phase of the cell cycle (known as **cell-cycle specific** or **CCS**). CCNS drugs kill the cell during the dividing and resting phases. CCS drugs are effective against rapidly growing cancer cells. Table 20-1 summarizes antineoplastic agents.

Growth fraction and doubling time are two factors that play a role in cancer cell response to antineoplastics. **Growth fraction** is the percentage of the cancer cells that are actively dividing. A high growth fraction is seen when the cells are dividing rapidly and a low growth fraction occurs when the cells are dividing slowly. In general, antineoplastics are more effective against cancer cells that have a high growth fraction. Leukemias and some lymphomas have high growth fractions and tend to respond better to antineoplastic treatment. Breast carcinomas tend to have a low growth fraction and thus tend to respond unfavorably to antineoplastic treatment. In general, small, early forming, and fast-growing tumors respond better to antineoplastic drugs. Solid tumors have a large percentage of cells in the G_0 phases, so these tumors have

TIP

All antineoplastic agents are cytotoxic (poisonous to cells) and therefore interfere with normal as well as neoplastic cells.

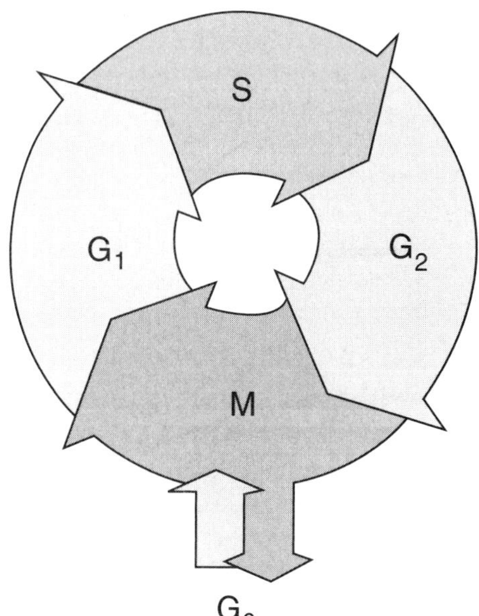

FIGURE 20-1 The cell cycle

TABLE 20-1 Antineoplastic Agents Used for Chemotherapy in Animals

Drug Type	Drug Category	Examples	Cell-Cycle Effect
CCNS	*Alkylating agents*: Cross link DNA to inhibit its replication	• cyclophosphamide (Cytoxan®, Neosar®) • cisplatin (Platinol®) • chlorambucil (Leukeran®) • melphalan (Alkeran®) • carboplatin (Paraplatin®)	• Work on all phases of the cell cycle, but are more effective in the G_1 and S phases • Tend to be used for Lymphoproliferative diseases, osteosarcoma, mast cell tumors, and carcinomas
	Antitumor Antibiotics: Inhibit DNA, RNA, and (in some cases) protein synthesis	• doxorubicin (Adriamycin®) • dactinomycin, also known as actinomycin D (Cosmegen®) • mitoxantrone (Novantrone®)	• Work on all phases of the cell cycle, but doxorubicin is more effective at the S phase • Tend to be used for lymphoproliferative diseases, sarcomas, and carcinomas
	Steroid drugs: Action may include anti-inflammatory effects, suppression of bone marrow cells, reduction of edema, and suppression of tumor growth	• corticosteroids • estrogens • progestins • androgens	• Work on all phases of the cell cycle, but more effective at the S and M phases • Tend to be used for lymphoproliferative diseases, reproductive cancers, mast cell tumors, and CNS tumors
CCS	*Antimetabolites*: Cell-cycle specific drugs that affect the S phase (involving DNA synthesis)	• methotrexate (Mexate®): inhibits folic acid synthesis • 5-fluorouracil (Adrucil®, 5-FU®): pyrimidine base analogs • cytarabine (Cytosar-U®): pyrimidine base analog • azathioprine (Imuran®): antagonizes purine base metabolism	• Work on S phase of the cell cycle. They either inhibit folic acid, which is needed for the synthesis of proteins and DNA, or by being an analog of pyrimidine or purine (which are bases occurring in DNA and RNA) and incorporating into the DNA/RNA molecule • Work for lymphoproliferative diseases and carcinomas
	Alkaloids: Chemicals derived from plants; stop cancer cell division	• vincristine (Oncovin®, Vincasar®) • vinblastine (Alkaban-AQ®, Velbane®)	• Work on the M phase of the cell cycle, inhibiting mitosis and causing cell death • Work for lymphoproliferative diseases, mast cell tumors, sarcomas, and carcinomas

a low growth fraction and tend to be less sensitive to antineoplastic drugs. **Doubling time** is the time required for the number of cancer cells to double. When tumors age and enlarge, their growth fraction decreases and their doubling time increases.

Administering Antineoplastics

Veterinarians often administer anticancer drugs in various protocols or therapeutic combinations—a procedure called *combination therapy*. The advantage to combination therapy is that a tumor is more effectively destroyed when bombarded with drugs that target different sites or act in different ways.

Calculation of antineoplastic drug doses usually is based on body surface area in square meters. Body surface area (BSA) is determined from body weight in kilograms using prepared charts (see Appendix B). A dosage for cisplatin is 20 $mg/m^2/day$. A dog's weight of 10 kg (22 lb) would translate into 0.46 m^2. To calculate the dose for this dog, you would take 20 $mg/m^2/day \times 0.46\ m^2 = 9.2\ mg/day$. The concentration of cisplatin is 1 mg/ml, so this dog would need $9.2\ mg/day \times 1\ ml/1\ mg = 9.2$ ml of cisplatin per day.

When preparing and administering antineoplastics, great care must be taken to ensure safe handling. Table 20-2 lists proper handling techniques for antineoplastic agents.

When administering antineoplastic agents IV, it is recommended to infuse unmedicated IV solution first through the same administration set/catheter, to check for blood return, pain, redness, or edema, before starting the solution containing the antineoplastic agent. After giving the antineoplastic agent, infuse unmedicated solution IV through the same administration set/catheter before withdrawing needles or IV sets, to ensure that residual antineoplastic agents do not remain in or on this equipment.

TIP

Some antineoplastic drugs need to be refrigerated before reconstitution; others do not. Always read labels to ensure proper handling and storage of these drugs.

TIP

Advise clients to report pain, redness or edema, near the injection site after treatment.

Immune Enhancers

Biologic response modifiers (BRMs) are agents used to enhance the body's immune system. These drugs are used in conjunction with antineoplastic protocols to help the animal mount an immune response against the tumor.

Interferons are a group of naturally occurring proteins that have antitumor and antiviral effects. Interferons regulate lymphocytes and block replication of viral cells. Three major types of interferon have been identified: alpha, beta, and gamma. Alpha interferons are the group used to treat tumors and have also been used to treat viral infections in cats, such as FeLV and FIP. Commercially available products include Roferon-A® injection, Intron A® injection, Alferon N® injection, and Actimmune® injection. These products must be stored refrigerated or frozen. They can also be diluted and frozen in the appropriate dose size. They are typically diluted in a solution of 3 million IU/ml in 1 liter of sterile saline. Side effects include fever and lethargy.

Colony stimulating factors (CSFs) stimulate the growth, maturation, and differentiation of bone marrow stem cells. They have been used to treat neutropenia in dogs and cats. Types include G-CSF, which stands for granulocyte colony stimulating factor; and GM-CSF, which stands for granulocyte macrophage colony stimulating factor. G-CSF is marketed as filgrastim (Neupogen®) and GM-CSF is marketed as sargramostim (Leukine®). Side effects of filgrastim include bone pain, diarrhea, and anorexia. Use with caution during pregnancy, lactation, or with myeloid cancers, as it acts as a growth factor for any tumor type. Side effects of sargramostim include hypotension.

Interleukins are a group of chemicals that play various roles in the immune system. Interleukin 2 promotes replication of antigen-specific T cells and has been used to treat some cancers in dogs and cats.

Acemannan is a chemical used in the treatment of fibrosarcomas and mast cell tumors in dogs and cats. It is a potent stimulator of macrophage activity. It is administered

TABLE 20-2 Guidelines for Handling Antineoplastic Agents

Guidelines for Handling Antineoplastic Agents
Antineoplastics should be handled by trained personnel.
Handling of antineoplastic agents by pregnant women is generally contraindicated.
Ideally, antineoplastic agents should be prepared under a laminar flow hood. If a laminar flow hood is not available, preparation should be done in a separate room in a work area away from cooling or heating vents and away from other people. The worktable should be covered with a disposable plastic liner.
Use latex gloves to protect the skin. Gloves made of polyvinyl chloride may be permeable to some cytotoxic agents.
Good handwashing before and after handling drugs is essential.
Prevent contact of antineoplastic agents with skin or mucous membranes. If this occurs, wash the area immediately with large volumes of water, document the contact, and seek medical assistance.
Before drug preparation, don a disposable, nonpermeable surgical gown with closed front and knit cuffs that completely cover the wrists.
Goggles (or eye shield) and particulate filtration masks are recommended to avoid ocular and respiratory contact.
If reconstituting the drug, vent the vial at the beginning of the procedure. Special vent needles are available and should be used. This lowers the internal pressure in the vial and reduces the risk of spilling or spraying the solution when the needle is withdrawn from the vial.
Wipe all external surfaces of syringes and bottles with alcohol. All disposable equipment should be placed in a separate disposable plastic bag specifically marked for chemotherapy and collected for incineration.
Wear latex gloves when disposing of vomit, urine, or feces from animals receiving antineoplastics.
Maintain a record of all exposure during preparation, administration, clean-up, and spills.

as a prelude to surgery and by concurrent intraperitoneal and intralesional injections weekly for a minimum of six treatments. Prior to use, it must be reconstituted with sterile diluent, and it is only good for four hours after rehydration.

Monoclonal antibodies are identical immunoglobulin molecules that have cytotoxic effects on tumor cells. They may be conjugated with antineoplastics and/or other agents to deliver them directly to the tumor cells.

Immune Dehancers

Immunosuppressive drugs work mainly by interfering with one of the stages of the cell cycle or by affecting cell messengers. They are used primarily in treating immune-mediated disorders in which the immune system is overactive. Examples include:

- cyclosporine (Optimmune® Ophthalmic Ointment, Sandimmune® oral and for injection, Atopica®), which inhibits the proliferation of T-lymphocytes. It is used in veterinary medicine for managing keratoconjunctivitis sicca (KCS) in dogs and for immune-mediated skin disorders. Side effects include nephrotoxicity and vomiting.

- azathioprine (Imuran® tablets and injectable), which affects cells in the S phase of the cell cycle (Table 20-1). It also inhibits T- and B-lymphocytes and is used mainly in dogs for immune-mediated diseases such as systemic lupus and immune-mediated hemolytic anemias. Side effects include bone marrow suppression. This drug is not recommended for use in cats or pregnant animals.
- cyclophosphamide (Cytoxan®), which interferes with DNA and RNA replication and thus disrupts nucleic acid function. It is used in conjunction with antineoplastic agents and for immune-mediated diseases such as autoimmune hemolytic anemia. Side effects include bone marrow suppression.

Enzymes

The use of an enzyme drug, L-asparaginase (Elspar®, Oncaspar®, Erwinase®), offers a different approach to treating tumor cells. L-asparaginase works by hydrolyzing asparagines into aspartic acid and ammonia. Malignant cells are dependent on an exogenous source of asparagine for survival. Normal cells are able to synthesize asparagine and thus are less affected by rapid depletion of this substance. By giving the enzyme L-asparaginase, asparagine is quickly converting to aspartic acid and ammonia, making it unavailable to the tumor cells. L-asparaginase is used for the treatment of lymphomas, mast cell tumors, and thrombocytopenia. Side effects include pain at the injection site, hypotension, and diarrhea.

Adverse Reactions

Antineoplastic agents cause adverse reactions on rapidly dividing normal cells, such as those in the gastrointestinal tract and blood. Table 20-3 lists the general adverse reactions caused by antineoplastic agents.

TABLE 20-3 General Adverse Reactions to Antineoplastic Agents

Adverse Reactions	Patient Considerations
Bone marrow suppression	• Low white blood cell counts increase the risk of infection • Low platelet counts may result in bleeding • Low red blood cell counts leading to anemia can cause lethargy
Gastrointestinal effects	• Anorexia can be due to taste left in the mouth from antineoplastic agents and/or nausea • Nausea and vomiting may be secondary to antineoplastic use • Diarrhea may be secondary to antineoplastic use and could lead to dehydration
Alopecia	• Hair loss varies; the animal should be kept out of the cold and should avoid excessive sunlight
Infertility	• Infertility secondary to antineoplastic use can be permanent
Cardiotoxicity	• Damage to the heart from antineoplastic use can be permanent

SUMMARY

Antineoplastic drugs work by blocking phases of the cell cycle. The groups of antineoplastic drugs are alkylating agents, antitumor antibiotics, antimetabolites, steroids, and alkaloids. Antineoplastic drugs are classified as either cell-cycle nonspecific (CCNS) or cell-cycle specific (CCS). They can be used singly or in combination with other antineoplastic drugs.

Biologic response modifiers (BRMs) are used to enhance the body's immune system. BRMs include interferon, colony stimulating factors, interleukin, acemannan, and monoclonal antibodies.

Immunosuppressive drugs are used to combat overactive immune responses that cause immune-mediated disease. Examples of immunosuppressive drugs are cyclosporine, azathioprine, and cyclophosphamide.

Enzymes work by speeding up reactions. In the treatment of tumors, L-asparaginase works by speeding up the conversion of a product needed by malignant cells to a byproduct that the malignant cells cannot use or metabolize.

Table 20-4 lists the drugs covered in this chapter.

TABLE 20-4 Drugs Covered in This Chapter

Category	Generic Name	Trade Name(s)
Alkylating agent	cyclophosphamide cisplatin chlorambucil melphalan carboplatin	Cytoxan®, Neosar® Platinol® Leukeran® Alkeran® Paraplatin®
Antitumor Antibiotic	doxorubicin dactinomycin (actinomycin D) mitoxantrone®	Adriamycin® Cosmegen® Novantrone®
Steroid Drug	corticosteroids: prednisone dexamethasone estrogens progestins androgens: danazol	 Meticorten®, Prednis® Azium®, Decadron® ECP®, Synovex C® Component E-C® Danocrine®, Cyclomen®
Antimetabolite	methotrexate 5-fluorouracil cytarabine azathioprine	Mexate® Adrucil®, 5-FU® Cytosar-U® Imuran®
Alkaloid	vincristine vinblastine	Oncovin®, Vincasar® Alkaban-AQ®
Biologic response modifier	interferon filgrastim	Roferon-A® injection, Intron A® injection, Alferon N® injection, Actimmune® injection Neupogen®

TABLE 20-4 *Continued*

Category	Generic Name	Trade Name(s)
Biologic response modifier, cont.	sargramostim interleukin acemannan	Leukine® Interleukin-2® Acemannan® Immunostimulant monoclonal antibodies Canine Lymphoma Monoclonal Antibody 231®
Immunosuppressive	cyclosporine azathioprine cyclophosphamide	Optimmune® Ophthalmic Ointment, Sandimmune®, Atopica® Imuran® Cytoxan®
Enzyme	L-asparaginase	Elspar®, Oncaspar®, Erwinase®

CHAPTER REFERENCES

Delmar's A-Z NDR-97 nurses drug reference. Clifton Park, NY: Thomson Delmar Learning.

Veterinary pharmaceuticals and biologicals (12th ed.). (2001). Lenexa, KS: Veterinary Healthcare Communications.

Veterinary values (5th ed.). (1998). Lenexa, KS: Veterinary Healthcare Communications.

Wynn, R. (1999). *Veterinary pharmacology.* Cambridge, MA: Harcourt Learning Direct.

CHAPTER REVIEW

Matching

Match the drug name with its action.

1. ______ cisplatin
2. ______ doxorubicin
3. ______ interferon
4. ______ interleukins
5. ______ vincristine
6. ______ acemannan
7. ______ cyclophosphamide
8. ______ azothioprine
9. ______ cyclosporine
10. ______ methotrexate

a. CCS antineoplastic that is an antimetabolite
b. CCNS antineoplastic that is an antitumor antibiotic
c. BRM that is used to treat tumors and viral infections
d. BRM that is used to treat fibrosarcomas and mast cell tumors
e. immunosuppressive that inhibits T cell proliferation
f. CCNS antineoplastic that has immunosuppressive activity
g. immunosuppressive that inhibits B and T cells
h. CCNS antineoplastic that is an alkylating agent
i. group of BRMs that promotes replication of T cells
j. CCS antineoplastic that is an alkaloid

Multiple Choice

Choose the one best answer.

11. Which antineoplastic drug inhibits folic acid synthesis?
    a. methotrexate
    b. 5-fluorouracil
    c. cytarabine
    d. vincristine
12. The percentage of cancer cells that are actively dividing is called the
    a. doubling time.
    b. synthesis phase.
    c. malignant phase.
    d. growth fraction.
13. Drugs used to enhance the body's immune system are
    a. CCNS drugs.
    b. CCS drugs.
    c. BRMs.
    d. immunosuppressive drugs.
14. Which type of interferon is used to treat tumors and viral infections in cats?
    a. alpha
    b. beta
    c. gamma
    d. omega
15. Which immunosuppressive drug is used orally, as an injectable, and as an ophthalmic ointment?
    a. cyclophosphamide
    b. azathioprine
    c. cyclosporine
    d. monoclonal antibodies
16. Which category of CCNS antineoplastic drug inhibits DNA replication and is most effective in the G_1 and S phases of the cell cycle?
    a. alkylating agents
    b. antitumor antibiotics
    c. steroid drugs
    d. antimetabolites
17. Which category of antineoplastic drug consists of analogs of bases occurring in DNA and incorporates into the DNA molecule?
    a. antitumor antibiotics
    b. antimetabolites
    c. alkaloids
    d. steroid drugs
18. Calculation of antineoplastic drugs is based on
    a. conversion of weight in kg to body surface area in square centimeters.
    b. conversion of weight in kg to body surface area in square meters.
    c. conversion of weight in pounds to body surface area in square feet.
    d. conversion of weight in pounds to body surface area in square inches.

True/False

Circle a. for true or b. for false.

19. CCNS antineoplastics work on any phase of the cell cycle.
    a. true
    b. false
20. Antineoplastics are safe to handle using usual methods.
    a. true
    b. false

Case Studies

21. A 15-yr.-old F/S Beagle (25#) is presented to the clinic after being diagnosed with malignant lymphoma at a referral clinic. Part of the multidrug treatment protocol includes methotrexate 2.5 mg/m^2 po sid.
    a. Using the chart provided in Appendix B, calculate an oral dose of methotrexate for this beagle.
    b. Methotrexate comes in 2.5 mg scored tablets. How many pills will this dog take per dose?
    c. What do you want to advise the client to do when giving the pills to the dog?

CHAPTER 21

Vaccines

OBJECTIVES

Upon completion of this chapter, the learner should be able to

- differentiate between specific and nonspecific immunity.
- differentiate between active and passive immunity.
- differentiate between natural and artificial immunity.
- explain what a vaccine is and why they are used.
- differentiate between inactivated, modified live, live, and recombinant vaccines.
- describe the use of toxoids, antitoxins, and antisera.
- differentiate between autogenous, polyvalent, and monovalent vaccines.
- describe routine care and handling of vaccines.
- list various vaccines given to animals.

KEY TERMS

nonspecific immunity
specific immunity
active immunity
antigen
antibody
passive immunity
natural immunity
artificial immunity
vaccine
inactivated (killed) vaccine
adjuvant
bacterin
attenuated (modified-live) vaccine
live vaccine
recombinant vaccine
toxoid
antitoxin
antiserum
autogenous vaccine
polyvalent (multiple-antigen) vaccine
monovalent vaccine
antibody titer
core vaccine
noncore vaccine

INTRODUCTION

A cat owner receives a reminder card from your clinic, noting that her cats are due to receive their annual vaccinations. She makes an appointment and brings her cats to the clinic at the appointed time. She asks you about the safety of vaccines because she read in her cat magazine that some vaccines might cause cancer. Can you tell her how safe (or unsafe) vaccines are? Does one type of vaccine cause more adverse effects than other vaccines? Should she vaccinate her cats at all? Should she vaccinate them annually? Can you explain the importance of vaccination protocols and how they protect animals from disease?

PROTECTION AGAINST DISEASE

An animal's physiology can protect it against disease in a variety of ways. Some ways, classified as **nonspecific immunity**, include things like physical barriers (intact skin), mucus production, inflammation, fever, and phagocytosis of foreign material. Nonspecific defense mechanisms are directed against all pathogens and are the initial defense against invading agents. Other ways animals protect themselves are specific for a particular antigen. **Specific immunity** takes over when the nonspecific mechanisms fail. Some advantages to specific immunity are that it is targeted for a specific antigen and it has memory. Specific immunity arises from the B- and T-lymphocytes. Lymphocytes that mature in the thymus are called T-lymphocytes, whereas those that mature elsewhere are called B-lymphocytes. T-lymphocytes are responsible for *cell-mediated immunity*, in which T-lymphocytes directly attack the invading antigen. Cell-mediated immunity is important in protecting against intracellular bacterial or viral infections, fungal diseases, and protozoal diseases. After the first exposure to an antigen, the cell-mediated response takes about six days to yield optimal protection. B-lymphocytes are responsible for *antibody-mediated immunity*, in which activated B-lymphocytes produce antibodies that react with antigen. Antibody-mediated immunity (also known as *humoral immunity*) is important in protecting against extracellular phases of systemic viral and bacterial infections and for protection against endotoxin- and exotoxin-induced diseases. After the first exposure to an antigen, the humoral response takes about 14 days to reach optimal protection.

There are four ways to acquire specific immunity: actively, passively, naturally, or artificially. *Active* and *passive* refer to whose immune system reacts to the antigen; *natural* and *artificial* refer to how the immunity is obtained. Table 21-1 gives examples of the various types of immunity.

- **Active immunity** arises when an animal receives an **antigen** (anything that stimulates the immune response) that activates the B- and T-lymphocytes and causes that animal to produce **antibodies** (proteins made by B-lymphocytes to counteract antigens). The animal actively makes antibodies. Active immunity takes time to develop, but creates memory. *Memory* in this context is the capacity of the immune system to respond more rapidly and strongly to subsequent antigenic challenge from the same antigen, because of the formation of memory cells. Active immunity lasts a relatively long time. Active immunity can be natural or artificial.
- **Passive immunity** arises when an animal receives antibodies from another animal. Antibodies are produced in a donor animal by active immunization. Once the antibody level is high enough, blood is taken from the donor animal and the immunoglobulin portion is precipitated and purified. The immunoglobulin can then be given to similar species because it is not recognized as foreign. Passive immunity provides immediate onset of immunity; however, the animal is protected for a shorter time and does not create memory. Passive immunity can be natural or artificial.
- **Natural immunity** is any immunity acquired during normal biological experiences.
- **Artificial immunity** is any immunity acquired through medical procedures.

TABLE 21-1 Examples of Immunity

Example	Type of Immunity
Dog getting parvovirus from another dog	Naturally acquired active immunity
Dog gets vaccinated for parvovirus	Artificially acquired active immunity
Horse gets tetanus antitoxin	Artificially acquired passive immunity
Foal gets maternal antibodies through colostrum	Naturally acquired passive immunity

WHY VACCINATE?

Almost everyone has received vaccines, but did you know exactly why you were being vaccinated, and what you were receiving that kept you from contracting a particular illness? You probably remember receiving several vaccines, against several different diseases, in the course of your life. That's because the body's response to an antigen is specific to that particular antigen. Vaccines trigger specific immune responses to help fight future infections from a specific agent, so animals receiving a vaccine may develop immunity to the inducing antigen only.

Did you ever receive a vaccine after you had already contracted the illness? Or when you were first coming down with the illness? By then it was too late, because your body must be exposed to the antigen beforehand to build a defense against the next exposure to the antigen. The animals in your care, like you, are unlikely to benefit from a vaccine once they have been exposed to a particular virus. The point of a vaccine is to produce immunity without producing the disease.

TIP

Vaccines are biological frauds; they mimic infections by pathogens to trigger the immune response into mounting a reaction.

WHAT IS A VACCINE?

The term *vaccine* comes from the Latin word *vacca* (cow), because the cowpox virus was used in the first preparation from active immunization against smallpox. A **vaccine** is a suspension of weakened, live, or killed microorganisms administered to prevent, improve, or treat an infectious disease. Exposure to a weakened, live, or killed version of a harmful microorganism stimulates your immune system to develop immunity. An animal's immune system works the same way.

TIP

In general, it is easier to induce protection against viruses than against bacteria.

Types of Vaccines

A vaccine should be considered from the standpoints of antigen selection, effectiveness, ease of administration, safety, and cost. In natural immunity, an infectious agent stimulates B- and T-lymphocytes to create memory cells. With vaccination, the objective is to obtain the same result with a modified version of the microorganism or its parts.

Types of vaccine include the following:

TIP

Safe and effective vaccines should mimic the natural protective response, not cause disease, have long-lasting effects, and be easy to administer.

Inactivated (killed). Inactivated or killed vaccines are made from microorganisms, microorganism parts, or microorganism byproducts that have been chemically treated or heated to kill the microorganism. In this process the antigenic structure of the microorganism is kept intact so that it can stimulate an immune response. **Adjuvants** are substances that enhance the immune response by increasing the stability of the vaccine in the body, thus stimulating the immune system for a longer period of time. The advantages to inactivated or killed vaccines are that they are usually safe, stable, and unlikely to cause disease. Disadvantages to inactivated or killed vaccines include the need for repeated doses to ensure protection, resulting in increased client cost. Also, the adjuvant and/or preservatives may cause reactions. If bacteria are inactivated, the "vaccine" is known as a **bacterin**. If only part of the microorganism is used, it is called a *subunit vaccine*. Subunit vaccines were developed to isolate the most important part of the microorganism—the part needed to produce the desired immune response—and eliminate the parts that can cause adverse reactions or interfere with the immune response.

Attenuated (modified-live). When microorganisms go through a process of losing their virulence (called *attenuation*), they become modified-live virus vaccines (MLVs). The microorganisms have been altered but must be able to replicate within the patient to provide immunity. The body will mount an immune response to modified-live vaccines, but they do not usually produce disease. The immunity produced by MLV vaccines usually lasts longer than immunity produced by killed vaccines. MLVs also have better efficacy and quicker stimulation of cell-mediated immunity than killed vaccines. Disadvantages of MLVs include possible abortion when given to pregnant animals; some MLVs can produce mild forms of the disease and can be shed into the environment. Handling and storage of MLVs is critical and are discussed later in this chapter.

Modified-live microorganisms have been grown in media containing adjusted levels of chemicals that trigger and enhance mutations of that microorganism. These mutations change the microorganism's metabolism in a way that alters its ability to cause disease. Temperature-sensitive vaccine microorganisms develop this way. Temperature-sensitive microorganisms lose their ability to grow at the animal's normal body temperature, though they can grow at other temperatures (like those of the ocular or nasal mucosa). Chemically altered microorganisms are considered safer than modified-live microorganisms. They produce the same level of immunity, but the duration of the immunity is shorter.

TIP

MLVs offer quicker protection in the face of an outbreak than killed vaccines.

Live. Live vaccines are made from live microorganisms that may be fully virulent (able to cause disease) or avirulent. *Brucella abortus* RB51 is a live vaccine used in cattle and must be handled with great care. The advantages to live vaccines are that fewer doses are needed to achieve immunity, they last longer, adjuvants are not needed (thus eliminating possible reactions), and they are inexpensive. Disadvantages include possible residual virulence; hence the requirement of careful handling.

Recombinant. Recombinant vaccines are made in a variety of ways. The general principle of recombinant vaccines is that a gene or part of a microorganism is removed from one organism (usually the pathogen) and inserted into another microorganism. The microorganisms are "recombined" to make something new. The advantages to recombinant vaccines include fewer side effects, effective immunity, and varied routes of administration. Increased cost is a disadvantage. Here are some examples of different types of recombinant techniques.

- Recombinant genes coding for a surface or other molecule are isolated from the pathogen and inserted into a nonpathogenic cloning vector. This technique yields large amounts of antigen that can be produced and used in a vaccine.
- Genes may be deleted from a pathogenic microorganism that is then used to produce a vaccine. This vaccine stimulates the immune response, yet has a low risk of producing the disease.
- Microorganisms may have selected parts that stimulate the immune response. Genes that code for those selected parts can be inserted into a recombinant organism to produce a vaccine containing only the antigenic part.

TIP

Toxoid vs. Antitoxin. Toxoids are inactivated toxins used like a vaccine, while antitoxins are antibodies produced in response to a toxin and are used when exposure to a toxin is likely. Toxoids provide active immunity, while antitoxins provide passive immunity.

Toxoids. Toxoids are a special type of vaccine used against toxins instead of microorganisms. The toxin is deactivated by heat or chemicals but is still able to stimulate antibody production. Toxoids have shorter durations of effectiveness than other vaccines and may contain adjuvants and preservatives.

Antitoxins. Antitoxins are substances that contain antibodies (immunoglobulins) obtained from an animal that has been hypersensitized to neutralize toxins. The immunity produced is short-lived because the immunity rendered is passive. Antitoxins may also contain preservatives that can cause adverse reactions.

Antiserum. Antiserum is antibody-rich serum obtained from a hypersensitized or actually infected animal. The immunity produced is short-lived because the immunity rendered is passive. Antiserum may also contain preservatives that can cause adverse reactions.

Autogenous. Autogenous vaccines are produced for a specific disease problem in a specific area from a sick animal. For example, an organism may be isolated from a farm where an infectious disease outbreak is occurring and made into a vaccine for that specific farm. The microorganisms are grown in culture, killed, and mixed with an adjuvant. These vaccines may contain endotoxin and other byproducts found in the culture; therefore, they should be used with caution.

Multiple-antigen vaccines. Multiple-antigen vaccines contain more than one antigen and are referred to as **polyvalent**. Polyvalent vaccines contain a mixture of different antigens and can be convenient to administer because fewer injections are needed. For a polyvalent vaccine to be approved, the manufacturer must show that each part of the vaccine induces the same level of immunity as does the single-antigen (referred to as **monovalent**) vaccine. Table 21-2 compares polyvalent and monovalent vaccines.

TIP

The effectiveness of a polyvalent vaccine should be clinically indistinguishable from that of monovalent vaccines.

Other vaccine possibilities. Some other possible mechanisms for vaccine production include use of peptides (using chemically synthesized protein fragments that antibodies recognize and bind to); anti-idiotypes (an antigen-binding region of a given antibody can be antigenic in a different animal, causing that animal to produce antibodies to the original antibody); and DNA (injection of a piece of DNA into cells to stimulate an immune response). Some of these methods have been utilized to a degree; however, adverse effects have limited their use to date.

Table 21-3 summarizes the type of response stimulated by each vaccine type. Table 21-4 compares the advantages and disadvantages of various types of vaccines.

TABLE 21-2 Comparison of Polyvalent and Monovalent Vaccines

Vaccine Type	Advantages	Disadvantages
Polyvalent	• Convenient • Fewer injections to administer • Less expensive when giving multiple vaccine agents	• Rate of adverse reactions increases as the number of antigens increases
Monovalent	• Can select only the desired antigens to administer • If animal has had an adverse reaction in the past, can give different vaccines on different days or exclude certain ones	• Using several monovalent vaccines exposes patients to higher amounts of proteins and possibly adjuvants • Adverse reactions increase when giving many monovalant vaccines at a time (although this can be decreased by giving at different times)

TABLE 21-3 Immune Response Stimulated by Particular Vaccines

Response	Vaccine Type
Stimulates antibody response	• Killed (inactivated) • Subunit • Toxoid
Stimulates cellular response	• Live • Modified-live (attenuated) • Recombinant

Maternally Derived Antibodies

Vaccines are not effective when given in the presence of maternal antibodies that come primarily from colostrum (maternal antibodies may also be transmitted via the placenta). Young animals receive a small amount of natural immunity from their mother's milk, in the form of colostrum, ingested during the first few days of nursing. Maternally derived antibodies from colostrum provide disease resistance for a few weeks. This temporary maternal protection provides variable antibody levels among animals and decreases by six to nine weeks. To continue and enhance this protection, young animals should receive vaccinations and booster vaccinations to ensure that an appropriate immune response has occurred. Booster vaccinations are recommended in young animals, because effective vaccination varies among individuals, due to variable

TABLE 21-4 Advantages and Disadvantages of Vaccine Types

Killed Vaccines, Subunit Vaccines, and Toxoids	
Advantages	**Disadvantages**
• No risk of reverting to virulent form • No risk of vaccine microorganism spreading between animals • Decreased abortion risk • More stable storage • No mixing, which decreases the risk of contamination	• More likely to cause allergic reactions and postvaccination lumps • Two initial doses needed at least ten days apart • Slower onset of immunity • Immune response may not be as strong or as long as MLV • Tend to be more expensive than MLV
Modified-Live (Attenuated) Vaccines	
Advantages	**Disadvantages**
• One initial dose is usually sufficient for protection (additional boosters may be required) • More rapid protection • Less likely to cause allergic reactions or postvaccination lumps	• Could revert to the virulent form • Could produce disease in immunosuppressed animals • Could produce an excessive immune response • Some risk of abortion • Must be handled and mixed with additional care • Risk of contamination during mixing

levels of maternally derived antibodies; and because some vaccines must be administered multiple times to achieve a satisfactory level of immunity.

Vaccine Allergies

Always ask clients about allergies in their animals prior to immunization. Advise clients of what to look for in regard to vaccine reactions and what they should do if they observe these reactions. Typical vaccine reactions include local reactions at the injection site, fever, lethargy, vomiting, salivation, and difficulty breathing. More severe vaccine reactions include anaphylaxis and immune diseases, such as vaccine-associated sarcoma in cats and autoimmune hemolytic anemia in dogs. Previous vaccine reactions should be reported to the veterinary medical staff prior to vaccination and recorded in the medical record.

ISSUES OF VACCINE USE

Vaccines play an important role in controlling and preventing infectious diseases. Veterinarians have greatly reduced the incidence of various infectious diseases by establishing vaccination protocols and educating clients about the importance of vaccinating their animals. To ensure that vaccination protocols are successful, the following items should be addressed.

Care and Handling

Inappropriate handling of vaccines may lead to their inactivation. Vaccines (especially MLVs) are sensitive to sunlight, excessive heat, and freezing. Vaccines should be ordered from a reputable company that delivers them in shipments with cold packs. Upon delivery, vaccines should be unpacked and stored in the refrigerator. Frozen vaccines may have experienced cell death and overheated vaccines will have been rendered inactive.

Some vaccines must be reconstituted before administration. Use only the diluent provided by the manufacturer, as these diluents are pH-sensitive for each vaccine product. Vaccines that have to be reconstituted should be mixed as near to the time of administration as possible. Vaccine can be purchased in multiple- or single-dose vials. Multiple-dose vials carry the additional risk of contamination with each use. Table 21-5 lists vaccine handling guidelines.

Route of Administration

Vaccines should be given by the route identified by the manufacturer. The level of immune protection can be ensured only when given by the proper route; some vaccines deliver local and perhaps systemic immunity (as with intranasal vaccines) and some deliver systemic immunity (as with intramuscular and subcutaneous vaccines). Route of vaccine administration in food-producing animals is also important, as IM vaccines can cause more tissue damage (and hence damage to muscles used in meat processing) than SQ vaccines.

Using Vaccines

Mixing vaccines together to reduce the number of injections an animal receives is not recommended. The mixing of vaccines can result in one vaccine component interfering with another vaccine component, meaning that the animal does not receive an adequate dose. Vaccines should be given in different sites and those sites recorded in the

TABLE 21-5 Vaccine Handling Guidelines

Vaccine Handling Guidelines
• Protect vaccines from temperature extremes.
• Protect vaccines from ultraviolet light.
• Use only diluents supplied by the manufacturer for a specific product. These diluents are pH sensitive for a particular product.
• Do not mix vaccines in the same syringe unless recommended by the manufacturer.
• Vaccines should be used as soon as possible after reconstitution.
• Do not use chemically sterilized syringes when administering vaccines.
• Use the entire recommended dose of vaccine.
• Using multiple-dose vials may result in inadequate mixing and therefore unequal distribution of antigens and adjuvant.
• Multiple-dose vials have an increased risk of contamination.
• Use proper animal restraint when administering vaccines.
• Use the route of administration recommended by the manufacturer.
• Clean the injection site of any dirt or debris.
• Document vaccine administration in the animal's medical record. Documentation should include vaccine type, name, manufacturer, serial number, expiration date of vaccine, date of administration, route of administration, and administration site.

medical records in case of vaccine reaction. For food-producing animals, most vaccines have warnings regarding slaughter.

The use of partial doses of vaccines is not recommended. The dose of vaccine required to induce an immune response is species-specific, not based on the weight of the animal.

TIP

Any animal that receives less than the standard dose of vaccine, or receives a vaccine through a nonstandard route or at a nonstandard site, should be revaccinated.

Patient Considerations

Routine physical examination prior to vaccine administration is recommended. Animals in good health will respond well to vaccination. Patient factors to consider when vaccinating animals include:

- animal age at vaccination. Age recommendations for vaccination are provided by the manufacturer to avoid age-specific risks of infection and complications, to ensure that the animal has the ability to mount an appropriate immune response, and to account for potential interference by maternal antibodies.
- illness. Animals that are debilitated, malnourished, ill, or have a high fever should not be vaccinated in most circumstances. Animals that are not healthy cannot mount the proper immune response to achieve protection.
- use of concurrent medication. Animals that are receiving immunosuppressive drugs pose an interesting problem regarding vaccination. Animals on high doses of corticosteroids should not receive MLVs. Using killed vaccines in animals that are receiving high levels of corticosteroids may also be a problem, as the animal may not be able to mount an adequate immune response. Each case should be evaluated individually as to the dose of immunosuppressive drug, the duration of treatment, and the risk of not vaccinating the animal. The same theory applies to animals that are immunocompromised by infection, like FeLV in cats and various cancers in all animals.

- pregnancy. The use of MLVs in pregnant and nursing animals is not recommended, as microorganism particles can be shed and affect the offspring. Some MLVs can pass the microorganism to the fetus, resulting in disease in the offspring. There may be circumstances in which the benefits of vaccination outweigh these risks, so each case should be evaluated on its own.
- patient's environment. Animals that have the opportunity to be exposed to other animals, are living in clusters, and are living in an area where the infectious agent exists are at higher risk for acquiring that disease than animals that are not. Animals in these situations should be vaccinated; others should be assessed on a case-by-case basis. In other words, should an inside cat in a single-cat household receive every vaccine manufactured for cats? Does the benefit of vaccinating for every disease outweigh the vaccine risks?

TIP

Vaccination should be avoided in conjunction with other procedures that may induce immune system responses.

Vaccine Reactions

Most vaccines are safe and effective; however, adverse reactions have been reported with vaccines. These reactions include vaccination site reactions (local reactions such as pain or swelling at the site that usually resolve in two to four weeks), systemic reactions (allergic reactions ranging from mild to severe anaphylaxis, hives, wheezing, swelling, difficulty breathing, hypotension, and shock), and sarcomas in some species (*sarcomas* are tumors of connective-tissue origin). Some vaccine reactions are associated with the agent being given. Modified-live canine distemper vaccine can cause some dogs to develop neurological signs (such as seizures) after being vaccinated.

VACCINE PROTOCOLS

The practice of annual vaccination has been a hallmark of veterinary practice for many years. The concept of annual booster vaccinations is universally credited to the manufacturers of veterinary vaccines. The manufacturers included the "recommendation" to revaccinate annually on the product information sheets accompanying the vaccines. It is important to note that this recommendation was not based entirely on scientific research. The original source of the recommendation is not known, but may have been extrapolated from early experimental evidence on duration of immunity for rabies vaccines. No studies have ever been undertaken by veterinary biologics manufacturers to determine the duration of immunity bestowed by their products, excluding rabies vaccines, because this was not required until recently for product approval. To receive a license to produce a specific vaccine for sale, the manufacturer is required only to produce evidence that the product is safe and effective in providing immunity against a specific invader. With the exception of rabies vaccines, manufacturers were under no obligation to prove that this annual revaccination interval is necessary to provide immunity. This means there is really no data regarding how long immunity lasts in a vaccinated animal (with the exception of the rabies vaccine).

TIP

Until recently, only manufacturers of rabies vaccines were required by the USDA to evaluate duration of immunity after vaccination.

One way to assess when revaccination is necessary is the **antibody titer**. When an animal is exposed to a foreign protein (such as the surface of a virus or bacteria), a specific immune response against that particular protein (the antigen) occurs. This specific immune response results in the production of antibodies. It takes time for the animal to mount this specific immune response on its first exposure to the antigen. During this first exposure to the antigen, the animal produces memory cells. Memory cells have the ability to make antibody against that specific antigen in a much shorter time frame upon subsequent exposures. Immunity against a particular antigen due to memory cell development can last a very long time, even a lifetime.

TIP

By tradition, manufacturers of veterinary biologics have recommended annual revaccination; however, this is only a recommendation.

Antibody titers are serum tests that express the level of antibody to a particular antigen in a particular individual. Antibody titers are expressed as 1:2, 1:4, and so on, which represents the dilution at which an immune response is still adequate. Antibody titers that fall below the accepted dilution are generally considered inadequate. However, the correlation between antibody titer and protection against challenge exposure to the disease has not been thoroughly investigated in veterinary medicine. A low titer may not necessarily mean that the animal lacks protection to subsequent exposure to the antigen. Until further research on antibody titers is done, positive antibody titers should be interpreted as the animal having developed memory cells to that particular antigen.

Core versus Noncore

Vaccines for dogs and cats are now being described as belonging to one of two categories: core vaccines and noncore vaccines. **Core vaccines** are recommended for all individual animals because the consequences of infection are severe, infection poses a substantial zoonotic potential, disease prevalence is high, the organism is easily transmitted to others of its species, and/or the vaccine is safe and efficacious.

Noncore vaccines are recommended only for individual animals deemed to be at high risk for contact with the organism. Vaccination with noncore vaccines is based on evaluation of all risk factors, including vaccine safety and efficacy.

VACCINE EXAMPLES

The following is a list of some of the more common animal vaccines used in the United States for disease prevention.

Vaccines Available for Dogs

- DA_2PLP (canine distemper, canine adenovirus type 2 (infectious canine hepatitis), canine parainfluenza, leptospira, and canine parvovirus)
- CV (coronavirus)
- lyme
- rabies
- *Bordetella bronchiseptica* (kennel cough)
- canine herpes virus (CAV-1)
- *Giardia lamblia*

Vaccines Available for Cats

- FVRCP (feline viral rhinotracheitis, feline calicivirus, feline panleukopenia)
- C (*Chlamydia psittaci*); this may be found in a FVRCP-C vaccine
- FeLV (feline leukemia)
- rabies
- FIP (feline infectious peritonitis)
- *Giardia lamblia*
- dermatophyte (ringworm)
- FIV

Vaccines Available for Cattle

- rotavirus
- coronavirus
- *E. coli*

- *Brucella abortus* RB51
- clostridial disease (*C. perfringens* types B, C, and D; *C. chauvoei, C. novyi, C. septicum*, and *C. sordelli*); may also contain *C. hemolyticum*, making the vaccine an eight-way actor
- IBR (infectious bovine rhinotracheitis)
- PI-3 (parainfluenza-3)
- BVD (bovine viral diarrhea)
- BRSV (bovine respiratory syncytial virus)
- leptospirosis
- tetanus
- campylobacter

Vaccines Available for Horses

- EEE/WEE/VEE (Eastern, Western, and Venezuelan equine encephalitis)
- equine viral rhinopneumonitis
- equine influenza
- strangles (*Streptococcus equi*)
- equine viral arteritis
- equine monocytic ehrlichiosis (Potomac horse fever)
- rabies
- tetanus toxoid (tetanus antitoxin is used in horses with wounds and no history of or expired tetanus toxoid vaccine)
- anthrax
- equine protozoal myelitis
- West Nile virus

Vaccines Available for Ferrets

- canine distemper (make sure it is labeled for ferrets)
- rabies

SUMMARY

Vaccines are an important part of veterinary practice to prevent infectious diseases in animals. Vaccines confer artificially acquired active immunity. A vaccine is a suspension of weakened, live, or killed microorganisms administered to prevent, improve, or treat an infectious disease. Types of vaccines include inactivated (killed), modified-live (attenuated), live, recombinant, toxoid, antitoxins, and antisera. Vaccines are also described as autogenous, polyvalent, or monovalent.

Vaccines must be handled properly to ensure their efficacy. Vaccines are sensitive to sunlight, excessive heat, and freezing. Mixing vaccines together to reduce the number of injections may result in one vaccine component interfering with another vaccine component. Vaccines are given by a variety of routes, including IM, SQ, and IN.

Many patient factors should be considered when developing a vaccination plan for animals. These include potential for exposure to the infectious agent, age of the animal, health status of the animal, and environmental conditions.

CHAPTER REFERENCES

Blakeslee, D. (1996). *Vaccines and the immune system*, JAMA HIV/AIDS Resource Center, http://www.amc_assn.org/special/hiv/newsline/briefing/vacc.htm

Cooper, G. (2000). Considering the innocuous vaccine. Madison Area Technical College, Madison, WI.

Elston, T. (1998, January 15). AAFP/AFM vaccination guidelines. *JAVMA, 212*(2): 227–241.

Immunostimulation. (1986). *The Merck veterinary manual*, (6th ed.). Rahway, NJ: Merck & Co, Inc.

Understanding vaccines. (2002, January). NebGuide Nebraska Coorperative Extension G02-1445, http://www.ianr.unl.edu/pubs/animaldisease/g1445.htm

CHAPTER REVIEW

Matching

Match the vaccine type with its properties.

1. ______ toxoid
2. ______ live
3. ______ polyvalent
4. ______ inactivated
5. ______ autogenous
6. ______ recombinant
7. ______ modified-live
8. ______ antitoxin
9. ______ monovalent
10. ______ antisera

a. vaccine with only one antigen
b. vaccine with a mixture of antigens
c. vaccine in which microorganisms have been chemically treated or heated to kill the microorganism
d. vaccine in which genes or parts of a microorganism are moved from one organism to another
e. vaccine made from live microorganisms that may be virulent or avirulent
f. vaccine against toxins
g. vaccine in which microorganisms go through a process of losing their virulence (are attenuated)
h. vaccine that contains antibodies to neutralize toxins
i. vaccine that is antibody-rich serum from a hypersensitized animal
j. vaccine produced for a specific disease in a specific area

Multiple Choice

Choose the one best answer.

11. A substance added to a vaccine that enhances the immune response is a/an
    a. bacterin.
    b. recombinant.
    c. adjuvant.
    d. attenuation.

12. The process of a microorganism losing its virulence is termed
    a. recombinant.
    b. attenuation.
    c. avirulence.
    d. hypopathogenicity.

13. A farmer has an outbreak of salmonella on his farm and a specific vaccine is made for those bacteria. This type of vaccine is a/an
    a. antitoxin.
    b. antisera.
    c. autogenous.
    d. autovalent.
14. A vaccine made from bacteria is called a/an
    a. bacterin.
    b. prokaryotin.
    c. attenuated.
    d. modified.
15. If you get sick from hepatitis A food poisoning, you are given immunoglobulins to provide immediate protection. This is an example of
    a. naturally acquire active immunity.
    b. artificially acquired active immunity.
    c. artificially acquired passive immunity.
    d. naturally acquired passive immunity.

True/False

Circle a. for true or b. for false.

16. FVRCP is an example of a polyvalent vaccine.
    a. true
    b. false
17. Vaccines are made to protect against viruses and viral diseases.
    a. true
    b. false
18. A horse with a puncture wound and no history of tetanus prophylaxis would receive tetanus toxoid.
    a. true
    b. false
19. One advantage to live vaccines is that they do not need adjuvants.
    a. true
    b. false
20. Rabies vaccine is an example of a monovalent vaccine.
    a. true
    b. false

CHAPTER 22

Behavior-Modifying Drugs

OBJECTIVES

Upon completion of this chapter, the learner should be able to

- outline the various forms of behavior modification.
- discuss why behavior-modifying drugs are not a simple solution to behavior modification.
- list the various categories of behavior-modifying drugs.
- describe the factors considered when selecting which category of drug to use to treat a behavior disorder.

KEY TERMS

anti-anxiety drug
antidepressant drug

INTRODUCTION

The owner of a male German Shepherd dog comes into your clinic to get your opinion about his dog, which has been very protective of him lately and will not let other people or dogs near him. His dog is particularly aggressive toward other male dogs. This owner is very concerned and wants to know what he can do to prevent his dog from injuring someone or their animal. He suggests filing the dog's teeth or extracting them altogether, building an enclosure for the dog, or surrendering the dog to the animal shelter. He thinks that if he could get some medication to calm his dog down, the dog would not be so aggressive. What suggestions can you offer this owner? Is medication the best answer for management of this dog's behavior? What are some ways to avoid behavior problems? What do you do in behavior cases where rapid resolution of the animal's problem behavior is needed?

BEHAVIOR PROBLEMS IN ANIMALS

Behavior problems and solutions are now a rapidly expanding area of interest and knowledge in veterinary medicine. The use of drugs to treat problem behaviors is only a small part of treating animal behavior problems.

Correctly diagnosing the condition, examining social conditions that affect the situation, and altering external stimuli are all parts of behavior treatment.

Although drug therapy to treat behavior problems may seem simple and convenient, it is not the first or only choice the veterinary community has for modifying an animal's behavior. Owners need to know about the potential problems that behavior modification drugs may cause. Owners of animals with diagnosed behavior problems, who are considering the use of behavioral drugs, need to be made aware of potential side effects of long-term drug use—the most notable are the development of liver, kidney, and cardiovascular problems. Periodic monitoring of blood values (CBC, serum chemistries) and ECGs are needed, because animals on behavior-modifying drugs may not show signs of disease until they have progressed to a dangerously unhealthy state.

Clients should also be aware that many behavior-modifying drugs are used extra-label. Extra-label drug use requires a veterinarian/client/patient relationship and compliance with the Animal Medicinal Drug Use Clarification Act of 1994 (see Chapter 1). Clear treatment plans for using behavioral drugs must be formulated. Continual monitoring of the animal's health and assessment of whether the behavior-modifying drug is helping the animal must also be done.

It may seem easy for the owner of an animal with diagnosed behavior problems to reach for pills to change the animal's behavior. However, it should be emphasized that behavior modification includes retraining the animal and eliminating external stressors for the animal.

TIP

With many behavior-modifying drugs, it takes six to eight weeks of treatment before any change is seen.

CLASSES OF BEHAVIOR-MODIFYING DRUGS

The types of drugs used for behavior modification include anti-anxiety agents, antidepressants, and hormones.

Anti-anxiety Drugs

Anxiety manifests in multiple forms in animals. Some examples of problems associated with anxiety include separation anxiety, excessive vocalization, whining, whimpering, and inappropriate urination. **Anti-anxiety drugs** attempt to decrease or eliminate these behaviors.

Antihistamines Antihistamines inhibit the H_1 receptor site from binding histamine, a natural substance in the body that is released in response to tissue damage. Most antihistamines also produce some degree of sedation because they suppress the central nervous system. The use of antihistamines as behavior-modifying drugs centers on their side effect of CNS depression. Antihistamines have been used in the treatment of anxiety and the behaviors associated with anxiety (for example, inappropriate urination, pacing with vocalization, and travel-related anxiety). Antihistamines are also used to control pruritus (itching). Animals may develop anxiety associated with pruritus and the antipruritic effects of antihistamines appear to lessen this anxiety. Antipruritic effects of antihistamines are achieved at higher doses than are the antianxiety effects. Many classes of antihistamines have anticholinergic effects, so side effects may include urinary retention, increased heart rate, and increased respiratory rate. Antihistamines used in behavior modification include:

- hydroxyzine (Atarax®), which produces greater sedative effects than other antihistamines and can last four to six hours. Hydroxyzine has minimal anticholinergic effects.
- diphenhydramine (Benadryl®) has lesser sedative effects and may be used with some success for pruritus. It is also used in cases in which mild sedation may be helpful, such as travel-related anxiety.

TIP

Many behavior-modifying drugs found in different categories have similar functions and mechanisms of action. Therefore, their specific effect(s) may depend on the dosage used.

TIP

Benzodiazepines are not dependent on their ability to cause sedation in order to modify behavior.

TIP

When discontinuing the use of behavior-modifying drugs, withdrawal should generally be gradual, to avoid potential return of behavior problems or the development of side effects.

Benzodiazepines Benzodiazepines are chemically related compounds that are used to relieve anxiety. This group of drugs causes little drowsiness at normal therapeutic doses, does not readily cause the development of tolerance to their anti-anxiety effect, and does not interfere significantly with metabolism of other drugs. Benzodiazepines appear to work on the limbic system of the brain by potentiating the inhibitory action of gamma-aminobutyric acid (GABA), an amino acid that helps mediate nerve impulse transmission in the CNS. Benzodiazepines bind to specific sites in the brain and this binding appears to produce sedation and relieve anxiety. Benzodiazepines cause muscle relaxation that is independent of its sedative effect. At low dosages, benzodiazepines have a sedative effect; at moderate dosages, they have anti-anxiety effects; at high dosages they facilitate sleep. Benzodiazepines have been used to treat some forms of aggression (especially in cats), urine spraying, and noise phobias. Side effects include tolerance to the sedative effects of the drug and increased muscle spasticity. Examples in this group include:

- diazepam (Valium®), a C-IV controlled substance.
- chlordiazepoxide (Librium®), a C-IV controlled substance.
- lorazepam (Ativan®)
- flumazenil (Mazicon®)
- alprazolam (Xanax®)

Phenothiazines Phenothiazines have similar chemical structures and work by antagonism of dopamine. The neurotransmitter dopamine is unequally distributed in the brain, but is found in high concentrations in the limbic system. The limbic system is involved in control of body metabolism, body temperature regulation, wakefulness, vomiting, and hormonal balance. Increased dopamine levels are associated with some psychotic diseases in humans, notably schizophrenia and Parkinson's disease.

When phenothiazines are used as anti-anxiety drugs in animals, they suppress both normal and abnormal behavior. They have been used to treat aggression, but this treatment is now under question. Animals given phenothiazines may be more reactive to noises, thus making an aggressive animal's reactions more unpredictable. Side effects of phenothiazines include CNS (sedation, increased likelihood of seizure development), anticholinergic (dry mouth, constipation) and cardiovascular (hypotension, increased heart rate, and arrhythmia) problems. Examples of phenothiazines include:

- chlorpromazine (Thorazine®)
- acepromazine (PromAce®)
- promazine (Sparine®)
- perphenazine (Trilafon®)
- prochlorperazine (Compazine®)

Azapirones Buspirone (BuSpar®) is an azapirone anti-anxiety drug that is chemically different from the other anti-anxiety drugs, and does not cause sedation like most of the other anti-anxiety drugs. Buspirone is believed to work by blocking serotonin. Buspirone creates no convulsant or withdrawal problems. It is used to treat urine spraying in cats and anxiety-associated aggression. It usually takes weeks of treatment before a clinical response is seen. Side effects include gastrointestinal problems.

TIP

Long-term treatment with phenobarbital requires blood monitoring to avoid hepatotoxicity.

Barbiturates The anti-anxiety action of barbiturates was once attributed to their ability to cause CNS depression (sedation), but it is currently thought to be due to their effects on the GABA receptor. Barbiturates can cause liver problems (as it induces liver enzymes) and heavy tranquilization. These problems have led to decreased use. Long-term use of barbiturates for behavior modification is questionable because of variable and unpredictable results. Barbiturates used to modify behavior include phenobarbital (Luminal®), which has been used to control vocalization in cats in small doses; and carbamazepine (Tegretol®), which has been used to treat seizure-like anxiety in dogs.

Antidepressants

Antidepressant drugs are used in veterinary medicine to treat various mood changes, including aggression, and cognitive dysfunction in animals. Normally, the transmission of nerve impulses between two nerves or between a nerve and tissue takes place via the release of neurotransmitters from their storage sites at the nerve terminal (Figure 22-1). After the neurotransmitter combines with the appropriate receptors, there are several mechanisms that can reduce the concentration of neurotransmitter in the synaptic space. One mechanism involves reuptake of the neurotransmitter by the nerve terminal from which it was released. Another mechanism involves destruction of the neurotransmitter by the enzyme monoamine oxidase (MAO). The main neurotransmitters involved in behavioral disorders are summarized in Table 22-1.

Tricyclic Antidepressants Tricyclic antidepressants (TCAs) have a three-ring (tricyclic) structure and work by interfering with the reuptake of neurotransmitter by the presynaptic nerve cell. This interference increases the concentration of neurotransmitter at postsynaptic receptors in the CNS. TCAs inhibit serotonin reuptake, while TCA metabolites inhibit norepinephrine reuptake. TCAs have been used to treat separation anxiety, pruritic conditions, and compulsive disorders in animals. If side effects occur with TCAs, the use of TCA metabolites may give favorable results. Side effects of TCAs include anticholinergic effects (dry mouth, constipation, urinary reten-

Cats are more sensitive to TCAs than dogs.

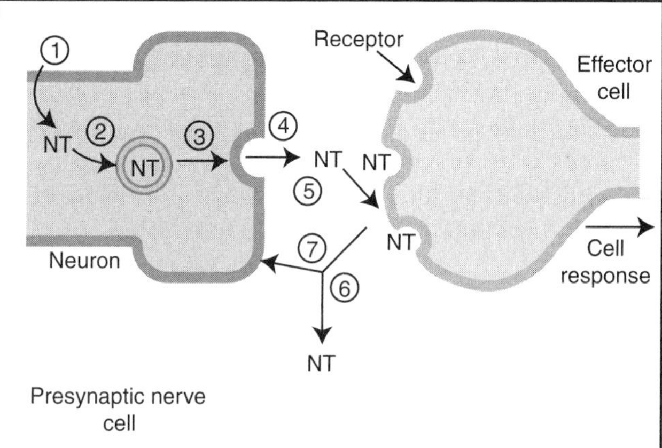

1. Neurotransmitter (NT) is synthesized from substances that enter or may enter neuron in intact form.
2. Neurotransmitter is stored in storage vesicles.
3. When neuron is stimulated, some storage vesicles fuse with neuron cell membranes and
4. Discharge neurotransmitter into the synaptic space.
5. Some neurotransmitter diffuses to the receptor in the effector cell and produces a cell response.
6. Neurotransmitter remaining in the synaptic space is either degraded by enzymes (e.g., monoamine oxidase) or
7. It reenters the neuron and is again stored by storage vesicles.

FIGURE 22-1 Nerve impulse transmission

TABLE 22-1 Neurotransmitters Involved in Behavior Disorders

• acetylcholine
• norepinephrine
• dopamine
• serotonin
• gamma-aminobutyric acid (GABA)

tion, and increased heart rate), liver problems, and thyroid effects (sick euthyroid condition and interference with thyroid medication). Examples of TCAs are amitriptyline (Elavil®), imipramine (Tofranil®), clomipramine (Clomicalm®, Anafranil®), and doxepin (Sinequan®).

TIP

Hypertensive problems can occur in animals given both MAOIs and food rich in tyramine (aged cheese, yogurt, and chicken liver).

Monoamine Oxidase Inhibitors Monoamine oxidase inhibitors (MAOIs) work by inhibiting the enzyme monoamine oxidase, thus reducing the destruction of neurotransmitters such as dopamine, norepinephrine, and epinephrine and increasing their free level in the CNS. MAOIs irreversibly inhibit MAO. It may take weeks to months to see the antidepressant effect of MAOIs. The only MAOI approved for use in dogs is selegiline (Anipryl®). Selegiline blocks the reuptake of dopamine and is used to treat cognitive dysfunction in aging dogs. Side effects include hypotension, drowsiness, and anticholinergic effects (dry mouth, increased heart rate, and constipation). MAOIs that may be used extra-label in animals include phenelzine (Nardil®), isocarboxazid (Marplan®), and tranylcypromine sulfate (Parnate®).

Selective Serotonin Reuptake Inhibitors Selective serotonin reuptake inhibitors (SSRIs) are as effective as TCAs, but have fewer side effects and tolerance problems. SSRIs selectively inhibit serotonin reuptake, resulting in increased serotonin neurotransmission. This group is structurally diverse; uses include treating depression, aggression, anxiety, phobias, and compulsive disorders. Examples of SSRIs are fluoxetine (Prozac®), sertraline (Zoloft®), and paroxetine (Paxil®).

Hormones

The hormones used or altered in treatment of behavior problems are the reproductive hormones progestins, estrogens, and testosterone.

Progestins and estrogens are used as behavior-modifying drugs mainly because of their calming effects, which are created by their suppression of the excitatory effects of glutamine (an amino acid) and their suppression of male-like behaviors. Side effects of these drugs include mammary gland hyperplasia, endometrial hyperplasia, pyometra, bone marrow suppression, and endocrine disorders. Drugs in this category include:

- diethylstilbestrol (DES), a synthetic estrogen that has a different chemical structure than natural estrogens but similar pharmacological effects. It is a second-line drug used to treat urinary incontinence. Other drugs, such as bethanecol and phenylpropanolamine, are the preferred, first-line treatment.
- medroxyprogesterone acetate (Depo-Provera®, Provera®), a synthetic progestin used to treat aggression, because of its calming and/or feminizing actions. It is also used to treat male-like behaviors of mounting, territorial marking (urination), inter-male aggression, and roaming.

- megestrol acetate (Ovaban®, Megace®), a synthetic progestin approved by the FDA to postpone estrus and alleviate false pregnancies in dogs. It is used extra-label to treat urine spraying, anxiety, and aggression in cats. It is also used in cats to treat eosinophilic ulcers and dermatitis conditions. Megestrol acetate may cause endocrine problems (iatrogenic Addison's disease and diabetes mellitus) that are usually resolved when treatment is discontinued.

Testosterone is an androgenic steroid responsible for secondary sex characteristics of the male, and it produces male-like behaviors. Drugs that inhibit testosterone production or block enzymes that convert testosterone to dihydrotestosterone (its potent form) have been used to treat aggressive behavior in male dogs. An androgen antagonist, delmadinone (Terdak®), has also been used to treat aggression in male dogs. Finasteride (Proscar®) is an androgen hormone inhibitor that inhibits the enzyme that converts testosterone to its potent form. Finasteride has been used in humans to treat benign prostatic hyperplasia, and may play a future role in veterinary medicine.

Table 22-2 summarizes behavior disorders in which drug therapies are used, and Table 22-3 lists some behavior-modifying drugs.

TIP

Because of the number and seriousness of side effects associated with hormonal drugs, they should be the last drugs tried in treating animal behavioral problems.

TIP

Hormonal drugs that are reproductive in nature should not be used in pregnant animals.

TABLE 22-2 Behavior Disorders and Some of Their Drug Therapies

Behavioral Disorder	Drugs Used for This Disorder
Urine spraying/marking	• benzodiazepines such as alprazolam, diazepam, and chlordiazepoxide • buspirone • TCAs such as clomipramine • SSRIs such as paroxetine and fluoxetine
Aggression	• TCAs such as amitriptyline • benzodiazepines such as diazepam • SSRIs such as fluoxetine • hormones such as medroxyprogesterone acetate and megestrol acetate
Obsessive-compulsive disorders	• TCAs such as amitriptyline • antihistamines (especially when due to pruritus) • benzodiazepines such as diazepam
Anxiety	• phenothiazines such as acepromazine • TCAs such as amitriptyline • buspirone • antihistamines (especially when due to pruritus) • benzodiazepines such as diazepam
Noise phobias	• benzodiazepines such as alprazolam • buspirone • TCAs such as clomipramine
Fear aggression	• TCAs such as amitriptyline and clomipramine • buspirone • benzodiazepines such as diazepam

TABLE 22-3 Behavior-Modifying Drugs

Drug Category	Drug	Function	Examples
Anti-anxiety agents	benzodiazepines	• Promote the inhibitor neurotransmitter GABA in the brain • Used to treat separation anxiety, phobias, aggression, and urine spraying	• diazepam (Valium®) • lorazepam (Ativan®) • alprazolam (Xanax®) • chlordiazepoxide (Librium®) • flumazenil (Mazicon®)
	antihistamines	• Cause CNS depression, resulting in sedation	• hydroxyzine (Atarax®) • diphenhydramine (Benadryl®)
	phenothiazines	• Work as dopamine antagonists	• chlorpromazine (Thorazine®) • acepromazine (PromAce®) • promazine (Sparine®) • perphenazine (Trilafon®) • prochlorperazine (Compazine®)
	azapirones	• Serotonin blockers • Used to treat urine spraying	• buspirone (BuSpar®)
Antidepressants	tricyclics	• Prevent uptake of neurotransmitters • Used to treat separation anxiety, obsessive licking, hypervocalization, urine spraying	• amitriptyline (Elavil®) • clomipramine (Clomicalm®, Anafranil®) • imipramine (Tofranil®) • doxepin (Sinequan®)
	serotonin reuptake inhibitors	• Inhibit serotonin uptake from the synapse • Used to treat obsessive licking, phobias, separation anxiety, aggression	• fluoxetine (Prozac®) • sertraline (Zoloft®) • paroxetine (Paxil®)
	monoamine oxidase inhibitors	• Block dopamine uptake from the synapse • Used to treat canine cognitive dysfunction or dementia	• selegiline (Anipryl®) • phenelzine (Nardil®) • isocarboxazid (Marplan®) • tranylcypromine sulfate (Parnate®)
Hormones	synthetic progestins	• Believed to correct hormonal imbalance	• megestrol acetate (Megace®, Ovaban®) • medroxy-progesterone (Depo-Provera®)
	synthetic estrogens	• Believed to correct hormonal imbalance	• diethylstilbestrol (DES) (Stilphostrol®)
	testosterone inhibitors	• Inhibit testosterone production or conversion to its potent form	• delmadinone (Terdak®) • finasteride (Proscar®)

SUMMARY

The use of drugs to treat problem behaviors is an expanding area of veterinary medicine. Remember that only a small part of treating animals with behavior problems involves drug therapy. Correctly diagnosing the condition, examining social conditions that affect the situation, and altering external stimuli are all parts of behavior modification.

When behavior modification drugs are used, owners need to be made aware of the potential problems these drugs may cause. Clients should also be told that many behavior-modifying drugs are used extra-label. It should be stressed to owners of animals with diagnosed behavior problems that behavior modification includes retraining the animal and eliminating external stressors for the animal, in addition to the seemingly easy drug treatment.

Classes of drugs used in the treatment of behavior disorders include anti-anxiety drugs, antidepressant drugs, and hormonal drugs. Many behavior-modifying drugs fall into multiple categories, especially when used at varying dosages.

CHAPTER REFERENCES

Delmar's A-Z NDR-97 Nurses drug reference. Clifton Park, NY: Thomson Delmar Learning.

Reiss, B., & Evans, M. (2002), *Pharmacological aspects of nursing care* (6th ed.). Clifton Park, NY: Thomson Delmar Learning.

Veterinary pharmaceuticals and biologicals (12th ed.). (2001). Lenexa, KS: Veterinary Healthcare Communications.

Veterinary values (5th ed.). (1998). Lenexa, KS: Veterinary Healthcare Communications.

CHAPTER REVIEW

Matching

Match the behavior-modifying drug with its classification.

1. ______ megestrol acetate
2. ______ diazepam
3. ______ acepromazine
4. ______ buspirone
5. ______ amitriptyline
6. ______ clomipramine
7. ______ fluoxetine
8. ______ hydroxyzine
9. ______ chlordiazepoxide
10. ______ selegiline

a. MAOI
b. SSRI
c. benzodiazepine
d. antihistamine
e. phenothiazine
f. progestin hormone
g. azapirone
h. TCA
i. barbiturate
j. estrogen hormone

Multiple Choice

Choose the one best answer.

11. Which class of behavior-modifying drugs works by potentiating the inhibitory action of GABA?

a. antihistamines
b. benzodiazepines
c. phenothiazines
d. azapirones

12. Which drug inhibits the reuptake of serotonin?

a. fluoxetine
b. selegiline
c. phenobarbital
d. amitriptyline

13. Which group of drugs works by inhibiting the enzyme that destroys neurotransmitters?

a. SSRI
b. TCA
c. MAOI
d. GABA

14. Which drug can cause diabetes mellitus during use?

a. diethylstilbestrol
b. medroxyprogesterone acetate
c. megestrol acetate
d. delmadinone

15. Which group of drugs works by dopamine antagonism?

a. azapirones
b. phenothiazines
c. progestins
d. TCAs

16. Which group of drugs is both anti-anxiolytic and antipruritic?

a. antihistamines
b. benzodiazepines
c. phenothiazines
d. SSRIs

True/False

Circle a. for true or b. for false.

17. The use of behavior-modifying drugs is the most effective way to treat behavior problems in animals.

a. true
b. false

18. Most behavior-modifying drugs are approved by the FDA for animal use.

a. true
b. false

19. Most behavior-modifying drugs cause sedation.

a. true
b. false

20. Most behavior-modifying drugs start affecting an animal's behavior immediately.

a. true
b. false

Case Studies

21. A 4-yr.-old M/N Pug (Figure 22-2) is presented for aggression toward its owner for the past year. The dog's history includes neutering at six months of age and routine vaccination, deworming, and heartworm testing. The owner states that the aggression started about a year ago: the dog bit her on the hand while she was brushing him. Since then, the owner has taken the dog to a groomer. The dog then began biting if he was not fed or petted on demand. The owner can no longer pick up the dog and is afraid to move the dog off the bed or chair or out of her lap. On PE, the dog is difficult to restrain and has to be muzzled. The results of the PE and baseline laboratory work (CBC, chemistry panel, UA) are normal. The dog is diagnosed with dominance aggression.

a. Is the first treatment option for this dog a behavior-modification drug?
b. Can you think of some reasons why this dog feels dominant over his owner?
c. Can you make any suggestions regarding this case?

FIGURE 22-2 Pug.
Photo by Isabell Francais

CHAPTER 23

Herbal Therapeutics

OBJECTIVES

Upon completion of this chapter, the learner should be able to

- outline the various forms of alternative medicine.
- differentiate between Western and Chinese herbal medicine.
- describe quality-control concerns about herbal supplements.
- describe the different forms and routes of administration of herbs.
- describe properties of commonly used herbs.
- explain ways to ensure proper client education regarding herb use in animals.

KEY TERMS

alternative medicine
complementary medicine
acupuncture
chiropractic
physical therapy
homeopathy
botanical medicine
nutraceutical medicine
holistic veterinary medicine
Western herbal medicine
Chinese traditional herbal medicine
poultice

INTRODUCTION

A client comes into the clinic and says she is interested in maintaining her cat's health by using herbal supplements. She uses a variety of herbal supplements herself and wants to know if she can use some of these on her cat. You check her file and see that her cat has had a heart murmur ausculted on PE for the past few years, but has not had a cardiac workup yet. You know that people can buy herbal supplements without a prescription and without their doctor's advice. Can they really cause any damage? Do you know for sure? What herbal supplements could you recommend (or discourage) based on this cat's history?

ALTERNATIVES

The use of alternative and complementary therapies has been well received by some members of the veterinary community and caused great concern to other members. Regardless of the veterinary viewpoint on these therapies, the numbers of animal owners using alternative therapies have increased over the years, many times without the knowledge of the veterinarian treating the animal. The use of the Internet, especially Website message boards on which pet owners praise the benefits of these therapies, has been a contributing factor in the growth of these therapies.

WHAT'S IN A NAME?

Alternative and complementary medicine consist of multiple diverse disciplines. The term **alternative medicine** applies to treatments or therapies that are outside accepted conventional medicine. **Complementary medicine** implies that these therapies can be used with or in addition to conventional treatment. The use of both terms gives a more inclusive definition to this group of therapies.

The American Veterinary Medical Association (AVMA) has set up guidelines for veterinary alternative and complementary medicine. Included in this field are:

- veterinary **acupuncture** and acutherapy, which consists of the examination and stimulation of body points by use of acupuncture needles, injections, and other techniques for the diagnosis and treatment of numerous conditions in animals.
- veterinary **chiropractic**, which is the examination, diagnosis, and treatment of animals through manipulation and adjustments of specific joints and cranial structures.
- veterinary **physical therapy**, which is the use of noninvasive techniques for the rehabilitation of animal injuries. This includes stretching, range-of-motion exercise, hydrotherapy, massage therapy, and application of heat and cold (Figure 23-1).
- veterinary **homeopathy**, a medical discipline in which animal conditions are treated by administration of substances that are capable of producing clinical signs in healthy animals. These substances are used therapeutically in very small doses. The theory behind homeopathy is that the signs caused by a substance (such as increased mucus production or coughing) are needed clinically to resolve the condition. If these substances are given in small doses, the signs will be produced and help treat the condition.

FIGURE 23-1 Dog receiving physical therapy. *Source:* Kris Jensen, MS, PT, SCS

TIP

Some veterinarian's concerns about the use of alternative and complementary therapies center around their belief in the lack of objective documentation for the safety and efficacy regarding these therapies.

TIP

Some herbs that are classified as alternative treatments have been used for many years. Consider the use of psyllium (bulk laxative), witch hazel (astringent), and aloe vera (burn treatment).

- veterinary **botanical medicine**, which uses plants and plant derivatives as therapeutic agents.
- **nutraceutical medicine**, which uses micronutrients, macronutrients, and other nutritional supplements as therapeutic agents. Examples of nutraceutical substances include glucosamine, chondroitin sulfate, and coenzyme Q.
- **holistic veterinary medicine** is a comprehensive approach to health care using both alternative and conventional diagnostic techniques and therapeutic approaches. Holistic veterinarians may use a combination of methods (including those previously listed) to treat a patient's condition.

The rest of this chapter covers the use of herbs in veterinary medicine.

WEST VERSUS EAST

The use of herbs in human and veterinary medicine can be divided into two main groups: Western and Chinese (also known as Eastern). The central tenet of **Western herbal medicine** is that individuals have an inner force that works to maintain physical, emotional, and mental health. Herbalists who practice Western herbal medicine believe that many diseases occur because an individual's inner force or natural immune system is out of balance.

Western herbal remedies have been used as far back as 3000 B.C. in ancient Egypt. Ancient Greece, Rome, and the Middle East also used herbal remedies. When Columbus arrived in America, New World plants became available to Europeans and blended plant use (New World and European plants) increased. For centuries, most Western medicine consisted of herbal medicine. Only since World War II has medicine relied less on plants and more on synthetic drugs.

Chinese traditional herbal medicine is based on a holistic philosophy of life that emphasizes the relationship among the mental, emotional, and physical components of each individual, as well as the importance of harmony between individuals, their social groups, and the greater population. Chinese traditional herbal medicine attempts to restore health through corrections of imbalances within a patient's body or between the patient and natural order.

Chinese traditional medicine, which includes the use of herbs and medicinal plants in combination, can be traced to three Chinese emperors: Fu Si (2852 B.C.), Shen Nong (3494 B.C.), and Huang Di (2697 B.C.). The principles are based on Taoism, a philosophy that emphasizes following the right path (Tao) in order to find one's place within the universe. This philosophy emphasizes herbal medicine and daily dietary habits. Prescriptions for herbal medicines were formulated to correct excesses of *yin* (cold, moisture, dimness, inward movement, quietness, and slowing) and *yang* (heat, dryness, brightness, outward movement, forceful action, and speed), blockages in the flow of *qi* (the universal life force), localized organ disorders, and emotional problems associated with physical illness.

Why Herbs?

Herbal supplements are one of the fastest growing segments of the dietary supplement market in the United States. It is not clear what is driving this trend, but there appear to be some consistent patterns. One reason may be the desire for a more holistic approach to health care, especially by older people who have more chronic ailments. This rationale extends the holistic approach to animals, especially as they age.

A second reason may be that people believe conventional treatments have real or perceived limitations. They may perceive a higher incidence of adverse effects with chemically based products than with "natural" products. Another reason for this trend

TIP

Some of the most potent and toxic chemicals come from plants. Herbs are considered "natural," but that does not mean that they never have serious side effects.

is that use of the Internet, combined with the advertising practices of the companies that produce herbal supplements, have influenced people's buying behavior. If people use herbal supplements for themselves, they tend to want to use them on their animals as well.

Another factor that may affect people's desire to use herbs in their animals is that herbs have been used for a long time. Human ailments have been treated for thousands of years with herbs. In addition, many conventional drugs are derived from herbs. The assumption is that herbs are safe and do not have the side effects seen in some conventional drugs.

However, keep in mind that herbal supplements can cause problems in some patients. Herbs can interact with conventional medications in many ways, including enhancing the effect of the medicine. Herbs can also increase or reduce the bioavailability of prescription drugs. Herbs that are safe for humans may not be safe for animals. For example, garlic is used in humans as an antimicrobial and cholesterol-lowering agent. In animals, garlic may cause anemia if given for long periods of time.

Quality Control

Herbal supplements do not require FDA approval because, under the 1994 Dietary Supplement Health and Education Act, they are considered food supplements. Safety and efficacy are not guaranteed and no outside monitoring programs are currently available to identify and judge the potency of these herbs. Because herbal products are unregulated, the concentrations of active ingredients can vary between dose forms (ginseng products have been found to vary 200-fold between capsule and liquid forms). Currently the FDA is working with several trade organizations to develop guidelines for herbal supplements. Some manufacturers have devised their own parameters for quality control. Table 23-1 gives information on factors affecting herb quality.

TABLE 23-1 Factors Affecting Herb Quality

Factor	Considerations
Environment	• Is the herb wild-grown or commercially grown? • Were fertilizers used for growing the herb? • What was the climate and rainfall during the growing season of the herb?
Plant	• Was the correct part of the plant used (seeds, stalks, or roots)? • Was the plant part harvested at the correct time? • Were the plants mature enough or young enough to yield the desired active ingredients of the herb? • Is the correct genus and species of the herb being used?
Handling	• After harvest was the herb properly handled? • After harvest, were temperature, sunlight, and humidity maintained in the proper range?

Herb Forms and Administration

Herbs can be given to patients in a variety of forms: capsules and tablets, ointments, extracts, **poultices**, compresses, teas, and in bulk. Table 23-2 summarizes forms of herbs.

Herbal treatments are administered in many different ways. They can be taken orally as a drink; applied to the skin as a cream, ointment, or poultice; taken internally in tablet or capsule form; and/or added to water for bathing.

Sometimes herbs are given in combination with each other. The belief is that a combination of herbs creates an additive effect of the advantages of the different herbs, while diluting the toxic effects of one herb with the benefits of other herbs in the combination.

Herbal Countdown

The following are some of the more popular herbal supplements used by people. Clients may be interested in using these same herbs for their animals.

TABLE 23-2 Herb Forms

Herb Form	Properties
Capsules and tablets	• Effective and convenient to give • Variety of sizes available
Ointments	• Topical application for specific areas • Should be applied to intact skin
Extracts	• Liquids come in water, alcohol tincture, and glycerin forms • Powders come in capsules and tablets • Standardized commercially to provide a certain number of milligrams of active ingredient per dose • Efficacy is more consistent
Poultices and compresses	• A poultice is made by boiling fresh or dried herbs, squeezing out excess liquid, cooling the herb, and applying it to the skin. The herb is then wrapped with gauze to hold the material in place. This is left on for a few minutes to hours. • A compress is made with an herbal extract (not the plant part) by soaking a cloth in the extract. The cloth is then applied to the skin.
Teas	• Best-known form for human patients • Most active ingredients of herbs are water-soluble and therefore available to the patient in tea form • Commercial teas usually do not have adequate amount of herb to be effective
Bulk	• Herb purchased in loose form • Form may be chopped, powdered, or fresh

Ginkgo Ginkgo, *Ginkgo biloba*, is the earth's oldest living tree species; many individual trees live for more than 1000 years (Figure 23-2). Gingko is the only tree to survive the atomic blasts of Hiroshima and Nagasaki. It has been used for centuries in China to help improve memory and to treat respiratory problems in humans.

The active ingredients in ginkgo are ginkgo flavone glycosides and the terpene lactones. It has been used in cases of vascular insufficiency, to enhance the utilization of oxygen and glucose by the brain, and to reestablish perfusion in areas of ischemia (deficiency of blood flow). It has also been used to treat depression. In animals, gingko has been used to reduce aging effects on the nervous system, to reduce hypertension, and as a general tonic in geriatric patients.

An ingredient in gingko is a potent inhibitor of platelet aggregating factor. Animals using aspirin, NSAIDs, and heparin should be given gingko with care. Gingko should be stopped at least one week prior to surgery. Gingko may also inhibit cytochrome P450 enzyme and may induce hypoglycemia. Animals with diabetes mellitus should not take gingko.

FIGURE 23-2 Ginko (*Ginkgo biloba*)

St. John's Wort St. John's Wort, *Hypericum perforatum* (Figure 23-3), got its common name because it is believed that the flowers of St. John's Wort have their brightest appearance on June 24th, the birthday of John the Baptist. In Europe people use St. John's Wort as a food coloring and flavor additive and topically for the treatment of burns and wounds.

About 10 groups of components are active ingredients in St. John's Wort. One component, hyperforin, regulates the effects of serotonin. In humans, St. John's Wort is used to treat mild to moderate depression. It has been extensively studied in humans and appears to work well in some patients. In animals, St. John's Wort has been used to treat behavior disorders such as lick granulomas, aggression, separation anxiety, and obsessive-compulsive disorders.

St. John's Wort can induce cytochrome P450 and alter the pharmacodynamics of other drugs. It should not be used with other antidepressant medications. It can also affect blood pressure and cause photosensitivity (especially in livestock) in higher doses.

Ginseng Ginseng, *Panax ginseng* (Figure 23-4), has been part of traditional Chinese medicine for more than 2000 years. It has been used to treat almost every ailment, but is popular as a general tonic and as a medication to help cope with stress.

FIGURE 23-3 St. John's Wort (*Hypericum perforatum*)

FIGURE 23-4 Ginseng (*Panax ginseng*)

FIGURE 23-5 Garlic (*Allium sativum*)

The main active ingredients are ginsenosides (13 have been identified) that are thought to be responsible for increasing energy, countering stress, and enhancing physical performance. Ginseng also seems to stimulate natural killer cell activity, resulting in improved ability to fight infection. Ginsenosides also interfere with coagulation and platelet function. In animals, ginseng has been used to treat extreme weight loss (such as that secondary to cancer), anorexia, systemic infections, and feline leukemia.

Dosage should be standardized to the ginsenoside content of the product being used. One problem with dosing ginseng is the great variability in commercial ginseng products and variation of the ingredient concentration from what is represented on the label. Ginseng seems to be well tolerated, but it is recommended not to give more to an animal in an attempt to get better results. The inhibition of platelet aggregation has raised some safety concerns with increased dosages or increased frequency of use. Ginseng use should be discontinued at least one week prior to elective surgery. Ginseng can also affect blood sugar levels in nondiabetic and diabetic animals, increase blood pressure and heart rate, increase gastrointestinal motility, and stimulate the brain to the point of seizures (when large doses are used).

Garlic Garlic, *Allium sativum* (Figure 23-5), has been used for more than 5000 years in humans, for a variety of conditions including parasitic infections, respiratory problems, and poor digestion. Louis Pasteur confirmed garlic's antimicrobial action and Albert Schweitzer used it in the treatment of amoebic dysentery.

Garlic's activity is due to its allicin content. Alliin is an odorless amino acid that, when exposed to air (when the garlic is broken), will produce allicin. Allicin is an unstable compound that gives garlic its characteristic odor. Therefore, any garlic preparation that is odorless is probably worthless. Allicin is converted to other sulfur compounds, such as ajoene, which are the active ingredients of garlic.

Garlic is used for numerous purposes, including reduction of cholesterol and triglyceride levels and hypertension. It is believed that garlic has anti-carcinogenic properties, especially against colon, stomach, and prostate cancers. It also has an inhibitory effect on *Heliobacter pylori*, which has been linked to the development of ulcers. Garlic has also been used to treat parasitic (roundworm, tapeworm, and hookworm infections) and fungal infections (ringworm). In animals, it has been used for flea prevention (this use has not been clinically proven), respiratory problems, and to acidify urine (testing has not proven this latter claim). Garlic is high in potassium and should be used with caution in animals with electrolyte imbalances.

Dosages of garlic vary with the preparation and desired effect. Precautions concerning garlic use revolve around inhibition of platelet aggregation and prolongation of bleeding times. Garlic should be discontinued at least one week prior to any surgical procedure. In high doses garlic can cause gastrointestinal upset, clotting problems, and Heinz body anemia. Because garlic enters the milk, these gastrointestinal problems can be passed to suckling young.

FIGURE 23-6 Echinacea (*Echinacea purpurea*)

Echinacea *Echinacea purpurea* (Figure 23-6) is a wildflower native to North America, commonly known as the purple coneflower. Historically, it has been used for a variety of conditions such as infections, joint pain, abscesses, and burns. Echinacea was part of the *National Formulary of the United States* until 1950, when antibiotic use increased and interest in echinacea waned. With the increasing problems of antibiotic resistance, echinacea's benefits have been reexamined.

The actions of echinacea are due to polysaccharides called fructofuranosides. These compounds contribute to tissue regeneration, regulation of the inflammatory response, and a mild cortisone-like effect. Echinacea stimulates phagocytosis and natural killer cell activity. Most evidence shows that echinacea is effective in limiting the severity of and shortening the duration of some infections.

Dosage of echinacea varies with the preparation of the herb. It is usually given at higher doses early in the course of disease for better results. Minimal side effects are seen with this herb. Echinacea should not be used in animals with autoimmune disorders or immunosuppressive disease such as FeLV or FIV.

Saw Palmetto Saw palmetto, *Serenoa repens* (Figure 23-7). was used by Native Americans in the treatment of urinary tract infections. By the early twentieth century, it was being used to treat benign prostatic hypertrophy in humans.

The active ingredients of saw palmetto are fatty acids from the berries, which produce an enzyme to prevent the conversion of testosterone to dihydrotestosterone (DHT). Saw palmetto does not alter the overall size of the prostate, but shrinks the inner prostatic epithelium. Saw palmetto has been used to treat benign prostatic hyperplasia, to stimulate appetite, and as a mild diuretic. Saw palmetto extract is well tolerated and few adverse side effects are noted other than mild gastrointestinal problems.

FIGURE 23-7 Saw palmetto (*Serenoa repens*)

Evening Primrose Evening primrose, *Oenothera biennis* (Figure 23-8), contains gamma-linolenic acid, which is a fatty acid in the omega-6 family. Evening primrose oil has been used in humans for treating premenstrual syndrome, rheumatoid arthritis, diabetic neuropathy, and eczema.

Dosage of evening primrose oil varies with the condition being treated. It is recommended that evening primrose oil be taken with food to increase absorption. Benefits may not be seen for several months in treating chronic conditions. Because this is an oil supplement, higher doses can lead to loose stools and abdominal cramps. Evening primrose oil may lower the seizure threshold in some people on phenothiazine medication, so use caution when giving it to animals with a history of seizures.

Goldenseal Goldenseal, *Hydrastis canadensis* (Figure 23-9), gets its name from the fact that when its stem is broken near the base, the scar resembles the gold wax used to seal letters. Native Americans used goldenseal as a dye and medicinally for skin and eye conditions and diarrhea. Because of overharvesting in the 1980s, goldenseal is considered an endangered species.

The active ingredients of goldenseal are the isoquinoline alkaloids—especially a substance called berberine. Berberine is found in the stem bark of the plant as well as in the roots. Goldenseal is used mainly as an antibacterial and antiparasitic drug. Its antiparasitic activity is primarily against amoebae, such as *Entamoeba sp.*, and the protozoan *Giardia*.

Dosage for oral use of goldenseal varies, but it should be taken only for a limited time (usually two to three weeks). High doses of goldenseal for prolonged periods of time can cause cardiac problems and central nervous stimulation. Goldenseal should not be used in pregnant animals because it stimulates uterine contractions.

Cranberry Cranberry, *Vaccinium macrocarpon* (Figure 23-10), is a North American heritage herb that has been used for food and as medicine mainly for the treatment of urinary tract infections. Its name comes from the Pilgrims who called it *crane berry* because the plant's stem looks like the neck and head of a crane.

It was believed that cranberry acidified urine, but that is now known to be false. The primary effect of cranberry is interference with the attachment of urinary pathogens (such as *E. coli*) to the urinary bladder wall. Cranberry has been used to treat urinary tract infections in cats. Cranberry compounds may also protect the stomach from adhesion of *H. pylori* and prevent the development of ulcers. High levels of cranberry consumption can lead to diarrhea and stomach problems. Prolonged use can lead to the formation of kidney stones.

FIGURE 23-8 Evening primrose (*Oenothera biennis*)

Valerian Valerian, *Valeriana officinalis* (Figure 23-11), is a plant native to Europe and North America. Valerian produces small, rose-colored flowers that bloom on a four-foot-high stem. Historically, it has been used as a sedative and anti-anxiety drug.

TIP

Catnip, part of the mint family, may be considered a herbal medicine. The active ingredient in catnip is nepatalactone, a chemical that closely resembles a chemical structure found in female cat urine. This may be why unneutered male cats have more of a reaction to catnip than do female or neutered male cats.

Valerian's effects have been attributed to several volatile oils, including valeric acid, valerenic acid, and valepotriates. Valerenic acid inhibits the breakdown of gamma-aminobutyric acid (GABA). Because GABA is inhibitory, hindering its breakdown results in higher levels of GABA, which decreases central nervous system activity. Valerian is primarily used as a sleep aid, to calm hyperactivity in dogs, and for mild tranquilization. Adverse reactions to valerian are mild, but may include hepatotoxicity with long-term use. Care should be taken if giving valerian with other depressant drugs.

Advice to Clients

Given the increasing public awareness and use of herbs and other complementary and alternative treatment methods, the veterinary profession needs to provide reliable information to clients so that responsible choices can be made. One way to provide reliable information is by developing regulation of herbal supplements. The National Animal Supplement Council has developed a Compliance Plus program, a nonregulatory program that applies to non-food-producing animals and the reporting of adverse effects of complementary and alternative treatments. Members of the National Animal Supplement Council would also be required to implement labeling, manufacturing, and quality control standards for their industry. This may help with the development of standards for the herbal supplement and product industry.

Another way to provide reliable information is by educating clients about herb use and possible side effects or interactions. The following are some guidelines for dealing with patients in a clinical setting.

- Ask each client whether they give herbs or other supplements to their animals.
- Inform clients that herb-drug interactions exist.
- Encourage the use of standardized products from respected manufacturers.
- Use herbal therapies in recommended doses; more is not better.
- Avoid herbs with known toxicities.
- Herb use in pregnant or nursing animals, the very young, or the very old may not be wise.

FIGURE 23-9 Goldenseal (*Hydrastic canadensis*)

FIGURE 23-10 Cranberry (*Vaccinicum macrocarpon*)

FIGURE 23-11 Valerian (*Valeriana officinalis*)

- Accurate diagnosis of the animal's condition is essential to evaluate all therapeutic options.
- Document all herb or supplement use in the animal's medical record.

SUMMARY

The use of complementary and alternative medicine and therapies has increased in recent years. Multiple disciplines of complementary and alternative medicine are recognized by the American Veterinary Medical Association.

Herbal supplements have been used for thousands of years for a variety of conditions. Examples include gingko, St. John's Wort, garlic, and ginseng. It is important to use a preparation with the correct part of the plant (root, stem, or seed) when using herbs, and to be aware of possible herb-drug interactions. Some herbs have side effects and may have to be discontinued prior to surgery. Always question clients about whether they use herbal supplements for their animals.

Table 23-3 lists the herbs discussed in this chapter.

CHAPTER REFERENCES

AltVetMed (complementary and alternative veterinary medicine): <http://www.altvetmed.com>.

Guidelines on alternative and complementary therapies. (1996). Schaumburg, IL: American Veterinary Medical Association.

Libster, M. (2002). *Delmar's integrative herb guide for nurses.* Clifton Park, NY: Thomson Delmar Learning.

Nolen, R. Facing crackdown, dietary supplement companies promise changes. (2002, August 15). *JAVMA, 221*(4): 479–483.

PDR for herbal medicines (2d ed.). (2000). Montvale, NJ: Thomson Medical Economics.

PDR for nutritional supplements. (2001). Montvale, NJ: Thomson Medical Economics.

Pet Education.com: <http://www.peteducation.com>.

Ramey, D. (2003, June 15). Regulatory aspects of complementary and alternative veterinary medicine. *JAVMA, 222*(12): 1679–1682.

Watkins, R. (2002, March). Herbal therapeutics: The top 12 remedies—Part I. *Emergency Medicine Journal, 34*(3): 12–19.

TABLE 23-3 Herbs Covered in This Chapter

Herb	Body System Category	Effect	Cautions	Form Used
Ginkgo	• Nervous • Cardiovascular	• Reduces aging-associated brain problems • Anticoagulant • Neutralizes free radicals that damage cells	• Do not use long term in animals with clotting problems	Tinctures, tea infusions, capsules, oil used internally and externally

(continues)

TABLE 23-3 *Continued*

Herb	Body System Category	Effect	Cautions	Form Used
St. John's Wort	Nervous	• Anti-anxiety • Antidepressant • Anti-inflammatory	• Photosensitization	Dried leaves and flowering tops used for tea infusion and tincture (internally), oil (externally)
Ginseng	Immune	• Immunostimulant that prevents disease • Stimulant of nervous system • Antifatigue drug	• Gastrointestinal problems if used long term	Roots used for internal administration with capsules and teas
Garlic	Cardiovascular	• Antibiotic activity to some species of gram + and gram − bacteria • Cholesterol and blood pressure reduction • Antiparasitic • Antifungal against ringworm	• Clotting problems may develop • Larger doses can cause gastrointestinal problems • Heinz body anemia may be seen with long-term use	Fresh pulp and capsules are given internally
Echinacea	Immune	• Immunostimulant • Anti-inflammatory	• None known	Roots are used for capsules, tinctures, and compresses
Saw palmetto	Urogenital	• Anti-inflammatory • Appetite stimulant	• Diarrhea in large amounts	Berries are used in an extract given internally
Evening primrose	Skin Mucous membranes	• Atopic eczema	• Avoid in animals with blood-clotting disorders	Extracted oil is used in ointments and compresses
Goldenseal	Gastrointestinal Urinary	• Antibiotic activity • Antiparasitic against amoeba and giardia • Less commonly used for urinary infections and gastroenteritis	• Do not use in pregnant animals (stimulates the uterus) • Diarrhea in large doses • Nervous system effects	Dried roots are used in tincture, capsule, and tea forms
Cranberry	Urogenital	• Prevent and treat urinary infections • Reduces urine calcium to help prevent stones	• None known	Berries and dried powder are used for juice and capsules
Valerian	Nervous	• Tranquilizer • Less commonly used as a diuretic and antispasmotic	• None known	Root is used dried or fresh in tincture form

Watkins, R. (2002, March). Herbal therapeutics: The top 12 remedies—Part II. *Emergency Medicine Journal, 34*(4): 43–54.

Watkins, R. (2002, March). Herbal therapeutics: The top 12 remedies—Part III. *Emergency Medicine Journal, 34*(5): 33–39.

CHAPTER REVIEW

Multiple Choice

Choose the one best answer.

1. The belief of which alternative medicine is that individuals have an inner force that works to maintain physical, emotional, and mental health?
   a. Western herbal medicine
   b. veterinary botanical medicine
   c. nutraceutical medicine
   d. Chinese traditional herbal medicine

2. Which type of alternative medicine is based on a holistic philosophy of life that emphasizes the relationship among the mental, emotional, and physical components of each individual, as well as the importance of harmony between individuals, their social groups, and the population at large?
   a. Western herbal medicine
   b. veterinary botanical medicine
   c. nutraceutical medicine
   d. Chinese traditional herbal medicine

3. What herb form involves boiling fresh or dried herbs, squeezing out excess liquid, cooling the herb, and applying it to the skin?
   a. dried root
   b. poultice
   c. compress
   d. dried stem

4. Which herb has been used to treat urinary tract infections in cats because it inhibits binding of some bacteria to the urinary bladder wall?
   a. catnip
   b. valerian
   c. cranberry
   d. saw palmetto

5. Which herbal supplement may cause photosensitivity?
   a. gingko
   b. ginseng
   c. cranberry
   d. St. John's Wort

6. Which herbal supplement has been used as a mild tranquilizer in dogs?
   a. valerian
   b. saw palmetto
   c. ginseng
   d. St. John's Wort

7. What is the active ingredient in catnip?
   a. nepatalactone
   b. ginsenoside
   c. allicin
   d. fructofuranoside

True/False

Circle a. for true or b. for false.

8. Because herbal supplements are used in humans and animals, they are regulated by the FDA.
   a. true
   b. false

9. Herbs are natural; therefore they can be used safely in pregnant and nursing animals.
   a. true
   b. false

10. All parts of herbs have equal potency and effectiveness.
   a. true
   b. false

Case Study

A 10-yr.-old DSH cat is brought into the clinic for lethargy. Her owner states that the cat has been acting "tired" and does not seem to want to eat. On PE, the cat's TPR are normal, a large amount of gas is palpated in the intestines, and some loose feces are seen in the perineal region. Blood is collected from this cat for a CBC, chemistry screen, and thyroid level. The cat's CBC results show that this cat is anemic, which is probably the cause of her lethargy. The cat's fecal exam reveals that the cat also has roundworms and hookworms. When told of the test results, the cat's owner says they cannot be correct, because she has been treating the cat with garlic supplements to get rid of the parasitic infection.

a. What can you tell this owner about garlic supplementation?
b. What additional information would you like to know from this client regarding her cat?
c. Could the garlic be causing any of the cat's health problems? Why?

APPENDIX A
Unit Conversions and Math References

PREFIXES FOR METRIC UNITS

1,000,000 = mega- = M

1,000= kilo- = k

100 = hecto- = h

10 = deka- = dk

0.1 = deci- = d

0.01 = centi- = c

0.001= milli- = m

0.000001= micro- = μ

0.000000001= nano- = n

0.000000000001 = pico- = p

METRIC-TO-METRIC CONVERSIONS

Linear measure: base unit is meters (m)

1 m = 100 centimeters (cm)

1 m = 1000 millimeters (mm)

1 m = 1,000,000 micrometers or microns (mcm or μ)

1000 m = 1 kilometer (km) or 0.001 km = 1 m

Volume measure: base unit is liter (l)

1 l = 100 centiliters (cl)

1 l = 1000 milliliters (ml)

1000 l = 1 kiloliter (kl)

Weight measure: base unit is gram (g)

1 g = 100 centigrams (cg)

1 g = 1000 milligrams (mg)

1 g = 1,000,000 micrograms or 0.000001 g = 1 microgram (mcg or μg)

0.001 mg = 1 mcg or 1 μg (a gram is 1000 times a milligram; a milligram is 1000 times a microgram); 1 mg = 1000 mcg or μu

1000 g = 1 kg

METRIC, HOUSEHOLD, AND APOTHECARY CONVERSIONS

Length

Metric base unit = meter

1 meter = 1.0936 yards

1 centimeter = 0.39370 inch

1 inch = 2.54 centimeters

1 kilometer = 0.62137 mile

1 mile = 5280 feet or 1.6093 kilometers

1 foot = 0.3048 meter

Mass

Metric base unit = gram

1 kilogram = 2.2 pounds

1 pound = 453.59 grams

1 pound = 16 ounces

1 grain = 65 milligrams

15 grains = 1 gram

1 dram = 3.888 grams

1 ounce = 28.35 grams

1 ton = 2000 pounds

1 gram = 0.035274 ounces

Volume

Metric base unit = liter

1 liter = 1.0567 quarts

1 gallon = 4 quarts

1 gallon = 8 pints

1 pint = 2 cups = 16 fluid ounces

1 cup = 8 ounces

1 gallon = 3.7854 liters

1 quart = 32 fluid ounces

1 quart = 0.94633 liter

1 minim = 0.06 milliliter

1 fluid dram = 3.7 milliliters

1 fluid ounce = approximately 30 milliliters

1 milliliter = 1 cubic centimeter

Temperature

$°F = 1.8°C + 32$

$°C = \frac{°F - 32}{1.8}$

DOSE CALCULATIONS

To calculate a dose:

1. Convert the animal's weight from pounds to kilograms (unless the dosage is given in pounds).
2. Multiply the animal's weight in kg by the dosage = dose.
3. Divide the dose by the drug concentration, with the unit of drug to administer in the numerator (set up units so that they cancel).

APPENDIX B
Conversion of Body Weight in Kilograms to Body Surface Area in Meters (for Dogs and Cats)

DOGS

kg	m^2	kg	m^2
0.5	0.06	26.0	0.88
1.0	0.10	27.0	0.90
2.0	0.15	28.0	0.92
3.0	0.20	29.0	0.94
4.0	0.25	30.0	0.96
5.0	0.29	31.0	0.99
6.0	0.33	32.0	1.01
7.0	0.36	33.0	1.03
8.0	0.40	34.0	1.05
9.0	0.43	35.0	1.07
10.0	0.46	36.0	1.09
11.0	0.49	37.0	1.11
12.0	0.52	38.0	1.13
13.0	0.55	39.0	1.15
14.0	0.58	40.0	1.17
15.0	0.60	41.0	1.19
16.0	0.63	42.0	1.21
17.0	0.66	43.0	1.23
18.0	0.69	44.0	1.25
19.0	0.71	45.0	1.26
20.0	0.74	46.0	1.28
21.0	0.76	47.0	1.30
22.0	0.78	48.0	1.32
23.0	0.81	49.0	1.34
24.0	0.83	50.0	1.36
25.0	0.85		

Although the above chart is compiled for dogs, it can also be used for cats. A formula for more precise values is

$$\text{BSA in m}^2 = \frac{K \times W^{2/3}}{10^4}$$

BSA = Body surface area

m^2 = square meters

W = weight in grams

K = 10.1 (dogs)
10.0 (cats)

Table adapted from *Veterinary Values* (5th ed.). (1998). Lenexa, KS: Veterinary Healthcare Communications.

APPENDIX C
Metric Conversions Guide

When doing metric conversions, use this table to make it easier.

	3		2		1	
kg		g		cg		mg
kl		l		cl		ml
km		m		cm		mm

Each time you need to convert something to a larger or smaller measurement, move left or right on the table. When you cross a line with a number, move the decimal place that many places to the left or right, depending on your conversion. If you pass through more than one line, add up the numbers to know the number of places to move the decimal. For example, how many cm are there in 5 m? You move from m to cm crossing only the line with the number 2; hence, you add 2 decimal places or zeros in this case (in the direction of your move). How many km are in 20 mm? You move from mm to km crossing 3 lines (a line with the number 1, a line with the number 2, and a line with the number 3). Adding 1 + 2 + 3 = 6. You move the decimal places 6 places to the left (the direction of your move) and your answer is 0.000020 km.

Be sure to validate your answer by making sure the larger metric unit has the smaller numerical value.

APPENDIX D
Vitamin Supplements

WATER-SOLUBLE VITAMINS

Vitamin	Function	Example	Administration Route
vitamin C	• antioxidant • needed for collagen formation • maintains capillaries, bones and teeth	• ascorbic acid parenteral injection • OTC brands	IM, SQ, IV, oral
thiamine (vitamin B_1)	• part of coenzyme needed for cellular respiration • helps support nervous system	• Vita-Jec® Thiamine • generic • part of B complex • OTC brands	IM, SQ, IV, orally
riboflavin (vitamin B_2)	• part of coenzymes needed for cellular respiration • involved in oxidation of fat and protein	• Part of B complex product • OTC brands	IM, SQ, IV, orally
niacin (nicotinic acid)	• part of coenzymes needed for cellular respiration • involved in oxidation of fat and protein	• part of B complex product • OTC brands	IM, SQ, IV, orally
folacin (folic acid)	• coenzyme needed for production of hemoglobin • helps in formation of DNA	• part of B complex product • OTC brands	IM, SQ, IV, orally
vitamin B_6	• coenzyme used for synthesis of hormones and hemoglobin • CNS control	• part of B complex product • OTC brands	IM, SQ, IV, orally
pantothenic acid (*d*-panthenol)	• part of coenzyme needed for oxidation of carbohydrates and fats • Aids in formation of hormones and some neurotransmitters	• part of B complex product • OTC brands	IM, SQ, IV, orally

(continues)

WATER-SOLUBLE VITAMINS (CONT.)

Vitamin	Function	Example	Administration Route
vitamin B_{12} (cyanocobalamin)	• part of coenzyme needed for synthesis of nucleic acids and myelin • involved in RBC production	• vitamin B_{12} injection • Part of B complex product • OTC brands	IM, SQ, IV, orally
biotin	• coenzyme needed for amino acid and fatty acid metabolism	• part of B complex product • OTC brands	IM, SQ, IV, orally
choline	• part of coenzyme needed for metabolism	• part of B complex product • OTC brands	IM, SQ, IV, orally

FAT-SOLUBLE VITAMINS

Vitamin	Function	Example	Administration Route
vitamin A	• antioxidant synthesized from beta-carotene • needed for healthy eyes, skin, hair, mucous membranes, and bone growth	• A-D Injection® • Vitamin AD Injection® • OTC brands	IM, SQ, orally
vitamin D	• a group of steroid vitamins needed for development and maintenance of bone and teeth	• A-D Injection® • Vitamin AD Injection® • OTC brands	IM, SQ, orally
vitamin E	• antioxidant that prevents oxidation of vitamin A and polyunsaturated fatty acids • prevents white muscle disease in sheep and calves	• vitamin E/selenium combinations (Bo-Se®, L-Se®, Mu-Se®, E-Se®, Seletoc®) • Vitamin E injectable • OTC brands	IM, SQ, orally
vitamin K	• needed for synthesis of blood clotting • also used in the treatment of rodenticide poisoning	• AquaMEPHYTON®, Mephyton®, Konakion® (all of these products are vitamin K_1)	IM, SQ, IV, orally

APPENDIX E
Inventory Management

WHY KEEP A DRUG INVENTORY?

- It tells you what is physically available for use and what is on order.
- It is a source of manufacturers/distributors with addresses, telephone numbers, and representatives' names.
- It involves receiving, unpacking, listing, and pricing of drugs and supplies.

TYPES OF INVENTORY MANAGEMENT

- Alphabetical: all drugs and supplies arranged from A to Z
- Classification: drugs are categorized by use
- Categoric: capsules/tablets in one area, injectables in another, and so on
- Numeric: each drug is assigned a number and stored accordingly

MONITORING CURRENT INVENTORY

- Should at least be done annually
- Computer programs provide instant inventory
- Hard-copy methods include ledgers and inventory cards
- Should include:
  - item
  - supplier
  - expiration dates
  - cost per unit
  - quantity purchased
  - quantity on hand
  - date of current inventory

CONTROLLED SUBSTANCE INVENTORY

- Purchase records for schedule II–V drugs (schedule I drugs are not utilized in veterinary clinics) must be kept for at least five years.
- Schedule II drugs must be logged separately.
- Inventory follow-up must be done every two years.
- Storage areas must be locked (preferably double-locked).
- Records must be easily retrievable.

RECEIVING SHIPMENTS

- Check packing slip to make sure you received the correct items, quantity of items, size and strength of medication, and so on. Make sure the package was sent to the correct address.
- Call the distributor immediately if there is a problem.
- Refuse shipment of damaged items. Remember to retain the shipping slip and a copy.
- Make sure the shipment was handled correctly (was not frozen, left in heat, etc.)
- Once the shipment has been checked, file the packing slip for back orders, individual companies, and so on.
- When unpacking, make sure to rotate stock so that items that will expire first are used first and any outdated items can be returned to the supplier for a possible credit.
- Check dating and make sure items can be used within the expiration dates (you may need to request longer dating if an item is not used frequently).
- Wait for back orders or cancel them and order through another company.

APPENDIX F
Managing Animal Toxicities

I: INITIATE LIFE SUPPORT MEASURES

Life support measures are aimed toward maintaining normal respiratory, cardiovascular, and neurologic function.

Clinical Sign	Drug
bradycardia	atropine
CNS depression	naloxone
hyperactivity	diazepam
metabolic acidosis	sodium bicarbonate
seizures	diazepam, phenobarbital

II: DECONTAMINATION

The goal of any decontamination process is to reduce further exposure of the animal to the toxicant. Animals must be continually monitored even after decontamination to make sure they do not develop signs of toxicity from other routes of exposure. The following table lists ways to decontaminate animals.

Application	Drug	Comments
adsorbents	• activated charcoal	Used with recent ingestions to adsorb the toxins from the body. It is used at a 10:1 activated charcoal-to-toxicant ratio. It is contraindicated with caustic agents.
bathing	• detergent or shampoo	Removes most dermal exposures.
cathartics	• sorbitol • sodium sulfate • magnesium sulfate	Causes diarrhea to help move toxicants out of the body. May be combined with activated charcoal.
emetics	• syrup of ipecac • 3% hydrogen peroxide • apomorphine • xylazine • salt	Good for recent exposures. Contraindicated with corrosive agents, in animals with altered mental status, and in rabbits/rodents.

III: CHELATION THERAPY

Chelators have the ability to form chelator-metal complexes that enhance the elimination of metal. Chelators work best soon after metal exposure. Any possible metal source should be removed to avoid further contamination.

Chelator	Metal(s) Chelated	Comments
EDTA	• lead • zinc	Given in 5% dextrose; dosage may have to be adjusted based on renal function. Multiple treatments may be needed, but should not exceed 5 days.
Deferoxamine mesylate	• iron	Usually supplemented by gastric lavage with magnesium hydroxide
DMSA (2,3-dimercaptosuccinic acid)	• arsenic • lead • mercury	May reduce gastrointestinal absorption of lead, with limited side effects
diphenylthiocarbazone	• thallium	Renal function should be monitored and dose/treatment adjusted if needed
d-penicillamine	• lead • copper • mercury	Side effects include proteinuria and hematuria; may see vomiting, anorexia, or CNS depression with this drug
Prussian blue (ferric-cyanoferrate)	• thallium	Stains feces blue

IV: ANTIDOTAL THERAPIES

Antidotal therapies are used to counteract the effect of toxicants in various ways. Some stabilize vital signs, decrease exposure, facilitate toxicant removal, or antagonize the toxicant at its site of action. Supportive therapy should also be considered in these cases, as the patient will benefit from a combined approach to toxin treatment.

If Poisoned By	Use Drug
acetaminophen	• acetylcysteine
anticoagulant rodenticides	• vitamin K_1
atropine	• physostigmine
carbamate insecticides	• atropine
carbon monoxide	• oxygen
cholecalciferol (vitamin D)	• calcitonin
cyanide	• sodium nitrite, sodium thiosulfate
digoxin, digitoxin	• sheep-origin antibody fragments with digoxin
ethylene glycol	• alcohol dehydrogenase inhibitors such as 4-methylpyrazole and ethanol
fluoride	• calcium borogluconate
heparin	• protamine sulfate
NSAIDs	• sucralfate, metoclopramide
opiates	• naloxone
organophosphates	• atropine
organophosphate insecticides	• pralidoxime chloride (2-PAM)
snakebite (rattlesnake, copperhead, water moccasin)	• antivenin

Source: Poison Management. In *Veterinary Values* (5th ed.). (1998). Lenexa, KS: Veterinary Healthcare Communications.

APPENDIX G
Web Sites of Veterinary Pharmaceutical Companies

United States Food and Drug Administration: http://www.fda.gov

3M Animal Care Products: http://www.mmm.com
Abbott Labs: http://abbott.com
Activon Products: http://www.activon.com
Addison Biological Laboratory: http://www.addisonlabs.com
Agri Laboratories: http://www.agrilabs.com
Airex Laboratories: http://www.bullenairx.com
Air-Tite Products Co.: http://www.air-tite.com
Alcide Corporation: http://www.alcide.com
Alex C. Fergusson Inc.: http://www.afco.net
Allergan: http://www.allergan.com
Allied Monitor: http://www.alliedmonitor.com
Alpharma Inc.: http://www.alpharma.com
American Animal Health, Inc.: http://www.aahinc.com
Amgen: http://wwwext.Amgen.com
Anchor Division of Boehringer Ingelheim Vetmedica, Inc.: http://www.bi-vetmedica.com
Animal Blood Bank: http://www.io.com/~tradew.abb
Antec International: http://www.antecint.com
Aqua Health Ltd.: http://www.cobequin.com
Argent Chemical Laboratories: http://www.argent-labs.com
Aristavet Pharmaceuticals LLC: http://www.aristavet.com
Balchem Corporation: http://www.balchem.com
Bayer: http://www.bayer.co.uk/products/animalhealth.html and http://www.bayerus.com/ah
Bimeda, Inc: http://www.bimeda.com
Bio-ceutic Division of Boehringer Ingelheim Vetmedica: http://www.bi-vetmedica.com
Biocor Animal Health: http://www.biocorah.com
Bio-Derm Laboratories, Inc.: http://www.biogroom.com
Biopure Corporation: http://www.biopure.com
Bio-Tek Industries, Inc.: http://www.bio-tekdisinfectant.com
Biovance Technologies: http://www.biovance.com
Bioyme Incorporated: http://www.biozymeinc.com

Blue Ridge Pharmaceuticals, Inc: http://www.brpharma.com
Boehringer Ingelheim Animal Health:
http://www.boehringer-ingelheim.com/biahi/
Bou-Matic: http://www.boumatic.com
Bristol-Myers Squibb: http://www.bms.com
Butler Company: http://www.wabutler.com
Champion Protector/Performer Brand: http://www.agrilabs.com
Colorado Serum Company: http://www.colorado-serum.com
Combe, Inc.: http://www.combe.com
Conklin Company, Inc.: http://www.conklin.com
Davis Veterinary Products: http://www.davismfginc.com
Dermapet: http://www.dermapet.com
DMS Laboratories: http://www.rapidvet.com
Durvet, Inc: http://www.durvet.com
DVM Formula: http://www.dvmformula.com
DVM Pharmaceuticals, Inc.: http://www.DVMPharmaceuticals.com
Elanco Animal Health: http://www.elanco.com/
Eli Lilly: http://www.lilly.com
Equicare Products: http://www.farnam.com
EVSCO Pharmaceuticals: http://www.evscopharm.com
Farnam Companies: http://www.farnam.com
First Priority, Inc.: http://www.prioritycare.com
Fort Dodge Animal Health:
http://www.wyeth.com/divisions/fort_dodge.asp
GlaxoSmithKlein (GlaxoWellcome + Smith Klein Beecham):
http://www.gsk.com
Grand Laboratories, Inc.: http://www.grandlab.com
Halocarbon Laboratories: http://www.halocarbon.com
Happy Jack, Inc.: http://www.happyjackinc.com
Hawthorne Products, Inc.: http://www.hawthorne-products.com
Heska Corporation: http://www.heska.com/
IDEXX Laboratories: http://www.idexx.com
Immucell Corporation: http://www.immucell.com
Immvac, Inc: http://www.immvac.com
IVAX: http://www.ivax.com
Johnson & Johnson: http://www.johnsonandjohnson.com/home.html
Jorgensen Labs, Inc.: http://www.jorvet.com
JPI Jones Animal Health: http://www.soloxine.com
Loveland Industries, Inc.: http://www.lovelandindustries.com
Luitpold Pharmaceuticals, Inc.: http://www.luitpold.com
Maine Biological Laboratories: http://www.mainebiolab.com
Med-Pharmex, Inc.: http://www.med-pharmex.com
Merial (Merck + Aventis): http://www.merial.com/main.html
Monsanto Company: http://www.monsanto.com/dairy/
Moorman's Inc.: http://www.moormans.com

Neogen Corporation: http://www.neogen.com
Novartis Animal Health: http://www.novartisah.com/
Pet-Ag, Inc.: http://www.petag.com
Pfizer Animal Health: http://www.pfizer.com
Pharmacia & Upjohn Company: http://www.pnuanimalhealth.com
Pharmaderm: http://www.pharmaderm.com
Phoenix Pharmaceutical, Inc.: http://www.phoenixpharmaceutical.com
Professional Biological Company: http://www.colorado-serum.com
Prolabs Ltd.: http://www.agrilabs.com
Rhone-Poulenc: http://www.rpan.com/
Schering-Plough: http://sp-animalhealth.com
Sergeant's Pet Products: http://www.sergeants.com
Synbiotics Corporation: http://www.synbiotics.com
Techmix, Inc.: http://www.techmixinc.com
Vedco, Inc.: http://www.vedcom.com/
Vet-A-Mix: http://www.lloydinc.com
Veterinary Dynamics, Inc.: http://www.thegrid.net/vdi
Veterinary Products Laboratories: http://www.vpl.com
Vetlife: http://www.vetlife.com
Vet Solutions: http://www.vetsolutions.com
Vets Plus, Inc.: http://www.vets-plus.com
Vet Tek, Inc.: http://www.durvet.com
Vetus Animal Health c/o Burns Veterinary Supply: http://www.burnsvet.com
Vineland Laboratories: http://www.vinelandlabs.com
Warner Lambert: http://www.warner-lambert.com
Wildlife Pharmaceuticals, Inc.: http://www.wildpharm.com
Zinpro Corporation: http://www.zinpro.com

APPENDIX H
Proper Use of Needles and Syringes

- Syringes are calibrated to allow precise measurement of medication. Make sure you know if the calibrations represent metric units, apothecary units, or international units.
- Use the syringe volume that is closest to the volume of medicine to be delivered for the greatest accuracy. For example, do not use a 60 cc syringe to measure 2 cc of fluid volume. The best choice in this case would be a 3 cc syringe.
- Disposable syringes are sterilized, prepackaged, nontoxic, nonpyrogenic, and ready for use. These syringes are preferred for use over nondisposable glass syringes.
- Selection of needle size is based on the amount and viscosity of the medication. Thick, oily medications are given through needles with larger lumens; thinner medications are given through needles with smaller lumens.
- Needle size is inversely proportional to its number. For example, 16-gauge needles have large lumens and 25-gauge needles have small lumens.
- Needle size selection is also based on route of administration. Intradermal injections are given with small-lumen needles (like 25-gauge) and subcutaneous injections in large animals are given with large-lumen needles (like 16- or 18-gauge).
- Needle size selection is also based on species and age of the animal.
- Needle length is based on route of administration. To give a deep muscle injection, a needle of appropriate length must be used.
- Needles must be disposed of in a rigid, puncture-resistant container (Figure H-1).

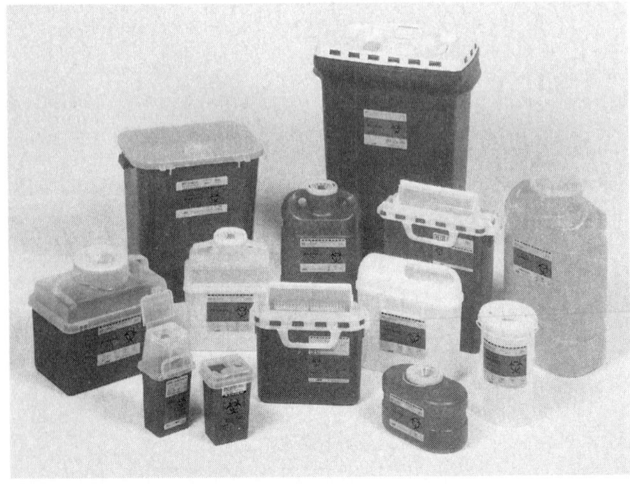

FIGURE H-1 Examples of sharps containers. Courtesy of Becton, Dickinson and Company

APPENDIX I
Withdrawing (Aspirating) Medication from a Vial

STANDARD PRECAUTIONS:

PURPOSE:

Medication is supplied in a variety of packaging.

Medication from a vial must be aspirated into a syringe for parenteral injection.

EQUIPMENT/SUPPLIES:

Appropriate syringe and needle with cover

Vial of medication

Antiseptic wipes or sponges

Disposable gloves

Sharps container

PROCEDURE STEPS:

1. Read the veterinarian's order and assemble equipment.
2. Wash hands. Apply gloves if needed.
3. Select the proper-sized needle and syringe for the medication and the route. If necessary, attach the needle to the syringe.
4. Check the vial label against the veterinarian's order.
5. Remove the metal or plastic cap from the vial. If the vial has been opened previously, clean the rubber stopper by applying a disinfectant wipe in a circular motion (Figure I-1A).
6. Remove the needle cover—pull it straight off.
7. Inject air into the vial as follows:
   a. Hold the syringe pointed upward at eye level. Pull back the plunger to take in a quantity of air equal to the ordered dose of medication.
   b. Hold the vial upright (inverted) according to personal preference. Take care not to touch the rubber stopper.
   c. Insert the needle through the rubber stopper of the vial. Inject the air by pushing in the plunger (Figure I-1B).

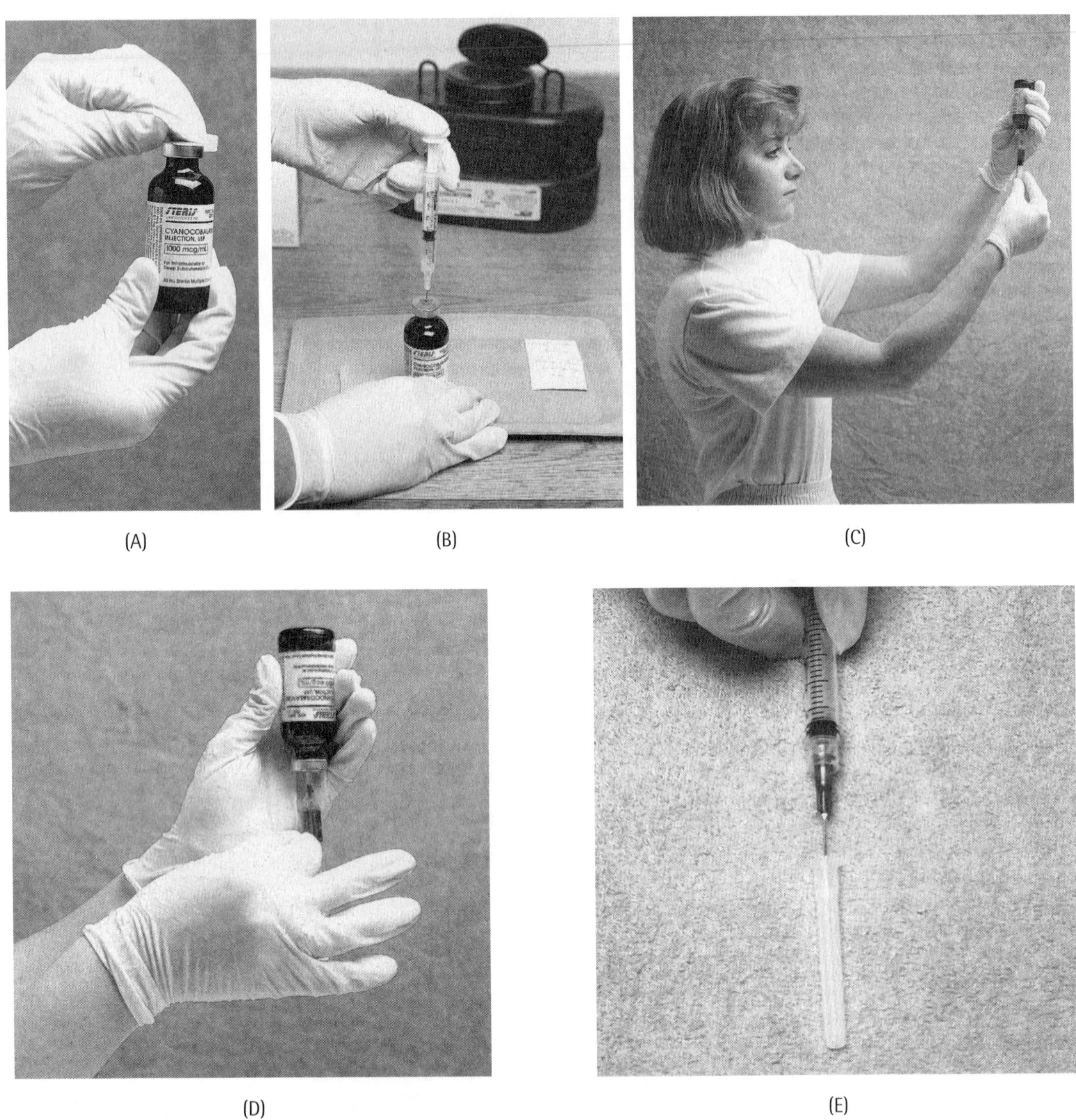

(A) (B) (C) (D) (E)

FIGURE I-1 (A) Disinfect the rubber stopper on the medication vial with an alcohol swab. (B) Keeping the bevel of the needle above the fluid level, inject an amount of air equal to the medication quantity to be withdrawn. (C) Hold the syringe pointed upward at eye level and keep the bevel of the needle in the medication. Pull back the plunger and aspirate the quantity of medication ordered. (D) Tap the syringe to eliminate air bubbles. Your hand should hold the syringe while you tap it. (E) After the correct dose has been withdrawn, recover the sterile needle. To safely recap needles, the "one-hand" technique can be used. First, place the cap on a flat surface and remove your hand from the cap. With one hand, hold the syringe and use the needle to "scoop up" the cap. When the cap covers the needle completely, use the other hand to secure the cap with the needle hub. Be careful to handle the cap at the bottom only (near the hub).

8. Withdraw the medication: Hold the vial and the syringe steady. Pull back on the plunger to withdraw the measured dose of medication. Measure accurately. Keep the tip of the needle below the surface of the liquid; otherwise, air will enter the syringe. Keep the syringe at eye level (Figure I-1C).
9. Check the syringe for air bubbles. Remove them by tapping sharply on the syringe (Figure I-1D).
10. Remove the needle from the vial. Replace the sterile needle cover using the "one hand" technique described in Figure I-E.
11. Check the vial label against the veterinarian's order.
12. The dose is now ready for injection.
13. Return a multiple-dose vial to the proper storage area (cabinet or refrigerator). Dispose of unused medication in a single-dose vial according to facility procedure. (Remember, disposal of a controlled substance must be witnessed and the proper forms signed.)
14. Discard used syringe-needle unit immediately after use in a sharps container (Figure I-2).

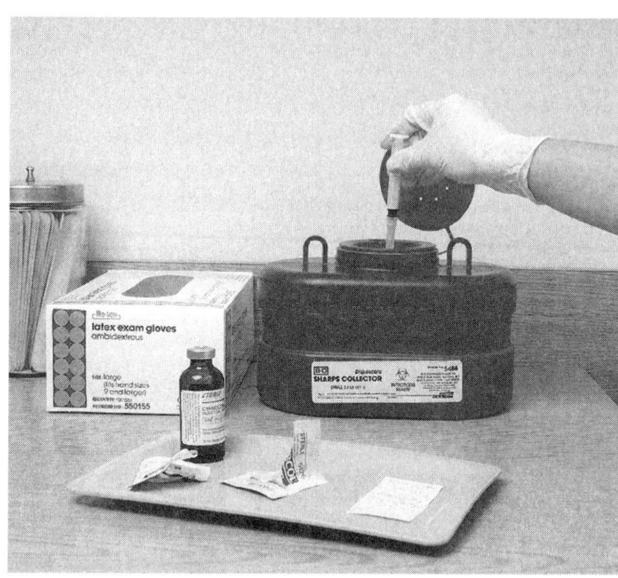

FIGURE I-2 Dispose of the used syringe-needle unit in a sharps container.

APPENDIX J
Withdrawing (Aspirating) Medication from an Ampule

STANDARD PRECAUTIONS:

PURPOSE:

Medication is supplied in a variety of packaging.

An ampule is a sterile, glass, single-dose container of liquid medication. It is aspirated into a syringe for parenteral injection.

EQUIPMENT/SUPPLIES:

Ampule of medication
Alcohol wipes
Sterile gauze sponges
Sharps container
Sterile syringe-needle unit
Disposable gloves

PROCEDURE STEPS:

1. Check the veterinarian's order.
2. Wash hands and gather equipment. Put on gloves.
3. Obtain ampule of medicine. Check for correct medication, dose, route, and time. Check medication expiration date.
4. Flick ampule of medication (medication will often get "trapped" above the neck of the ampule). A sharp flick of the wrist will help force all the medication down below the neck of the ampule into the body of the ampule (Figure J-1A).

   RATIONALE: This is important to ensure that all medication is available in the body of the ampule for calculation of the correct dose. If some of the medication remains trapped above the neck in the top of the ampule, some medication will not be available for use and it is possible to give an incorrect dose, especially if the client is to receive the entire contents of the ampule.

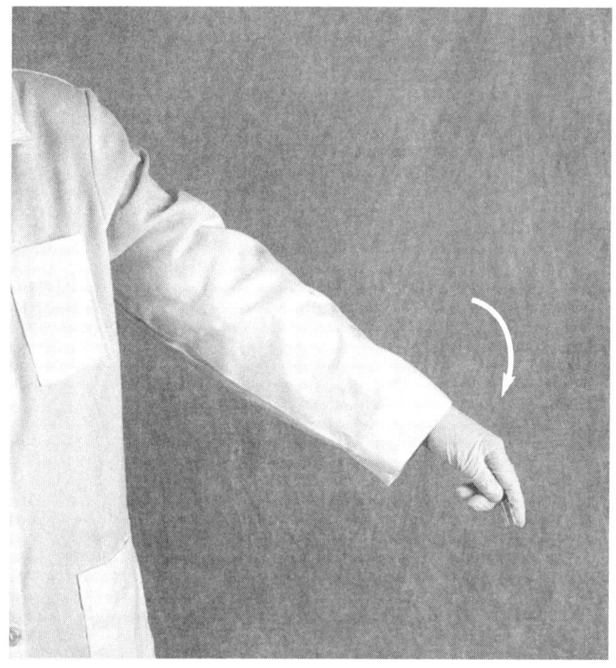

(A)

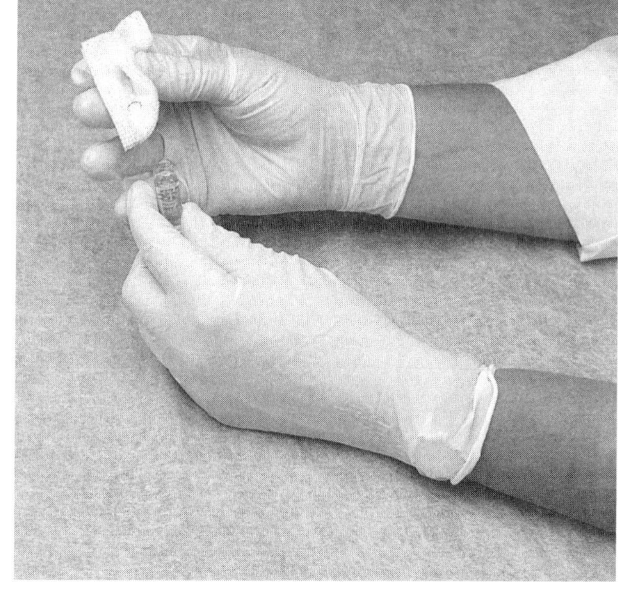

(B)

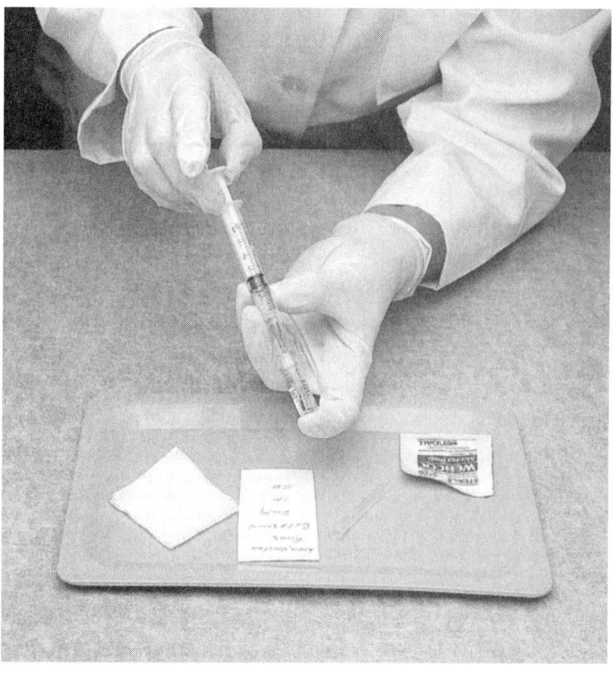

(C)

FIGURE J-1 (A) Hold ampule by the top and force all the medication into the bottom of the ampule by a snap of the arm and wrist. (B) Remove the top from the ampule. Turn your hand up and out simultaneously. (C) Aspirate the required dose into the syringe.

5. Thoroughly disinfect the ampule neck with an alcohol swab. Check label (second time).

   RATIONALE: The needle will enter the opening of the ampule. Wiping the neck of the ampule prior to removal of the top ensures disinfection of the neck or opening of the ampule.

6. With a sterile gauze, wipe dry the neck of the ampule. Completely surround the ampule with the gauze and forcefully snap off the top of the ampule by pushing the top away from you (Figure J-1B). Discard top in sharps container.

   RATIONALE: Ensures safety from possible injury from broken glass.

7. Place opened ampule down on medicine tray.

8. With a prepared sterile syringe-needle unit, aspirate the required dose into the syringe (Figure J-1C).

APPENDIX K
Reconstituting a Powder Medication for Administration

STANDARD PRECAUTIONS:

PURPOSE:

Drugs for injection may be supplied in a powdered (dry) form and must be reconstituted to a liquid for injection. A diluent (usually sterile saline) is added to the powder, mixed well, and the appropriate dose drawn up to be administered.

EQUIPMENT/SUPPLIES:

Diluent
2 appropriately sized syringe-needle units
Antiseptic swabs
Disposable gloves
Sharps container

PROCEDURE STEPS:

1. Wash hands.
2. Prepare the syringe-needle unit before reconstituting powder medication (Figure K-1A).
3. Remove tops from diluent and powder medication containers and wipe with alcohol swabs (Figure K-1B).
4. Insert the needle of a sterile syringe-needle unit through the rubber stopper on the vial of diluent that has been cleansed with an antiseptic swab. The syringe-needle unit should have an amount of air in it equal to the amount of diluent to be withdrawn (Figures K1-C and D).
5. Withdraw the appropriate amount of diluent to be added to the powder medication (Figures K-1E and F). Cover the sterile needle on the syringe containing the appropriate amount of diluent.
6. Add this liquid to the powder medication container that has been cleansed with an antiseptic swab (Figure K-1G).

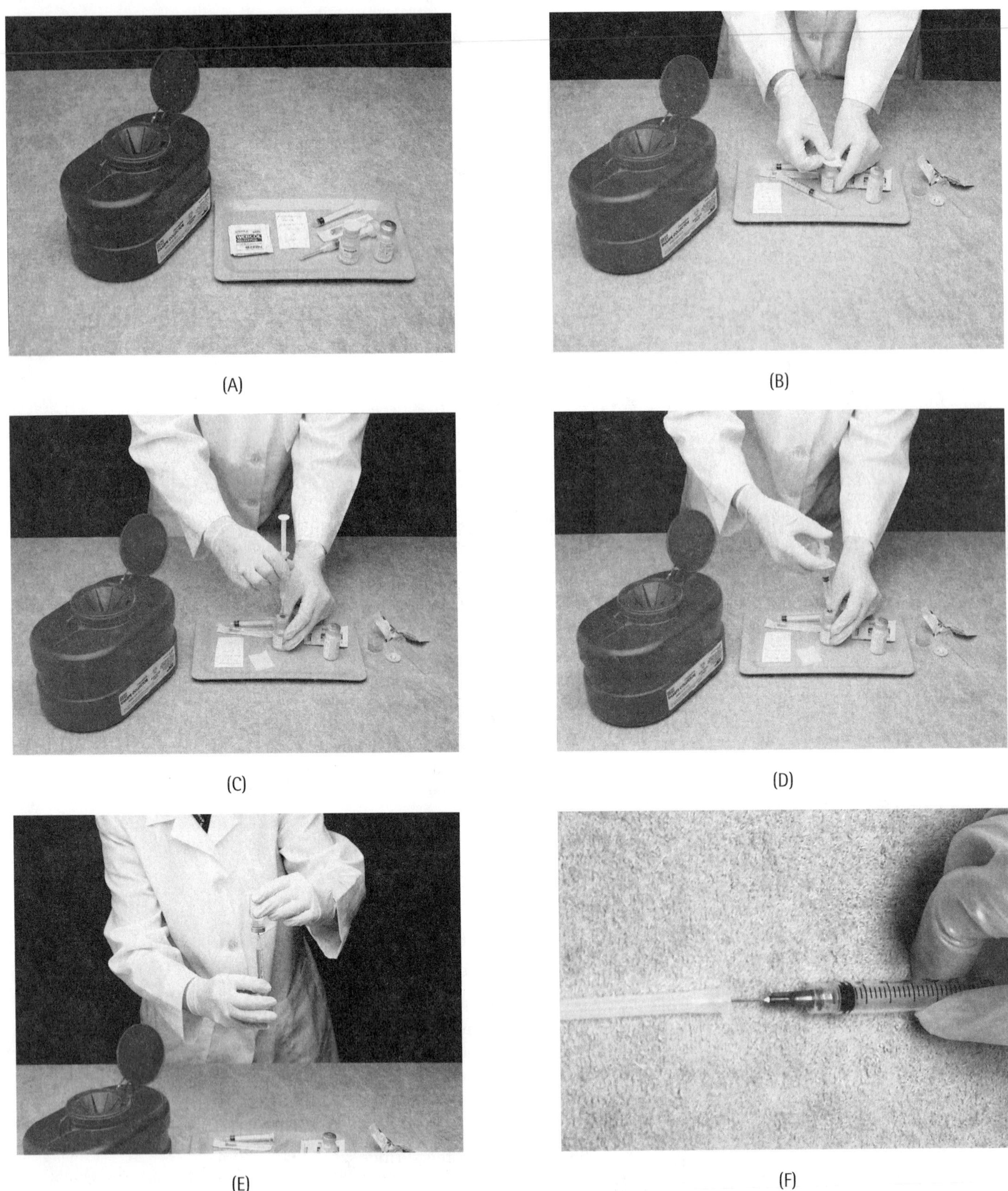

(A) (B) (C) (D) (E) (F)

FIGURE K-1 (A) Supplies for reconstituting powder medication. (B) Remove tops from diluent and powdered medication containers. Wipe top of each with an alcohol swab. (C) Prepare to inject air in an amount equal to the amount of diluent being removed from the diluent vial. (D) Inject air into the diluent vial. (E) Prepare to separate the vial from the syringe-needle unit after withdrawing diluent. (F) Cover the sterile needle on the syringe containing diluent, using the "one hand" technique described in Appendix I.

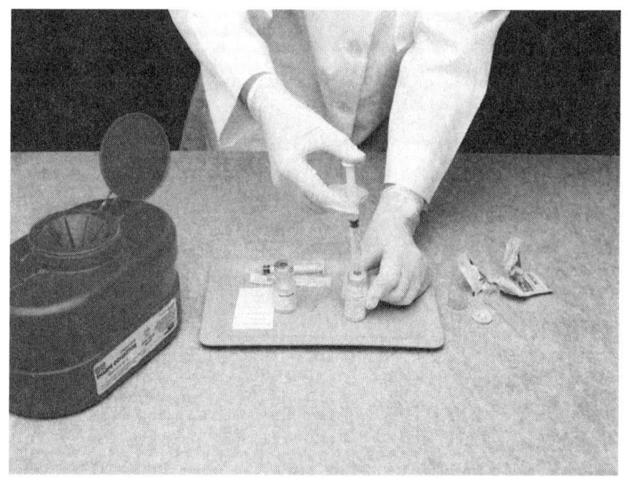

(G)

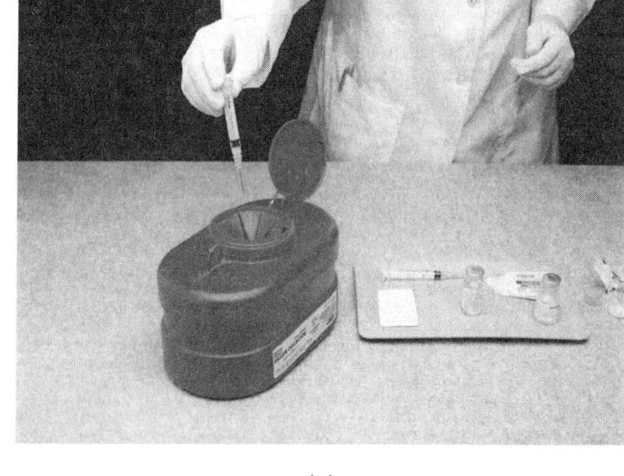

(H)

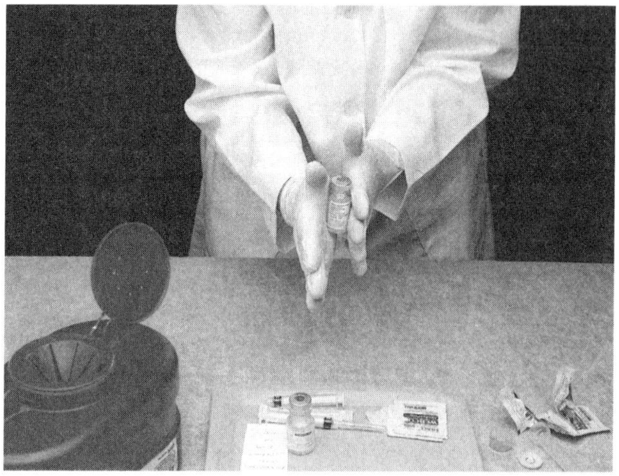

(I)

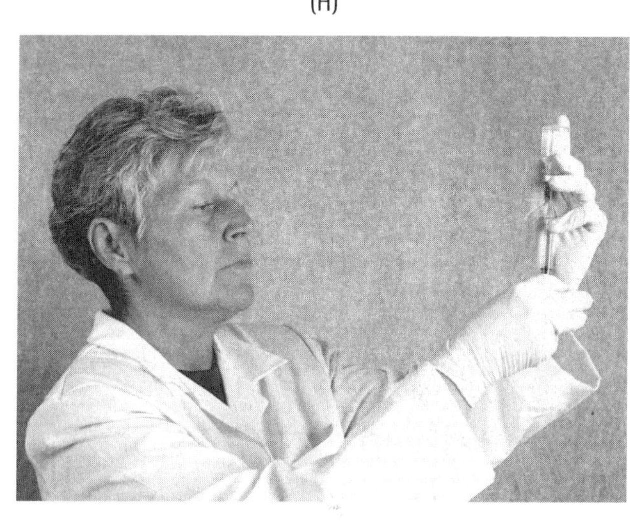

(J)

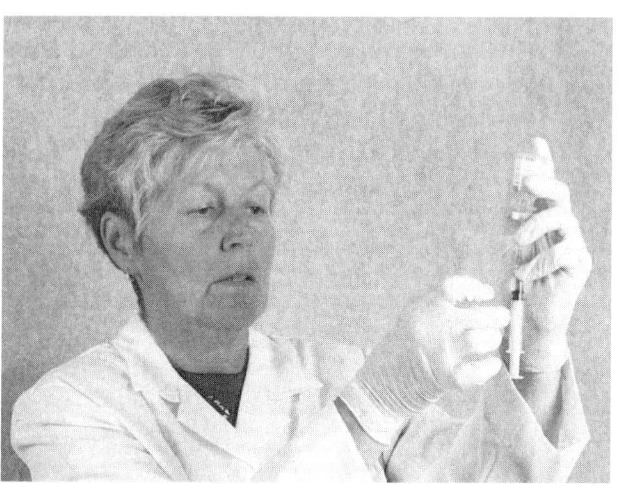

(K)

FIGURE K-1 (G) Inject diluent into the vial containing powdered medication. Before injecting, clean the top of the vial again with an alcohol swab. (H) Discard the syringe-needle unit used for diluent in a sharps container. (I) Roll the vial of powdered medication with diluent between the palms of your hands to mix it well. Label the vial with the date, amount of diluent added, strength of dilution, time mixed, your initials, and the expiration date. (J) Use a second sterile syringe-needle unit to draw up the prescribed dose of medication ordered by the veterinarian. (K) Flick away any air bubbles that cling to the side of the syringe.

7. Remove needle and syringe from vial with powder medication and diluent and discard them into a sharps container (Figure K-1H).

8. Roll the vial between the palms of the hands to completely mix together the powder and diluent (Figure K-1I). Label the multiple-dose vial with the dilution or strength of the medication prepared, the date and time, your initials, and the expiration date.

9. With a second sterile syringe and needle, withdraw the desired amount of medication (Figure K-1J).

10. Flick away any air bubbles that cling to the side of the syringe (Figure K-1K).

APPENDIX L
Drugs Listed Alphabetically and Classified by Category

Drug	Drug Category/Treatment
Acemannan	Immunomodulator
Acepromazine	Antiemetic
	Sedative
	Preanesthetic
	Restraint
Acetaminophen	Analgesic
Acetazolamide	Carbonic anhydrase inhibitor (glaucoma treatment)
	Metabolic acidosis
Acetylcysteine	Mucolytic
	Acetaminophen toxicity
Acetylsalicyclic acid (Aspirin)	NSAID
	Antiplatelet
	Antipyretic
Actinomycin-D	Antineoplastic Antibiotic
Activated charcoal	GI adsorbant
Albendazole	Antinematodal
Albuterol	Bronchodilator
Alcuronium	Neuromusuclar blocker
Allopurinol	Urate urolithiasis
Alprazolam	Behavioral disorders
Altrenogest	Pregnancy maintenance
	Estrus suppression
Aluminum hydroxide	Antacid
	Phosphate binder
Aluminum carbonate	Antacid
Amikacin	Antibiotic
Amiloride	Diuretic
Aminopentamide	Antidiarrheal
	Antiemetic
Aminophylline	Bronchodilator
Aminopromazine	Smooth muscle relaxant
Amiodarone	Antiarrhythmic
Amitraz	Demodex and sarcoptic mange
	Ear mites
Amitriptyline	Anxiety separation
	Behavior (urine spraying)
Ammonium chloride	Urinary acidifer
Amoxicillin	Antibiotic
Amoxicillin/Clavulanic acid	Antibiotic
Amphotericin B	Antifungal
Ampicillin	Antibiotic
Amprolium	Anticoccidial
Apomorphine	Emetic
Asparaginase	Antineoplastic
	Immune mediated thrombocytopenia
Atenolol	Hypertension reducer
	Anti-arrhythmic
Atipamezole	Reversal agent for medetomidine
Atracurium	Neuromuscular blocker
Atropine	Increases heartrate
	Agent for cholinergic toxins
	Antidiarrheal
	Antiemetic
	Mydriatic (ocular)
	Cycloplegic
Azathioprine	Antineoplastic antimetabolite
	Immune mediated diseases
Bacitracin	Antibiotic
	Growth promotant
Baclofen	Urethral obstruction
Beclomethasone dipropionate	Anti-inflammatory
Bendroflumethiazide	Diuretic
Benzoyl peroxide	Antiseborrheic
Betamethasone	Anti-inflammatory
Bethanechol chloride	Cholinergic for urinary bladder atony
Bisacodyl	Stool softener
Bismuth subsalicylate	GI protectant
Boldenone undecylenate	Anabolic steroid
Bretylium	Ventricular fibrillation
Bromocriptine mesylate	Hyperadrenocorticism
	Pseudo-pregnancy
Buparvaquone	Antiprotozoal
Bupivacaine	Local anesthetic
Buprenorphine	Analgesic

Drug	Drug Category/Treatment
Buspirone	Anxiety relieving
Butorphanol tartrate	Antitussive
	Analgesic
	Preanesthetic
	Sedation
Butoxypolypropylene glycol	Ectoparasite repellent
Calcitonin	Hypercalcemia treatment
	Cholecalciferol toxicosis treatment
Calcitrol	Parathyroid
Calcium carbonate	Antiulcer
Calcium chloride	Calcium supplement
	Ventricular systole
Calcium gluconate	Calcium supplement
	Ventricular systole
Captopril	Vasodilator
Carbachol	Miotic (ocular)
Carbadox	Antibiotic
	Growth promotant
Carbamazepine	Anticonvulsant
Carbaryl	Ectoparasitic
Carbenicillin	Antibiotic
Carbimazole	Hypterthyroidism
Carboplatin	Antineoplastic alkalinizing agent
Carnitine (l-carnitine)	Nutritional supplement for cardiomyopathy
Carprofen	NSAID
Catecholamines	Cardiac contractility enhancers
Cefaclor	Antibiotic
Cefadroxil	Antibiotic
Cefamandole	Antibiotic
Cefazolin sodium	Antibiotic
Cefataxime	Antibiotic
Cefonicid	Antibiotic
Cefoperazone	Antibiotic
Cefotaxime	Antibiotic
Cefotetan	Antibiotic
Cefoxitin sodium	Antibiotic
Ceftazidime	Antibiotic
Ceftizoxime	Antibiotic
Ceftriaxone	Antibiotic
Cefuroxime	Antibiotic
Ceftiofur	Antibiotic
Cephalexin	Antibiotic
Cephalosporin	Antibiotic
Cephalothin	Antibiotic
Cephapirin	Antibiotic
Cephradine	Antibiotic
Chloral hydrate	Restraint
	Sedation
	Preanesthetic
Chlorambucil	Antineoplastic alkylinizing agent
Chloramphenicol	Antibiotic
Chlordiazepoxide	Irritable colon syndrome
	Anti-anxiety
Chlorinated hydrocarbons	Ectoparasitic
Chlorothiazide	Diuretic
	Udder edema
Chlorpheniramine	Antihistine
	Antipruritic
Chlorpromazine	Antiemetic
	Tranquilizer
Chlorpyrifos	Ectoparasitic (OP)
Chlortetracycline	Antibiotic
	Growth promotant
Choline chloride	Hepatic lipidosis
Chondroitin	Cartilage supplement
Chorionic gonadotropin	Luteinizer of follicular cysts
	Ovulation inducer
	Induces descent of inguinal testis
Cimetidine	Antiulcer
Ciprofloxacin	Antibiotic
Cisapride	Antiemetic
	Gastrointestinal ileus
Cisplatin	Antineoplastic alkylinizing agent
Clindamycin	Antibiotic
Clomipramine	Antianxiety
	Lick granuloma prevention
Clonazepam	Anti-anxiety
Cloprostenol sodium	Abortion
	Induction of parturition
Clorazepate	Phobia prevention
	Anticonvulsant
Clorsulon	Antitrematodal
Closantel	Antiparasitic
Clotrimazole	Antifungal
Cloxacillin	Antibiotic
Coal tar	Antiseborrheic
Codeine	Antitussive
	Analgesic
Copper naphthenate	Antifungal
Corticosteroids	Anti-inflammatory
Coumaphos	Ectoparasitic
Cyanocobalamin	Vitamin B_{12}
Cyclophosphamide	Antineoplastic alkylinizing agent
	Immune mediated disease
Cyclosporine	Immunosuppressant
	Ophthalmic form for keratoconjunctivitis sicca
Cypionate	Androgen (testosterone)
Cyproheptadine	Appetite stimulant
	Bronchospasm control
Cytarabine	Antimetabolite antineoplastic

Drug	Drug Category/Treatment
Cythioate	Ectoparasitic (OP)
Cytosine arabinoside	Antineoplastic
Dactinomycin	Antineoplastic antibiotic
Danazol	Synthetic hormone
	Thrombocytopenia
Dantrolene	Functional urethral obstruction
	Malignant hyperthermia
Decoquinate	Anticoccidal
Delmadinone	Testosterone inhibitor
Desmopressine acetate	Central diabetes insipidus
	Von Willebrand's disease
Desoxycorticosterone acetate (DOCA, DOCP)	Adrenal cortex drug
Detomidine	Sedation
	Analgesia
Dexamethasone	Anti-inflammatory
	CNS trauma
	Adrenocortical collapse
Dexpanthenol	Intestinal atony
	Prokinetic agent
Dextran 40 and dextran 70	Plasma expanders for shock
Dextromethorphan	Antitussive
Diazepam	Sedative
	Restraint
	Anticonvulsant
	Appetite stimulant
	Muscle relaxant
Diazinon	Ectoparasitic (OP)
Dichlorphenamide	Carbonic anhydrase inhibitor
Dichlorvos	Hook-, whip- and roundworm antiparasitic
Diclofenac	NSAID
Dicloxacillin	Antibiotic
Difloxacin	Antibiotic
Diethylcarbamazine	Heartworm prophylaxis
Diethylstilbestrol	Estrogen used for hormone-responsive urinary incontinence
	Mismating
Diphenoxylate	Antidiarrheal
Digitoxin	Cardiac contractility enhancer (heart failure)
	Tachyarrhythmias
Digoxin	Cardiac contractility enhancer (heart failure)
	Tachyarrhytmias
Dihydrostroptomycin	Antibiotic
Diltiazem	Antiarrhythmic
Dimenhydrinate	Motion sickness
Dimercaprol	Chelating agent for arsenic
Dimethylsulfoxide	Anti-inflammatory
	Vehicle for drug transport

Drug	Drug Category/Treatment
Dioctyl sulfosuccinate	Stool softener
Diphenhydramine	Antihistamine
	Antipruritic
Diphenoxylate	Antidiarrheal for irritable colon
Dobutamine hydrochloride	Cardiac contractility enhancer
Docusate calcium	Stool softener
Docusate sodium succinate	Stool softener
Dolasetron	Antiemetic
Domperidone	Prokinetic agent
Dopamine	Cardiac contractility enhancer
	Renal vasodilator
Doxapram	Respiratory stimulant
Doxarubicin	Antineoplastic antibiotic
Doxepin	Behavior modifier
Doxycycline	Antibiotic
Doxylamine	Antihistamine
d-pencillamine	Chelating agent
Droperidol	Sedation
Edrophonium chloride	Cholinergic (used to diagnose myasthenia gravis)
EDTA	Anticoagulant
	Chelating agent
Enalapril	Vasodilator
Enanthate	Androgen (testosterone)
Enflurane	Inhalant anesthetic
Enrofloxacin	Antibiotic
Ephedrine	Adrenergic (used to treat urinary sphincter incontinence)
	Bronchodilator
Epinephrine	Cardiac contractility enhancer (cardiac arrest)
	Anaphylaxis treatment
	Bronchodilator
	Adrenergic agonist (ocular use)
Eprinomectin	Ectoparasitic
Epsiprantel	Anticestodal
Ergonovine maleate	Uterine involution
Erythromycin	Antibiotic
Erythropoietin	Stimulates RBC production
Estradiol cypionate	Mismating
	Urinary incontinence
	Induction of estrus
	Pyometra
Ethacrynic acid	Diuretic
Ethanol 20%	Ethylene glycol toxicity
Etodalac	NSAID
Etorphine	Narcotic analgesic
Famotidine	Antiulcer
Febantel	Antiparasitic
Febendazole	Antiparasitic (nematodes, cestodes, flukes)

Drug	Drug Category/Treatment
Fenoterol	Bronchodilator
Fentanyl	Narcotic analgesic
Fentanyl citrate	Analgesic
	Preanesthetic
Fenthion	Ectoparasitic
Ferrous sulfate	Anti-anemia (iron supplement)
Filgrastim	Biologic response modifier
Finasteride	Testosterone inhibitor
Fipronil	Ectoparasitic
Florfenicol	Antibiotic
Flucinolone	Anti-inflammatory
	Immunosuppressant
Fluconazole	Antifungal
Flucytosine	Antifungal
Fludrocortisone	Adrenal cortex drug
Flumazenil	Benzodiazepine antagonist
	Anti-anxiety
Flumethasone	Anti-inflammatory
Flunixin meglumine	NSAID
	Endotoxic shock
Fluorescein	Ocular stain
Fluoroquinolones	Antibiotics
Fluorouracil	Antineoplastic antimetabolite
Fluoxetine	Behavior modifier
Flurbiprofen sodium	NSAID
Folic acid	Dietary supplement
Follicle-stimulating hormone	Induction of estrus
	Male hypogonadism treatment
Fomepizole	Alcohol dehydrogenase inhibitor for ethylene glycol toxicity
Furazolidone	Antibiotic/antiprotozoal
Furosemide	Diuretic
	Udder edema
	Antihypertensive
Gallamine	Neuromuscular blocker
Gentamicin sulfate	Antibiotic
Glipizide	Oral hypoglycemic agent
Glucosamine	Cartilage supplement
Glycerin	Glaucoma
	Osmotic diuretic
Glycopyrrolate	Anticholinergic (for bradycardia/ bronchodilation)
Gonadorelin	Gonadotropic drug
	Ovarian cysts
	Induction of ovulation
Griseofulvin	Antifungal
Guaifenesin	Skeletal muscle relaxant
	Expectorant
Halothane	Inhalant anesthetic
hCG	Gonadotropin
Heparin	Anticoagulant
Hetacillin	Antibiotic
Hetastarch	Plasma volume expander
Homatropine	Mydriatic
Hyaluronate sodium	Synovitis
Hydralazine	Vasodilator
Hydrochlorothiazide	Diuretic
	Udder edema
	Antihypertensive
	Calcium oxalate uroliths
Hydrocodone bitartrate	Cough suppressant
Hydrocortisone	Anti-inflammatory
	Replacement therapy
Hydrocortisone sodium succinate	Shock
	Hypoadrenocortical crisis
Hydroflumethiazide	Diuretic
Hydrogen peroxide	Emetic
Hydromorphone	Narcotic analgesic
Hydroxyzine	Antihistamine
Hypertonic saline	Shock (plasma volume expander)
Idoxuridine	Herpesviral keratitis
Imidacloprid	Ectoparasitic
Imidazole	Antifungal
Imidocarb dipropionate	Babesiosis
Imipramine	Separation anxiety
	Urethral sphincter incontinence
	Narcolepsy
Insulin	Diabetes mellitus (lowers blood glucose)
Interferon	Immunomodulating
	Antiviral
Ipecac syrup	Emetic
Iron dextran	Iron deficiency anemia
Isocarboxazid	Behavior modifier (MAOI)
Isoflupredone	Anti-inflammatory ketosis
Isoflurane	Inhalant anesthetic
Isoniazid	Antibiotic
Isopropamide	Sinus bradycardia
	Antidiarrheal
Isoproteronol	Cardiac contractility enhancer
	Bradycardia
Isoxsuprine hydrochloride	Laminitis
	Navicular disease
Itraconazole	Antifungal
Ivermectin	Antinematodal
	Ectoparasitic
Kanamycin	Antibiotic
Kaolin-pectin	GI protectant
Ketamine hydrochloride	Chemical restraint
	Anesthesia
Ketoconazole	Antifungal
	Hyperadrenocorticism
Ketoprofen	NSAID
Ketorolac tromethoamine	NSAID
L-asparaginase	Antineoplastic enzyme

Drug	Drug Category/Treatment
Lactulose	Laxative
	Hepatic encephalopathy
Lansoprazole	Antiulcer
	Esophageal reflux
Lasalocid	Growth promotant
	Anticoccidial
Levamisole	Antiparasitic
	Immune stimulation
Levothyroxine sodium (T_4)	Thyroid supplement
Lidocaine	Local anesthetic
	Epidural anesthetic
	Anti-arrhythmic
	Gastrointestinal ileus
Lime sulfur	Ectoparasitic
Lincomycin	Antibiotic
	Growth promotant
Liothyronine (T_3)	Thyroid supplement
Lisinopril	Vasodilator
Loperamide	Antidiarrheal
Lorazepam	Sedation
Lufenuron	Ectoparasitic
Luteinizing hormone (LH)	Ovulation stimulator
Magnesium gluconate	Hypomagnesemia
Magnesium hydroxide	Antacid
	Laxative
Magnesium oxide	Hypomagnesemia
Magnesium sulfate	Osmotic cathartic
	Ventricular tachyarrhythmias
	Hypomagnesemia
Mannitol 20%	Diuretic
	Glaucoma treatment
	Oliguric renal failure
Marbofloxacin	Antibiotic
Mebendazole	Antinematodal
Meclizine	Antihistamine
	Motion sickness
Meclofenamic acid	Analgesic
Medetomidine hydrochloride	Chemical restraint
	Sedation
Medroxyprogesterone	Progestin
	Prevents abortion
	Aggressive behavior
	Skin conditions
Megestrol acetate	Progestin
	Prevents estrus
	Treats inflammatory conditions (eosinophilic placques)
Melarsomine	Antinematodal (for Dirofilariasis)
Meloxicam	Anti-inflammatory
Melphalan	Antineoplastic
Meperidine hydrochloride	Analgesic
	Sedative
Mepivacaine	Local anesthetic

Drug	Drug Category/Treatment
Metaproterenol sulfate	Bronchodilation
Methadone hydrochloride	Analgesia
Methazolamide	Carbon anhydrase inhibitor
Methicillin	Antibiotic
Methimazole	Thyroid drug (hyperthyroid treatment)
Methionine	Urinary acidifer
	Laminitis
Methocarbamol	Muscle relaxant
	Controls toxicosis
	Anticonvulsant
Methohexital	Anesthesia induction
Methoprene	Ectoparasitic (IGR)
Methotrexate	Antineoplastic antimetabolite
Methoxamine	Adrenergic
Methoxyflurane	Inhalant anesthetic
Methscopalamine	Antidiarrheal
Methylcellulose	Laxative
Methylene blue	Nitrate/nitrite toxicities
	Cyanide toxicities
Methylprednisolone	Anti-inflammatory
Methylprednisolone sodium succinate	Shock treatment
	Spinal cord trauma
4-methylpyrazole	Alcohol dehydrogenase inhibitor (ethylene glycol toxicity)
Methyltestosterone	Androgen
Methyxanthines	Bronchodilators
Metoclopramide	Antiemetic
	Gastrointestinal ileus
	Prokinetic agent
Metronidazole	Antibiotic
	Antiprotozoal
Mexiletine	Antiarrhythmic
Mibolerone	Androgen that prevents estrus
Miconazole	Antifungal (dermatophyte)
Midazolam	Sedative
	Preanesthetic
Milbemycin oxime	Prophylactic for heart-, hook-, round- and whipworms
Mineral oil	Laxative
Misopristol	Antiulcer
Mitotane (*o,p'*-DDD)	Adrenal cortex drug
Mitoxantrone	Antineoplastic antibiotic
Monensin	Growth promotant
	Anticoccidial
Morphine sulfate	Analgesia
	Preanesthetic
Moxidectin	Antiparasitic
Mupirocin	Antibiotic
Mycobacterium cell wall fraction	Immunostimulant
	Sarcoid treatment
Nafcillin	Antibiotic
Naloxone	Respiratory stimulant
	Reversal agent for morphine

Drug	Drug Category/Treatment	Drug	Drug Category/Treatment
Naltrexone	Behavior modification	Penicillin	Antibiotic
	Reversal agent for morphine		Growth promotant
Nandrolone deconate	Anabolic steroid	Pentazocine	Analgesia
	Bone marrow stimulant	Pentobarbital	Anticonvulsant
Naproxen	Anti-inflammatory		Anesthetic
Natamycin	Antifungal		Euthanasia solution
Neomycin	Antibiotic	Permetrin	Ectoparasitic
Neostigmine	Cholinergic (myasthenia gravis)	Perphenazine	Fescue toxicity
	Gastrointestinal atony		Antiemetic
Nicotinamide	Ketosis		Anti-anxiety
Nitenpyram	Ectoparasitic	Phenelzine	Behavior modifier (MAOI)
Nitrofurantion	Antibiotic	Phenobarbital	Anticonvulsant
Nitrofurazone	Antibiotic	Phenoxybenzamine	Adrenergic (urinary bladder atony, laminitis, antidiarrheal)
Nitroglycerin	Vasodilator		
	Laminitis		
Norepinephrine	Resuscitation	Phenylbutazone	NSAID
Norfloxacin	Antibiotic	Phenylephrine	Vasodilator
Norgestomet	Synchronized breeding	Phenylpropanolomine	Adrenergic (urinary sphincter enhancer)
Nortriptyline	Behavior modification		
Novobiocin sodium	Antibiotic	Phenytoin	Anticonvulsant
Nystatin	Antifungal	Phosmet	Ectoparasitic (OP)
Ofloxacin	Antibiotic	Physostigmine	Cholinergic
Omeprazole	Antiulcer		Atropine toxicity
	Esophageal reflux	Phytonadine	Hemostatic drug
Ondansetron	Antiemetic	Pilocarpine	Glaucoma
Opium	Antidiarrheal		Miotic
	Narcotic analgesic	Piperacillin	Antibiotic
Orbifloxacin	Antibiotic	Piperazine	Antinematodal (roundworms)
Organophosphate	Ectoparasitic	Piperonyl butoxide	Ectoparasitic
Orgotein	Analgesic (osteoarthritis)	Pirlimycin	Antibiotic
Oxacillin	Antibiotic	Piroxicam	NSAID
Oxazepam	Appetite stimulant	Poloxalene	Frothy bloat
Oxfendazole	Antiparasitic	Polycarbophil	Laxative
Oxibendazole	Antiparasitic	Polymerized methyl silicone	Antifoaming agent
Oxymorphone	Sedative	Polymyxin	Antibiotic
	Restraint	Polysulfated glycosaminoglycans	Cartilage protectant
	Preanesthetic		
	Analgesic	Potassium bromide	Anticonvulsant
Oxytetracycline	Antibiotic	Potassium chloride	Potassium supplement
	Growth promotant	Potassium citrate	Calcium oxide urolithiasis
Oxytocin	Hormone (enhances uterine contractions)		Urinary alkalinizer
		Potassium iodide	Antifungal
	Stimulates milk let-down	Pralidoxime chloride (2-PAM)	Organophosphate toxicity
	Mastitis, metritis	Praziquantel	Anticestodal
Pancrealipase	Enzyme	Prazosin	Vasodilator
Pancuronium bromide	Paralytic agent	Prednisolone	Anti-inflammatory
Paramethasone	Anti-inflammatory		Immunosuppressant
Paregoric	Antidiarrheal		Shock
Paromomycin	Antiprotozoal		Hypoadrenocorticism
Paroxetine	Behavior modifier	Prednisone	Anti-inflammatory
d-penicillamine	Lead poisoning		Immunosuppressant
	Copper hepatopathy		Shock
	Cystine uroliths		Hypoadrenocorticism

Drug	Drug Category/Treatment
Primidone	Anticonvulsant
Procainamide	Antiarrhythmic
Prochlorperzine/isopropamide	Antiemetic
Progesterone	Suppresses estrus
	Maintains pregnancy
Promazine hydrochloride	Sedation
	Preanesthetic
Propantheline	Antidiarrheal
	Sinus bradycardia
	Rectal relaxation
Proparacaine	Topical anesthetic (ocular)
Propionate	Androgen (testosterone)
Propofol	Anesthetic
Propranolol	Vasodilator
	Treats tachycardia
Propylene glycol	Ketosis
Propylthiouracil	Hyperthyroidism
Prostaglandins	Muscle stimulant
	Abortion
Protamine sulfate	Heparin antagonist
Pseudoephedrine	Nasal decongestant
Psyllium	Laxative
Pyrantel	Antinematodal (round- and hookworms)
Pyrethrins	Ectoparasitics
Pyrilamine maleate	Antihistamine
Pyrimethamine	Toxoplasmosis
Pyriproxyfen	Ectoparasitic (IGR)
Quinidine	Antiarrhythmic
Ranitidine	Antiulcer
Reserpine	Fescue toxicity
Rifampin	Antibiotic
Robenidine	Anticoccidial
Rotenone	Ear mitocide
Roxarsone	Growth promotant
	Swine dysentery
Salicyclic acid	Antiseborrheic
Salmeterol	Bronchodilator
Sarafloxacin	Antibiotic
Selamectin	Ectoparastic
Selegiline	Adrenal cortex drug (pituitary dependent)
	Behavior modifier
Selenium and alpha tocopherol	Selenium/Vitamin E deficiency
Sertraline	Behavior modifier
Sevoflurane	Inhalant anesthetic
Silver sulfadiazine	Antibiotic
	Antifungal
Sodium bicarbonate	Hyperkalemia
	Renal failure
	Acidosis
Sodium iodide	Antibiotic
	Antifungal
Sodium nitrite	Cyanide toxicity
Sodium sulfate	Osmotic cathartic
Sodium thiosulfate	Cyanide toxicity
	Arsenic toxicity
Spectinomycin	Antibiotic
Spironolactone	Diuretic
Stanozolol	Anabolic steroid
	Anemia
Staphage lysate	Immune stimulant
Streptomycin	Antibiotic
Succinylcholine	Paralytic
Sucralfate	Antiulcer
Sulfadiazine	Antibiotic
Sulfasozine	Antibiotic
Sulfadimethoxine	Anticoccidial
Sulfadimethoxine/ormetoprim	Antibiotic
Sulfamethoxazole/ trimethoprim	Antibiotic
Sulfaquinoxaline	Anticoccidial
Sulfasalazine	Antibiotic
	Inflammatory bowel disease
Sulfathiazole	Antibiotic
	Antiprotozoal
Sulfisoxazole	Antibiotic
Sulfonamides	Antibiotic
Sulfur	Antiseborrheic
Syrup of ipecac	Emetic
Tepoxalin	Anti-inflammatory
Terbutaline	Bronchodilator
Terfenadine	Antihistamine
Testosterone	Androgen
	Produce teaser animal
	Growth promotant
Tetanus antitoxin	Tetanus treatment
Tetracaine	Local analgesic
Tetracycline	Antibiotic
Theophylline	Bronchodilator
Thiabendazole	Antiparasitic
Thiacetarsamide	Heartworm adulticide
Thiamine hydrochloride	Lead poisoning
	Thiamine deficiency
Thiamylal sodium	Anesthetic induction
	General anesthesia (large animals)
	Tranquilization
Thiopental sodium (thiopentone)	Anesthetic induction
	General anesthesia (large animals)
Thyroid-releasing hormone (protirelin)	TRH stimulation test
Thyroid-stimulating hormone (thyrotropin)	TSH stimulation test

Drug	Drug Category/Treatment
Thyroxine (T_4)	Hypothyroidism
Ticarcillin	Antibiotic
Ticarcillin/clavulanic acid	Antibiotic
Tiletamine-zolazepam	Restraint Anesthetic
Tilmicosin	Antibiotic
Tiopronin	Cysteine urinary calculi
Tobramycin	Antibiotic
Tocainide	Antiarrhythmic
Tolazoline	Reversal for xylazine
Tranexamic acid	Hemorrhage
Tranylcypromine sulfate	Behavior modifier (MAOI)
Trenbolone acetate	Growth promotant
Triamcinolone	Anti-inflammatory
Triamcinolone acetate	Anti-inflammatory
Trichlormethiazide	Diuretic
Trifluridine	Herpesviral keratitis
Triiodothyronine (T_3)	Hypothyroidism
Trimeprazine	Antihistamine
Trimethobenzamide	Antihistamine
Trimterene	Diuretic
Tripelennamine	Antihistamine CNS stimulant

Drug	Drug Category/Treatment
Tropicamide	Mydriatic
Tylosin	Antidiarrheal Antibiotic Growth promotant
Valproic acid	Seizures
Vancomycin	Antibiotic
Vasopressin	Diabetes insipidus drug
Verapamil	Calcium channel blocker (anti-arrhythmic)
Vidarabine	Antiviral
Vinblastine	Antineoplastic
Vincristine	Antineoplastic alkaloid
Virginiamycin	Growth promotant Antibiotic
Warfarin	Anticoagulant
Xylazine	Emetic Anesthetic
Yohimbine	Neurologic stimulant Reversal agent for xylazine
Zeranol	Growth promotant
Zolazepam	Anti-anxiety

GLOSSARY

A

absorption: the movement of drug from the site of administration into the fluids of the body that will carry it to the site(s) of drug action

acetylcholine: chemical that transmits nerve impulses

acetylcholinesterase: an enzyme that destroys acetylcholine

active immunity: a reaction that occurs when an animal receives an antigen that activates the B- and T-lymphocytes and produces memory

active transport: movement of particles against the concentration gradient, from a region of low concentration to a region of high concentration

acupuncture: examination and stimulation of body points by use of acupuncture needles, injections, and other techniques

adjuvants: substance that enhances the immune response by increasing the stability of a vaccine in the body

adrenal cortex: the outer part of the adrenal gland

adrenergic blocking agents: drugs that block the effects of the adrenergic neurotransmitters; also called *sympatholytics*

adrenergic drugs: drugs that simulate the sympathetic nervous system; also called *sympathomimetics*

adrenocorticotropic: stimulating to the adrenal cortex

adsorbents: drugs that bind substances such as bacteria, digestive enzymes, and/or toxins to protect the intestinal mucosa from damage

adult administration set: device consisting of a set of tubing, drip chamber, and clamp used to administer fluid that typically delivers 15 drops per milliliter

adulticides: drugs that kill adult parasites, especially heartworms

affinity: strength of binding between a drug and its receptor

afterload: the force needed to push blood out of the ventricles; the impedance to ventricular emptying presented by aortic pressure

agonists: drugs that bind to a cell receptor and cause action

alternative: treatments and/or therapies that are outside accepted conventional medicine

aminoglycosides: a group of broad-spectrum antibiotics that affect protein synthesis of bacteria (primarily gram-negative bacteria)

anabolic: tissue-building

anabolic steroids: drugs with the steroid structure that aid in tissue building

analgesics: pain relievers

androgen: any steroid hormone that promotes male characteristics

anesthetics: drugs that interfere with the conduction of nerve impulses and are used to produce loss of sensation

anestrus: period of the estrous cycle when the animal is sexually quiet

angiotensin: a vasoconstrictive substance found in blood; its inactive form (**angiotensin I**) is converted to the active form (**angiotensin II**) by the enzyme renin

antacid: drug that neutralizes HCl and reduces pepsin activity

antagonists: drugs that inhibit or block the response of a cell when bound to its receptor

anterior: toward the front (of a body part) (term is not used extensively in veterinary medicine other than to describe organs such as the pituitary gland or eye chambers)

anthelmintics: drugs that kill worm parasites in the gastrointestinal tract, liver, lungs, and circulatory system

anthrax: a fast-killing bacterial infection caused by *Bacillus anthracis*

anti-anxiety drugs: agents that lessen anxiety but do not make the animal drowsy

antiarrhythmics: drugs used to treat arrhythmias or abnormal variations in heart rhythm

antibiotic: a chemical substance produced by a microorganism that has the capacity, in dilute solutions, to kill (bactericidal activity) or inhibit the growth (bacteriostatic activity) of bacteria

antibiotic resistance: condition in which bacteria continue to multiply despite administration of a particular antibiotic

antibiotic residue: presence of an antibiotic or its metabolites in animal tissues or food products
antibody: protein made by activated B-lymphocytes to counteract a specfic antigen
antibody titer: a serum test that expresses the level of antibody to a particular antigen in a particular individual
anticancer drugs: drugs that stop the cancerous activity of malignant cells without disturbing the chemistry of healthy cells; also called *antineoplastic agents* and *chemotherapeutic agents*
anticestodal drugs: drugs that kill or inhibit cestode worms (tapeworms)
anticholinergic drugs: drugs that treat vomiting, diarrhea, and excess gastric secretion by blocking parasympathetic nerve impulses; also called *parasympatholytics*
anticoagulants: agents that prevent blood clotting
anticoccidial drugs: drugs that inhibit coccidia organisms; also called *coccidiostats*
anticonvulsants: drugs that inhibit convulsions
antidepressant drugs: drugs used to counteract depression in humans; may be used to treat various mood changes and behavior problems such as aggression and cognitive dysfunction in animals
antidiarrheals: drugs that decrease peristalsis to reverse diarrhea
anti-emetics: substances that prevent vomiting
antifoaming agents: drugs that reduce or prevent the formation of foam
antifungals: drugs that prevent or treat disease spread by fungi
antigen: anything that stimulates immune response
antihistamines: drugs that block the effects of histamine
anti-inflammatory drugs: drugs that reduce inflammation
anti-infectives: drugs that counteract infectious agents
antimicrobials: substances made in the chemical laboratory that kill microorganisms or suppress their multiplication or growth
antinematodal drugs: agents that kill or inhibit nematodal worms (roundworms)
antineoplastic agents: see *anticancer drugs*
antiprotozoal drugs: drugs that kill or inhibit protozoal organisms
antipruritic: substance that controls itching
antipyretic: fever-reducing
antiseborrheic: substance that counteracts abnormal flaking and/or scaling of the outermost layer of the epidermis
antiseptics: substances that kill or inhibit the growth of microorganisms on living tissue
antiserum: antibody-rich serum obtained from a hypersensitized or actually infected animal
antisialogues: drugs that decrease salivary flow
antitoxins: substances that neutralize toxins by use of antibodies obtained from a hypersensitized animal
antitrematodal drugs: drugs that kill or inhibit trematodes (flukes)
antitussives: drugs that suppress coughing
antiulcer drugs: drugs that prevent erosions of the gastrointestinal mucosa
antivirals: drugs that prevent viral penetration of the host cell, or inhibit the virus's production of DNA or RNA
apothecary system: a system of liquid units of measure used chiefly by pharmacists; also called the "common system"
aqueous: prepared in water
arrhythmia: a variation from the normal beating of the heart; often causes reduced cardiac output due to the abnormal pumping activity
artificial immunity: immunity acquired through medical procedures
astringent: agent that constricts tissues
atrophy: wasting; decreasing in size
attenuated vaccine: see *modified-live vaccine*
autogenous vaccine: vaccine produced for a specific disease in a specific area from a sick animal

B

bacilli: rod-shaped bacteria
bacitracin: a polypeptide antibiotic used primarily as a topical medication for skin, mucous membranes, and eyes
bactericidal: agent that kills bacteria
bacterin: inactivated bacteria that serve as a vaccine against bacteria
bacteriostatic: agent that inhibits or slows the growth of bacteria
beta-adrenergic blocker: drug that inhibits beta-adrenergic receptors
beta-2-adrenergic agonists: drugs that stimulate beta receptors
beta-lactamase: enzyme produced by some bacteria that destroys the beta-lactam ring of some antibiotics (such as penicillin and cephalosporin)
bioavailability: the degree to which a drug is absorbed and reaches the circulation
bioequivalency: similar ability to be absorbed and to produce similar blood levels after absorption
biologic: a medicinal preparation made from living organisms or their products; examples include vaccines and serums
biologic response modifiers (BRMs): agents used to enhance the body's immune system
biotransformation: the chemical alteration of drug molecules into metabolites by the body cells of animals; also called *metabolism*
blood-enhancing drugs: drugs that affect red blood cells and their production
bolus: a concentrated mass of pharmaceutical preparation or mass of food

bolus administration: injection of a drug in a minute amount of fluid, with a syringe and needle only
botanical medicine: discipline that uses plants and plant derivatives as therapeutic agents
brand name: the name of a drug, usually capitalized, which establishes proprietary recognition of a particular manufacturer's product; also called *proprietary name, trade name*
***British Pharmacopoeia* (*BP*)**: the United Kingdom counterpart of the *USP*
British Veterinary Codex: a compendium of British veterinary drug standards
BRMs: see *biologic response modifiers*
broad-spectrum antibiotics: drugs that act on both gram-positive and most gram-negative bacteria
bronchi: air passages located between the trachea and bronchioles
bronchodilators: drugs which widen the airway for better breathing
bulk laxatives: drugs that absorb water and swell to form an emollient gel that stimulates peristalsis and fecal elimination

C

calcium-channel blockers: drugs that block the influx of calcium ions into the myocardial cells
carbonic anhydrase inhibitor: category of drugs that interfere with the production of carbonic acid, which leads to decreased aqueous humor formation
carcinogenicity: the ability or tendency to produce cancer
cardiac glycosides: drugs used to increase the strength of cardiac contractions
catecholamines: a group of chemically related compounds having sympathomimetic action (e.g., epinephrine and norepinephrine)
cathartics: harsh laxatives that result in a soft to watery stool and abdominal cramping
cc: see *cubic centimeter*
cell-cycle nonspecific (CCNS): anticancer drug that works on any phase of the cell cycle
cell-cycle specific (CCS): anticancer drug that works on a specific phase of the cell cycle
cell-mediated: affected by cellular rather than chemical elements
centi-: prefix meaning one hundredth of a unit
central nervous system (CNS): the portion of the nervous system consisting of the brain and spinal cord
cephalosporins: broad-spectrum antibiotics effective against most gram-positive and many gram-negative organisms
chemotherapeutic agents: see *anticancer drugs*
Chinese traditional herbal medicine: discipline based on a holistic philosophy of life that emphasizes the relationship among the mental, emotional, and physical components of the individual, as well as the importance of harmony among individuals, their social groups, and the general population
chiropractic: examination, diagnosis, and treatment of animals through manipulation and adjustments of specific joints and cranial structures
chloramphenicol: a broad-spectrum antibiotic not considered first-line because of its toxicity
cholinergic blocking agent: see *anticholinergic drugs*
cholinergic drugs: agents that mimic the action of the parasympathetic nervous system; also called *parasympathomimetics*
cholinergic receptors: sites stimulated by the nerve terminal's release of acetylcholine
chronic study: see *long-term toxicity test*
clindamycin: an antibiotic that is a derivative of lincomycin
clinical trials: tests done in the target species to determine if the drug is safe and effective in that species
CNS: see *central nervous system*
CNS stimulant: drug that reverses CNS depression
coccidiosis: a protozoan infection that causes various intestinal disorders, some serious and even fatal, in various species
coccidiostats: see *anticoccidial drugs*
colloid solution: fluids with large molecules that enhance the oncotic force of blood, causing fluid to move from the interstitial and intracellular spaces into the vascular spaces; also referral to as colloids
combination therapy: the administration of various combinations of anticancer drugs
Compendium of Veterinary Products (CVP): a compendium of United States veterinary standards
competitive nondepolarizers: drugs that block activity at the neuromuscular junction; also called *curarizing agents*
complementary medicine: therapies that can be used with or in addition to conventional treatment
controlled substances: drugs considered dangerous because of their potential for abuse or misuse and whose use is monitored by the Drug Enforcement Administration
core vaccine: vaccine recommended for all individual animals because the consequences of infection are severe, infection poses a substantial zoonotic risk, disease prevalence is high, the organism is easily transmitted to others of its species, and or the vaccine is safe and efficacious
corpus luteum: a "yellow body" that produces progesterone in pregnant animals
corticosteroids: hormones produced by the adrenal cortex
crystalloid solution: sodium-based electrolyte solution or solution of glucose in water that has a composition similar to that of plasma fluid; also referred to as crystalloids

cubic centimeter (cc): the liquid volume that would fill a one-centimeter-square cube
curare alkaloids: toxic botanical extracts originally used as arrow poisons in South America
curarizing agents: see *competitive nondepolarizers*
cycloplegic: drug that paralyzes the ciliary muscles of the eye and minimizes pain due to ciliary spasm

D

DEA: see *United States Drug Enforcement Administration*
decongestant: drug that decreases congestion of the nasal passages by reducing swelling
depot preparation: see *repository medication*
dewaxing agent: substance used to remove debris and wax (usually from the ear)
diabetes insipidus: a disease characterized by increased thirst and urination, caused by failure of the posterior pituitary to release sufficient antidiuretic hormone
diabetes mellitus: a broad term denoting a group of diseases characterized by insulin deficiency
diestrus: period of the estrous cycle after metestrus; a short phase of inactivity in polyestrous animals
diffusion: movement of atoms, ions, or molecules from an area of high concentration to an area of low concentration
digestive enzyme: enzyme supplement used in animals in which the pancreas is not producing sufficient enzymes for digestion
digitalis: a derivative of the foxglove plant used to treat cardiac failure
digitalis drugs: see *cardiac glycosides*
direct marketing: purchase of a drug directly from the company that makes it
disinfectant: chemical that kills or inhibits the growth of microorganisms on inanimate objects
distribution: physiological movement of drugs from the systemic circulation to the tissues
distributors: agencies that purchase drugs from the manufacturing companies and resell the drugs to veterinarians; also called *wholesalers*
diuresis: increased excretion of urine
diuretics: drugs that increase the volume of urine excreted by the kidneys and thus promote release of water from the tissues
dosage: amount of drug per an animal species' body weight or measure (e.g., 100 mg/kg)
dosage interval: how frequently the dosage is given (e.g., bid)
dosage regimen: the dosage and the dosage interval together (e.g., 100 mg/kg bid)
dose: the amount of a drug that is administered at one time to achieve the desired effect
doubling time: time required for the number of cancer cells to double in mass
dram: a fluid volume measure in the apothecary system; consists of 60 minims
dressing: substance applied to an area to draw out fluid or relieve itching
drug abuse: the illicit use of an illegal drug or the improper use of a legal prescription drug
Drug Enforcement Administration: see *United States Drug Enforcement Administration*
drying agent: product that reduces moisture (usually used for the ear)

E

ECF: see *extracellular fluid*
ectoparasite: a parasite living on the surface of the host's body
edema: abnormal intercellular fluid accumulation
effective dose-50 (ED_{50}): the dose of a test drug that causes a defined effect in 50 percent of the animals that take it
elimination: removal of a drug from the body; also called *excretion*
emetic: drug that induces vomiting
emulsion: a mixture of two immiscible liquids, one being dispersed throughout the other in small droplets
endoparasite: a parasite that lives within the body of the host
epidural: within the epidural space (also known as the "subarachnoid space")
ergot: derivative of rye fungus used to induce labor and treat migraines
erythromycin: A broad-spectrum antibiotic produced by a strain of the bacterium *Streptomyces erythreus*
erythropoietin: protein made in the kidneys that stimulates red blood cell formation from bone marrow stem cells
estrogen: a hormone that promotes female sex characteristics
estrous cycle: the entire sexual cycle in the female
estrus: heat (in a sexual or reproductive sense); the period of the estrous cycle in which the female is receptive to the male
euthanasia solution: chemicals used to humanely end an animal's life
excretion: see *elimination*
expectorants: drugs that enable or increase the coughing-up and swallowing of material from the lungs
expiration date: the date before which a drug meets all specifications and after which the drug can no longer be used
extracellular fluid (ECF): fluid found in the intravascular and interstitial spaces
extra-label drug: drug used in a manner not specifically described on the FDA-approved label

F

facilitated diffusion: diffusion that utilizes a carrier molecule
FDA: see *United States Food and Drug Administration*

fecal softeners: drugs that soften and lubricate the fecal mass
feed conversion efficiency: rate at which food is converted to tissue
first-line drug: the veterinarian's first drug of choice for a particular use
fluid dram: 60 minims
fluid ounce: a common standard unit used to measure liquid volume
follicular phase: stage of the estrous cycle in which the graafian follicle is present and produces estrogen
Food and Drug Administration: see *United States Food and Drug Administration*
fungicidal: chemical that kills fungi

G
gastroenteritis: inflammation of the lining of the stomach and intestines
generic company: business that sells, under its own company name, drugs that are no longer under patent protection
generic equivalent: drug determined to be the therapeutic equal of a brand-name drug for which the patent has expired
generic name: the official identifying name of a drug agreed upon by the FDA and the company of origin; also called *nonproprietary name*
germicide: chemical that kills microorganisms
glaucoma: group of diseases resulting in increased intraocular pressure
glucagon: a hormone that increases blood glucose levels by promoting the breakdown of liver glycogen into glucose, which exits the liver and enters the bloodstream
glucocortisoids: corticosteroids that raise the concentration of liver glycogen and blood sugar in the body; also used as anti-inflammatory agents and to terminate late-stage cattle pregnancies
glycogen: the chief carbohydrate storage material in animals
gonadotropins: hormones that stimulate the gonads (testes in males and ovaries in females)
grain: the basic unit of weight measurement in the apothecary system
gram (g): the metric standard unit of weight measurement
gram-negative bacteria: strains of bacteria that are decolorized by the Gram's stain process; they have cell walls more complex than those of gram-positive bacteria
gram-positive antibiotics with an extended spectrum: drugs that act on both gram-positive and some gram-negative bacteria
gram-positive bacteria: strains of bacteria that resist decolorization by the Gram's stain process; they have cell walls less complex than those of gram-negative bacteria
Gram's stain: a staining and decolorizing procedure that divides bacteria into gram-positive and gram-negative strains
growth fraction: the percentage of cancer cells that are actively dividing
growth promotants: chemical compounds that can improve growth; also called "anabolic agents"

H
half-life: the time required for the amount of drug in the body to be reduced by half of its original level
hemostatic drug: agent that helps promote clotting
Henle's loop: a U-shaped renal tubule that is part of the nephron; also known as the *loop of Henle*
heparin: anticoagulant drug
holistic medicine: comprehensive approach to health care that uses both alternative and conventional diagnostic techniques and therapeutic approaches
homeopathy: medical discipline in which animal conditions are treated by administration of very small doses of substances that are capable of producing clinical signs in healthy animals
hormone: a chemical substance produced in one part of the body that is transported to another part of the body, where it influences and regulates cellular and organ activity
household system: system of measure in which an approximate dose is acceptable; based on 1 pound (lb) containing 16 ounces (oz)
hydrophilic: water-loving
hyperadrenocorticism: disease caused by hyperactivity of the adrenal cortex; may be caused by tumors of the pituitary gland, tumors of the adrenal cortex, or excessive treatment with corticosteriods
hyperthyroidism: excessive functional activity of the thyroid gland
hypertonic: having osmolality greater than that of blood and extracellular water
hypogonadism: the decreased function and retarded growth and sexual development of the gonads (testes in males and ovaries in females)
hypokalemia: potassium deficiency
hypothyroidism: the underproduction of thyroxine and the effects of that underproduction
hypotonic: having osmolality lower than that of blood and extracellular water

I
iatrogenic: caused by treatment
IC: see *intracardiac*
ICF: see *intracellular fluid*
ID: see *intradermal*
IM: see *intramuscular*
immiscible: incapable of mixing

immunomodulating agent: drug that adjusts the immune response to the desired level

immunosuppressive drug: drug that decreases the immune response by interfering with one of the stages of the cell cycle or by affecting cell messengers to the immune system

inactivated vaccine: vaccine made from microorganisms, microorganism parts, or microorganism byproducts that have been chemically treated or heated to kill the microorganisms; also known as *killed vaccine*

indwelling catheter: a catheter specially designed to remain in place

induce: to increase the rate of metabolism of an enzyme system

inflammation: a natural response of living tissue to injury and infection; involves vascular changes and release of chemicals that help the body destroy harmful agents at the injury site

infusion of fluid: administration of large volumes of fluid continuously over extended periods of time

inhalation: taking in by breathing

inotropy: the force of contraction

insulin: a hormone formed in the pancreas that responds primarily to a rise in blood glucose and promotes storage of glucose in the liver as glycogen

interferon: a protein substance with multiple roles in the body's natural defenses, chief among them stimulating noninfected cells to produce antiviral proteins

intermediate: describes the ability of bacteria to sometimes be affected and sometimes not be affected by a particular antibiotic

intermittent therapy: treatment done by diluting a drug dose in a small volume of fluid and administering it via an indwelling catheter for a period of 30 to 60 minutes

intra-arterial: within an artery

intra-articular: within the joint

intracardiac (IC): within the heart

intracellular fluid (ICF): fluid within the cell

intradermal (ID): within the dermis

intramammary: within or into the mammary gland

intramedullary: within the marrow cavity of a bone

intramuscular (IM): within the substance of a muscle

intraosseous (IO): within a bone

intraperitoneal (IP): within the peritoneal or abdominal cavity

intrathecal: within the subarachnoid space of the meninges

intravenous (IV): within a vein

ionized: charged

ion trapping: a condition in which a drug molecule changes from its ionized (charged) form to the nonionized (uncharged) form as it moves from one body compartment to another body compartment

islets of Langerhans: the pancreatic cells that secrete insulin

isotonic: having osmolality the same as blood and extracellular water

IV: see *intravenous*

K

keratoconjunctivitis sicca (KCS): disease in which tear production is decreased resulting in persistent mucopurulent conjunctivitis and corneal scarring and ulceration; also known as "dry eye"

keratolytic: substance that removes excess keratin and promotes loosening of the outer layers of the epidermis

killed vaccine: see *inactivated vaccine*

kilo-: metric prefix meaning one thousand times the unit

kinetics: the scientific study of motion

L

lacrimogenic: drug that increases tear production

laxative: medicine that loosens the bowel contents and encourages evacuation

lethal dose (LD_{50}): the dose of a test drug that kills 50 percent of the animals that take it

lincomycin: an antibiotic produced by a strain of *Streptomyces* that works by inhibiting protein synthesis

lipophilic: fat-loving

liter (l): the metric standard unit used to measure liquid volume

live vaccine: vaccine made from live microorganisms, which may be fully virulent or avirulent

loading dose: initial dose of a drug given to get the drug concentration up to the therapeutic range in a very short period of time

long-term toxicity test: a toxicity test of a drug that lasts from three months to two years; also called *chronic study*

loop diuretics: diuretics that block sodium reabsorption in Henle's loop

loop of Henle: see *Henle's loop*

luteal phase: stage of the estrous cycle in which the corpus luteum is present and produces progesterone

M

macrolides: antibiotics with large molecular structure and many-membered rings

maintenance dose: dose of drug that maintains or keeps the drug in the therapeutic range

maintenance volume: fluid volume needed by the animal on a daily basis to maintain body function

margin of safety: the difference in magnitude between a drug's effective dose and its lethal dose; LD_{50}/ED_{50}

materia medica: the study of the physical and chemical characteristics of materials used as medicines

MIC: see *minimum inhibitory concentration*

measurement: the use of standard units to determine the final weight or volume of substances

metabolism: the chemical alteration of drug molecules into metabolites by the body cells of animals; also known as *biotransformation*
meter: the metric standard unit used to measure length
metestrus: period of the estrous cycle after sexual receptivity
metric system: system of measure developed by the French that is based on factors of 10
micro-: prefix meaning one millionth of a unit
microfiliaricide: drug that kills microfiliaria of *Dirofilaria immitis*
microgram (mcg): a unit of metric weight equivalent to one-thousandth of a milligram or one-millionth of a gram
microsomal enzymes: enzymes found primarily in the liver that are involved in the biotransformation (metabolism) of many drugs
milli-: a Latin prefix meaning one thousandth of a unit
milliliter (ml): a metric value of liquid volume equal to one thousandth of a liter
mineralocorticoids: corticosteroids that function primarily to help the body retain sodium and the kidneys to secrete potassium
minim: the liquid volume of a drop of water from a standard medicine dropper
minimum inhibitory concentration (MIC): the lowest concentration of a particular antibiotic that visually inhibits the growth of bacteria
miotic: drug that constricts the pupil
miscible: able to be mixed
modified-live vaccine (MLV): vaccine made with microorganisms that have undergone a process to nullify or reduce their virulence; also known as *attenuated vaccine*
monograph: a written account about a single drug
monovalent vaccine: single-antigen vaccine
mucolytics: drugs designed to liquify thick mucous secretions for better removal from the lungs
mucosal protective drug: drug that inhibits pepsin; may or may not cover an ulcer to protect it
mydriatic: drug that dilates the pupil

N

narcotic antagonist: see *opioid antagonist*
narrow-spectrum antibiotics: drugs that act specifically on the gram-positive or the gram-negative families of bacteria
natural immunity: immunity acquired during normal biological experiences
nebulize: administer as a fine spray for inhalation
neuroleptanalgesics: agents that are a combination of an opioid and a tranquilizer or sedative
neuromuscular blockers: drugs that produce paralysis
neuromuscular junction: the point where a nerve fiber meets the muscle it affects
noncore vaccine: vaccine recommended only for individual animals deemed to be at high-risk for contact with the organism
nonionized: uncharged
nonparenteral: administered through the gastrointestinal tract (orally)
nonproprietary name: see *generic name*
nonspecific immunity: defense mechanisms that are directed against all pathogens and are the initial defense against invading agents
nonsteroidal anti-inflammatory drug (NSAID): drug that reduces inflammation by inhibiting prostaglandin production (e.g., aspirin and phenylbutazone)
nutraceutical medicine: discipline that uses micronutrients, macronutrients, and other nutritional supplements as therapeutic agents

O

ongoing fluid loss: fluid loss due to vomiting, diarrhea, or other pathology
ophthalmic drugs: drugs formulated as eye drops and eye ointments to be instilled into the eye or applied directly to the eyelid
opioid antagonist: drugs that block the binding of opioids to their receptors; also known as narcotic antagonists
oral: by mouth
osmolality: concentration of a solution in terms of osmols of solutes per kilogram of solvent
osmotic diuretics: drugs that consist of large molecules that have limited capability of being reabsorbed into the blood from the kidney tubule, resulting in large amounts of fluid lost through urine
osmotic pressure: the ability of particles to attract water
OTC: see *over-the-counter drug*
otitis externa: outer ear infection
otitis interna: inner ear infection
otitis media: middle ear infection
ovarian follicle: the ovum and its encasing cells
over-the-counter (OTC) drug: drug for which no prescription is needed
oxytocic: labor-producing

P

package insert: information provided with the drug bottle or vial by the manufacturer to clarify the drug's properties and uses
parameter: the intensity of effect measured on a subjective scale of either 1 through 3 or 1 through 10
parasympatholytics: see *anticholinergic drugs*
parasympathomimetics: see *cholinergic drugs*
parenteral: administered by routes other than the gastrointestinal tract
particulates: composed of several particles
passive diffusion: movement of atoms, ions, or molecules from an area of high concentration to an area of low concentration

passive immunity: protection conferred when an animal receives antibody from another animal

pediatric administration set: device consisting of a set of tubing, drip chamber, and clamp used to administer fluid that typically delivers 60 drops per milliliter

peristalsis: a wavelike contraction of longitudinal and circular muscle fibers that moves food through the gut

phagocytosis: engulfing solids into the cell by surrounding the particle and forming a vesicle

pharmacodynamics: the study of a drug's mechanisms of action and its biological and physiological effects

pharmacokinetics: the study of the absorption, distribution, biotransformation, and excretion of drugs

pharmacology: study of the history, sources, physical and chemical properties of drugs; also includes study of how drugs affect living systems

pharmacotherapeutics: field that examines the treatment of disease with medicines

pharmacotherapy: treatment of disease with medicine

physical therapy: use of noninvasive techniques for the rehabilitation of animal injuries

***Physician's Desk Reference* (PDR)**: a compendium of United States human drug standards

pinocytosis: engulfing fluids into the cell by surrounding the particles and forming a vesicle

polymyxin B: a polypeptide antibiotic effective against gram-negative bacteria only

polypeptide: a chain of amino acids

polypeptide antibiotics: agents composed of polypeptides that are effective against bacteria

polyuria: the formation and excretion of a large volume of urine

polyvalent vaccine: multiple-antigen vaccine

posterior: toward the rear (of a body part): term is not used frequently in veterinary medicine other than to describe organs such as pituitary gland and organ parts such as eye chambers

poultice: a form of herbal treatment made by boiling fresh or dried herbs, squeezing out excess liquid, cooling the herb, and applying it to the skin

powder: a dry, granulated version of a drug mixed with inert bulking and flavoring agents to allow dilution of the drug

preclinical studies: series of tests performed on laboratory animals to determine the safety and effectiveness of a drug

preliminary studies: tests performed to determine if a product will perform as expected

preload: the volume of blood entering the right side of the heart; the ventricular end-diastolic volume

prescription: an order, written by a licensed veterinarian, to a pharmacist to prepare the prescribed medicine, to affix the directions, and to sell the preparation to the client

prescription drug: drug that is limited to use under the supervision of a veterinarian or physician

probiotic: substance that seeds the gastrointestinal tract with beneficial bacteria

proctoscopy: examining the rectum with a proctoscope

proestrus: period of the estrous cycle before sexual receptivity

progesterone: a female sex hormone produced and secreted by the corpus luteum

prokinetic agents: gastrointestinal drugs that increase the motility of parts of the GI tract to enhance movement of material through it

proprietary name: see *brand name*

prostaglandins: group of naturally occurring, chemically related chemicals that stimulate uterine contractions and other smooth muscle

protectant: drug that coats the gastrointestinal mucosa

protocols: therapeutic combinations

pruritus: itching

psychoactive drug: a drug that affects the mind or behavior

purgatives: harsh cathartics that cause watery stool and abdominal cramping

Q

quinolones: synthetic, broad-spectrum antibiotics

R

receptor: specialized cell binding component that is a three-dimensional protein or glycoprotein located on the cell surface, in the cytoplasm, or within the nucleus of the cell

recombinant vaccine: vaccine containing a gene or part of a microorganism that was removed from one organism and inserted into another organism, thus "recombining" the genetics of both to make something new

rehydration volume: fluid volume used to correct dehydration

renal tubules: tubes within the nephron of the kidney

renin: an enzyme released by the kidney that activates angiotensin (a potent vasoconstrictor)

repository medication: a drug injected with a substance that delays absorption; also called *depot preparation*

resistant: the ability of a microorganism to withstand doses of a drug that will kill or inhibit the growth of most members of its species

respiratory stimulants: drugs that stimulate breathing

ruminant: an animal with one glandular stomach and three esophageal outpouches that regurgitates and chews undigested food (e.g., cattle, sheep, goat)

S

SC: see *subcutaneous*

scours: diarrhea (in livestock)

seborrhea: a skin condition characterized by abnormal flaking and/or scaling of the outermost layer of the epidermis

seborrhea oleosa: seborrhea with increased production of sebum (oil)
seborrhea sicca: seborrhea with decreased production of sebum (oil)
second-line drugs: drugs that may have toxic side effects or are used only when the first-line choice cannot be used or is ineffective
sedatives: drugs that allay irritability and excitement
sensitive: the ability of a microorganism to be killed or its growth inhibited by an antibiotic
shelf life: the time during which a drug remains stable and effective for use
short-term tests: tests that occur in the hours following a drug test dose; checks for adverse reactions
short-term toxicity tests: toxicity test of a drug that lasts from the start of the test to about three months; used to assess adverse drug reactions that occur during short-term use of the drug
soak: substance applied to an area to draw out fluid or relieve itching
solute: the dissolved substance of a solution
solution: water that holds a dissolved substance; also called the *vehicle*
solvent: the dissolving substance of a solution
spasmolytic: drug that relieves muscle spasticity
special tests: tests done to determine specific outcomes, such as carcinogenicity or teratogenicity
specific immunity: defense mechanisms against a particular antigen that arise from the B- and T-lymphocytes
spectrum of action: the range of bacteria that an antibiotic or antimicrobial will eradicate
sporicidal: chemical that kills spores, which are especially resistant to chemicals
SQ: see *subcutaneous*
status epilepticus: ongoing seizures
steady state: point at which drug accumulation and elimination are balanced
steroidal anti-inflammatory drugs: group of anti-inflammatory drugs that have the steroid configuration of carbon atoms in four interlocking rings (e.g., glucocorticoids such as cortisol)
strangles: an equine streptococcal infection of the nose and throat
streptococcal bovine mastitis: a mammary strepococcal infection in cows
Streptomyces erythreus: a bacterium that produces erythromycin
subcutaneous: beneath the layers of the skin; abbreviated SC, SQ, or subQ
subdural: between the dura mater and the arachnoid membrane of the meninges; below the dura mater
sulfonamides: a group of drugs that inhibit the synthesis of folic acid to block the growth of a wide variety of bacteria
suspension: a finely divided, undissolved substance dispersed in water
subQ: see *subcutaneous*
sympatholytics: see *adrenergic blocking agents*
sympathomimetics: see *adrenergic drugs*
systems-oriented screen: a test of a drug's effect on a particular physiological system

T

tablet: medicine mixed with an inert binder and molded or compressed into a hard mass
teratogenicity: the capacity to cause birth defects
testosterone: the primary male sex hormone
tetracyclines: broad-spectrum antibiotics effective against many gram-negative and gram-positive bacteria
therapeutic index: an expression synonymous with margin of safety
therapeutic range: the drug concentration in the body that produces the desired effect in the animal with minimal or no signs of toxicity
therapy: treatment of disease
thiazides: diuretics that act directly on the renal tubules to block sodium reabsorption and promote chloride ion excretion
thyroid hormone: a collective term for two active hormones found in the thyroid gland: *thyroxine* and *tri-iodothyronine*
thyroid replacement therapy: the use of thyroid hormone to treat hypothyroidism
thyroxine (T_4): a thyroid hormone that acts as a catalyst in the body and influences many metabolic, growth, reproductive, and immune effects
tolerance: reduced effect from the use of a substance resulting from repeated use
tonicity: a property based on an osmolality; describes the amount of solutes in a solution
topical: a medication that goes on the skin or mucous membranes
topical otic dewaxing agent: see *dewaxing agent*
topical otic drying agent: see *drying agent*
total daily dose: the entire amount of drug delivered in 24 hours
total dose: the total amount of a drug given to an animal in one administration
toxicity evaluation: an evaluation to see if a drug causes convulsions, seizures, or changes in blood pressure, heart rate, respiration rate, sleep, or muscle tension
toxoid: vaccine against a toxin
trade name: see *brand name*
tranquilizers: drugs that calm anxiety
tri-iodothyronine (T_3): a thyroid hormone that acts as a catalyst in the body and influences many metabolic, growth, reproductive, and immune effects

U

United States Drug Enforcement Administration (DEA): the federal agency (in the United States) that enforces the controlled substances laws and regulations of the

United States and brings to the criminal and civil justice system those involved in violating the laws regarding the use, manufacture, or distribution of controlled substances

United States Food and Drug Administration (FDA): the federal agency (in the United States) formed in 1906 that approves or disapproves drugs to ensure the safety, purity, and effectiveness of drugs produced and used in the United States.

***United States Pharmacopeia* (*USP*)**: a publication that is the recognized legal drug standard of and for the United States

urinary acidifying drugs: drugs that lower the pH of urine

urinary alkalizing drugs: drugs that raise the pH of urine

urinary incontinence: loss of voluntary control of micturition

urolith: a stone in the urinary system

V

vaccine: a suspension of weakened, live, or killed microorganisms administered to prevent, improve, or treat an infectious disease

vasoconstriction: narrowing of the blood vessels

vasodilators: drugs used to widen arteries and/or veins, which alleviates vessel constriction and improves cardiac output

vehicle: see *solution*

veterinarian/client/patient relationship: a relationship between a veterinarian and client for which the veterinarian assumes the responsibility for making clinical judgments regarding the health of the animal(s) (patients) and the need for medical treatment, and the client agreees to follow the veterinarian's instructions

***Veterinary Pharmaceuticals and Biologicals* (*VPB*)**: a compendium of United States veterinary drug standards

veterinary pharmacology: the study and use of drugs in animal health care

virucidal: chemical that kills viruses

volatilize: rapidly evaporate

W

Western herbal medicine: discipline based on belief that individuals have an inner force that works to maintain physical, emotional, and mental health

wholesalers: see *distributors*

withdrawal time: period of time after drug administration during which the animal cannot be sent to market for slaughter and the eggs or milk must be discarded, because of the potential for drug residues to persist in these products

INDEX

ANSWER KEY TO TEXT REVIEW EXERCISES

Chapter Review: 1

2. b
4. g
6. c
8. h
10. i
12. c
14. c
16. a
18. b
20. b

Chapter Review: 2

2. d
4. e
6. a
8. i
10. h
12. d
14. a
16. c
18. b
20. a

Chapter Review: 3

2. b
4. c
6. a
8. d
10. f
12. c
14. b
16. d
18. b
20. a

Chapter Review: 4

2. d
4. a
6. b
8. e
10. i
12. a
14. a
16. a
18. a
20. d

Chapter Review: 5

2. d
4. i
6. f
8. c
10. h
12. b
14. generic, trade
16. nothing orally
18. three times daily
20. orally

Chapter Review: 6

2. h
4. i
6. d
8. b
10. f
12. b
14. b
16. d
18. c
20. d
22. b
24. b
26. a
28. 1550
30. 2550
32. 0.015
34. 0.400
36. 15
38. 1
40. 20.08
42. 4.54
44. 3125
46. split it into quarters
48. 0.98
50. 10
52. 38.9°C
54. 77°C

Chapter Review: 7

2. i
4. d
6. a
8. f
10. g
12. b
14. a
16. b
18. c
20. b

Chapter Review: 8

2. d
4. j
6. h
8. i
10. c
12. d
14. b
16. a
18. b
20. d

Chapter Review: 9

2. b
4. e
6. f
8. h
10. j
12. d
14. c
16. c
18. b
20. c

Chapter Review: 10

2. d
4. g
6. b
8. f
10. j
12. a
14. d
16. a
18. b
20. b

Chapter Review: 11

2. a
4. j
6. d
8. b
10. i
12. c
14. a
16. c
18. b
20. b

Chapter Review: 12

2. i
4. b
6. c
8. f
10. d
12. a
14. a
16. a
18. d
20. d

Chapter Review: 13

2. d
4. a
6. b
8. d
10. e
12. b
14. d

Chapter Review: 14

2. c
4. a
6. i
8. e
10. j
12. b
14. b
16. d
18. c
20. b

Chapter Review: 15

2. f
4. h
6. e
8. b
10. a
12. b
14. b
16. c
18. d
20. b

Chapter Review: 16

2. b
4. a
6. a
8. b
10. c
12. b
14. c
16. c
18. b
20. a

Chapter Review: 17

2. d
4. h
6. j
8. b
10. e
12. a
14. b
16. c
18. a
20. b

Chapter Review: 18

2. j
4. i
6. b
8. e
10. c
12. b
14. a
16. d
18. c
20. b

Chapter Review: 19

2. d
4. c
6. b
8. c
10. a
12. a
14. b
16. b
18. a
20. b

Chapter Review: 20

2. b
4. i
6. d
8. g
10. a
12. d
14. a
16. a
18. b
20. b

Chapter Review: 21

2. e
4. c
6. d
8. h
10. i
12. b
14. a
16. a
18. b
20. a

Chapter Review: 22

2. c
4. g
6. h
8. d
10. a
12. a
14. c
16. a
18. b
20. b

Chapter Review: 23

2. d
4. c
6. a
8. b
10. b

SOFTWARE SYSTEM REQUIREMENTS AND INSTALLATION INSTRUCTIONS

DELMAR'S VETERINARY DOSAGE ACCU-CALC® (2004 UPDATE) on CD-ROM

System Requirements

- Pentium, 100 MHz (or faster)
- 32 MB RAM
- 800 × 600 SVGA with 24-bit color display
  640 × 480 display recommended
- CD-ROM drive (4x or faster)
- Sound card and speakers
- Windows® 98 SE, Windows® ME, Windows® 2000, Windows® XP or later Windows® versions
- Direct X, Version 5 (or later)
- 2.5 MB free disk space for program files
- 1 MB free disk space for data files

Installation Notes

1. Run SETUP.EXE from the CD-ROM disk.
2. Select a program folder or copy to the default folder: C:\Program Files\AccuCalc.®
3. The AccuCalc® CD-ROM disk must be in the drive to run the program.
4. Data files are saved to the user's folder on systems with Windows® 2000 or Windows® XP. Default data folder: C:\Documents and Settings\User\Application Data\AccuCalc®.
5. On systems with older versions of Windows®, data is saved to the Windows data folder: C:\Windows\Application Data\AccuCalc®.

DELMAR'S FLUIDS and ELECTROLYTES for VETERINARY TECHNICIANS on CD-ROM

System Requirements

- Pentium, 100 MHz (or faster)
- 32 MB RAM
- 800 × 600 SVGA with 24-bit color display
- CD-ROM drive (4x or faster)
- Sound card and speakers
- Windows® 95 or later Windows® versions

Installation Notes

Before you run the installation program, check the system requirements. This program will take approximately 10 megabytes of disk space.

Windows™

1. Insert the disk in the CD-ROM drive. Select "Start" and "Run."
2. Type "D: \Setup" or "E: \Setup" and press OK. The drive letter indicated depends on the drive designation of your computer's CD-ROM drive.
3. Follow the on-screen prompts.